# Handbook of Research on Emerging Rule-Based Languages and Technologies:
## Open Solutions and Approaches


Adrian Giurca
*Brandenburgische Technische Universität Cottbus, Deutschland*

Dragan Gašević
*Athabasca University, Canada*

Kuldar Taveter
*Tallinn University of Technology, Estonia*


Volume II


Information Science REFERENCE

**INFORMATION SCIENCE REFERENCE**

Hershey · New York

Director of Editorial Content: Kristin Klinger
Senior Managing Editor: Jamie Snavely
Managing Editor: Jeff Ash
Assistant Managing Editor: Carole Coulson
Typesetter: Jeff Ash and Jennifer Johnson
Cover Design: Lisa Tosheff
Printed at: Yurchak Printing Inc.

Published in the United States of America by
Information Science Reference (an imprint of IGI Global)
701 E. Chocolate Avenue
Hershey PA 17033
Tel: 717-533-8845
Fax: 717-533-8661
E-mail: cust@igi-global.com
Web site: http://www.igi-global.com/reference

and in the United Kingdom by
Information Science Reference (an imprint of IGI Global)
3 Henrietta Street
Covent Garden
London WC2E 8LU
Tel: 44 20 7240 0856
Fax: 44 20 7379 0609
Web site: http://www.eurospanbookstore.com



Library of Congress Cataloging-in-Publication Data

Handbook of research on emerging rule-based languages and technologies : open solutions and approaches / Adrian Giurca, Dragan Gašević, and Kuldar Taveter, editors.
p. cm.

Includes bibliographical references and index.
Summary: "This book provides a comprehensive collection of state-of-the-art advancements in rule languages"--Provided by publisher.

ISBN 978-1-60566-402-6 (hardcover) -- ISBN 978-1-60566-403-3 (ebook) 1. Expert systems (Computer science) 2. Rule-based programming. I. Giurca, Adrian, 1966- II. Gašević, Dragan. III. Taveter, Kuldar.

QA76.76.E95.H3554 2009
006.3'3--dc22

2008055311

British Cataloguing in Publication Data
A Cataloguing in Publication record for this book is available from the British Library.

All work contributed to this book is new, previously-unpublished material. The views expressed in this book are those of the authors, but not necessarily of the publisher.

## Editorial Advisory Board



## List of Reviewers

# List of Contributors

# Table of Contents

**Section III**
**Rule Applications**

# Detailed Table of Contents

NKRL is a semantic language expressly designed to deal with all sort of 'narratives', in particular with those ('non-fictional narratives') of an economic interest. From a knowledge representation point of view, its main characteristics consists in the use of two different sorts of ontologies, a standard, binary ontology of concepts, and an ontology of n-ary templates, where each template corresponds to the formal representation of a class of elementary events. Rules in NKRL correspond to high-level reasoning paradigms like the search for causal relationships or the use of analogical reasoning. The chapter describes the main characteristics of the NKRL inference rules.



This chapter describes various graphical notations for rule modeling. Rule modeling methodologies, empowered with graphical notations, play an important role in helping business experts and rule engineers to represent business rules formally for further deployment into a rule execution system. Rules, represented graphically, can be easier understood by business people and by technicians without intensive technical learning. In this chapter the authors mainly focus on three graphical notations for rules: UML/OCL, URML and ORM.



This chapter addresses the problems of modeling and interchanging ontological commitments in order to support ontology based decision making. The authors model and visualize ontological commitments by means of an extension to Object Role Modeling Language (ORM), which was called ORM Plus (ORM+) and is now named Semantic Decision Rule Language (SDRule-L). SDRule-L is a commitment language for modeling dynamic and non-monotonic decision rules. SDRule-L models are further stored in an XML-based markup language called Semantic Rule Markup Language (SDRule ML), which is a hybrid language of Rule Markup Language (Rule-ML) and Object Role Modeling Markup Language (ORM-ML).

Agent-Oriented Programming (AOP) is a relatively new programming paradigm, proposed by Yoav Shoham, which views software systems as consisting of a set of agents that interact with one another to solve problems beyond their individual capabilities. Since the inception of the paradigm, a number of AOP languages have been proposed. This chapter focuses on one such language, the Agent Factory Agent Programming Language (AFAPL), a practical rule-based language that has been applied to a wide range of problem domains including robotics, virtual and mixed reality environments, and mobile computing.



This chapter presents a number of model transformations for round-trip engineering between Web services and reaction rules. In addition it describes the mappings of models of Web services with reaction rules into the Drools rule language, thus enabling the run time execution semantics for our rule-based models.

**Section III**
**Rule Applications**



This chapter presents the ViDRE system that bridges the gap between business rules and service-oriented computing by exposing business rules as Web services. In contrast to existing rule engines, our approach supports distributed rule execution using meta-rules which includes automatic transformation of rules on both client and service side.

# Preface

Rules have a history from the early days of computer science, including computational models, compiler technology, artificial intelligence, databases, and logic programming. They included backward chaining rules (or derivation rules), forward chaining rules (or production rules), Event-Condition-Action Rules (or reaction rules) while many researchers also consider integrity constraints to be rules as well.

The history of rule-based programming and rule-based systems started in the 1960s and 1970s with theorem proving, parsing, AI reasoning, and Prolog, a logical and a declarative programming language. During that time, a number of other rule languages were created. Between the 1970s and 1990s, rule languages were used mainly in academia in artificial intelligence applications, notably expert systems, and logic programming. However, since the late 1990s, many practical applications of rules in various problem domains have been developed.

Rules became increasingly important in business modeling and requirements engineering, and as a high level programming paradigm especially in the engineering of e-business applications and Semantic Web applications. In each of these fields different rule languages and tools are being used. Rules are used in applications to model and manage some parts of the application business logic. They are best used in applications with dynamic business logic (e.g., applications) where changes in the business logic are frequent and they need to be immediately reflected in the application behavior such as insurance (insurance rating), financial services (loans, claims routing and management, fraud detection), government (tax calculations), telecom customers (care and billing), and electronic commerce (personalizing the user's experience, recommender systems, and auctions). Moreover, rules have continued to play an important role in AI shells for knowledge-based systems and in intelligent agents, today both needing a Web interchange format, and such XML/RDF-standardized rules are now also usable for the declarative specification of Web services.

A number of companies became involved in rule engines design and production (IBM, ILOG, Fair Isaac, HP and others). Later open source projects appeared (e.g., JBoss Drools and Jena Rules) and other software vendors (e.g., Oracle, TIBCO and SAP) started building their own tools for developing rule-based systems. In the last decade appeared standardization efforts pioneered by the Rule Markup Initiative (RuleML), and continued by various parties including the W3C Working Group on Rule Interchange Format (RIF).

This handbook aims to provide a comprehensive state-of-the-art of rule languages, methodologies for modeling, designing, and implementing rules, rule interoperability/interchange, and rule-based applications. The target readers are practitioners dealing with concrete problems, researchers looking for challenging research topics, and users and managers willing to advance their current business practices and software solutions. The book helps researchers to learn more about the current state of the art in the area.

In addition, the book fills a gap in the literature by relating the rule applications to Model Driven Architecture and ontology development on the Semantic Web. It gives an overview of the major issues researchers, developers, and users of rule-based systems may face while researching, developing, and using rule-based systems. This topics are related to: (a) business rules' representation methods; (b) rule history and lifecycle management; (c) rule types and forms of reasoning; (d) rule verbalization, visualization, and documentation; (e) interoperability with external (data) sources; (f) testing and debugging tools; and (g) authoring environments (IDE vs. Browser vs. Custom). The developers will accommodate rule languages and methodologies to designing and building rule-based applications. The users and managers will learn more about rule-based systems, so that choosing solutions may become easier.

## ORGANIZATION AND STRUCTURE

The handbook features chapters authored by 54 leading experts offering an in-depth description of key terms and concepts related to rule technologies, issues and trends in business rules' management, and rule modeling and rule interoperability technologies in modern organizations worldwide. The handbook contains 27 peer-reviewed contributions organized into three main sections.

Section I covers the topics of rule languages and rule engines. Particularly, this explores various rule formalisms (such as description logic and logic programming), domain specific rule languages, various aspects of rule-based reasoning, modular rule languages, graphical notations for rule languages, and the rule languages for the Web.

Section II is devoted to standards, methodologies, and tools for rules and rule-based systems. This includes topics such as agile development of rule systems, verification and debugging principles used in rule bases, development of Web applications, services, and rule usage in agent-oriented systems.

Finally, Section III looks into rule applications, which spans across various application domains such as business process management, high-level agent architectures, service-oriented systems, ontology-driven applications, policy and rule languages, rules and model-driven software development techniques, and tools for rule mining and modeling.

The handbook contains many practical examples, case studies, and references to external sources containing many useful tools, systems, and engines, which allow the readers to experiment with the presented research results and potentially use them in their current research and practice.

*Adrian Giurca, Brandenburgische Technische Universität Cottbus, Deutschland*
*Dragan Gašević, Athabasca University, Canada*
*Kuldar Taveter, Tallinn University of Technology, Estonia*

February 2009

# Acknowledgment

This handbook would not have been possible without the significant contributions of many individuals and organizations. We are very grateful to the members of the editorial advisory board and reviewers for their dedication in reviewing the papers and providing the authors with substantial feedback. We also thank the authors for their efforts in writing and revising their papers, and we thank IGI Global for publishing the handbook and for a great collaboration throughout all stages of the work on this special issue.

*Adrian Giurca, Brandenburgische Technische Universität Cottbus, Deutschland*
*Dragan Gašević, Athabasca University, Canada*
*Kuldar Taveter, Tallinn University of Technology, Estonia*

# Chapter XV
# The Agent Object Relationship Simulation as a Business Process

**Emilian Pascalau**
*Brandenburg University of Technology, Germany*

**Adrian Giurca**
*Brandenburgische Technische Universität Cottbus, Deutschland*

**Gerd Wagner**
*Brandenburg University of Technology, Germany*

## ABSTRACT

*The use of agent-based simulation models is growing and attracted a lot of attention recently both for researchers and business management. Agent-Object Relationship (AOR) is an agent-based simulation paradigm that uses reaction rules to model agents' behavior. The goal of this chapter, besides exemplifying the AOR concepts by means of a use case, is to investigate the use of business process modeling notation (BPMN) to model the AOR simulation process. Moreover it discusses aspects of a distributed architecture for an AOR simulation system. The chapter concludes with the fact that BPMN is well suited to model the AOR simulation process.*

## INTRODUCTION

This chapter has two main pillars on which it stands on: agent-based simulation and modeling and business process modeling. The use of agent-based simulation models is growing and attracted a lot of attention recently both for researchers and business management. Business process modeling is on an ascending curve of interest and research.

This chapter tries to bind together these major paradigms of nowadays research efforts, agent based simulation modeling and business process modeling.

The Agent-Object-Relationship (AOR), an agent based simulation paradigm, is used to illustrate the agent based simulation concepts.

AOR employs the concept of reaction rules for modeling behavior of organization information systems. AOR provides an agent based simulation platform based on a high level ontological meta-model and offers compared to other simulation platforms almost a complete list of features specific to agent based simulation systems: it is independent from a specific programming language, it is model driven, uses rule based (declarative) behavior modeling, supports a concept of space, supports the distinction between objects and agents, includes a cognitive model of perceptions, supports the distinction between facts and beliefs.

The chapter starts with an overview of AOR paradigm. The use case that follows the AOR overview illustrates the AOR concepts and also exemplifies the AOR modeling capabilities. The AOR simulation principles are then narrated and explained. The cornerstone of the chapter is the introduction of a distributed architecture for AOR and the modeling as a business process of the AOR simulation process. The modeling as a business process is a starting point for future research that includes modeling of agent behavior as business processes using BPMN notation.

## RELATED WORK

Before going into the middle of the problem it is not with out reason to have a short look over the main topic of the current chapter.

Nowadays agents are encountered almost every where, or at least the term agent. There are several types of agents: intelligent agents, information agents, mobile agents, personal assistant agents and other types of agents.

It is claimed that agent technologies facilitate software development based on their interaction's high level of abstractions.

There might be raised several possible questions such: What is an agent? Do agents have something in common? And so on.

In Russell & Norvig, 1995 it is argued that agents are entities perceiving their environment through sensors and acting upon that environment through effectors. Agents are not isolated entities but they are able to communicate and collaborate with other entities. There are several group efforts towards the standardization of multi agents systems such as: Foundation for intelligent Physical Agents (FIPA), the Object Management Group, the Knowledge-able Agent-oriented System (KAoS).

The Multi Agents Systems (MAS) have generated many works and many systems were implemented in order to put in practice theory behind MAS. MAS' base concepts are introduced in Wooldridge, 2000.

Bradshaw, 1997 presents a list of common agent attributes:

- **Adaptivity:** Being able to learn and improve with experience
- **Autonomy:** Goal oriented, proactive and self-starting behavior
- **Collaborative behavior:** Working together with other agents to achieve common goal
- **Inferential Capability:** The ability to act on abstract task specification using prior knowledge of general goals
- **"Knowledge-level" communication ability:** The ability to communicate with other agents
- **Mobility:** Being able to migrate in a self oriented way from one host platform to another
- **Personality:** The capability of manifesting the attributes of a believable character
- **Reactivity:** Ability to sense and act selectively

- **Temporal continuity:** Persistence of identity and state over time

As stated in Railsback & Lytinen & Jackson, 2006 most of today's Agent Based Simulation (ABS) systems, such as Swarm, Repast, SeSam, Mason follow the so called framework and library approach, providing a framework – a set of standards concepts for designing and describing agent base simulation systems along with a software library, for simulation, implementing the framework concepts. Usually in most of the ABS systems the simulation model is directly encoded in the programming language chosen for implementing the library.

SeSAm (Shell for Simulated Agent Systems) system introduced in Kluegl & Puppe, 1998, provides a generic environment for agent-based systems modeling and experimentation. Kluegl & Herrler & Oechslein, 2004 presents how can SeSAm be used for software development. SeSAm allows visual agent modeling as well as experimenting and analyzing agent-based simulation experiments.

Repast (Recursive Porous Agent Simulation Toolkit) was introduced in North & Howe & Collier & Vos, 2005a and 2005b. Repast introduces a new approach to agent-based simulation development and execution; it brings advanced features for agent behavioral specification and dynamic model self-assembly. Tatara & North & Howe & Collier & Vos, 2006 provides an introduction to Repast modeling by means of a simple predator-prey example.

Swarm as stated in Minar & Burkhart & Langton & Askenazi, 1996 is a multi-agent software platform for the simulation of complex adaptive systems. In the Swarm system the basic simulation unit is the swarm. A swarm is a collection of agents executing a schedule of actions. Swarms supports hierarchical modeling approaches where agents can be composed of swarms of other agents in nested structures.

Another platform that follows the framework and library approach is MASON. MASON stands for Multi Agent Simulator of Neighborhoods. Based on Luke & Balan & Panait & Cioffi-Revilla & Paus, 2003 and Luke & Cioffi-Revilla & Panait & Sullivan, 2004 MASON is a fast, easily extendable, discrete event multi-agent simulation toolkit written in Java; designed for simulating a wide range of multi agent simulation tasks ranging from swarm robotics to machine learning to social complexity environments.

NetLogo based on Wilensky, 1999 is a multi-agent programming language and modeling environment for simulating complex phenomena. The NetLogo dedicated simulation language is more high-level and simpler to use than ordinary programming languages.

NetLogo, Swarm and Repast are considered to be the most widely used system as stated in Railsback & Litynen & Jackson, 2006.

## AOR

### AOR Overview

Although there are several platforms that try to realize the general idea of agent systems there is no generally accepted standard and also there are few approaches that try to define the basic concepts in a rigorous way. One of these approaches is Agent-Object-Relationship (AOR), introduced in Wagner, 2003 and Wagner, 2000. Another one is AUML introduced in Odell & van Dyke Parunak & Bauer, 2000.

AOR offers a range of high-level abstractions that facilitate modeling, expressing the concepts and interactions between agents and also for effectively simulating simulation scenarios.

It was created with the scope of modeling and simulating information systems based on the agent-object relationship, rather than for software engineering. It is based on some fundamental paradigms such as Entity Relationships modeling

(ER), Unified Modeling Language (UML), and Logic Programming (LP). As stated in Taveter & Wagner, 2001a Agent-Orientation is highly significant for business information systems since business processes are driven by and directed towards agents (or actors), and hence has to comply with the physical and social dynamics of interacting individuals and institutions. Wagner, 2003 argues that "semantics of *business processes* may be more adequately captured if the specific business agents associated with the involved events and actions are explicitly represented in the information systems in addition to passive business objects". Agent-Object-Relationship emphasizes: the fundamental role of agents and their mental state; interaction and communication between agents; allows capturing dynamic and deontic semantics of enterprise modeling better than object-oriented approaches such as UML.

According to Tavater & Wagner, 2001b business rules that define interactive behavior of business agents are best formalized as reaction rules.

AOR creation was influenced by the work from Wagner, 1997 and Shoham, 1993. AOR comprise in addition to the reactive agent part of the work from Wagner, 1997 a modeling language for agents.

AOR comes with the ontological distinguish between agents and objects. Agents are considered active entities and Objects are passive entities. The ontological distinguish is developed in a detailed level in the Unified Foundational Ontology (UFO) (Guizzardi & Wagner, 2004). UFO combines General Formal Ontology (GFO) (Degen, & Heller & Herre & Smith, 2001) and OntoClean (Welty and Guarino, 2001) ontology. It is used as a *foundation of agent concepts* and for *evaluating agent-oriented modeling methods*.

In AOR an entity is an agent, an event, an action, a claim, a commitment, or an ordinary object and fundamental associations, compositions and generalizations are supplemented by special relationships between agents and events, actions, claims and commitments.

The main differences between agents and objects as they were introduced by the Agent-Oriented programming paradigm in Shoham, 1993 are also retained by AOR. As stated in Shoham, 1993 at p. 56, agent oriented programming (AOP) can be viewed as a specialization of object oriented programming. The comparison between OOP and AOP is based upon the following criteria: basic unit, parameters defining state of basic unit, process of computations, types of messages, and constraints on methods. While in OOP the state of an object has no particular structure and is only given by the object attributes in AOP an agent state is given by its metal state that usually comprise mental components such as: beliefs, commitments, capabilities, choices…

Agents in AOR are based on the principle of internal or *subjective view* of the world. This conducts to the fact that there is no notion of *objective time* or the distinctions between knowledge and beliefs. The environment that comprises objects and agents is the one that knows everything through the fact that any communication is performed through the environment. In AOR everything is a typed entity. The state of such an entity is denoted by its set of properties. AOR entities are decoupled in different epistemic categories such as agents, events, actions, commitments, claims and objects. AOR emphasizes distinction between communicative and non-communicative actions and events. Actions create events but not all events are created by actions (e.g. ExogenousEvents).

The agents modeled by AOR are reactive agents. Such an agent has its state represented by a knowledge base and its behavior modeled by means of actions and reaction rules. AOR is based on list of 19 fundamental ontological principles described in detail in Wagner, 2003.

As the reader already knows in AOR the fundamental concept is the Entity. An Entity is sub classed by passive (objects) and active entities (agents), events and messages.

AOR has two main types of events, EnvironmentEvents that belong, as the name says, only to the environment and InternalEvents which are agent internal events.

EnvironmentEvents are ExogenousEvents, CausedEvents and ActionEvents. ExegenousEvents are the so called non action events. These are events that are not controlled by the environment. For example the *wind blows* because *it blows*.

An agent is a processor for internal events. An agent sends OutMessageEvents and internally deals with ActualInMessages. An ActualInMessage is the internal mapping of a PotentialInMessage. An agent is actor for ActionEvents. ActionEvents are non-communicative action events. OutMessageEvents are communicative action events, as expressed by the subClassOf relationship.

An AOR belief property is expressed like in logic programming in the form of belief triple, containing subject, object and value. An AOR agent can have three different belief types: objective beliefs, self belief properties and beliefs about other entities.

## Modeling with AOR

AOR modeling language introduces two basic types of AOR models: external and internal models. In an external model is adopted the perspective of an external observer looking at the prototypical agents and their interaction inside of the problem domain. The internal model adopts the first person view of an agent. More clearly an external model models agents and behavior (expressed with environment rules) that takes place in the environment, while the internal model focuses on a specific agent type and its internal behavior (modeled with agent rules). Often these two types of models are condensed in a single one. Figure 1 depicts the core elements of an external AOR model.

An external AOR model may comprise one or more of the following diagrams:

- Agent Diagrams depict the agent types of the domain, relevant object types and the relationship between them.
- Interaction Frame Diagrams depict the action event types and commitment/claim types that determine interactions between agent types
- Interaction Sequence Diagrams depict prototypical instances of interaction processes.
- Interaction Pattern Diagrams focus on general interaction patterns expressed by means of reaction rules

These types of diagrams are detailed in Wagner, 2003.

*Figure 1. AORML – External model core elements*

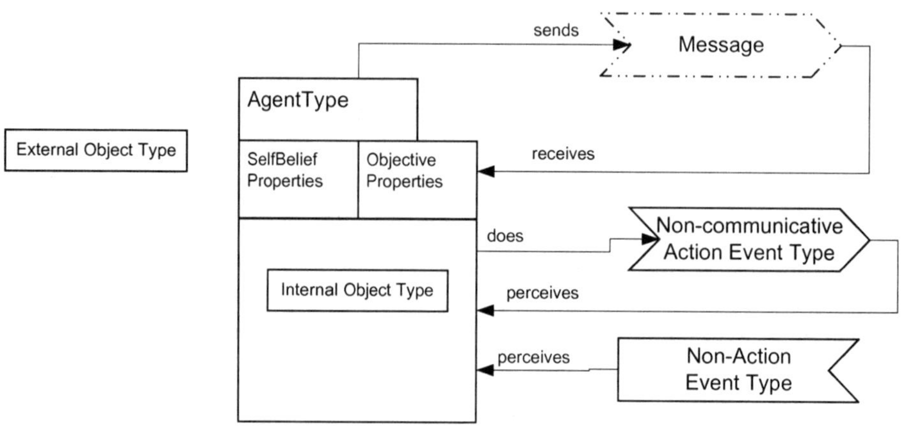

AOR models can be casted to UML models. In this way those UML tools that provide extension mechanism in the form of stereo types and tagged values can be used to define AOR models. A complete mapping from the AOR meta-model to their base UML classes is presented in Wagner, 2002.

## WHY IS AOR DIFFERENT FROM OTHER APPROACHES?

There are several aspects that differentiate AOR from other approaches. However it is based on several approaches like ER, UML and LP, and allows capturing more domain specific semantics, as argued in Wagner, 2003. The modeling language of AOR allows integration of state and behavior modeling in one diagram. Agent-Object-Relationship Modeling Language (AORML), introduced in Wagner, 2003 allows the deontic concepts of rights and duties. AOR seems to be the first approach that employs modeling behavior with reaction rules. Compared directly to other agent based platforms like SWARM, Repast, SeSam, NetLogo, MASON, AOR:

- Is independent from a specific programming language. AORSL (AOR Simulation Language) is XML based
- It is model driven. AOR is based on meta-model.
- It is rule-based (declarative) behavior modeling. Behavior of entities is modeled by means of reaction rules.
- Supports a concept of space. In AOR objects and agents are subclasses in terms of UML of a physical object. Such a physical object has a series of properties with respect to space and physical appearance.
- Supports the distinction between objects and agents. AOR distinguishes between passive (objects) and active (agents) entities.
- Includes a cognitive model of perceptions. AOR takes into account the idea of potential and actual perceptions.
- Supports the distinction between facts and beliefs. AOR distinguishes between objective and self belief properties.

At this moment AOR does not:

- Support beliefs about other entities. Future versions envision beliefs of other entities in the form of belief triples.
- Include the concept of goal. This point is highly coupled with the next sub-point of proactive behavior.
- Allow to model or simulate proactive behavior. AOR models only reagents or reactive agent but future versions will probably include and will allow modeling and simulation of proactive behavior, with respects to goals and sub goals.

### Example

This section presents a simulation scenario that will help the reader understand: how to model a scenario with AORML; how scenario entities are modeled with UML based on the AOR meta-model.

#### Scenario

On a track of a given length, there are a number of cars, trucks and cyclists. The track length is 30km. Cars and trucks have to respect a regular speed specific to their type if they are inside a radar range. Maximum allowed velocity for trucks is 50km/h and for cars is 60km/h. This is also the city regular velocity. They all can reach a maximum speed, and they can accelerate or decelerate at some given rates. They accelerate as long as they are far enough from the next car or they are below the regular speed and within the radar's range. If they are close to a cyclist they have to slow down to city regular speed until they have overtaken the cyclist. When a car or truck is too close from a car or truck, it must speed up to overtake the car or truck in front. For ease we have only 1 instance of each agent type. The simulation takes place 3600 cycles (steps). The duration of one step is 1s (second).

CarAgent and TruckAgent have the same list of properties. They just define different types, so we will present just the CarAgent. BicycleType has only one property.

CarAgent has a list of predefined properties – objective properties. A part of these properties are inherited from PhysicalObject, some from Entity because in AOR simulation everything is an Entity and some from Agent.

The properties inherited from PhysicalObject are: x, y, z, vx, vy, vz, vxx, vyy, vzz, v, a.

x, y, z are used to express the position of the objective part of the agent inside the environment. vx, vy, vz express the velocity on each of the 3 coordinates axes of a 3D Cartesian system. vxx, vyy and vzz are also the velocities on each of the 3 coordinates axes of a 3D Cartesian system. The necessity of having two velocity variables for each axis comes from Newton's kinematics equations. They are used to express the initial and final velocity. Based on them all other unknown variables from Newton's kinematics equations can be derived. This is why PhysicalObject has a – acceleration, as a derived property. Newton's kinematics equations can be used to compute the values for acceleration, initial velocity, final velocity, initial location or final location. Yet in the AOR case initial location expressed by x, y, z, initial velocity expressed by vx, vy, vz and final velocity expressed vxx, vyy, vzz might be always known. And all the other unknown variables can be computed but not vice versa.

perceptionRadius is inherited from Agent. This one is another objective property and its meaning is to express a perception radius that the objective part of the AOR agent can have.

It is time to explain that an objective property is important only to the environment. In AOR simulations there is a difference between the objective agent and the subjective agent. The difference comes from the separation between "objective" (referring to physical properties) and "artificial" (internal properties).

The subjective part of our CarAgent has several self belief properties: myAcceleration, myPossibleAcceleration, myVelocity and myTargetVelocity. Both these properties are needed as self belief properties because they are related to one another. They are used to actually compute the new position of the agent. A belief property is expressed as a belief triple. Triple structure is: subject, predicate, object. A self belief triple is a triple in which the subject is the agent itself. We consider that cars move with a uniform acceleration. The value for acceleration that a car may have is denoted by the self belief property: myPossibleAcceleration. myAcceleration is the current acceleration of a car or a truck. This property can have two possible values: myPossibleAcceleration or -myPossibleAcceleration when the car or truck de-accelerates. maxVelocity is the maximum velocity that a CarAgent can have. As a comparison with the real world this would be the maximum speed that you see on your car dashboard for example 200km/h. myVelocity on the other hand is the actual velocity or the velocity at a given moment of time. These two myVelocity and maxVelocity have an objective correspondent. Normally this would have been v. Yet in the current example for ease and also because the space is one dimensional vx is used. And the last self belief property is myTargetVelocity. This property is used when the car needs to accelerate or de-accelerate to some specific velocity.

BicycleType has only one self belief property named velocity. It is considered that bicycles moves uniformly with the same velocity always, they do not accelerate nor de-accelerate.

Passive and active entities from our example are also denoted in UML notation in Figure 2.

Now moving on to objects two types of objects are used in the example: SpeedLimitSign and SpeedLimitCancelationSign.

SpeedLimitSign is an object of the environment. Its role is to signal the start of a speed limit range. Also as a real speed limit sign, it also tells the regular or the admissible velocity in its range. This range has as "lower" limit the sign itself and as superior limits a SpeedLimitCancelationSign. To express the maximum admissible velocity this object has a property named carAdmMaxVel and a truckAdmMaxVel. These are objective properties and are known only to the environment. No other entity can access it. Opposed to agents, objects can have only objective properties. They do not posses self belief properties neither belief properties. Such an object extends also PhysicalObject. In this way each object has or can have position, velocity and acceleration.

SpeedLimitCancelationSign is just a simple AOR object that has no objective properties other than those inherited from PhysicalObject. The role of this particual object is just to signal the end of the speed limit range, and no properties other than its physical position are needed. To set its position the PhysicalObject built in properties are used, specifically x property.

Events are depicted in the Figure 3. The reader should notice that ActionEvents are modeled in the meta-model as sub classes of Environment-Events. A fundamental tenet of the foundations of agent-orientation is that an action is an intentional event and, hence, can only be only be performed by an entity with intentionality. However this approach has been taken because in AOR actions performed by agents have effect only in the environment. Besides these it is important to recall that in AOR environment is the one that actually modifies its state. By state altering here we understand modification of object or agent's position and so forth. These are the reasons why actions in AOR are subclasses of environment events. We must underline here that the environment is in no way considered as an agent.

CarAgent and TruckAgent can perform one of the following actions: StartSlowDown, End-SlowDown, StartSpeedUp, EndSpeedUp.

The subjective agent decides to take the action and the result (effect) of that action is seen in the environment. To compare with real world a car speeds up with acceleration. The speedup action is performed from car's interior using the gas pedal. Yet the effect is seen in the environment because the car changes position – objective

*Figure 2. Example Entities in UML notation*

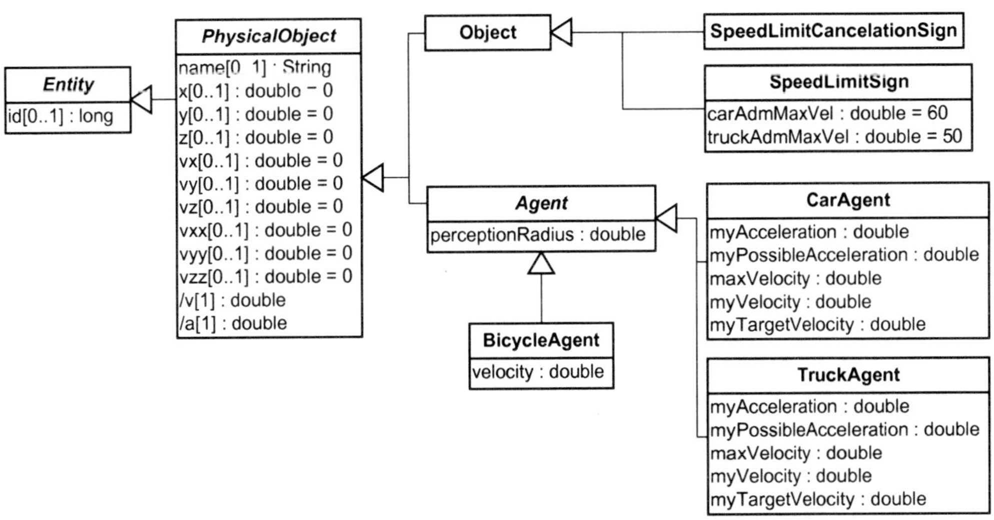

*Figure 3. Example – Events in UML notation*

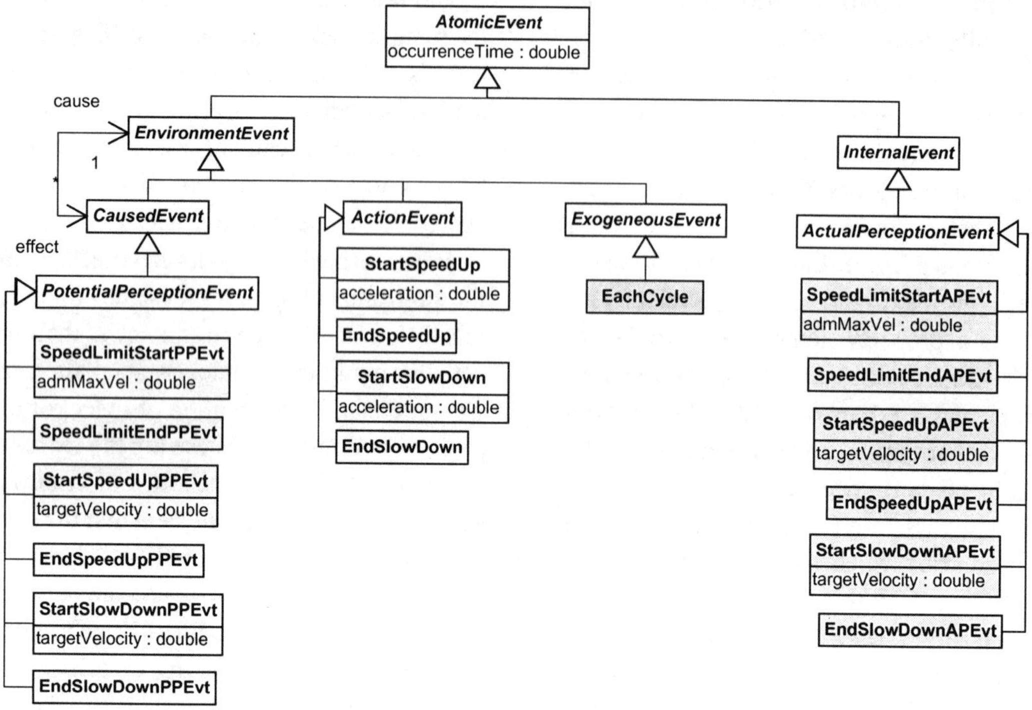


property, faster. Environment uses the acceleration transported as a property of StartSpeedUp action event and uses it to compute the new objective velocity vx. The acceleration used in the current example is the value of myPossibleAcceleration property.

EndSpeedUp action is used to stop the process of speeding up.

StartSlowDown and EndSlowDown work in the same way.

Potential perception events are environment events that are mapped to agent internal events – actual perception events. Subjective agents deal only with actual perception events, so they have to map their perceptions (PotentialPerceptionEvents) into ActualPerceptionEvents.

SpeedLimitStartPPEvt is sent by the environment and shall be perceived or not by the agents. A PotentialPerceptionEvent has some predefined properties: perceiver, perceiverType, perceivers, perceiverTypes. perceiverTypes is a special property that specifies all possible agent types that can perceive such an event. The most used one is perceiver and is used to set the id of the agent that has to perceive this PotentialPerceptionEvent. It has one property admMaxVel. This property has two possible values: SpeedLimitSign.carAdmMaxVel or SpeedLimitSign.truckAdmMaxVel.

The corresponding ActualPerceptionEvent is SpeedLimitStartAPEvt. An agent uses a mapping rule to create such an event from a PotentialPerceptionEvent.

EachCycle is an ExogenousEvent or a non-action event. Each event in AOR has a predefined occurrenceTime property. This property stands for the time when the event must be consumed. This event occurs at each second. To express the periodicity of such an event the periodicity property has to be used. For specifying the stop time of such an event the stopCondition property must be used. In the current scenario this property has the value 3600, exactly as the end of whole simulation.

*Figure 4. Example – Rules in UML notation*

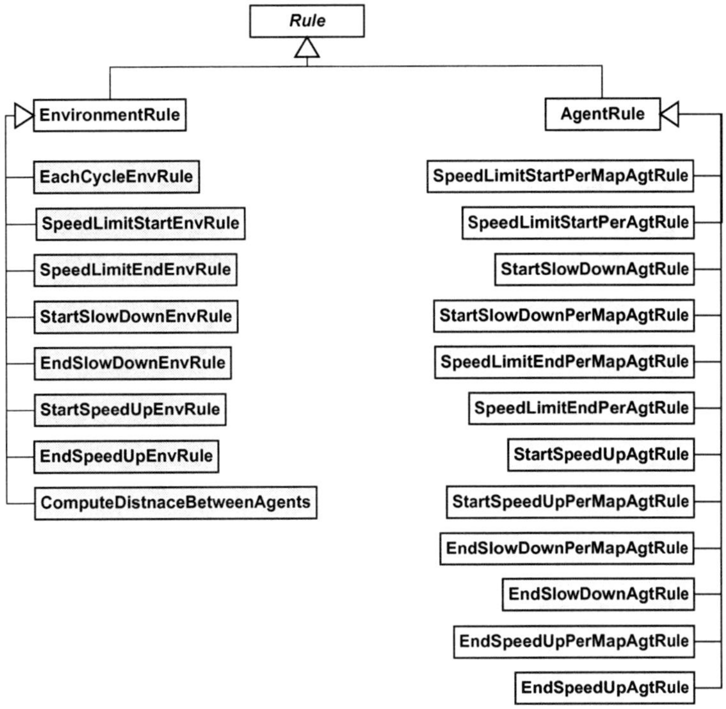


Rule modeling in AORSL is explained in the next lines, and rules modeling the example are depicted in Figure 4.

SpeedLimitStartPerMapAgtRule is a mapping rule. It maps the PotentialPerceptionEvent SpeedLimitStartPPEvt to SpeedLimitStartAPEvt. An agent rule is triggered by an event; it may involve passive or active entities; it can have a condition expression; a resulting event and it can perform a state change. SpeedLimitStartPerMapAgtRule mapping rule is triggered by the SpeedLimitStartPPEvt and has as resulting event a SpeedLimitStartAPEvt.

SpeedLimitStartAgtRule is the agent rule that is triggered by the SpeedLimitStartAPEvt. Note the environment rule that creates the SpeedLimiStartPPEvt is called SpeedLimitStartEvnRule. This rule is triggered by the EachCycle exogenous event, and it checks based on the actual position and the agent objective property perceptionRadius if the agent is in the range of the SpeedLimitSign.

Figure 5 depicts a small part of the current example modeled with AORML. In the figure it is modeled the whole process of a car entering in the range of a SpeedLimitSign. The rule with id: SLSER which stands for SpeedLimitStartEnvRule is triggered by the EachCycle exogenous event. This is expressed with an event arrow that points towards the rule. The rule involves two entities: a car agent and the SpeedLimitSign object. So there are 2 condition arrows pointing towards the rule. The condition arrows are annotated with the conditions. From the rule pointing towards the StartSpeedLimitPPEvt there is a post condition arrow. This one is annotated with the state effect of the rule: the admMaxVel property of the event is assigned the value of sign.carAdmMaxVel. Following the same process inside of the agent we have the mapping rule SpeedLimitStartPerMapAgtRule. The mapping rule has as resulting event the actual perception event named SpeedLimitStartAPEvt. This one is a trigger for the SpeedLimitStartAgtRule expressed in the

*Figure 5. Example – AORML modeling*

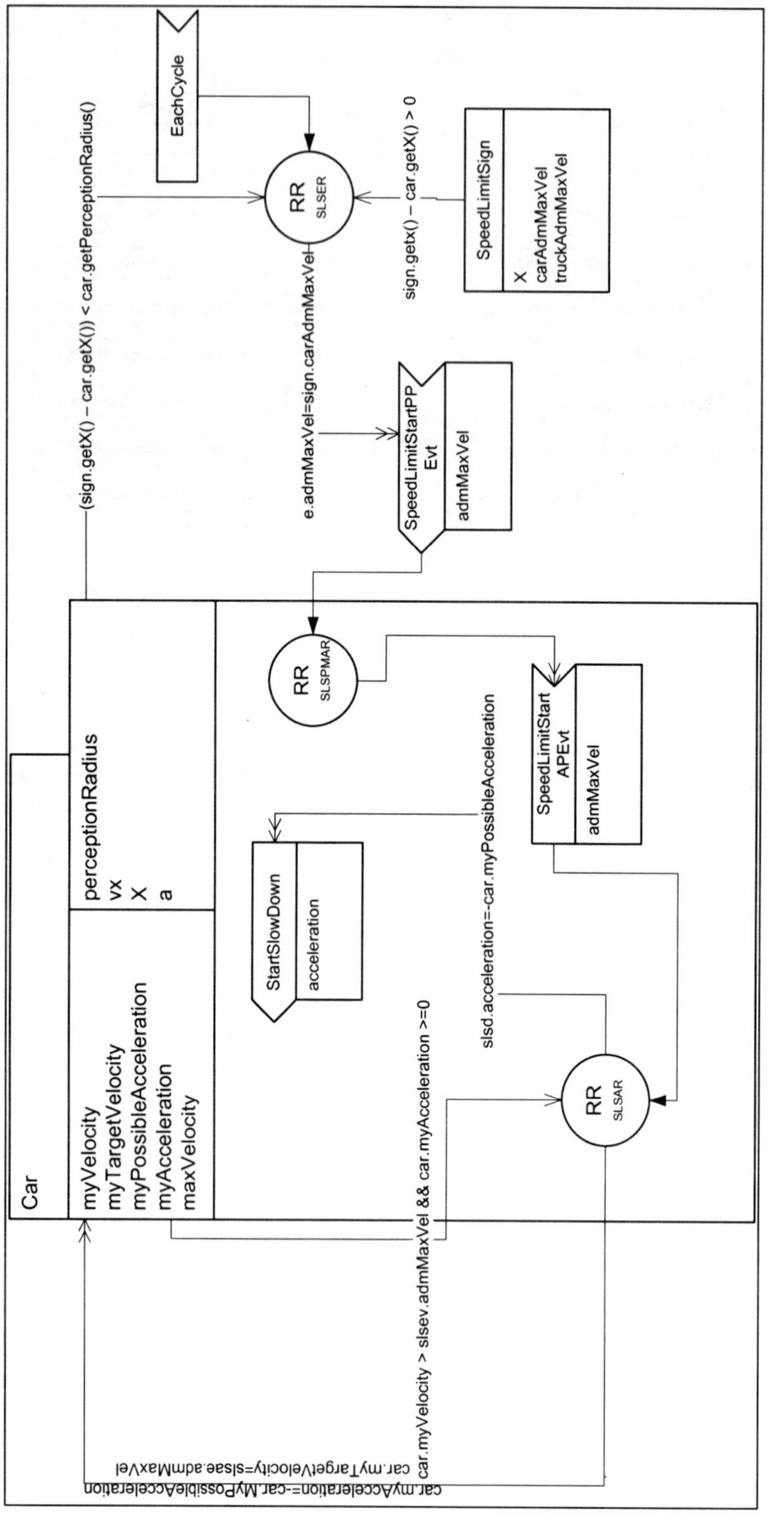

visual model with the id SLSAR. Other simulation scenarios modeled with AOR are the MIT Beer Game in van Luin & Tulba & Wagner, 2004, and in Guizzardi & Aroyo & Wagner, 2003.

## WHAT ARE THE AORS SIMULATION PRINCIPLES?

Agent-based simulation is a new paradigm applied for simulation problems.

AOR simulations can be categorized under the title "agent-based simulation" (ABS), but opposed too many other implemented systems like SWARM or Mason, AOR has a theoretical foundation, as the reader could have already noticed, in the form of a meta-model. AOR simulation scenarios are simulated as multi agent systems. In ABS agents can perform actions, perceive the environment and act upon it and react to changes in it. Agents have also a mental state comprised of knowledge/beliefs, goals, memories and commitments, and this complies with the knowledge representation of Artificial Intelligence (AI). According to Wagner, 2004 Agent-Object-Relationship Simulation (AORS) aims to be established as an ABS framework based on a high-level declarative specification language with a UML-based visual syntax and an underlying simulation meta-model as well as an abstract simulator architecture and execution model.

According to Wagner, 2004 AORS is an Agent-Based Discrete Event Simulation (ABDES) framework. In a discrete event simulation the operation of a multi agent system is represented as a chronological sequence of events. Events occur at some moment in time and their occurrence marks a change of state in the system. As stated in Ball, 1996 Discrete Event Simulation (DES) time advances either by using *event-based time progression* (also called *next event*), meaning that time is advanced based on the occurrence time of the next event, or using *incremental time progression* (also called *time slicing*), where simulated time advances in regular time steps.

AORS takes advantage of both of these methods such that it has a simulation time clock with a cycle of 100ms but the effective state change of the system is performed on the occurrence of a next event. Event occurrence time is based upon simulation time.

To develop the AORS platform standard ABDES is extended by adopting a number of ontological distinctions from AORML:

- Endurant (as they are called in UFO) entities of the system are classified in agents and objects.
- Agents have beliefs about the state of the environment and process percepts (referring to events)
- Events are either *actions events* or *non-action events* (exogenous events).
- Action events can be communicative events (messages) or non-communicative action

Exogenous events from AORML are non-action events because they are not caused by other events or external actions events. They have no real actor inside of the simulation and they are generated periodically either randomly or as successive events.

AORS is partitioned in two *conceptual subsystems*:

- The environment simulator – is the manager of all physical objects and physical state of an agent. There is only one environment simulator in the system.
- The agent simulator – manager for the internal state of an agent.

Several explanations are necessary:

- An agent in AOR has two different states: a physical one accessible *only in* the environment simulator and *only by* the environment simulator such that this particular state of an agent is changeable only by the environment simulator.

- For each agent its mental state is simulated with an agent simulator.
- There exists a logical association between the mental state of an agent and its physical state.

The state of an AORS system is composed of:

- The simulated time *t* – it represents a cycle counter. At each simulation start are automatically generated a *start* and *end date-time calendar.* Based on the cycle counter *t*, a cycle duration constant, which by default has the value of 100ms, and also based on start date-time calendar the simulated time can be transformed into date-time calendar.
- The environment simulator state consisting of:
    - Passive entities – objects (i.e. geographic position and any other user defined properties they may have)
    - Physical state of all agents (i.e. geographic position)
- Internal agent states comprising perceptions, beliefs, memory, goals and so on
- A list of future events

AORS execution cycle flow has it roots in the execution of reagents, based on Wagner, 1997, and it is a *perception-reaction* cycle.

The reactive behavior of an agent is specified by means of reaction rules grouped in sets. Mainly the cycle involves:

1. Getting the next message from a queue; check if it triggers any reaction rules; if not then repeat 1, otherwise continue
2. Check conditions for each triggered rules; if they hold then assimilate the epistemic effect of the triggered action, if any, in the knowledge base; in case it is
    a. A physical action then execute it
    b. A communicative action then send the message
3. Repeat from 1

Running an AOR simulation cycle, based on Wagner 2004, comprises the followings:

1. Environment simulator computes (determines) all current events. Current events are those events from the future events list which have occurrence time the actual time.
2. Environment simulator computes the new state based on the current state and based on current events
    a. New environment state
    b. A set of new caused events. To these events the occurrence time is set. They are created based on the environment rules and they include:
        i. New caused *environment events*
        ii. *Receive message events* that have to be received next step. Their occurrence time is $t+1$. They are derived from the corresponding current *send message events*
        iii. Next step perception events. Their occurrence time would be $t+1$. They are computed from the current *environments events* and from the current *non-communicative action events.*
    c. The environment simulator hands over to each agent simulator the current *receive message events* and *perception events* belonging to the agent being simulated.
3. Each agent simulator, based on its current internal state, based on its list of current received messages and perception events, also based on its current periodic time events and reminder events, computes:
    a. A new internal state
    b. A set of new action events. Their occurrence time will be set to $t+1$. They are computed based on internal agent rules.

4. The list of future events is updated by deleting all processed events and by adding all newly computed caused events from step 2b and new action events from step 3b.
5. If the list of future events is not empty or it was not reached the simulation end date-time then the environment simulator sets the simulated time *t*
   a. to *t+1* if it is in the *time slicing* mode
   b. to the occurrence time of the next event from the list of future events if it is *next event* mode
6. Repeat from step 1

It is assumed that simulating communication between agents and the act of perceive done by an agent might be performed with information lose. Because of the environment through which the communication takes place, information might be altered on the way. This means that in general for each external event (from the environment) and each agent the simulation system has to compute a *potential perception PP*. From this potential perception each internal agent (agent simulator) must compute a corresponding *actual perception AP*. For simplicity when the perception is done with out information lose then a PP event and the AP event are the same. This can be expressed in the following way: *E=PP=AP*. When it comes to messages, an agent sends an out message, and the simulation system has to compute a *potential in message PIM*. The receiver agent must map this to an *actual in message* event *AIM*.

In Wagner, 2004 it is argued that the abstract simulation cycle presented can be implemented in any programming language. To exemplify this in Wagner, 2004 it is introduced a Prolog prototype of an AOR Simulation System.

## Architecture of an AOR Based Simulator

This section presents the component view of an AOR based simulator. The Simulation Engine (Core) is the component that strictly follows the AOR meta-model and AOR simulation principles.

The architecture envisions the fulfillment of the following goals:

- Decoupling of components
- Ease in maintenance
- Ease in extensibility
- General enough to allow implementation on different platforms such as C++, C#, Java, PHP5

Figure 6 depicts the main components of an AOR based simulator.

These components in no particular order are: GUI, Controller, XSLT Generator, Simulation Engine (Core), and Logger.

### Controller

The Controller is the system coordinator and it must take care mainly of the following tasks:

- Communicates with GUI
- Performs open, save , run simulation and any other similar operations
- Creates and uses the XSLT Generator
- Creates and configures the Logger
- Requests reports from logger
- Communicates with other different components such as viewers or players.

### XSLT Generator

This component is a special one. It receives AORSL description files from the controller and based on them it generates simulation files. It also sends status of the process to the Controller. This component receives at input XML files in AORSL format and generates using XSLT processing instructions the simulation files in a platform dependent format such as: Java, C++, and so on.

*Figure 6. AORS system architecture – components view*

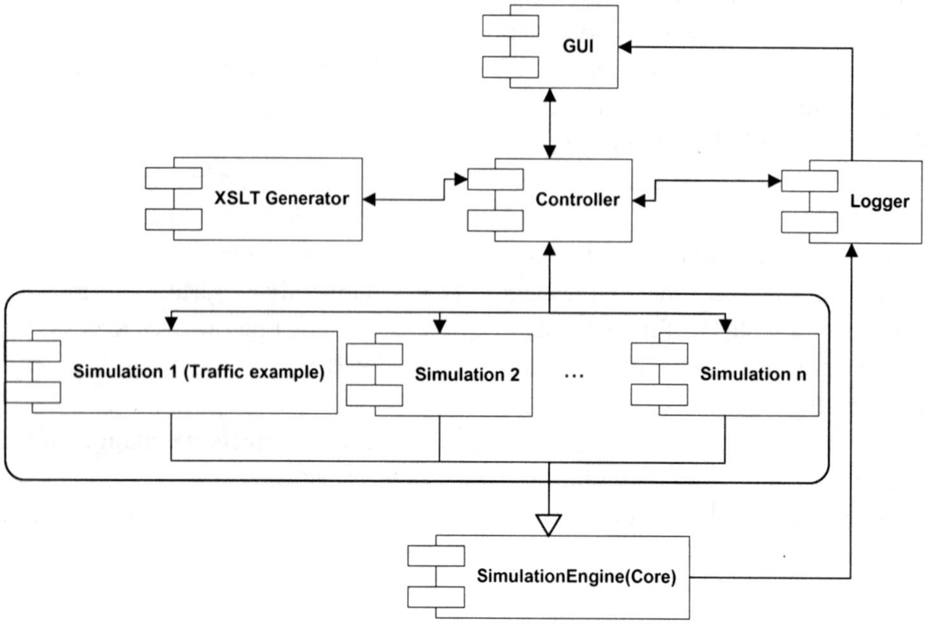


## GUI

The main functionality is of interfacing between the user and the system coordinator – Controller. Through it the user controls the system. It also displays information received either from the Controller or from the Logger component. Basic functionality such as: open, save, run is forwarded directly to the Controller. GUI should have no connection with the Simulation Engine, and changes regarding simulations should be received through the Logger.

## Logger

It is mainly a *data controller* and the source for all kind of reports. Receives all changes from the Simulation Engine, and communicates specific changes to different other components or sub components such as GUI, report managers and so on. It must also be capable of data serialization in different formats and to different location types such as: XML, DB, console etc. "Cache-like" functionality are also envisioned.

## Simulation Engine

It is the core of the system that actually performs the simulations. It follows the AOR meta-model and AOR simulation principles. The files generated by the XSLT generator extend the Simulation Engine; they add concrete implementations to abstract methods from Simulation Engine. The generated classes are identified in Figure 6. by the name "Simulation no.". The core can not run without the generated classes. In fact the controller runs a simulation from the generated classes.

Next section introduces the distributed architecture of an AOR Simulation system.

# AOR Distributed Architecture

The distributed architecture of an AOR simulation system is depicted in Figure 7.

The AOR distributed architecture is composed of the AOR system and the subjective agents that run remotely. The decomposition in this specific way is not random. It is based on the conceptual decomposition, introduced by AOR, between the objective world and the subjective world. The AORS system comprises two main components SimulationEngine and EnvironmentSimulator.

The SimulationEngine encapsulates the communication component and a fact base containing the environmentEvents. Internally environmentEvents are exchanged between the communication component and the SimulationEngine fact base. The SimulationEngine fact base exchanges environmentEvents with the InferenceEngine of the EnvironmentSimulator.

The EnvironmentSimulator has three main logical components: ProductionMemory, InferenceEngine, WorkingMemory.

ProductionMemory comprises all EnvironmentSimulator rules. Against them facts stored in WorkingMemory are matched. The WorkingMemory contains the agents' objectiveState.

The InferenceEngine of the EnvironmentSimulator is an event based one. Mainly its functionality comprises the event detection performed by the EventMatcher. Following the event detection is rule selection, and then PatternMatching or condition evaluation based on WorkingMemory facts. The result of rule execution is either a change in the WorkingMemory or the apparition of new events, or both.

A remote SubjectiveAgent has a similar architecture as illustrated in Figure 7. A remote SubjectiveAgent's WorkingMemory contains beside beliefs also internalEvents. Between the InferenceEngine and CommunicationComponent PercepetionEvents and ActionEvents are exchanged. PerceptionEvents are sent towards the InferenceEngine, coming from the remote AOR system. ActionEvents are sent to CommonicationComponent from the InferenceEngine.

The communication between remote SubjectiveAgents and AOR system is done based on no specific communication protocol, such as HTTP.

Communication assumes mainly the following: communication establishing, message production and message consumption.

## AOR SIMULATION AS BUSINESS PROCESS

The Business Process Management Initiative (BPMI) has developed a standard Business Process Modeling Notation (BPMN), OMG, 2008. The primary goal of BPMN is to provide a notation that is readily understandable by all business users such that business processes can be illustrated using a standard notation understandable also to business analysts that create the initial drafts of the processes, continuing towards technical developers responsible for implementing the technology that will perform those processes, and finally going on, towards the business people who will manage and monitor those processes.

A more elaborated and refined introduction to business processes is given in Weske, 2007. The book tries as the author recalls, mainly to narrow the gap between two different points of view underlined by two different communities: researchers with a background in formal methods interested in investigating structural properties of processes and the second group consisting of software community interested in providing robust and scalable software systems.

Recalling from Taveter and Wagner, 2001a Agent Orientation is highly significant for business information systems since business processes are driven by and directed towards agents (or actors), and hence has to comply with the physical and social dynamics of interacting individuals and institutions. Beside these reasons there are also several that are raised from the definitions given to business processes in Weske, 2007 and agent definition from Wooldridge, 2002. In Weske, 2007 was defined a business process management system as "a generic software system that

*Figure 7. AORS system distributed architecture*

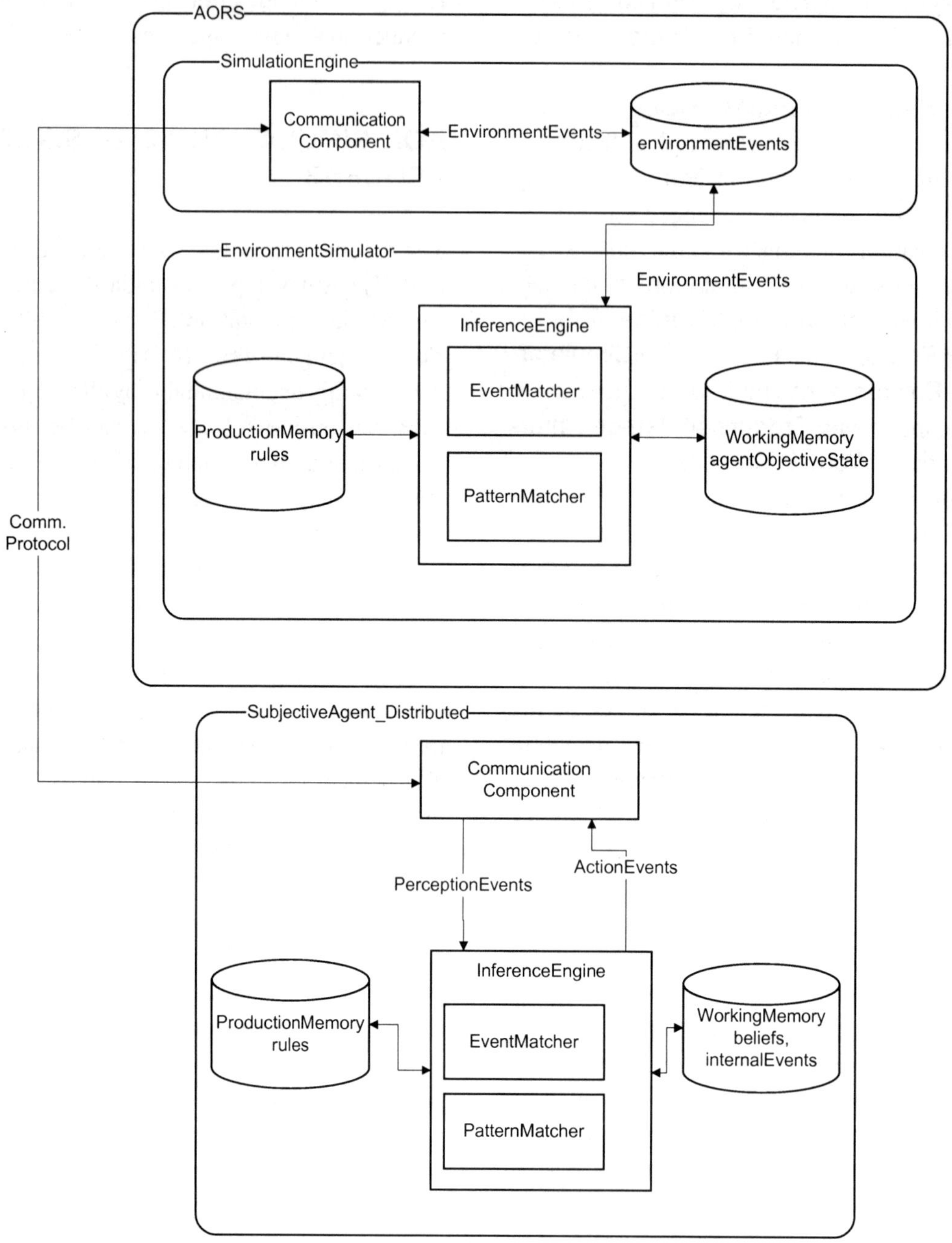


is driven by explicit process representations to coordinate the enactment of business processes." On the other side in Wooldridge, 2002 was argued that "an agent is a computer system that is capable of independent action on behalf of its user or owner".

AOR simulation cycle that was presented in "Which are the AOR simulation principles" section is modeled in Figure 8 and Figure 9 as business process.

The AORS System as depicted in Figure 8 comprises two lanes: SimulationEngine and

EnvironmentSimulator. The SimulationEngine is the coordinator of the simulation process. The simulation process starts in the SimulationEngine lane. Then the whole system is initialized and initial events are created and also the EnvironmentSimulator is initialized and inside the EnvironmentSimulator lane the agents ObjectiveState is initialized. Before starting the simulation process the startTime is saved. The process continues and the simulation cycling comes next. A simulation cycle starts by sending the AOR events that need to be processed by the EnvironmentSimulator. One by one the AOR events are verified if they trigger any EnvironmentRule. The triggered rules are executed. When all AOR events are exhausted then the new computed AOR Events are sent to the SimulationEngine. The addEvents task is in charge with this. The Figure 8 illustrates the fact that addEvents task from SimulationEngine performs an update on events object. Now the process continues divided with two parallel activities. There is produceMessages task. This task produces corresponding messages for the distributed SubjectiveAgents. The Figure 8 illustrates by means of a complex event based gateway followed by throwing messages the activity of sending asynchronous messages to the remote SubjectiveAgents.

The second activity that runs in parallel with produceMessages is receiveActionEvents task. This task is a special one because is time based. The system listens for incoming ActionEvents a predefined time period, for example 10 seconds. As illustrated in Figure 8 even though the produceMessages finishes, the process does not continue until the time quanta has not expired. It is important to notice that the receiveActionEvents listens for incoming ActionEvents from remotely SubjectiveAgents. Also a keen observation is that if for particular reasons a remote SubjectiveAgent does not succeed in providing ActionEvents in the specified time quanta then those ActionEvents will be lost. The receiveActionEvents task updates the events object with the received ActionEvents. After the expiration of the time quanta then the process goes on with computeNewExogenousEvents task and deleteProcessedEvents task. These two tasks can not be performed in parallel because both depend on event's occurrenceTime property. Running them in parallel will obviously conduct in information lose.

This cycle is run as long as there are any simulationSteps left. If no simulationSteps have to be simulated then the simulation endTime is saved and the simulation ends.

The business process that drives all the distributed SubjectiveAgents is depicted in Figure 9. Distributed SubjectiveAgents are started independently by the AORS System, as is denoted through the use of the start event. initializeSubjectiveAgent task follows immediately after the start event. It is in charge with creating the initial state of the system: creation of initial beliefs and initial internal events. consumeMessage follows the initialization task. A distributed SubjectiveAgent waits for the incoming of a message from the remote AORS System. The received message is consumed, meaning by that the extraction of AORS Event that has been transported. It is then checked if the AORS received message triggers any rule. If yes then the rule is executed. The result of a rule execution may conduct in an internal state change (update) of beliefs and also to the change of internalEvents. It means that new internal events can be created. Also ActionEvents can be created. The newly created ActionEvents are sent back to the remote AORS System. This is depicted with the help of an intermediate message event. The process goes further either after the rule has been executed or the new ActionEvents have been sent if the received message triggered any rule, or in the case the event didn't trigger any rule immediately. All internalEvents are to be processed next. The events comprised in the internalEvents list are checked one by one if they trigger any rule. If yes, then like previously explained, all new ActionEvents that are

*Figure 8. AOR system - business process*

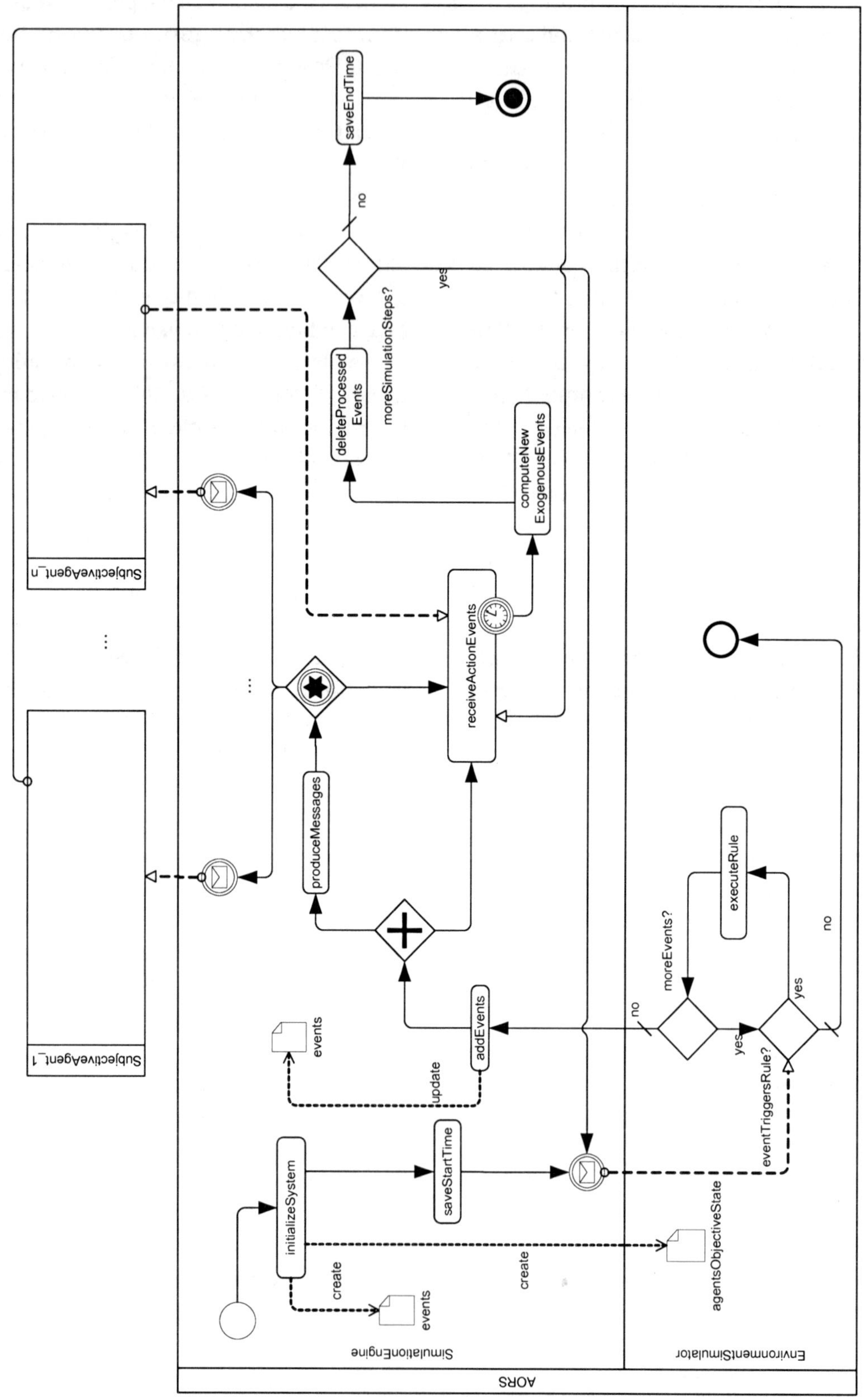

*Figure 9. Distributed SubjectiveAgent - business process*

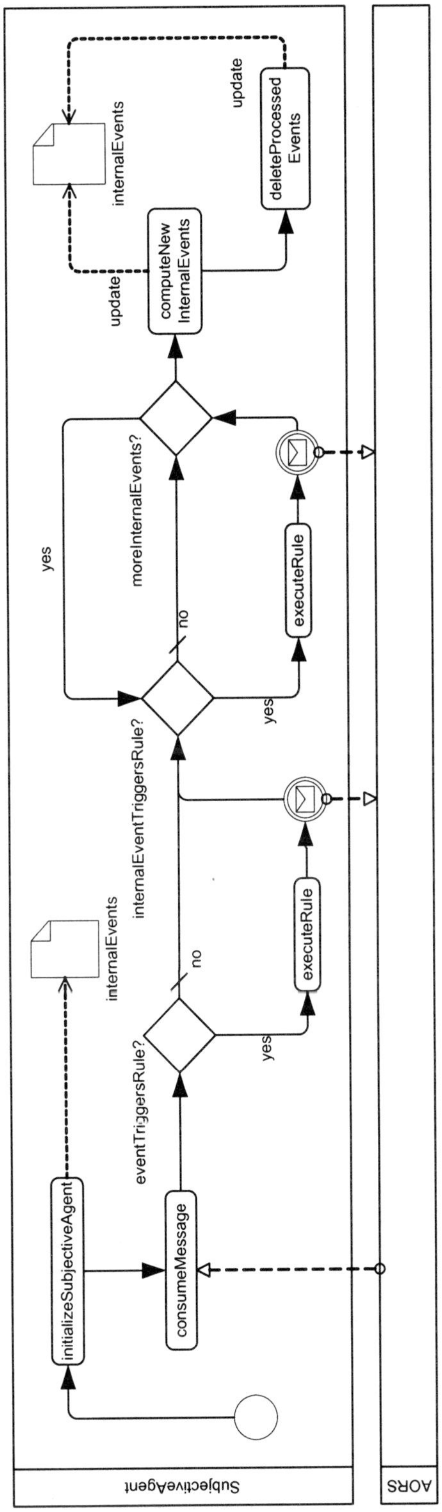


computed are sent back to the remote AORS System. Also all necessary updates to the internal state are performed. The process continues with two other tasks: computeNewInternalEvents and deleteProcessedEvents. They all update internalEvents object. Here a question can be raised: Why these two tasks can not be performed in parallel? Well the answer comes from the AORS simulation principles. New events are computed based on the occurrenceTime (which is in fact simulationStep) of the old events. The deleteProcessedEvents task is based also on the occurrenceTime of events. Running these tasks in parallel obviously conducts to information lose. The process of a distributed Subjective Agent is not finished programmatically, after the simulation ends it remains running waiting for new messages from the remote AORS System.

The section presented the AOR simulation as a business process. Future work envisions a more important and complex problem, the one of mapping AOR fundamental concepts to business processes concepts.

## WHERE IS AOR HEADING TO AND AOR OPEN PROBLEMS?

Right now the AOR system is a reactive agent-oriented system, as the reader already knows. Yet AOR evolving curve is not finished, because the final destination point is the one of modeling and simulating vivid agents Wagner, 1997. According to Wagner, 1997 a *vivid agent* is a software-controlled system whose state is represented by a knowledge base, and whose behavior is represented by means of *action* and *reaction rules.*

Vivid agents' behavior is a combination of reactive and proactive acting through a nondeterministic interleaving of perceptions, reactions, planning and plan execution. Vivid agents are also called in Wagner, 1997 *full agents* that can generate and execute plans for solving certain tasks. Such an agent consists of a *mental state*, an *event queue* that records perception events, a *set of action rules* and a *set of reaction rules.* Actions are deliberately planned to solve a certain task while reactions may be independent from the current believe state.

The mental state comprises a *knowledge base* representing agent's beliefs, a *set of tasks the agent has accepted* to perform, a *set of tasks the agent has to perform* continuously and a *set of current intentions* i.e. goals/plan pairs.

Vivid agent execution cycle comprise *planning for permanent tasks, planning for temporary tasks, execution of the last action of a plan, plan execution* and *re-planning.*

The state of an agent, according to Wagner, 1997 page 29, is transformed by five kinds of transitions: event perception, reaction in response to an event, action, re-planning and planning.

Actually AOR does not provide any means for simulating commitments and claims and also does not define semantics for them. Rights and duties are in the same situations.

AOR supports only simple beliefs and future versions envision the complete belief feature in the form of belief triples. It also has in plan to extend the AOR meta-modeling with activities and goals.

## CONCLUSION

The chapter has given an overview of the Agent-Object Relationship (AOR) paradigm that comes with the ontological distinction between active and passive entities that helps in capturing more semantics about the natural and artificial systems in the form of information systems and their interrelationships. The main concepts of AOR have been exemplified with the help of an example. The chapter introduces architectural design of a distributed AOR system, and last but not the less the AOR simulation process is modeled as a business process. The modeling as a business process is a starting point for future research that includes modeling of agent behavior as business processes using BPMN notation.

## REFERENCES

Ball, P. (1996). *Introduction to discrete event simulation.* University of Strathclyde.

Bradshaw, J. M. (1997). Introduction to software agents. In J.M. Bradshaw (Ed.). *Software agents,* (pp. 3-46). AAAI Press.

Degen, W., Heller, B., Herre, H., & Smith, B. (2001). GOL: Towards an axiomatized upper level ontology. In B. Smith & N. Guarino (Eds.), In *Proceedings of FOIS'01.* Ogunquit, Maine, USA, October 2001. ACM Press.

Guizzardi, G., & Wagner, G. (2004, June 8).Towards ontological foundations for agent modeling concepts using the Unified Foundational Ontology (UFO). In P. Bresciani, P. Giorgini, B. Henderson-Sellers, G. Low & M. Winikoff (Eds.). *Revised Selected Papers of the 6th International Bi-Conference Workshop on Agent-Oriented Information Systems* (LNCS 3508, pp. 110-124).

Guizzardi, R.S.S., Aroyo, L., & Wagner, G. (2003). Agent-oriented knowledge management in learning environments: A peer-to-peer help-desk case study. In *Agent Mediated Knowledge Management, International Symposium AMKM* (LNCS 2926, pp. 57-72).

Kluegl, F., & Puppe, F. (1998). The multi-agent simulation environment SeSAm. In *Proceedings of Workshop "Simulation in Knowledge-based Systems".* Report tr-ri-98-194, Reihe Informatik, Universität Paderborn.

Kluegl, F., Herrler, R., & Oechslein, C. (2004). From simulated to real environments: How to use SeSAm for software development. In *Multi Agent System Technologies* (LNCS 2831, pp. 13-24).

Luke, S., Balan, G.C., Panait, L., Cioffi-Revilla, C., & Paus, S. (2003). MASON: A Java multi-agent simulation library. In *Proceedings of the Agent 2003 Conference.*

Luke, S., Cioffi-Revilla, C., Panait, L., & Sullivan, K. (2004). MASON: A new multi-agent simulation toolkit. In *Proceedings of the 2004 SwarmFest Workshop.*

van Luin, J., Tulba, F., & Wagner, G. (2004, May). Remodeling the beer game as an agent-object-relationship simulation. In C. Espinasse (Ed.). *Proceedings of the 5th workshop on Agent-Based Simulation (ABS2004)* (pp. 5-10). Lisbon, Portugal.

Minar, N., Burkhart, R., Langton, C., & Askenazi, M. (1996). *The swarm simulation system: A toolkit for building multi-agent simulations.* SFI Working Paper 96-06-042.

North, M.J., Howe, T., Collier, N., & Vos, J.R. (2005a). The repast simphony runtime system. In *Proceedings of the Agent 2005 Conference on Generative Social Processes, Models, and Mechanisms, ANL/DIS-06-5* (pp. 1-8). Argonne National Laboratory and The University of Chicago.

North, M.J., Howe, T., Collier, N., & Vos, J.R. (2005b). The repast simphony development environment. In *Proceedings of the Agent 2005 Conference on Generative Social Processes, Models, and Mechanisms, ANL/DIS-06-5* (pp. 159-166). Argonne National Laboratory and The University of Chicago.

Odell, J., van Dyke Parunak, H., & Bauer, B. (2000). Extending UML for agents. In G. Wagner, Y. Lesperance, & E. Yu (Eds.), *Proceedings of the 2nd International Workshop on Agent-Oriented Information Systems.* Berlin, Germany: iCue Publishing.

OMG (2008). *Business process modeling notation.* V1.1, OMG

Railsback, S.F., Lytinen, S.L., & Jackson, S.K (2006). Agent-based simulation platforms: Review and development recommendations. *Simulation, 82*(9), 609-623.

Russell, S. & Norvig, P. (1995). *Artificial intelligence: A modern approach.* Englewood Cliffs, NJ: Prentice Hall.

Shoham, Y. (1993). Agent-oriented programming. *Artificial Intelligence, 60*, 51-92.

Tatara, Eric & North, Michael J. & Collier, Nick & Vos, Jerry R. (2006). An Introduction to Repast Modeling by Using a Simple Predator-Prey Example. In *Proceedings of the Agent 2006 Conference on Social Agents: Results and Prospects* (pp. 1-11). Argonne National Laboratory and The University of Chicago.

Taveter, K., & Wagner, G. (2001a, September). Agent-oriented enterprise modeling based on business rules. In *Proceedings of International Workshop on Open Enterprise Solutions: Systems, Experiences and Organizations (OES-SEO2001).* Rome.

Taveter, K., & Wagner, G. (2001b, November). Agent-oriented enterprise modeling based on business rules. In *Proceedings of the 20th International Conference on Conceptual Modeling (ER2001)* (LNCS 2224, pp. 527-540).

Wagner, G. (1997). Vivid agents - How they deliberate, how they react, how they are verified. (Extended version of G. Wagner, Logical and operational model of scalable knowledge and perception-based agents). In W. Van de Velde and J.W. Perram (Eds.), *Agents breaking away* (LNCS 1038).

Wagner, G. (2000, May). Agent-oriented analysis and design of organizational information systems. In J. Barzdins & A. Caplinskas (Eds.). *Proceedings of 4th International Workshop on Databases and Information Systems.* Vilnius, Lituania.

Wagner, G. (2002, July). A UML profile for external AOR models. In *Proceedings of Third International Workshop on Agent-Oriented Software Engineering (AOSE-2002)* (Vol. 2585). Bologna, Italy: Springer-Verlag.

Wagner, G. (2003, July). The agent-object-relationship meta-model: Towards a unified view of state and behavior. In *Information Systems, 28*(5), 475-504.

Wagner, G. (2004). AOR modeling and simulation: Towards a general architecture for

agent-based discrete event simulation. In P. Giorgini, B. Henderson Sellers & M. Winikoff (Eds.), *Proceedings of Agent-Oriented Information Systems, 5th International Bi-Conference Workshop, AOIS 2003, Melbourne, Australia and Chicago, IL, USA* (LNCS 3030, pp. 174-188).

Welty, C., & Guarino, N (2001). Supporting ontological analysis of taxonomic relationships. *Data and Knowledge Engineering, 39*(1), 51-74.

Weske, M. (2007). *Business process management: Concepts, languages, architectures.* Springer.

Wilensky, U. (1999). *NetLogo: Center for connected learning and computer based modeling.* Northwestern University. Evanston, IL, USA.

Wooldridge, M. (2002). *An introduction to multi agent systems.* John Wiley & Sons.

## KEY TERMS AND DEFINITIONS

**ABS:** Agent based simulation

**Agent:** Entity perceiving the environment through sensors and acting upon that environment through effectors

**AOR Modeling:** Agent oriented relationship modeling

**AORS:** Agent oriented relationship simulation

**Business Rules:** Business rules or business rule sets describe the operations, definitions and constraints that apply to an organization in achieving its goals.

**DES:** Discrete event simulation

**MAS:** Multi agent systems

# Chapter XVI
# Developing Rule-Based Applications for the Web: Methodologies and Tools

**Vassilis Papataxiarhis**
*National and Kapodistrian University of Athens, Greece*

**Vassileios Tsetsos**
*National and Kapodistrian University of Athens, Greece*

**Isambo Karali**
*National and Kapodistrian University of Athens, Greece*

**Panagiotis Stamatopoulos**
*National and Kapodistrian University of Athens, Greece*

**Stathes Hadjiefthymiades**
*National and Kapodistrian University of Athens, Greece*

## ABSTRACT

*Embedding rules into Web applications, and distributed applications in general, seems to constitute a significant task in order to accommodate desired expressivity features in such environments. Various methodologies and reasoning modules have been proposed to manage rules and knowledge on the Web. The main objective of the chapter is to survey related work in this area and discuss relevant theories, methodologies and tools that can be used to develop rule-based applications for the Web. The chapter deals with both ways that have been formally defined for modeling a domain of interest: the first based on standard logics while the second one stemmed from the logic programming perspective. Furthermore, a comparative study that evaluates the reasoning engines and the various knowledge representation methodologies, focusing on rules, is presented.*

## 1. INTRODUCTION AND MOTIVATION

Nowadays, with the evolution of traditional web of documents to a more complex web of services, an increasing demand for embedding intelligence to Web applications arises. In this context, the efficient management of knowledge seems to play a key role in order to achieve smart behavior of Web applications and to overcome several issues of such environment (e.g., information integration). Ontologies, mainly written with Semantic Web technologies, constitute a well-established paradigm for representing knowledge on the Web. Though, current efforts are focused on extending ontologies with more expressive forms of knowledge like rules. In fact, given the state-of-the-art in the realization of the Semantic Web vision, rules constitute the next prominent challenge. Since the ontology layer of the Semantic Web architecture stack has reached a sufficient degree of maturity through Web Ontology Language (OWL) (Dean et al., 2004), the next step of progress involves the integration of rules with ontologies, most of them based on subsets of First Order Logic (FOL).

Rules are capable of extending the expressiveness provided by ontology languages through the definition of more complex relationships between individuals. Additionally, as a modular form of knowledge, they fit well in domains like personalization, policies and business-to-business (B2B) interaction. However, it has been shown that extending ontologies even with simple forms of rules can lead to undecidability of key inference problems.

On the other hand, many business-logic applications have extensively taken advantage of existing rule management systems or solvers (Jess, 2008; ILOG, 2008; Drools, 2008), aiming at facilitating the knowledge management process. As a result, the success of rules in non-Web applications moved Web researchers to use traditional rule engines on the Web.

However, the aforementioned stable rule systems have not been originally created for open and heterogeneous environments like the Web. Such platforms have adopted different knowledge representation formalisms, mainly based on principles of logic programming, instead of classical logic. As a consequence, they differ from recent Semantic Web technologies in many aspects, including representational features and reasoning functionality, as well. Hence, building a rule-based application for the Web with existing rule technologies is not a straightforward task.

In the rest of this chapter, we provide foundational knowledge on this topic together with implementation issues, techniques and design patterns. Section 2 briefly describes how the things have gone so far in the area of Web knowledge representation formalisms. In Section 3, various knowledge representation methodologies and tools are discussed. Specifically, Section 3.1 demonstrates the different languages and formalisms, derived from both classical logic and logic programming view, while Section 3.2 focuses on various engines able to reason over such knowledge bases. Section 4.1 gives the main requirements for rule-based web applications. The evaluation presented includes both a qualitative comparison (Section 4.2) of the existing approaches and a performance analysis (Section 4.3) of current ontology reasoners and rule engines. Finally, several future trends and open issues are identified in Section 5. Hence, this chapter aims at becoming a helpful guide for applying rules to Web applications.

## 2. THE STORY SO FAR

The knowledge representation languages proposed (see Section 3.1) for representing knowledge on the Web are based either on the Classical Logic (CL) perspective or on Logic Programming[1] (LP). As a result, a debate was started between the Database community and AI researchers, respectively, in order to determine the more suitable of the two approaches in the formalization of Web knowledge. Additionally, different languages of the same perspective, providing various degrees of expressivity, have been proposed. Hence, the integration of knowledge with Web applications was more complicated. Recently, with the evolution of Semantic Web technologies, these modeling paradigms have been extensively discussed by researchers (Motik, Horrocks, Rosati, & Sattler, 2006; Patel-Schneider & Horrocks, 2006; Eiter, Ianni, Polleres, Schindlauer, & Tompits, 2006a; Donini, Lenzerini, Nardi, & Schaerf, 1998; Antoniou, et al., 2005; Boley, Kifer, Patranjan, & Polleres, 2007).

Clearly, this confrontation also affected rules. Although, significant efforts have been devoted in order to develop appropriate rule languages for the Web (see Section 3.1), a debate between proponents of the different perspectives has appeared. The main argument involves the degree that rules would be combined with ontologies. Initially, as formulated by Tim Berners-Lee, it had been commonly accepted that Semantic Web should be structured in a hierarchy of layers that seamlessly interoperate (Antoniou & van Harmelen, 2004). However, some proposals (mainly stemmed from the database community) preferred to keep the rules and the ontology layer separated, in order to preserve expressiveness and decidability of reasoning process, as mentioned in (Eiter et al., 2006a). Specifically, in (Kifer, de Bruijn, Boley, & Fensel, 2005) the authors argue that the rules layer of Semantic Web architecture should be placed next to the ontology layer of the Semantic Web stack in order to take advantage of important rule languages based on logic programming. However, in (Horrocks, Parsia, Patel-Schneider, & Hendler, 2005) the authors criticize the aforementioned approach and denote that such a distinction would lead to two Semantic Webs based on different semantics. These discussions along with the Semantic Web Activity workgroup have led to several modifications of the stack. Figure 1 presents the initial and the current form of the Semantic Web stack diagram specified by the W3C Semantic Web Activity Group (W3C Semantic Web Activity, 2008).

The aforementioned facts gave rise to confusion inside the Web community. If the experts (i.e., the logicians stemming from both AI and databases) could not decide and recommend a unified framework for formalizing knowledge, how Web developers and users should choose the best language according to their needs? The current solutions are either to study the capabilities provided by all these languages (i.e., available expressiveness, support by existing reasoning

*Figure 1. a) Tim Berners-Lee Semantic Web layer "cake" and b) the latest form of Semantic Web stack diagram (adapted from W3C Semantic Web Activity, 2008)*

a)

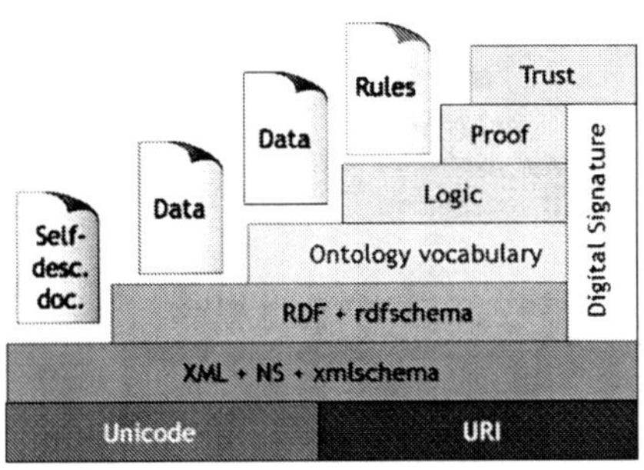


b)

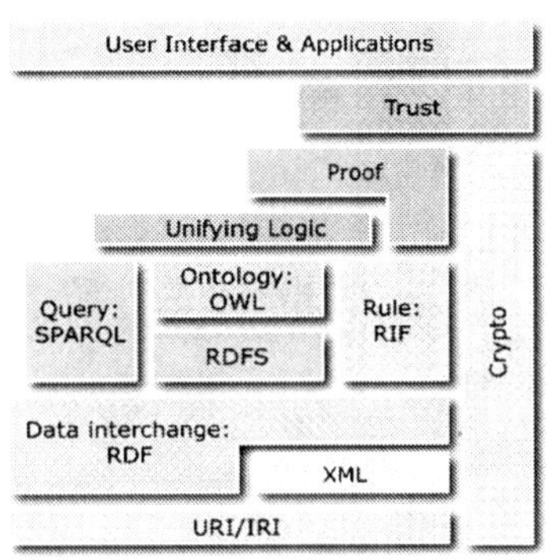

engines, tractability etc.) or to exclude knowledge from their applications.

## 3. KNOWLEDGE REPRESENTATION METHODOLOGIES AND TOOLS

This section surveys different knowledge representation methodologies along with reasoning modules that have been proposed for managing knowledge on the Web. After an overview of Description Logics that comprise the main formalism for representing ontological knowledge in Web applications, we focus on rules.

### 3.1 Knowledge Representation Formalisms

#### Description Logics

Description Logics (DLs) (Baader, Calvanese, McGuiness, Nardi, & Patel-Schneider, 2003) are subsets of first-order-logic, originating from semantic networks and frame-based systems. They constitute a family of knowledge representation languages that aim at providing well-understood mechanisms in order to formalize knowledge that describes a domain. This way, DLs are equipped with formal, logic-based semantics, emphasizing on reasoning process. Typical reasoning tasks include consistency checking of the knowledge base, concept satisfiability, instance checking etc.

A DL-based knowledge base is composed of two components: TBox and ABox. TBox contains the vocabulary of the application domain, called *terminology,* as well as axioms based on that vocabulary. Practically, such vocabulary consists of *concepts* and *roles.* Concepts are generic descriptions of sets of individuals, while roles constitute binary predicates for defining properties of the individuals. On the other hand, ABox includes assertions of individuals that may refer to either concepts or roles. For example, a statement declaring that a specific individual is instance of a concept resides in ABox, while a statement denoting that "every human is mortal" belongs to TBox. Figure 2 shows the generic architecture of a DL knowledge representation system.

Each description logic language is determined by a set of constructs, enabling the use of atomic concepts and atomic roles in order to define complex ones. These constructs directly affect the expressive power of the language and, thus, the complexity of inference tasks. As a result, the selection of the appropriate description logic language in order to describe a specific domain includes the examination of the imposed requirements for representation expressiveness.

*Figure 2. Architecture of a Description Logics knowledge-based system*

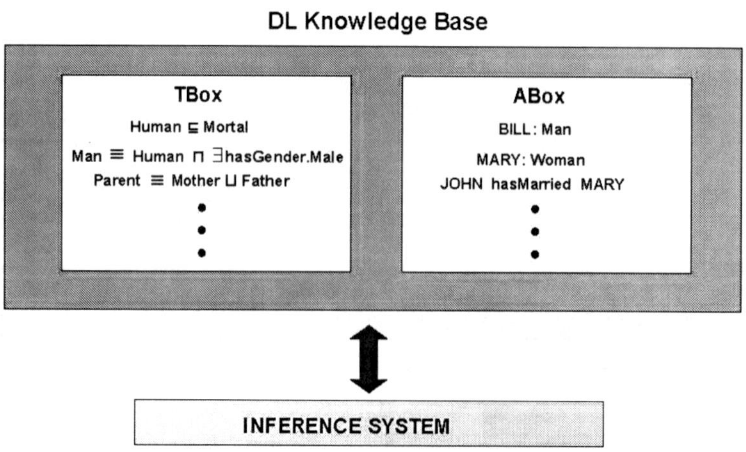

DLs set up the base for the definition of Resource Description Framework (RDF) (Klyne & Caroll, 2004), RDF Schema (Brickley & Guha, 2004) and Web Ontology Language (OWL) (Dean et al., 2004) that constitute W3C standards for representing knowledge on the Web. Specifically, RDF is a simple data model for representing information on the Web. RDF statements are expressed in the form of *(subject, predicate, object)* triples. A set of such statements can be viewed as a graph where the subjects and the objects of these statements constitute the graph nodes while the predicates correspond to the graph edges. RDFS is an extension of RDF for expressing simple taxonomies through the definition of class/property hierarchies and domain/range of properties. OWL is totally based on DLs and it comes in three species: OWL-Lite, OWL-DL and OWL-Full. OWL-Full is the most expressive OWL species, since it takes advantage of all the OWL language primitives. OWL-DL limits OWL-Full to a subset of OWL primitives so as to achieve efficiency of reasoning. Finally, OWL-Lite is a sublanguage of OWL-DL by restricting more its expressiveness.

## Description Logic Programs

The expressiveness of Description Logic Programs (DLP) approach (Grosof et al., 2003) corresponds to a fragment of OWL defined by the expressive intersection of Description Logics and logic programming. This approach intends to define a mapping from DL to logic programming (specifically, Horn programs in which no function symbols, negation and disjunction are allowed) and vice versa. An instance of such a mapping in the case of conjunction follows:

$$C_1 \sqcap C_2 \sqsubseteq D \quad \equiv \quad D(x) \leftarrow C_1(x) \wedge C_2(x) \qquad (1)$$

However, for the sake of decidability, DLP offers limited expressiveness, since the aforementioned mapping covers only a few DL-constructs (in particular, conjunction, disjunction and quantification restrictions). For example, Description Logic Programs do not cover negation in class descriptions nor fully support cardinality restrictions.

## Semantic Web Rule Language

Semantic Web Rule Language (SWRL) (Horrocks et al., 2004b) is probably the most popular formalism in Web community for expressing knowledge in the form of rules. Specifically, SWRL is based on a combination of Web Ontology Language (OWL) (Dean, 2004) and Rule Markup Language (RuleML) (RuleML, 2008) and has been proposed as a W3C candidate standard for formalizing the expression of rules in Web context. Contrary to DLP, SWRL extends OWL-DL with a specific form of Horn-like rules.

The proposed rules are in the form of an implication between the body and the head of the rule. A typical SWRL rule can be of the following form:

$$a_1 \wedge a_2 \wedge \ldots \wedge a_n \rightarrow b_1 \wedge b_2 \wedge \ldots \wedge b_m \qquad (2)$$

where $a_i$ and $b_i$ are OWL atoms of the following forms:

- Concepts, e.g., C(x), where C is an OWL description, in general, and x is either a variable, an OWL individual or a data value.
- Object properties, e.g., P(x,y), where P is an OWL property and x, y are either variables, individuals or data values.
- Datatype properties, e.g., P(x,y), where P is an OWL property, x is variable or individual, while y is a data value.
- B(x1,x2,...), where B is a built-in relation and x1, x2, ... are either variables, individuals or data values.
- sameAs(x,y) or differentFrom(x,y) where x, y are either variables, individuals or data values.

The main advantage of SWRL is the simplicity it offers, while extending the expressiveness of OWL. Another benefit of SWRL is its compatibility with OWL syntax and semantics, since they are both combined in the same logical language. On the other hand, extending OWL-DL with SWRL rules leads to undecidability of simple inference problems. A possible solution of this problem is presented in (Motik, Sattler, & Studer, 2005) which introduces the notion of *DL-safe* rules. Specifically, that approach restricts the application of SWRL rules only to individuals of the ABox part of the DL knowledge base. Moreover, SWRL does not support negation (neither classical nor negation as failure - NAF) and disjunctions. Finally, there is no efficient support of first-order provers to execute reasoning over SWRL. Usually, the SWRL rules are translated to existing rule systems (e.g., Jess (O'Connor, Knublauch, Samson, & Musen, 2005)) that handle the reasoning tasks partially, since they are not aimed to manage knowledge expressed in terms of first-order logic or subsets.

In the case of a Web service composition paradigm, some SWRL example rules could be the following:

profile:hasOutput(?S1,?out) ∧ profile:hasInput(?S2,?in) ∧ process:parameterType(?in, ?cin) ∧ process:parameterType(?out, ?cout) ∧ (rdfs:subClassOf(?cout, ?cin) ∨ rdfs:subClassOf(?cin, ?cout)) → composableWith(?S1, ?S2)

(3)

The abovementioned rule captures the knowledge that a web service S1 is composable with a service S2, if an input of S1 is either subclass or superclass of at least one output of S2. This rule uses specific namespaces of OWL-S ontology like "profile" and "process".

## Answer Set Programming

Answer Set Programming (ASP) (Gelfond & Lifschitz, 1991) is a paradigm for knowledge representation and declarative programming. It has several advantages compared to other logic programming paradigms (e.g., Prolog) such as:

- *Full declarativity*: the order of rules in a program is not important.
- *ASP programs are in general decidable*
- *Non-monotonic inference*: both negation as failure (NAF) and strong negation are supported, thus enabling default reasoning and reasoning under the Closed World Assumption (CWA).
- *Availability of efficient solvers*: there are several ASP solvers that are scalable enough to deal with large knowledge bases.

A general ASP rule is of the following form:

$$a_1 \vee a_2 \vee \ldots \vee a_n \leftarrow b_1 \wedge \ldots \wedge b_k \wedge \text{not } b_{k+1} \wedge \ldots \wedge \text{not } b_m \quad (4)$$

where $a_i$ and $b_j$ are literals (atoms or strong negations of atoms) and *not* denotes NAF.

A set of such rules is an ASP program. What is interesting is the fact that these rules can have disjunctions in their head. This is a very important feature of ASP, since it introduces non-determinism in the inference process (i.e., an ASP program may have several models which are called *answer sets*).

In order to be able to use ASP on the Web, the ASP rules should be combined with Web knowledge. Since ontologies is the most common way to represent knowledge on the Web, an interaction between ASP programs and ontologies is deemed necessary. A solution to this integration problem is description-logic programs (or dl-programs). These consist of ASP rules that may contain queries to DL knowledge bases. For example, the following rule "brings" into the ASP program all instances of the class MovieTitle in the DL knowledge base:

movie(X) ← DL["Movie"](X). (5)

Several extensions to dl-programs were proposed (and implemented) so that they become more "suitable" for open environments like the Web, where information may be expressed in many diverse ways (e.g., multiple different ontologies). The most well known extension is HEX-programs (Eiter, Ianni, Schindlauer, & Tompits, 2005), which enable handling knowledge expressed in various formalisms, even with potentially different semantics (e.g., RDF(S) and OWL). HEX-programs contain several features (e.g., higher-order logic features) that enable more flexible integration with external knowledge bases. These extensions result in new syntax elements. For example, the atom *&rdf[u](s,p,o)* evaluates to true if <s p o> is an RDF triple asserted at URI u. Finally, another advantage of HEX-programs is that they allow using external data processing services that logic programming cannot handle (e.g., string processing). An interesting engine for HEX-programs is dlvhex[2] which is described in Section 3.2.

There have been proposed ASP-based several applications in the context of (Semantic) Web. One of the most promising is Web service composition (Rainer, 2005). The authors apply ASP techniques to "build" service compositions from available services that match a certain service request.

## Web Service Modeling Language (WSML)

WSML (de Bruijn et al., 2005) is a language of representation languages for the Semantic Web. These languages are based on several different formalisms such as Description Logics, Logic Programming and First-Order Logic. Some of the basic variants of WSML are the following:

- **WSML-Core:** A subset of a Description Logic which falls inside the Horn logic fragment of FOL. It supports subsumption reasoning and query answering.
- **WSML-Flight:** An extension of WSML-Core which also supports full Datalog rules, default negation and integrity constraints. It can provide query answering in the context of Logic Programming.
- **WSML-Rule:** An extension of WSML-Flight with support for function symbols and unsafe rules.

Reasoning for these WSML variants can be implemented by several Logic Programming engines. For some features of these languages, DL reasoners can also be used or First Order theorem provers. These languages have been extensively used in several European projects, mainly in the application domain of semantic web services. Hence, several APIs, tools and other facilities are available for building WSML-enabled applications.

## Defeasible Rules

Defeasible logic (Nute, 1994) is a rule-based, non-monotonic approach able to deal with incomplete knowledge and inconsistencies. These features have been widely remarked in the context of realizing Semantic Web vision, mostly in information integration areas (e.g., ontology merging). As a result, some efforts in research community (Antoniou, Billington, Governatori, & Maher, 2001) were devoted to carry the advantages of defeasible logic in the area of Semantic Web technologies.

The main idea behind defeasible logic reasoning systems is the ability to handle a number of additional features with regard to classical rules like priorities of rules, default inheritance, exceptions, etc. There are three different types of rules in a defeasible logic reasoning system: a) classical rules (called *strict rules*), b) *defeasible rules* that can be contradicted by other rules and c) *defeaters* used to specify exceptions of defeasible rules. This way, an important aspect achieved by such reasoning modules is their capability of resolving the possible conflicts that arise among defeasible rules.

### Classification of Approaches Integrating Ontologies and Rules

Although several approaches have been discussed for combining rules with Semantic Web ontologies (Horrocks et al., 2004b; Grosof, Horrocks, Volz, & Decker, 2003; Eiter, Lukasiewicz, Schindlauer, & Tompits, 2004; Bassiliades, Antoniou, & Vlahavas, 2006; Rosati, 2006a; Rosati, 2006b), there is no totally accepted solution in the field. The main topic of argumentation is the degree of integration between the ontology layer and the rules layer. In this section we intend to provide a brief classification of the proposed approaches.

Two main categories of integration approaches have been distinguished in this context:

a. **Homogeneous approaches.** These approaches suppose a tight semantic integration of the two layers. Specifically, both ontologies and rules are embedded in a common logical language, permitting predicate sharing in a coherent way. In such approaches, ontology concepts and properties may be defined through rules. The most typical homogeneous paradigm is the combination of SWRL rules with OWL ontologies. This is also the most In addition, Description Logic Programs (DLP) (Grosof et al., 2003) constitutes another similar approach.
b. **Hybrid approaches.** These approaches correspond to a strict semantic separation between ontologies and rules. In particular, this strict separation concerns the rule predicates and the ontology elements. Hence, the vocabulary (concepts and properties) offered by the ontologies is used as a conceptualization of the domain and rules cannot directly define ontology classes or properties. Many integration approaches adhere to this category, including Answer Set Programming (Gelfond & Lifschitz, 1991), *dl-programs* (Eiter et al., 2004) and *DL+log* (Rosati, 2006c).

The user may find more details on this topic in (Eiter et al., 2006a).

Table 1 summarizes the aforementioned knowledge representation languages and their basic features.

## 3.2 Reasoning Engines

In this section we intend to provide the reader with a comparative feature analysis of existing reasoning modules, including description logic reasoners and rule engines, as well.

*Jena2* (McBribe, 2002) is the second generation of Jena Semantic Web programming toolkit, which is a Java framework for developing applications based on Semantic Web technologies. Specifically, Jena provides an Application Programming Interface (API) for creating, storing, managing and querying RDF graphs as well as RDFS, OWL ontologies in various formats (RDF/XML, N3 and N-triples). The RDFS reasoner included in Jena framework does not support datatypes and blank node entailments. The built-in OWL reasoner is very limited, since it is a rule-based implementation of OWL-Lite. However, Jena is supplied with an interface which facilitates the connection and interoperation of the framework with any external reasoner that supports the DIG (DL-Implementation Group) standard (Bechhofer, Moller, & Crowther, 2003). Hence, the API provided by Jena could be integrated with most of the existing description logics reasoners. Furthermore, Jena provides a query engine in order to execute SPARQL (Prud'hommeaux & Seaborne, 2005) queries over RDF graphs.

*Table 1. Basic features of various knowledge representation languages*

| Feature / Language | Logical Foundation | Decidability* | Serialization formats |
|---|---|---|---|
| **OWL** | Classical Logic (FOL subset) | OWL-Lite: decidable<br>OWL-DL: decidable<br>OWL-Full: undecidable | XML/<br>N-triples (textual) |
| **OWL + SWRL** | Classical Logic (FOL subset) | undecidable | XML |
| **DLP** | DL and LP intersection | decidable | textual serialization (in terms of rules) |
| **ASP** | Extension of LP (disjunction in rule heads, DL queries) | decidable | textual serialization (in terms of rules) |
| ***DL+log*** | DLs + Datalog rules (disjunctive, non-monotonic) | decidable | textual serialization (in terms of rules) |

*Note: *regarding key inference problems (e.g., consistency of the knowledge base)*

*RacerPro system* (RacerPro, 2008) is the commercial[3] extension of Racer (Haarslev & Moller, 2001), which is probably the most popular reasoning engine for OWL ontologies to practitioners of Semantic Web technologies. RacerPro can be seen as a knowledge-based repository that can handle and it is a system for managing OWL ontologies, in particular. It implements the DIG interface and it offers an optimized tableau calculus for the description logic SHIQ(D). Additionally, it supports qualified cardinality restrictions as well as some extensions of OWL (e.g., OWL-E (Pan & Horrocks, 2004) except user defined XML datatype expressions). Moreover, the latest version of RacerPro includes a first implementation of an SWRL rule engine.

*Pellet* (Sirin, Parsia, Grau, Kalyanpur, & Katz, 2007) is a Java-based, open source reasoner capable of handling expressive OWL ontologies. It implements an optimized tableau algorithm, augmented with a number of additional features (e.g., support for Unique Name Assumption - UNA, closed world reasoning, SPARQL query answering). Pellet also provides an explanation service in order to facilitate the debugging of the ontology engineering (Parsia, Sirin, & Kalyanpur, 2005). Although typical reasoners detect the inconsistencies between concepts of the knowledge base (KB), Pellet can explain *why* a concept description led to unsatisfiability. This way, the reasoner provides user with additional knowledge (e.g., relevant axioms or restrictions) sufficient to understand the problem and reform the KB properly. Finally, Pellet allows ontologies to use XML-Schema built-in and user-defined datatypes that extend numeric and date/time types.

*Bossam* (Jang & Sohn, 2004) is a RETE-based, forward-chaining reasoning module for reasoning and querying over RDF(S) and OWL documents, while it also supports rules execution (SWRL rules are included). It is based on Logic Programming (LP), augmented with some expressiveness features stemming from First-Order-Logic (FOL). For example, a number of additional to LP features are provided by Bossam, including support for both classical negation and NAF and disjunctions in the body of rules. Hence, it does not support complete reasoning over OWL Ontologies. Furthermore, it facilitates the integration of rules with Java by supporting a procedural attachments mechanism for SWRL rules. Finally, Bossam provides an API for managing the engine, loading ontologies and rules, querying RDF(S)/OWL documents and giving explanations about derived facts. Currently, Bossam does not support SPARQL query answering, while the serialization of the knowledge base to a persistent store (e.g., file system) is another missing feature.

*FACT++* system (Tsarkov & Horrocks, 2006) is the descendant of FACT (Horrocks, 1998). Contrary to the lisp-based FACT system, FACT++ is an open source reasoner for SHOIQ(D) implemented in C++. It is based on tableaux algorithms in order to provide both TBox and ABox reasoning tasks and it can be accessed through the DIG interface. It also supports a number of additional features like handling enumerated classes (a.k.a. nominals) and it is a very efficient TBox reasoning engine. However, the main disadvantage of FACT++ is its inefficiency to support complete ABox reasoning. Hence, FACT++ is unsuitable to applications that call for instance classification and retrieval.

*KAON2* (KAON2, 2008) is the successor of KAON Project (KAON, 2008) and unlike pure description logic reasoners, KAON2 does not implement a tableaux algorithm. In fact, KAON2 is a hybrid reasoning module able to handle both ontologies expressed in description logics terms and Disjunctive Datalog programs. It implements algorithms that reduce description logic SHIQ(D) to Disjunctive Datalog (Hustadt, Motik, & Sattler, 2004), taking advantage and applying well-known practices stemming from deductive databases (e.g., magic sets) to DL reasoning. It can also handle SWRL, F-Logic ontologies and SPARQL query answering, as well. Moreover, it provides a Java API in order to accommodate the management and the integration of different knowledge formalisms (e.g., OWL ontologies with SWRL rules). However, KAON2 does not support reasoning about nominals and cannot handle a large number of cardinality restrictions. Hence, it cannot deal with the full OWL-DL expressiveness.

*dlvhex* (Eiter, Ianni, Schindlauer, & Tompits, 2006b) is the name of a prototype application for computing the models of so-called HEX-programs, which are an extension of Answer Set Programs towards integration of external computation sources. dlvhex can communicate with OWL and RDF knowledge bases and can also return the results in RuleML syntax. A strong point of dlvhex is that it enables developers to write and embed plug-ins to the core engine. In fact, support for RDF and OWL is implemented through plug-ins too. Recently, another plug-in for querying HEX models through the SPARQL language (Polleres & Schindlauer, 2007). Through this plug-in dlvhex can be used as a query engine by providing a rewriter from SPARQL to rules. The source code and binaries of dlvhex are publicly available.

*DR-DEVICE* (Bassiliades, et al., 2006) is a defeasible logic reasoner for the Web based on the CLIPS expert system shell (CLIPS, 2008) that intends to integrate Semantic Web standards (RDF metadata, XML-syntax of rules) with non-monotonicity (e.g., strong negation). Specifically, it provides reasoning services over RDF metadata by taking advantage of rules defined by defeasible logic (strict rules, defeasible rules and defeaters). It also claims for reasoning efficiency compared to other systems based on logic programming.

*Jess* (Jess, 2008) is a Java framework for editing and applying rules, since it contains a scripting environment and a rule engine, as well. It supports a CLIPS-like language suitable for developing applications based on declarative rules (a.k.a. expert systems). Jess also uses an optimized version of Rete algorithm (Forgy, 1982) tailored for Java, comprising a very efficient rule engine. Recently, the evolution of rule technologies on the Web has led Jess to rebound its practical value in the community of Web developers. Moreover, the fact that Jess is a Java-based system facilitates its integration with a number of Web programming paradigms like Java servlets or applets. Finally, it supports backward-chaining and some additional features like procedural attachments.

Table 2 presents the types of inference support provided by the aforementioned reasoning modules. In (Cardoso, 2007), an analysis of the current trends and the adoption of available reasoning engines by the Semantic Web community are presented in detail.

## 4. EVALUATION OF EXISTING APPROACHES

In this section, a comparative study across the aforementioned methodologies for introducing rules in Web applications is presented. Firstly, we give some fundamental requirements that have to be satisfied by such formalisms. Afterwards, we compare various aspects from the perspectives of classical logic and logic programming, giving indicative examples. Finally, some experimental results that examine the efficiency of current efforts in the combination of ontologies with rules are presented.

### 4.1 Rule-Based Web Applications are Still "Web Applications"

Before evaluating the existing approaches to introducing rules in Web applications, we should identify the main requirements for such approaches. These requirements mainly stem from the nature of the Web itself and its current status. Firstly, WWW is a ubiquitous and massive multi-user distributed environment. In fact, this massive characteristic is its strong point, and is a direct consequence of its architectural and technological simplicity. Web technologies such as HTTP, HTML and XML are very simple to learn and use, even for plain users who are not IT experts. Such simplicity should be taken as granted for any Web-related technology, and this applies to rule systems as well. In this context, rules for the Web should be written and used even by users/developers not familiar with advanced knowledge engineering concepts. For example, negation is a rather advanced topic in logic-based systems (with Horn rules). Hence, the semantics of rules should be such that can be easily understood by naïve users. However, this would affect their expressiveness and would limit the inference power of rule-based systems that may be required by more demanding applications (i.e., Web service discovery engines). It is worthy to mention that most of the existing rule-based applications for the Web have adopted SWRL approach (see Section 3.1) in order to express rules. SWRL is neither a highly expressive language (e.g., no negation is available) nor a decidable one, but it remains simple.

Another requirement is that the rule-based systems for the Web are compatible with existing (Semantic) Web standards, such as XML, RDF, OWL and SAWSDL (Farrell & Lausen, 2007). This means that direct support for URIs or XML syntax should be available. On the other hand, writing rules in XML (e.g., like the SWRL and RuleML approaches propose) is a cumbersome task. This is indeed a difficult decision that has to be taken by the Web architects.

Furthermore, the addition of rules in Web applications requires their compatibility with existing forms of Web knowledge that have already attained a certain maturity level. Most of the knowledge bases developed in the context of Web are expressed in the form of ontologies (mainly written in OWL or RDFS). As a result, rules have to be integrated with ontologies properly.

*Table 2. Types of inference support by various reasoning engines*

| Inference Support / Modules | TBox Reasoning | Abox Reasoning | Rules Reasoning |
|---|---|---|---|
| **Jena2** | Limited (incomplete RDFS/ OWL reasoning) | Limited (incomplete RDFS/ OWL reasoning) | JenaRules (forward/tabled backward chaining) |
| **RacerPro** | √ | √ | nRQL rules, first implementation of SWRL (forward chaining) |
| **Pellet** | √ | √ | DL-safe rules (SWRL subset) |
| **Bossam** | sound, incomplete | sound, incomplete | SWRL, Buchingae (forward chaining) |
| **FACT++** | √ | sound, incomplete | - |
| **KAON2** | √ (except nominals) | √ (except nominals) | DL-safe rules (SWRL subset) |
| **dlvhex** | Limited, through interface with external reasoner | Limited, through interface with external reasoner | DL-Rules (non-monotonic logic program rules with queries to DL KB) |
| **DR-DEVICE** | - | - | Defeasible rules |
| **Jess** | Limited* | Limited* | SWRL*, Jess rules (forward/backward chaining) |

** through appropriate transformations (Mei, Bontas & Lin, 2005; O'Connor et al., 2005)*

## 4.2 Qualitative Comparison of Classical Logic and Logic Programming Approaches

The aforementioned knowledge representation methodologies (see Section 3.1) are either based on classical logic (e.g., DLs and SWRL) or logic programming (e.g., answer-set programming and defeasible rules). Across the literature, several differences between classical logic and logic programming paradigms have already been identified. In this section, we intend to survey and discuss these efforts by presenting the main incompatibilities that impede the reconciliation of the two approaches.

### Monotonicity vs. Non-Monotonicity

Classical logic is based on standard model theoretic semantics and adheres to monotonicity of entailment. Informally, monotonicity means that the addition of new information to the knowledge base cannot invalidate conclusions inferred by current knowledge. As a result, classical logic is able to deal with incomplete information by nature.

Instead of classical logic, logic programming has non-monotonic features. It assumes complete knowledge and there is a unique model describing the state of the world. This way, addition of knowledge may reduce the inferences.

### Unique Name Assumption

Another difference between the two paradigms is that logic programming approaches typically deploy the *Unique Name Assumption* (UNA). This assumption supposes that different names represent different objects of the world. This fact imposes severe limitations on the Web context, since several distinct URIs[4] may refer to the same content or data. On the contrary, in classical logic there is no one-to-one correspondence between the names and the objects of the domain. Hence, equality between individuals represented by different names can be derived. Although it imposes a huge computational cost, most of the current description logic reasoners support reasoning with equality.

### Negation

Classical logic and logic programming face the aspect of negation from different perspectives. With respect to its monotonic nature, negation in classical logic allows inferring new information only if the truth or the falsehood of a statement is explicitly declared. This fact is related with the *Open World Assumption* (OWA) of classical logic theory where incompleteness of the knowledge base is considered.

On the other hand, *Negation-As-Failure* (NAF) adheres to the *Closed World Assumption* (CWA). Specifically, if a description is not known to be true, then the truth of the negated description is drawn. In that sense, absence of knowledge draws to negated knowledge and, thus, NAF has a non-monotonic behavior. Modeling the world according to CWA seems to be somewhat inappropriate for the Semantic Web. Since knowledge on the Web is not always available (e.g., web servers breakdowns), such an assumption could lead to incorrect inferences. However, the usefulness of different types of negation in rule-based Web applications is demonstrated in (Wagner, 2003; Alferes, Damasio, & Pereira, 2003).

### Constraints and Restrictions

Another aspect where the classical paradigm differs from logic programming is on the treatment of *constraints* and *restrictions*. Regardless of the apparent similarity between constraints and restrictions, there are important differences between them.

Restrictions are used in the classical logic paradigm and they constitute part of the logical theory. By adding restrictions to a knowledge base, the knowledge statements are also increased and, thus, the inference of additional knowledge is permitted.

Restrictions may be further classified to the following main categories:

- *Value restrictions* or *range restrictions*. They are used to restrict the values of a property and possibly infer new information according to their type (a.k.a. range).
- *Cardinality restrictions*. They are used for restricting the number of values that an

individual may have for a specific property. They can possibly infer the existence of new instances or equality between known individuals.

In terms of Description Logics, an example of cardinality restriction in the web service paradigm could be the following:

$$\text{Profile} \sqsubseteq \geq 1 \text{ hasInput} \quad (6)$$

Such rule denotes that every service profile should have at least one input.

On the other hand, constraints (also called *integrity constraints*) are mainly used in deductive databases and logic programming in order to check if the knowledge base is consistent with a number of specified conditions. As a result, constraints cannot draw inference of new information, but they may lead to inconsistencies in the case that some conditions specified by them are violated inside the knowledge base. Usually, constraints are rules without head (i.e., they have a "false" value in their head), denoting that the conditions stated in the body should not be satisfied concurrently by the knowledge base.

Similarly to restrictions, constraints may be divided to:

- *Value constraints* or *range constraints.* Their importance consists in checking the type of a property value.
- *Cardinality constraints.* They are used for checking the number of values that an individual may have for a specific property.

Similarly to (6), a cardinality constraint could be written in logic programming as:

$$\leftarrow \text{Profile(s)} \wedge \neg \text{hasInput(s,i)} \quad (7)$$

More details about integrity constraints and restrictions may be found in (de Bruijn, Polleres, Lara, & Fensel, 2005), where an alternative ontology language based on the logic programming subset of OWL is presented.

## Other Differences

A number of additional differences between the described approaches have also been discussed in the literature. (Eiter et al., 2006a) raises the point of non-ground entailment in the logic programming approach, declaring that, since the semantics of logic programming is defined in terms of sets of ground facts, the inference of non-factual knowledge is not allowed. Contrary to the logic programming, classical logic permits the entailment of non-factual knowledge. Furthermore, decidability issues are explored in the combination of the two perspectives. In addition, (Pattel-Schneider, 2006) describes the treatment of datatypes by classical logic approaches (e.g., DLs) and logic programming. Moreover, it examines the role of tools that support each methodology and how they can facilitate the modeling of knowledge.

The aforementioned modeling paradigms have been proposed for formalizing rules on the Web. Concerning the advantages and the disadvantages of each approach, we believe that both perspectives are useful in the context of Semantic Web.

The open nature of the classical logic seems to be more suitable for formalizing such a distributed environment. By this mean, Semantic Web technologies based on classical logic accomplish the integration of information stemming from different sources. Knowledge becomes shareable in the sense that it can be accessed without any obligation to adopt a specific model or schema. Hence, data are available to everyone through WWW so as to be retrieved and used in applications appropriately.

On the other hand, logic programming makes several assumptions in order to simplify tasks like reasoning or modeling of the knowledge base. There are several useful features of logic programming like high expressiveness, negation-as-failure and decidability of reasoning. Moreover, many of the reasoning engines targeting at managing knowledge expressed in classical logic formalisms have been recently extended to support such features (e.g., Pellet, RacerPro, Bossam). Additionally, both users and developers are very familiar with the world of databases and logic programming languages like Prolog. It is worthy to mention that even SPARQL engines return "no" in ASK query patterns with no solutions. However, "don't know" would fit better in an open environment where multiple solutions may exist outside the current knowledge base. This way, SPARQL takes advantage of the users' familiarity with common databases, deciding not to complicate the query answering process.

To summarize, last decade has shown to the developers of Web applications that different types of Semantic Web enabled applications requires different styles of modeling. As a result, the development of a framework that satisfy most of these needs seems to become unavoidable.

### 4.3 Quantitative Evaluation of Reasoning Modules

This section intends to quantify the total time required by popular reasoning modules in order to perform common reasoning tasks. Instead of (Pan, 2005; Gardiner, Horrocks, & Tsarkov, 2006), where the evaluation of description logic reasoners focuses on TBox reasoning tasks, here we examine both TBox and ABox reasoning as well as rules application. In particular, Figures 3 to 6 show the results of some indicative performance tests. These tests aim at demonstrating the efficiency of description logic reasoners and mainstream rule engines.

This quantitative evaluation involved the last stable versions of RacerPro (v1.9.0), Pellet (v1.5.1) and FACT++ (v1.1.11) description logic reasoners that were available for ontology reasoning as well as the Jess rule engine for SWRL rules application. Bossam was also tested as both ontology reasoner and rule engine in this evaluation. However, Bossam's results are not displayed in the figures, since such a comparison would be improper. This stems from the fact that although Bossam constitutes a sound reasoning module it is not a complete reasoner for OWL. However, we should mention that Bossam performed well (i.e., it accomplished smaller execution times than the other OWL reasoners) handling small-size ontologies, while it threw out-of-memory errors when provided with large-scale ontologies (i.e., containing 500 and 1000 instances). Moreover, all reasoners were managed through Protégé OWL API in Java except Bossam which provides its own API. Finally, Jess was accessed via SWRL-Jess bridge (O'Connor et al., 2005).

The knowledge base used in these tests stems from the Web service description paradigm and it consists of a domain ontology and OWL-S service descriptions. The domain ontology, which represents the domain of football (i.e., player/team details and statistics), includes concepts described through necessary and sufficient conditions. The elements of this ontology are used for the annotation of some Semantic Web services that provide search functionality to some underlying football statistics databases. Specifically, the service inputs and outputs are annotated through concepts of this domain ontology. The extract of the domain ontology used contains 37 concepts, 8 object properties and 3 datatype properties. Each service description involves 19 concept instances and 11 property instances.

Figure 3 depicts the performance of three description logic reasoners while checking the consistency of all classes over the aforementioned knowledge base. The execution times presented in this figure do not involve the time needed to load the models. The figure shows that these times increase as the number of instances contained in the knowledge base also increases. FACT++ and Pellet attained the best performance by achieving similar execution times. On the other hand, RacerPro performed worse than the others.

*Figure 3. Evaluation of consistency check (TBox)*

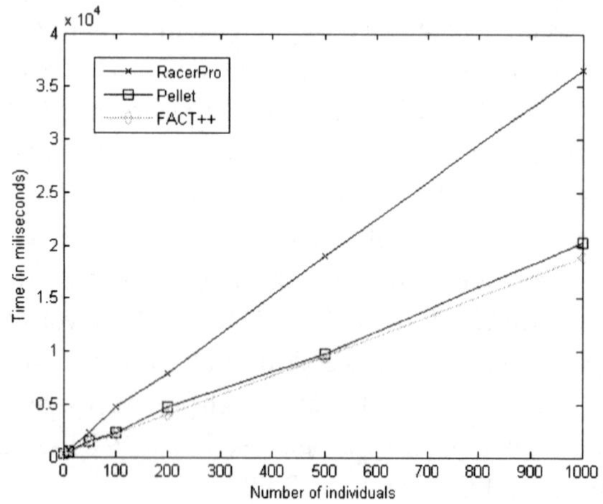


*Figure 4. Evaluation of classification (TBox)*

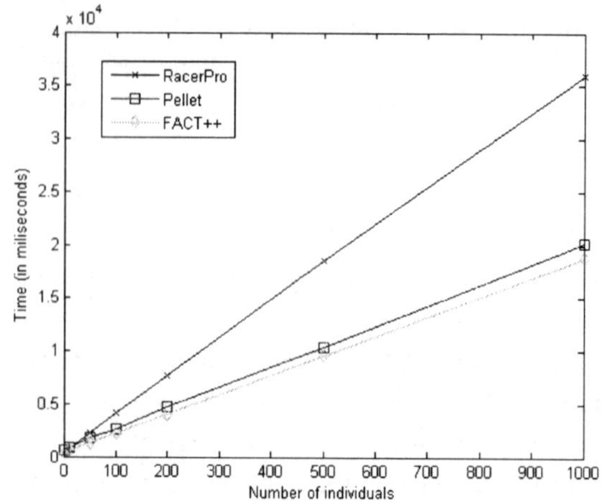


In Figure 4, the times required for classifying the hierarchy of the ontology are presented. Again, the times shown in the figure do not include the loading time of each model. The resulting performances are very similar to those of the consistency checking process. FACT++ has slightly better performance than Pellet, while RacerPro has the worst performance again.

Figure 5 shows the efficiency of RacerPro and Pellet reasoners in realizing reasoning over ABoxes. In this figure we do not display FACT++, since it provides incomplete ABox reasoning. In particular, the experiment concerns the identification of the types (i.e., classes) that each individual belongs to. The figure shows that even the reasoning over an ontology containing 100 instances using Pellet (which seems to be the most efficient reasoner in this reasoning task) requires about 20 seconds. Moreover, both reasoners failed to compute inferred individual types for ontologies containing 500 and 1000 individuals.

It is worthy to mention that version 1.9.2 Beta of RacerPro was also available during the experiments. This version seemed to remain slower than FACT++ and Pellet in TBox reasoning tasks (consistency check and classification of hierarchy) while it performed better in ABox reasoning. However, the results are not displayed in the figures above, since this is not a stable version of the reasoner.

Figure 6 shows the time required by Jess rule engine to apply SWRL rules over ontologies containing various numbers of instances. These rules represent service composition constraints that should apply to the inputs and outputs of the services. We applied two different sets of rules to the previously described knowledge bases: the first set consisted of 10 rules while the second of 50 rules. In both cases, the execution time increases proportionally to the number of instances

*Figure 5. ABox reasoning evaluation (computation of inferred individual types)*

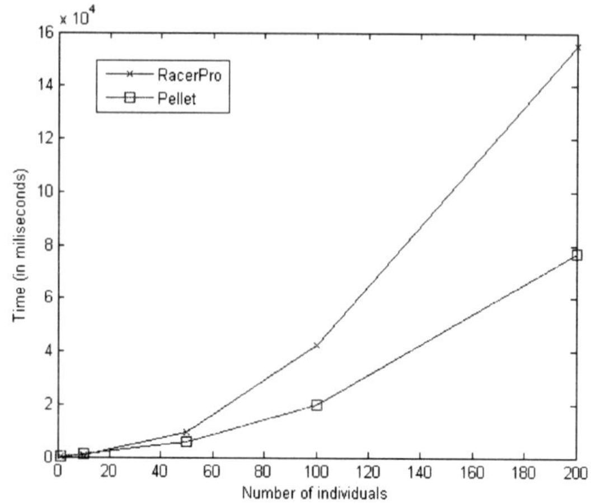

*Figure 6. Jess performance executing SWRL rules*

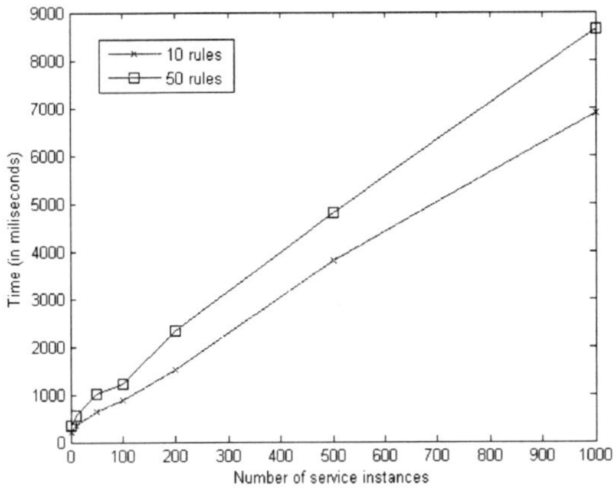


contained in the ontology. As expected, the delay imposed by the execution of rules was affected by the number of rules. However, the number of instances seems to affect execution times more than the number of rules. Hence, the inference process is complicated when real-time demands come up in distributed environments where applications have to support multiple individual instances.

To summarize, the time required to handle a large KB and execute reasoning in terms of Semantic Web technologies like DLs constitutes a limiting factor, even for the most efficient reasoners. This is a fundamental problem on the Web, since applications have to deal with a large number of instances representing data. Especially, when considering real time applications that call for small response times, the aforementioned results become unmanageable. This is the main reason why many researchers have recently focused on developing efficient and scalable reasoning applications over large individual sets. Instance Store (Horrocks, Li, Turi, & Bechhofer, 2004a) is such an approach that permits working in conjunction with any ontology reasoner that implements the DIG interface.

## 5. FUTURE TRENDS AND OPEN ISSUES

Several open issues that deserve further research efforts in the future have been identified throughout the chapter. Firstly, a Web standard for editing and embedding rules into Web applications should be specified. In our opinion, such a language should focus on its simplicity, instead of providing major expressiveness capabilities that will lead to a clumsy formalism. Rule Interchange Format (RIF) Working Group (RIF, 2008) works in this direction in order to produce W3C recommendations for enabling interchange of rules. Specifically, RIF does not intend to provide explicit mappings between various rule languages, but specify a mechanism capable of defining the meaning of the formulas of a rule language. This way, rules could be automatically translated across different formalisms.

Additionally, more tractable algorithms for Web-rules, TBox and ABox inference problems should be developed. These algorithms should be able to handle complex and heavy knowledge bases that combine rules and ontologies with a large number of concepts and relationships. Such knowledge bases are substantial, especially in the context of information integration where multiple domain vocabularies interact with upper-level ontologies. Furthermore, the Web is a large scale environment, where a huge set of resources are added every day. As a result, future knowledge bases will have to capture and describe more and more individuals. These facts should be taken into account by the developers that will target at designing new reasoning algorithms.

Finally, a more practical issue that should be investigated is the development of a unified reasoning framework capable of managing both ontologies and rules. Today, there is no efficient and easy-to-use integrated reasoning module that can reason over both formalisms. As a consequence, the developer should use at least two different reasoning modules to handle such an integrated knowledge base. This can result to unexpected situations. For example, the restrictions defined by the ontology can be violated by the application of rules, since the rule engines do not take into account these restrictions (e.g., disjointness of concepts). Similarly, the application of rules could produce knowledge that would be useful for further description logic inferences.

## 6. CONCLUSION

In this chapter we have discussed the application of rules to Web applications in order to achieve intelligent application behavior and efficient management of knowledge. We have described the main methodologies and technologies for integrating rules with ontologies, since the latter constitutes a mature knowledge technology on the Web. The chapter has also examined aspects of the different approaches. We also showed through several experiments that current reasoning modules are not efficient enough to manage knowledge stemming from large-scale environments like the Web, especially in the context of real-time applications which impose severe constraints in response times. Finally, more improvements should be made in the standardization of rules formalism on the Web and in the development of reasoning modules that can handle and reason over both ontologies and rules as integral knowledge.

## 7. REFERENCES

Alferes, J., Damasio, C., & Pereira, L. (2003). Semantic Web logic programming tools. (LNCS 2901, pp. 16-32).

Antoniou, G., Billington, D., Governatori G., & Maher, M., J. (2001). Representation results for defeasible logic. *ACM Transactions on Computational Logic, 2*(2), 255-287.

Antoniou, G., Damasio, C., V., Grosof, B., Horrocks, I., Kifer, M., Maluszynski, J., et al. (2005). *Combining rules and ontologies: A survey.* Deliverables I3-D3, REWERSE. Retrieved from, http://rewerse.net/deliverables/m12/i3-d3.pdf.

Antoniou, G., & van Harmelen, F. (2004). *A Semantic Web primer.* MIT Press.

Baader, F., Calvanese, D., McGuiness, D., Nardi, D., & Patel-Schneider, P. (2003). *The description logic handbook: Theory, implementation, and applications.* Cambridge: Cambridge University Press.

Bassiliades, N., Antoniou, G., Vlahavas, I. (2006). A defeasible logic reasoner for the Semantic Web, *International Journal on Semantic Web and Information Systems 2(1), 1-4.*

Bechhofer, S., Moller, R., & Crowther, P. (2003). The DIG description logic interface. In D. Calvanese, G. de Giacomo, & F. Franconi (Eds.), *Proc. of the 2003 Int. Workshop on Description Logics (DL 2003)* (Vol. 81 of CEUR Workshop Proceedings), Rome, Italy.

Boley, H., Kifer, M., Patranjan, P., L., & Polleres, A. (2007). Rule interchange on the Web. In *Reasoning Web 2007* (No. 4636, pp. 269-309). Springer.

Brickley, D., & Guha, R.V. (2004). *RDF vocabulary description language 1.0: RDF schema.* Recommendation 10 February 2004, W3C. Retrieved April 22, 2008, from http://www.w3.org/TR/rdf-schema/.

de Bruijn, J., Lausen, H., Krummenacher, R., Polleres, A., Predoiu, L., & Kifer, M., et al. (2005). The Web service modeling language WSML (Tech. Rep. WSML. WSML Final Draft D16.1v0.21). Retrieved from http://www.wsmo.org/TR/d16/d16.1/v0.21/.

de Bruijn, J., Polleres, A., Lara, R., & Fensel, D. (2005). OWL DL vs. OWL flight: Conceptual modeling and reasoning on the Semantic Web. In *Proc. WWW2005*, (pp. 623-632).

Cardoso, J. (2007). The Semantic Web vision: Where are we? *Intelligent Systems, 22*(5), 84-88.

*CLIPS: A tool for building expert systems.* (2008). Retrieved April 22, 2008, from http://clipsrules.sourceforge.net/.

Dean, M., Schreiber, G., Bechhofer, S., van Harmelen, F., Hendler, J., Horrocks, I., et al. (2004). *OWL Web ontology language reference.* W3C Recommendation 10 February 2004. Retrieved April 22, 2008, from http://www.w3.org/TR/owl-ref/.

Donini, M. F., Lenzerini, M., Nardi, D., & Schaerf, A. (1998). AL-log: Integrating datalog and description logics. *Journal of Intelligent Information Systems, 10(3)*, 227-252.

*Drools* (2008). Retrieved April 22, 2008, from http://www.jboss.org/drools/.

Eiter, T., Ianni, G., B., Polleres, A., Schindlauer, R., & Tompits, H. (2006a). Reasoning with rules and ontologies. In P. Barahona, F. Bry, E. Franconi, U. Sattler, & N. Henze (Eds.), *Reasoning Web, Second International Summer School 2005, Tutorial Lectures* (LNCS, pp. 93-127).

Eiter, T., Ianni, G., B., Schindlauer, R., & Tompits, H. (2006b). dlvhex: A system for integrating multiple semantics in an answer-set programming framework. In M. Fink, H. Tompits, & S. Woltran (Eds.), *Proceedings 20th Workshop on Logic Programming and Constraint Systems (WLP 06)*, (pp. 206-210).

Eiter, T., Ianni, G., B., Schindlauer, R., & Tompits, H. (2005). A uniform integration of higher-order reasoning and external evaluations in answer set programming. In *Proc. IJCAI 2005*. Morgan Kaufmann.

Eiter, T., Lukasiewicz, T., Schindlauer, R., & Tompits, H. (2004). Combining answer set programming with description logics for the Semantic Web. In *Proc. of the International Conference of Knowledge Representation and Reasoning* (KR04).

Farrell, J., & Lausen, H. (2007). *Semantic annotations for WSDL and XML schema.* W3C Recommendation, W3C, August 2007. Retrieved April 22, 2008, from http://www.w3.org/TR/sawsdl/.

Forgy, C. (1982). Rete: A fast algorithm for the many pattern/many object pattern match problem. *Artificial Intelligence, 19*, 17-37.

Gardiner, T., Horrocks, I., & Tsarkov, D. (2006). Automated benchmarking of description logic reasoners. In *Proc. of DL.*

Gelfond, M., & Lifschitz, V. (1991). Classical negation in logic programs and disjunctive databases. *New Generation Computing, 9*(3-4), 365-386.

Grosof, B. N., Horrocks, I., Volz, R., & Decker, S. (2003). Description logic programs: Combining logic programs with description logic. In *Proc. of the Twelfth International World Wide Web Conference (WWW 2003)*, (pp. 48-57). ACM.

Haarslev, V., & Moller, R. (2001). RACER system description. In R. Gor'e, A. Leitsch, and T. Nipkow (Eds.), *International Joint Conference on Automated Reasoning, IJCAR'2001, June 18-23, Siena, Italy*, (pp. 701–705). Springer-Verlag.

Horrocks, I. (1998). Using an expressive description logic: FaCT or fiction? In *Proceedings of the Sixth International Conference on Principles of Knowledge Representation and Reasoning (KR'98), Trento, Italy, June 2–5, 1998*, (pp. 636–647).

Horrocks, I., Li, L., Turi, D. & Bechhofer, S. (2004a). The instance store: DL reasoning with large numbers of individuals. In *Proc. of the 2004 Description Logic Workshop (DL~2004)*, (pp. 31-40).

Horrocks, I., Patel-Schneider, P.F., Boley, H., Tabet, S., Grosof, B., & Dean, M. (2004b). *SWRL: A Semantic Web rule language combining owl and RuleML*. W3C Member Submission, 21 May 2004. Retrieved April 22, 2008, from http://www.w3.org/Submission/SWRL/.

Horrocks, I., Parsia, B., Patel-Schneider, P., F., & Hendler, J. (2005). Semantic Web architecture: Stack or two towers? In *Proc. PPSWR 2005*, (pp. 37-41), Dagstuhl Castle, Germany.

Hustadt, U., Motik, B., & Sattler, U. (2004). Reducing SHIQ- description logic to disjunctive datalog programs. In *Proc. of the 9th International Conference on Knowledge Representation and Reasoning (KR2004)*, (pp. 152-162).

*ILOG, Business Rule Management Systems, Optimization Tools and Engines, Visualization Software Components, Supply Chain Applications* (2008). Retrieved April 22, 2008, from http://www.ilog.com/.

Jang, M., & Sohn, J-C. (2004). Bossam: An extended rule engine for OWL inferencing. In *Workshop on Rules and Rule Markup Languages for the Semantic Web at the 3rd International Semantic Web Conference* (LNCS 3323, pp. 128-138).

*Jess, The Rule Engine For the Java Platform* (2008). Retrieved April 22, 2008, from http://www.jessrules.com/jess/index.shtml

*KAON – The KArlsruhe ONtology and Semantic Web tool suite* (2008). Retrieved April 22, 2008, from http://kaon.semanticweb.org.

*KAON2 – Ontology Management for the Semantic Web* (2008). Retrieved April 22, 2008, from http://kaon2.semanticweb.org.

Kifer, M., de Bruijn, J., Boley, H., & Fensel, D. (2005). A realistic architecture for the Semantic Web. In *Proc. RuleML 2005* (pp. 17-29), Galway, Ireland.

Klyne, G., & Carroll, J.J. (2004). *Resource description framework (RDF): Concepts and abstract syntax*. Recommendation 10 February 2004, W3C. Retrieved April 22, 2008, from http://www.w3.org/TR/rdf-concepts/.

McBride, B. (2002). Jena: A Semantic Web toolkit, *IEEE Internet Computing, 6(6)*,55-59.

Mei, J, Bontas, E.P., & Lin, Z. (2005). OWL2Jess: A transformational implementation of the OWL semantics. In *Proceedings of International Workshops on ISPA* (LNCS 3759, pp. 599-608).

Motik, B., Sattler, U., & Studer, R. (2005). Query answering for OWL-dl with rules. *Journal of Web Semantics: Science, Services and Agents on the World Wide Web, 3*, 41-60.

Motik, B., Horrocks, I., Rosati, R., & Sattler, U. (2006). Can OWL and logic programming live together happily ever after? In *Proceedings ISWC-2006* (LNCS 4273, pp. 501–514).

Nute, D. (1994). Defeasible logic. In *Handbook of Logic in Artificial Intelligence and Logic Programming, Nonmonotonic Reasoning and Uncertain Reasoning* (Vol. 3). Oxford University Press.

O'Connor, M., Knublauch, H., Samson, T., & Musen, M. (2005). Writing rules for the Semantic Web using SWRL and Jess. *Protégé with Rules WS*. Madrid.

Pan, J. Z., & Horrocks, I. (2004). *OWL-E: Extending owl with expressive datatype expressions* (Technical report, IMG/2004/KR-SW-01/v1.0). Victoria University of Manchester.

Pan, J. (2005). Benchmarking DL reasoners using realistic ontologies. In *Proc. of the OWL: Experiences and Directions Workshop*.

Parsia, B., Sirin, E., & Kalyanpur, A. (2005). Debugging OWL Ontologies. In *Proceedings of the 14th International World Wide Web Conference (WWW2005), Chiba, Japan, May 2005*. Retrieved from http://www.mindswap.org/papers/debuggingOWL.pdf.

Patel-Schneider, P. F., & Horrocks, I. (2006). Position paper: a comparison of two modelling paradigms in the Semantic Web. In *Proceedings of the 15th International Conference on World Wide Web (Edinburgh, Scotland, May 23 - 26, 2006). WWW '06, New York, NY* (pp. 3-12). ACM Press.

Polleres, A., & Schindlauer, R. (2007). *DLVHEX-SPARQL: A SPARQL Compliant Query engine based on DLVHEX. In Proceedings of 2nd International Workshop on Applications of Logic Programming to the Web, Semantic Web and Semantic Web Services (ALPSWS2007), Porto, Portugal.*

Prud'hommeaux, E., & Seaborne, A. (2005). *SPARQL query language for RDF.* http://www.w3.org/TR/2005/WD-rdf-sparql-query-20050217/, 2005.

*RacerPro* (2008). Retrieved April 22, 2008, from http://www.racer-systems.com/products/racerpro/index.phtml.

Rainer. A. (2005). Web service composition under answer set programming. *KI-Workshop "Planen, Scheduling und Konfigurieren, Entwerfenl" (PuK).*

*RIF - Rule Interchange Format Working Group* (2008). Retrieved April 22, 2008, from http://www.w3.org/2005/rules/wiki/RIF_Working_Group

Rosati, R. (2006a). DL+log: Tight integration of description logics and disjunctive datalog. In P. Doherty, J. Mylopoulos, & C.A. Welty (Eds.), *Proceedings of the 10th International Conference on Principles of Knowledge Representation and Reasoning (KR 2006), Lake District, UK* (pp. 68-78). AAAI Press.

Rosati, R. (2006b). Integrating ontologies and rules: Semantic and computational issues. In P. Barahona, F. Bry, E. Franconi, N. Henze, & U. Sattler (Eds.), *Reasoning Web, Second International Summer School 2006, Lissabon, Portugal, September 4_8, 2006, Tutorial Lectures.* (LNCS 4126, pp. 128-151).

Rosati, R. (2006c). DL+log: Tight integration of description logics and disjunctive datalog. In *Proceedings of the Tenth International Conference on Principles of Knowledge Representation and Reasoning (KR 2006)* (pp. 68-78). AAAI Press.

*RuleML, Rule Markup Initiative* (2008). Retrieved April 22, 2008, from http://www.ruleml.org/.

Sirin, E., Parsia, B., Grau, B., C., Kalyanpur A., & Katz, Y. (2007). Pellet: A practical OWL-DL reasoner. *Journal of Web Semantics, 5*(2).

Tsarkov, D., & Horrocks, I. (2006). FaCT++ description logic reasoner: System description. In *Proc. of the International Joint Conference on Automated Reasoning (IJCAR 2006).*

*W3C Semantic Web Activity* (2008). Retrieved April 22, 2008, from http://www.w3.org/2001/sw/.

Wagner, G. (2003). Web rules need two kinds of negation (LNCS 2901, pp. 33-50).

## KEY TERMS AND DEFINITIONS

**Logic Programming:** It is a declarative paradigm of programming that is mainly based on first-order logic. Prolog is the most common logic programming language.

**Knowledge-Based System:** A system that exploits knowledge representation and reasoning techniques in order to achieve an intelligent behavior.

**Knowledge Representation:** It is the process of exploiting formal methodologies and languages (e.g., rules formalisms, ontologies) in order to capture and describe knowledge.

**Ontology:** A formal representation of a domain of discourse that describes concepts and relationships among them. This way, it provides a common vocabulary and allows for the inference of new knowledge.

**Reasoning:** The systematic process of inferring new knowledge by applying formal implication rules to a given knowledge base.

**Rule-Based Application:** An application that takes advantage of rules formalisms (e.g., Horn-clauses) in order to represent and infer new knowledge.

**Semantic Web:** It is an extension of the current Web where resources are described through formal syntax and semantics in order to be human- and machine-readable.

## ENDNOTES

1. Also called as Datalog view in the literature
2. http://con.fusion.at/dlvhex/
3. It offers free trials while educational and research licenses are available
4. Universal Resource Identifiers

# Chapter XVII
# Modeling and Programming with Commitment Rules in Agent Factory

**Rem Collier**
*University College Dublin, Ireland*

**Gregory M.P. O'Hare**
*University College Dublin, Ireland*

## ABSTRACT

*Agent-Oriented Programming (AOP) is a relatively new programming paradigm, proposed by Yoav Shoham, which views software systems as consisting of a set of agents that interact with one another to solve problems beyond their individual capabilities. Since the inception of the paradigm, a number of AOP languages have been proposed. This chapter focuses on one such language, the Agent Factory Agent Programming Language (AFAPL), a practical rule-based language that has been applied to a wide range of problem domains including robotics, virtual and mixed reality environments, and mobile computing. AFAPL is placed in context through a general introduction to the state-of-the-art in AOP. The chapter finishes with a discussion of some future trends for AOP and some concluding remarks.*

## INTRODUCTION

Agent-Oriented Programming (AOP) is a relatively new programming paradigm introduced by Yoav Shoham (1993) in which software systems are viewed as consisting of a set of agents that interact with one another to solve problems that are beyond their individual capabilities. More specifically, agents are viewed as high-level autonomous software entities that encapsulate a set of capabilities and whose internal state is comprised of a set of mental components such as beliefs, capabilities, choices and commitments. This view of agents as mentalistic entities is a common perspective within multi-agent systems research and underpins many of the most prominent agent theories (Cohen & Levesque, 1990; Rao & Georgeff, 1991; Wooldridge, 2000). These theories model the internal decision-making process of an agent in terms of the interplay between the constituent components of the underlying mental state. Their objective is to define how an agent is able to act in a rational goal-directed manner and to tease out various desirable properties that emerge from that action. Thus, the objective of AOP is to present a framework for developing a new class of programming languages that are derived from these theories.

This view of programs that consist of components whose state is defined by a set of mental qualities and in which computation is realized through speech act based message passing is not unique to Shoham. John McCarthy, who is widely acknowledged as being one of the first to associate mental qualities with machines (McCarthy, 1979) had, as early as 1990, written an draft proposal for a programming language he entitled Elephant 2000 (McCarthy, 1992). That said, McCarthy recognized at the time that his language could be implemented, and to this day, no implementation of Elephant 2000 exists (McCarthy, 2007). In contrast, a number of AOP languages have been proposed, implemented and successfully used to build a range of agent-oriented applications.

This chapter focuses on one such AOP language, entitled the Agent Factory Agent Programming Language (AFAPL) (Collier, 2001; Ross & Collier & O'Hare, 2004; Collier & Ross & O'Hare, 2005). AFAPL has its origins as part of a larger software engineering framework entitled *Agent Factory* (O'Hare, 1996; Collier, 1996; O'Hare & Collier & Conlon & Abbas, 1998; Collier, 2001; Collier & O'Hare & Lowen & Rooney, 2003). Both AFAPL and the associated framework have been designed specifically to support the fabrication of agent-oriented applications, and have been used extensively in the development of a number of prototype systems in areas such as: robotics (O'Hare & Duffy & Collier & Rooney & O'Donoghue, 1999; Dragone & Holz & O'Hare, 2007), mobile computing (O'Hare & O'Grady, 2003; Muldoon & O'Hare & Phelan & Strahan & Collier, 2003), virtual and mixed reality (Duffy & O'Hare & Campbell & Stafford & O'Grady, 2005), wireless sensor networks (Marsh & Tynan & O'Kane & O'Hare, 2004; O'Hare & O'Grady & Marsh & Ruzzelli & Tynan, 2006) and information retrieval (Peng & Collier & Mur & Lillis & Toolan & Dunnion, 2004; Lillis & Collier & Toolan & Dunnion, 2007).

The remainder of this chapter starts with an overview of Agent-Oriented Programming. This is followed by a general overview of the Agent Factory framework. The next section presents details of AFAPL, and after that some of the tools that have been developed to support its use are discussed. The penultimate section, discusses future trends in this area, and finally, some concluding remarks are presented.

## AGENT-ORIENTED PROGRAMMING

In his seminal paper, Shoham (1993) introduces the concept of Agent-Oriented Programming as apecialisation of Object-Oriented Programming (OOP). As is highlighted in Figure 1, the core computational unit becomes an agent rather than an object, and the state is constrained to mental components. The second significant restriction that is applied to AOP is the requirement that the only valid types of message are speech acts. Speech act theory (Searle, 1969) is a pragmatic theory of how humans communicate with one another, and has become the defacto standard within the multi-agent systems research community (Wooldridge & Jennings, 1995). Ultimately, the adoption of speech act theory has led to the specification of various Agent Communication Languages (ACLs), such as Knowledge Query Meta Language (KQML) (Finin & Labrou & Mayfield, 1997) and Foundation for Intelligent

Physical Agents (FIPA) ACL (FIPA, 2000) which form the basis for the communication infrastructure of most modern agent systems.

Based on this view of AOP, computation arises through a combination of message-passing and mental state update. Practically, this is achieved through the implementation of an agent interpreter that updates the agents' mental state and manages the passing of messages in accordance with some associated agent program. Specifics relating to the operation of this interpreter and the structure of the associated agent program are bound intimately to the particular agent theory that is chosen to underpin the corresponding programming language.

The first AOP language to be developed, entitled Agent-0, was intended as illustrative example (Shoham, 1993), and is described in more detail later. Agent-0 was followed closely by the PLAnning Communication Agents (PLACA) language (Thomas, 1995), which attempted to address the lack of support for plans and planning within Agent-0. Specifically, PLACA introduced capabilities as first class constructs within the language, and modeled the intentions of the agent rather than its commitments. Execution of the internal state thus involves matching capabilities to intentions through the formation of plans of action via a classical STRIPS (Fikes & Nilsson, 1971) style planning process. In contrast with Agent-0, which was made available for use, no prototype of PLACA was made public.

Around the same time, Arnand Rao proposed the AgentSpeak(L) language (Rao, 1996), which built on earlier experiences in working on the first Belief-Desire-Intention (BDI) agent architecture, known as the Procedural Reasoning System (PRS) (Ingrand & Georgeff & Rao, 1992), and subsequent theoretical and practical work carried out in this area (Rao & Georgeff, 1991) (Rao & Georgeff, 1995). While BDI-style agents had already been implemented Unfortunately, Rao's language was only a proposal and was not implemented until around 2004 when Jason, a Java based development environment for AgentSpeak(L) was created (Bordini & Hubner & Vieira, 2005). However, during this interim period, Hindriks & de Boer & van der Hoek & Meyer, (1999) developed 3APL, a rule-based programming language that is inspired by the BDI-architecture and which has many similarities with AgentSpeak(L). 3APL is the first AOP language whose syntax and semantics was formally specified using operational semantics (Plotkin, 1981), and while it is similar to AgentSpeak(L), Hindriks et al. (1999) point out that 3APL offers a richer set of rules than AgentSpeak(L) as it not only provides rules for matching goals to plans, but also provides support for modeling the modification and revision of goals. The 3APL interpreter has been implemented in Java and has been made available for use.

*Figure 1. Table showing OOP versus AOP [Reprinted from Artificial Intelligence, Volume 60, Issue 1, Yoav Shoham, Agent-oriented programming, pages 51-92, Copyright (1993), with permission from Elsevier]*

| | OOP | AOP |
|---|---|---|
| Basic Unit | Object | Agent |
| Parameters describing state of basic unit | Unconstrained | Beliefs, capabilities, choices, commitments, ... |
| Process of computation | Message passing and response methods | Message passing and response methods |
| Types of message | Unconstrained | Inform, request, offer, promise, decline, ... |
| Constraints on methods | None | Honesty, consistency, ... |

## Agent-0

In this section, we take an in-depth look at the first AOP language, Agent-0. Our motivation for doing this is two fold: firstly, Agent-0 is the first AOP language to be developed, and secondly, its design is centered around the notion of commitment to action as opposed to the more goal-driven BDI approach. Given that AFAPL also adopts a more commitment centric approach, it would seem natural and appropriate to focus on Agent-0.

Shoham formalizes the operation of Agent-0 through the definition defines a set of modalities that he uses to formally define the agent interpreter. These modalities act upon a first-order logic that is augmented with a time component. In this logic, Shoham augments both the base facts, and the modalities with times, for example: `finished(rem, chapter)`$^t$ represents that rem finished this chapter at time t; and `BEL`$^t_a$ `finished(rem, chapter)`$^{t'}$ represents that at time t, agent a believes that rem finished this chapter at time t'. Shoham justifies this representation of time by stating that: "we believe things both about different times and at different times, and the same is true of other modalities" (Shoham, 1993).

Shoham models actions through as being defined implicitly, so that, "rather than say that the robot took the action *raise-arm* at time t, we will say that the sentence *raise-arm(robot)* $^t$ is true" (Shoham, 1993). When conjoined with modalities this allows an agent to believe it is capable of raising its arm or to believe that it has raised its arm, or even allows the agent to believe such things about other agents.

Given this model of time and action, Shoham defines three fundamental modalities: belief, obligation, and capability. The first modality, which was introduced informally above, enables an agent to have beliefs about what it thinks the current state of the environment is, and potentially beliefs about those and other agents' beliefs. The second encompasses the notion of commitment to action, in that it represents the commitment of one agent to another at a given time to believe a certain state of affairs. Finally, the third modality defines states of affairs an agent can bring about at a given time. For example, an agent can have the capability at time 5 to open the door by time 8.

Two additional modalities, decision and able, are defined as extensions of obligation and capability respectively. Decision represents obligation to oneself, and able represents an "immediate" version of capability. By immediate Shoham means those capabilities that can be carried out at that time point, for example if an agent is able to open the door at time 5, then at time 5 it has the capability to open the door immediately.

Defining the above set of modalities provides a formal model of the agents' mental state. However, they provide little insight into the nature of the agent interpreter. This is provided through the ascription of various properties to this model, such as, internal consistency, introspection, and persistence of mental state.

Shoham implements this model through an agent interpreter. His design is illustrated in figure 2. The properties of the model are encoded within the update mental state phase of the interpreter cycle, and the representation of the mental state and capability encodes the modalities (in Agent-0, belief is represented by the B keyword, and obligation by the CMT keyword – representing commitment).

The behaviour of Agent-0 agents is programmed via a high-level programming language. This language defines syntax for the representation of positive and negative facts based upon standard LISP semantics, for example `(NOT (t (employee jones acme)))` represents the fact that jones is not an employee of acme at time t. Actions may be distinguished by type, private or communicative, and activation, conditional or unconditional. Thus actions types are distinguished by the keywords DO (for private), and INFORM, UNINFORM, REQUEST, UNREQUEST, or REFRAIN (for communicative). Unconditional

*Figure 2. Schematic of the Agent-0 Interpreter (Reprinted from Artificial Intelligence, Volume 60, Issue 1, Yoav Shoham, Agent-oriented programming, pages 51-92, Copyright (1993), with permission from Elsevier)*

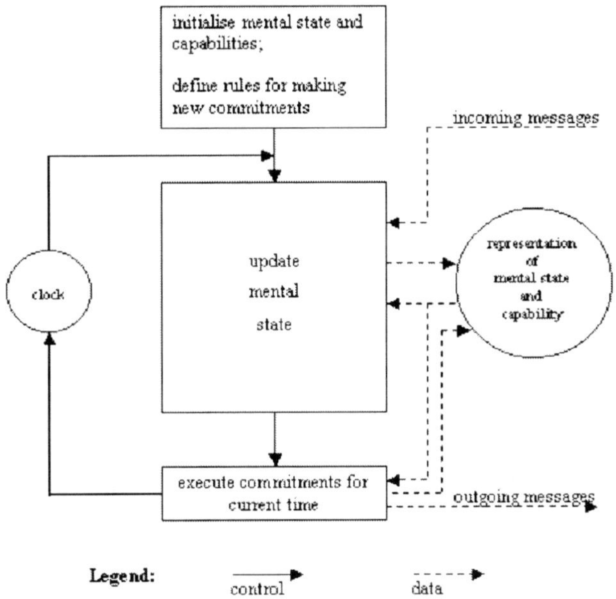


actions take the form (`DO t p-action`) which states that at time t, the agent should perform the private action p-action. Conditional actions are triggered through some mental condition. Mental conditions may be either a belief statement or a commitment (obligation) statement. Conditions in Agent-0 may include variable bindings.

Finally, the top layer of the Agent-0 language is a collection of commitment rules that define when and how agents should adopt new commitments. These combine some mental condition with some message condition and define a corresponding action that should be committed to should the conditions arise. As a result, it is these rules that define the overall behaviour of the agent and, while these behaviours are not defined in terms of goals that should be achieved, they do support the definition of undesirable situations in which the agent should act.

As Shoham acknowledges, Agent-0, was never intended to be a fully-fledged AOP language, but rather a prototype that could be used as a basis for other researchers. As a result, Shoham leveled a number of criticisms at both Agent-0 and AOP. We conclude this section be outlining some of the more main criticisms that are directed at AOP languages.

- Extensions to the Paradigm. The relation between AOP and OOP should be further investigated to consider additional properties of OOP such as inheritance, composition, and polymorphism. Informally, this would also implicitly require the introduction of some form of reuse mechanism.
- Extensive Tool Support. It is vital that developers wishing to use an AOP language have access to the same quality and variety of tools that they would have access to if developing an OOP program. This includes integration with established IDE's, such as Netbeans and Eclipse, and the development of equivalent tool sets for AOP languages,

- Clear Link to the Environment. Any practical AOP language must specify clearly how agents interact with their environment. Lack of clarity can result in confusion for developers, which will impact on the potential uptake of AOP languages.
- Integration with Current Best Practices for Agent Systems. Many of the earlier AOP languages were either not distributed, or made use of some form of bespoke infrastructure. Since the emergence of the FIPA Standards (FIPA, 2000) many of the agent toolkits have become, at least to some level, compliant. AOP languages must offer similar levels of compliance in order that any agents developed will be able to interoperate with other agents developed using other toolkits.

Armed with these general criticisms of AOP languages, the next three sections introduce the Agent Factory framework, a constituent AOP language, known as AFAPL, and some of the tools that have been developed to support the coding of agents using AFAPL.

## THE AGENT FACTORY FRAMEWORK

Agent Factory (AF) is an open source framework that has been developed to support the deployment of multi-agent systems (Collier et al., 2003). It is broadly compliant with the current FIPA standards and is implemented in the Java programming language. In its latest incarnation, the framework has been restructured to facilitate the deployment of applications that consist of multiple agents that employ a diverse range of agent architectures. In this the framework has become an enabling middleware layer that can easily be extended and adapted for different application domains.

As can be seen in the schematic of AF presented in Figure 3, the overall framework has been broken down into three parts:

- A distributed *Run-Time Environment (RTE)* that consists of a FIPA-compliant *agent platform* together with a number of prefabricated *agent system architectures* (software architectures);

*Figure 3. A Schematic of the Agent Factory framework*

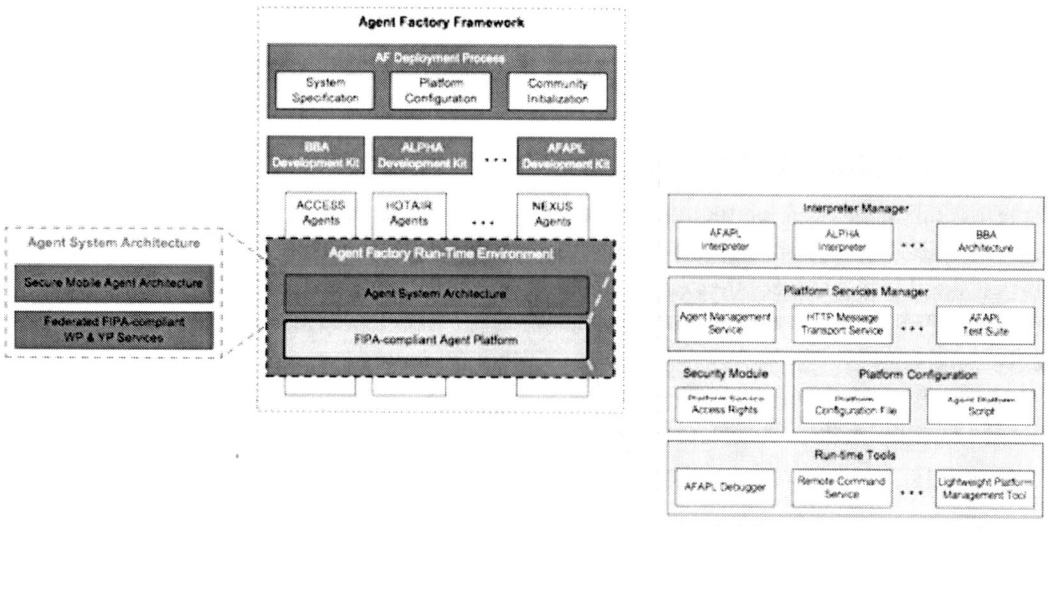

- A set of *Development Kits (DK)* that contain agent interpreter / architecture implementations together with relevant tool support; and
- A *Deployment Process* that provides a structured approach to the deployment of multi-agent systems using the framework.

Of these, the two core components are the agent platforms and the DKs. Specifically, each agent platform contains the machinery necessary to create and support agents that are developed using one or more of the DKs. This machinery includes:

- An *agent container* that holds the agents currently resident on the agent platform;
- A manager that supports the deployment of a set of *platform services* that implement shared resources that are available to some or all of the resident agents;
- A *security module* that controls what platform services each agent has access to;
- A module that manages the set of *agent interpreters / architectures* for the various types of agent that are currently deployable on the platform; and
- A module that creates and deploys any *run-time tools* that are required to provide necessary support for the visualization, management, profiling, and debugging of the resident agents.

The configuration of the agent platform and the specification of the initial agent community are supported via two deployment files: the *platform configuration file* (with a .cfg extension) is used to specify a configuration for each agent platform and the *agent platform script* (with a .aps extension) is used to specify the initial community of agents that are to be deployed on that platform. Examples of these two file types are given in the screenshot in Figure 4, which showcases a plugin for the Netbeans Integrated Development Environment. This plugin is also integrated with some of the tool support that comes as part of the AFAPL-DK (the development kit for AFAPL).

*Figure 4. Screenshot of the Netbeans IDE Plugin for Agent Factory. The upper pane shows an open platform configuration file, while the lower pane shows an open agent platform script file.*

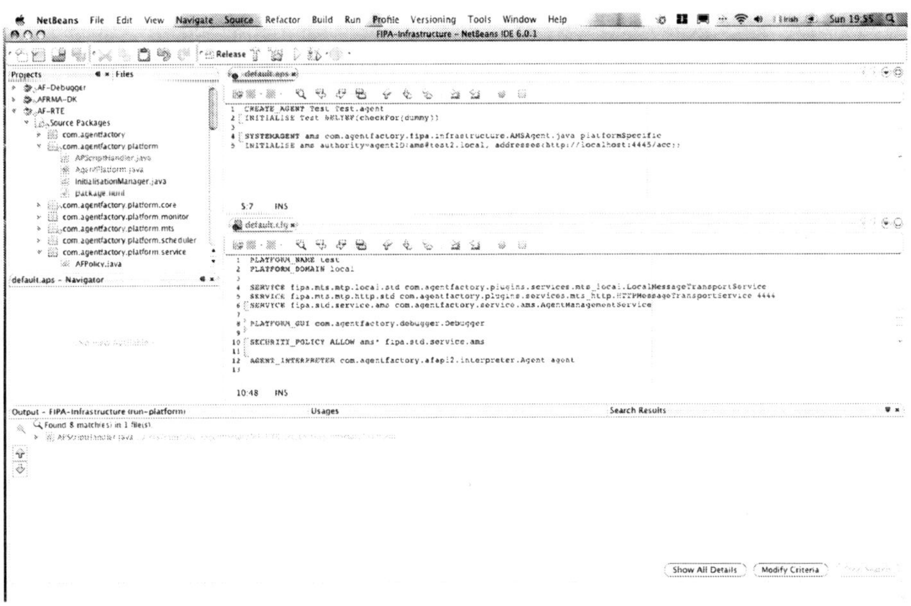

As can be seen in the example presented in Figure 4, each agent platform is given a unique name and is associated with a domain. In this case, the agent platform name is "test.local". Platform services are associated with a platform via the SERVICE command. As part of their declaration, each service is given a unique identifier (in the case of the HTTPMessageTransportService and AgentManagementService, this is specified in the associated FIPA specifications) together with the name of the class that implements the service. For configuration purposes, optional settings can also be specified after the class name. For example, the HTTPMessageTransportService in the above example is configured to work on port 4444.

Platform services are shared resources that are, by default, accessible by every agent that is currently residing on the platform, although agents must explicitly bind to a service in order to access it. One kind of platform service that is an exception to this is Message Transport Services (MTS). These services are used to support inter-agent communication, and deal with low-level issues of sending and receiving messages. All agents are automatically bound to all available MTS platform services when they are created. Currently, four MTS are provided as part of the framework: the *Local-MTS* provides intra-platform agent communication via Java message passing; the *HTTP-MTS* provides a HTTP-based message transport service that is based on the corresponding FIPA specification; the *UDP-MTS* provides a UDP-based message transport service that works on a multicast channel, but does not guarantee delivery; and the *MPP-MTS* provides a SMTP-style client-server based message transport service that can be used to circumvent firewalls.

The restriction of access to individual platform services is supported through the use of security policies. For example, in the above platform configuration file, a security policy is defined that restricts access to the AgentManagementService, a second FIPA service that provides support for the run-time management of agents (i.e. creation, suspension, resumption, and termination), so that only agents whose name beings with "ams" can access the service. Once a security policy is applied to a platform service, the default policy for the service becomes denial of access to that service.

Finally, the platform configuration file also allows multiple agent interpreters (or architectures) and run-time tools to be associated with an agent platform via the AGENT_INTERPRETER and PLATFORM_GUI keywords respectively. As with the SERVICE keyword, optional configuration information may be appended to the end of the mandatory set of parameters. This information is passed directly to the underlying Java implementation for the associated platform service, agent interpreter or run-time tool. Support for the development of these components is provided in the form of Java APIs, which include abstract base classes that may be extended to implement the new feature. Details of how to do this are provided via various development guides that are available on the Agent Factory website.

The agent platform script file is used to declare the initial agent community that will be deployed on an agent platform at startup. This file allows developers to define what agents will be created, to initialize the state of those agents, and where to start their execution. The script allows the creation of two kinds of agents: *system agents* and *application agents*. System agents are those agents that make up the system architecture, while application agents are those agents that contain the application logic. In the example presented in Figure 4, one system agent is created, with a platform specific name that begins with ams (platform specific names are generated by appending the specified name with an @ symbol followed by the full platform name). One application agent is created with the name Test. The main practical differences between system agents and application agents are that system agents are created first and have the chance to carry out initial configuration before the application agents are created, and system agents start executing by default, while application agents must be started explicitly via the START_AGENT command. Initial configuration information may be passed to either kind of agent via the INITIALISE keyword and the format of the data passed is specific to the underlying agent architecture.

The mapping between agent program and agent interpreter / architecture is based on the file extension used in when the agent is created. For example, in the above example, the Test agent is created using a file with a ".agent" extension which is used to identify its implementation as AFAPL target code). In cases where the architecture is implemented directly in Java, as is the case with the ams agent, the ".java" extension is used.

In summary, Agent Factory provides a modular and extensible framework for constructing and deploying multi-agent systems. It offers a set of APIs that can be used to implement and integrate agents that employ different architectures and can support multiple interpreters that realise high-level AOP languages via the creation of Development Kits. One such language, which is described in the next section, is the *Agent Factory Agent Programming Language*.

## THE AGENT FACTORY AGENT PROGRAMMING LANGUAGE (AFAPL)

Agent Factory (AF) provides support for the programming of agents through AFAPL, a purpose-built AOP language that is inspired by Agent-0. AFAPL builds on previous work presented in (O'Hare, 1996) and (Collier, 1996) in which VDM (Bjorner & Jones, 1978) was used to specify an agent interpreter for AF. Further, the specification presented in (O'Hare, 1996), was subsequently refined into an algebraic specification that was implemented using the UMIST Paradox System (Powell, 1989), while the specification presented in (Collier, 1996) was implemented in the Small-talk-80 programming language. Underlying the latter implementation was a purpose-built configuration language that supported the definition of agents that utilized the associated interpreter. Following on from these initial implementations, this configuration language has evolved into AFAPL. As part of this evolution, (Collier, 2001) formally defines the syntax and semantics using a multi-modal branching time first-order logic of commitment. The language itself combines the mental attitudes, whose syntax and semantics are defined via the logic, with additional programming constructs that support the specification of the interface between the agent and its environment. This has resulted in a practical AOP language that is implemented in Java, and has been successfully applied in a number of significant problem domains.

In contrast with many of the more recent AOP languages, AFAPL does not adopt the Belief-Desire-Intention (BDI) mental state architecture as its core. Instead, it employs a commitment-based mental state that is similar to the one used in the design of Agent-0. However, in contrast with Agent-0, AFAPL offers the following:

- Support for the definition of composite activities based on plan operators together with a mechanism for declaring, at design time, a set of partial plans (via a plan library).
- The implementation of a well-defined model of commitment management that specifies how commitments are adopted, refined, maintained and realised; and what to do in the event that a commitment fails.
- A set of tools, presented later, that provide comprehensive support for the development process.
- The provision of a well-defined interface between the agent and its environment that can be customised through the creation of *perceptors* (sensors) and *actuators* (effectors) and their subsequent declaration within the relevant AFAPL program.

- Support for the decomposition of agent programs into reusable units together with an import mechanism that is similar to #include in C.

In its current incarnation, AFAPL is a logic-based language in which an agent is viewed as a computational entity whose core mental state consists of *beliefs, plans* and *commitments*. Beliefs represent the agent's (possibly incorrect) knowledge of the state of the environment; plans are complex activities that may be performed by the agent, and which are combinations of plan operators and other (more primitive) activities; and commitments represent the activities that the agent has decided to perform. These three mental attitudes are joined by a set of *commitment rules*, which encode the situations in which the agent should adopt new commitments. In the following sections, we discuss these core concepts, the corresponding programming constructs and their associated managers in more detail. The syntax and semantics of AFAPL have been formalized via a multi-modal first-order branching-time logic (Collier, 2001). In this section, we present a detailed, but informal, overview of both the language and the operation of the associated aspects of the agent interpreter.

## Beliefs and Belief Management

The agent's beliefs represent the (possibly incorrect) knowledge that an agent has about itself and its environment. Knowledge takes the form of facts that are represented as well-formed formulae of an underlying domain language that is based on first-order logic. For example, the fact that Bob has finished ironing can be represented as: `finished(Bob, ironing)`. As is usual for such logics, variables may be introduced to specify unknowns, for example, the fact that someone has finished ironing can be represented as: `finished(?x, ironing)`. Building on this, the beliefs of an agent are represented as a set of grounded belief literals, which are themselves well-formed formulae of the underlying domain language. As is typical for such languages, these well-formed formulae are "wrapped" within an operator that classifies them as beliefs. For example, the belief that Bob has finished ironing is represented as: `BELIEF(finished(Bob, ironing))`.

A key decision made in the design of the belief model employed in AFAPL was to make beliefs transitory. That is, beliefs do not persist beyond the iteration of the interpreter cycle in which they are adopted. The motivation for this arises as a result of the use of *perceptors* as the core belief update mechanism. Perceptors are part of the environment interface of an agent and they are used to convert sensor data gathered from the environment into beliefs. For example, in the robot soccer domain, a perceptor may be used to generate a belief about the presence or absence of a ball within the robot's field of vision. This belief may be used to drive the overall strategy of the robot and result in a change in its underlying behaviour such as switching from a "defensive" behaviour to a "ball tracking" behaviour. Similarly, other agents may need to be aware of the current time, and as a result, may adopt beliefs about the current time. Rather than attempt to specify which kinds of beliefs should persist or to require that perceptors also manage the removal of existing beliefs; the approach adopted in AFAPL is to remove all beliefs at the beginning of the iteration and to build a new belief set via activation of the agents associated set of perceptors. This approach is inspired by the foundational approach to belief revision in that the perceptors generate a set of foundational beliefs at the start of each interpreter cycle. These foundational beliefs, which are termed the *current beliefs* of the agent, are then augmented through a set of *belief rules* that define a set of derived beliefs that may be inferred from the foundational beliefs.

As is highlighted in the upper part of Figure 5, perceptors are instances of concrete Java classes that extend the `com.agentfactory.core.agent.Perceptor` abstract base class, and implement the perceive() method. This abstract base class facilitates access to the associated agent implementation together with a number of helper methods that simplify access to shared resources (i.e. platform services), and simplify the adoption of beliefs. The example perceptor in Figure 5 generates beliefs about the agents name and the current iteration. When implemented, perceptors must be explicitly associated with an agent program through the PERCEPTOR construct. An example of a perceptor declaration is presented in the lower part of Figure 5.

By adopting an approach in which the current beliefs of the agent are directly generated from sensor data, the belief set that the agent uses to reason about how to act is, as far as possible, accurate with respect to that environment. However, the downside is that the beliefs of an agent become akin to a short-term memory, and information cannot be easily maintained over multiple interpreter cycles. In response to this, the AFAPL belief model includes a set of *temporal beliefs* that can be used to represent more persistent beliefs. Along with current beliefs and belief rules, this third component of the belief model holds beliefs that are associated with one of three temporal operators:

- the NEXT operator is used to define beliefs that will should be adopted at the start of the next interpreter cycle, for example, `NEXT(BELIEF(finished(Bob, ironing)))` represents the belief that Bob will have finished the ironing by the next iteration;
- the ALWAYS operator is used to define beliefs that persist for the lifetime of the temporal belief, for example, `ALWAYS(BELIEF(overworked(Bob)))` represent the belief that Bob is always overworked; and
- the UNTIL operator is used to define beliefs that persist until some associated condition arises, for example, the temporal belief `UNTIL(BELIEF(doing(Bob,ironing)), BELIEF(finished(?someone,ironing)))` represents the belief that Bob is doing the ironing, and this belief persists until the agent believes that someone (here represented by the variable) has finished ironing.

Temporal beliefs are kept separate from the current beliefs of the agent. The effect of adopting a temporal belief is that the associated belief is also adopted as a current belief. For example, if the agent adopts the temporal belief, `ALWAYS(BELIEF(overworked(Bob)))`

*Figure 5. An example Perceptor and the corresponding AFAPL declaration*

```
// The Java implementation
public class SelfPerceptor extends Perceptor {
    public void perceive() {
        adoptBelief("BELIEF(name(" + agent.getName() + "))");
        adoptBelief("BELIEF(iteration(" agent.getIteration() + "))");
    }
}

// The AFAPL declaration
PERCEPTOR self {
    CLASS SelfPerceptor;
}
```

*Figure 6. Belief management in AFAPL*

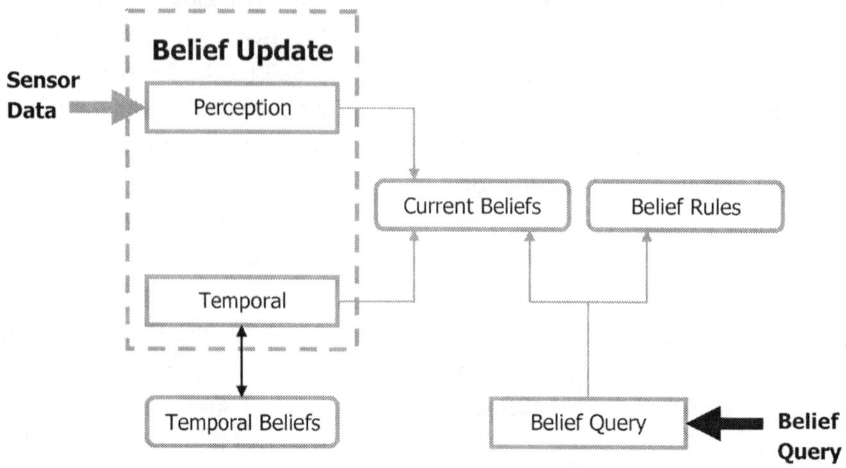


then the agent also adopts a current belief, `BELIEF(overworked(Bob))` on each iteration. As is illustrated in Figure 6, the adoption of current beliefs based on the temporal beliefs of the agent is managed via a temporal update step that is part of the belief update process. This step is preceded by a perception step in which the agent fires its perceptors. The perceptions generated by these perceptors are added to a *belief update set*, which also includes any beliefs adopted by actions that were performed during the previous interpreter cycle.

The current beliefs of the agent are used as a basis for making decisions about how the agent should act. This is supported through the concept of a situation. Informally, a situation is a conjunction of positive and negative belief literals that should be evaluated with respect to the current beliefs of the agent. For example, we can represent the situation that somebody is hungry and there is some food to eat as: BELIEF(hungry(?x)) & BELIEF(hasFood(?y)). Situations are evaluated through a belief query process (see Figure 6), which provides a resolution-based (Robinson, 1965) query mechanism through which the truth or falsity of the situation can be evaluated. The implementation of the belief query component is based on the reasoning mechanisms employed in Prolog interpreters, and is applied to both the current beliefs and the belief rules of the agent. In addition to assessing the truth or falsity of a situation, the belief query process can also return a set of all the variable bindings that, if applied, would cause the situation to be evaluated to true.

Finally, AFAPL includes additional support for defining valid terms in the underlying content language. This is done through the use of *ontologies*. Ontologies are specified using the ONTOLOGY construct and can be used to define a set of terms for the agent program. For example,

*Figure 7. An example AFAPL Ontology*

```
ONTOLOGY health {
    PREDICATE timeToCheck(?x, ?y);
    PREDICATE healthy(?x);
    PREDICATE failed(?x);
}
```

Figure 7 presents a simple ontology for a Health Monitoring agent that includes thee terms:

- `timeToCheck(?x, ?y)` is a term that is used to represent the fact that it is time to perform a health check on the agent with name ?x and addresses ?y;
- `healthy(?x)` is a term that is used to represent the fact that ?x is healthy; and
- `failed(?x)` is a term that is used to represent the fact the ?x has failed.

Initially, support for ontologies was introduced as a mechanism for detecting semantic errors during the compilation of agent programs. As such, a minimalist approach was adopted. It is anticipated that future versions of AFAPL will expand on this model to offer integration with more established ontology languages such as OWL.

## Actions and Plans

The actions and plans of an agent describe the set of activities that the agent is capable of performing. Actions refer to the primitive capabilities of the agent that are directly executable via an associated *actuator* unit, which forms the second part of the interface between the agent and its environment. In contrast, plans refer to the complex capabilities of the agent and consist of a set of plan operators that define orderings on an underlying set of activities, which may themselves also be either actions or plans.

Actuators implement the primitive capabilities of an agent. They are instances of Java classes that extend the abstract `com.agentfactory.core.agent.Actuator` base class, and implement the act(...) method. Again, this base class provides access to the agent implementation together with a number of helper methods that simplify access to shared resources, and the adoption of beliefs. Specifically, as was indicated in an earlier section, beliefs that are adopted via actuators are added to a *belief update set*, which forms the basis of the agents current beliefs for the next iteration.

To illustrate how actuators work, the upper part of Figure 8 presents an example actuator that can be used to dynamically create another agent at run-time. As can be seen, this actuator retrieves a reference to the AgentManagementService platform service, and invokes the createAgent(...) method on this service, taking the name and type of the agent to be created from the parameters of the action identifier, which is passed as a parameter to the act(...) method. The method then implements the action, returning a Boolean value to indicate whether or not it succeeded.

Similarly to perceptors, actuators must also be explicitly declared within the relevant AFAPL agent program. As is shown in Figure 8, this is achieved though the use of an ACTION statement. This statement defines an action in terms of an identifier (that is used to refer to the action in the context of commitments), a pre-condition that must be satisfied before the action may be executed, a post-condition that will be satisfied after the action is completed, and the class that implements that action. Informally, when an agent decides to perform do something it generates an appropriate activity identifier. This identifier is then matched against a corresponding action, and the agent uses this matched action to access and fire the underlying actuator unit. For example, if the agent decided to perform the activity `createAgent(Bob, mydesign.agent)`, then it would first match this activity identifier to the `createAgent` action. Based on this match, it would fire the associated `CreateAgent` actuator unit. When doing this, the original activity identifier is passed to the act(...) method, where the data stored within the arguments can be extracted.

In addition to action, AFAPL also includes support for the declaration of plans. A plan is basically a composite activity that combines a set of sub-activities whose order of realization depends upon an associated plan operator. Sub-

*Figure 8. An example Perceptor and the corresponding AFAPL declaration*

```
// The Java implementation
public class CreateAgent extends Actuator {
    public void act(FOS action) {
        String name = action.argAt(0).toString();
String type = action.argAt(1).toString();

AgentManagementService service = (AgentManagementService)
       getService("fipa.std.service.ams");
service.createAgent(name, type);
Service.resumeAgent(name);
        adoptBelief("BELIEF(created(" + name+ "))");
        return true;
    }
}

// The AFAPL declaration
ACTION createAgent(?name, ?type) {
    PRECONDITION BELIEF(true);
POSTCONDITION BELIEF(created(?name, ?type));

    CLASS CreateAgent;
}
```

*Figure 9. An example AFAPL plan for performing a health check*

```
PLAN healthCheck(?name, ?addr) {
      PRECONDITION BELIEF(true);
       POSTCONDITION BELIEF(true);

      BODY
           SEQ(request(agentID(?name, ?addr), ping),
         OR(DO_WHEN(BELIEF(fipaMessage(inform, sender(?name, ?ad), pong)),
                   adoptBelief(BELIEF(healthy(?name)))),
             SEQ(DELAY(milliseconds(5000)),
                adoptBelief(BELIEF(failed(?name))))));
  }
```

activities may themselves by either actions or plans. Currently, AFAPL supports the following plan operators:

- **SEQ($act_1$, ..., $act_n$)**: Defines a sequence of activities that should be performed. When realising this plan, the agent performs each activity consecutively, so $act_1$ is performed before $act_2$ and so on.
- **PAR($act_1$, ..., $act_n$)**: Defines a set of parallel activities that can be performed at the same time. When realising this plan, the agent tries to perform each activity at the same time. This may not happen in cases where the preconditions associated with a given activity are not met.
- **OR($act_1$, ..., $act_n$)**: Define a set of activities, only one of which must be completed successfully for the plan to be considered successful. For example, if the agent were to perform the plan **OR(A, B)**, then the agent need only complete either A or B. Once a

sub-activity is successful, all other attempts to complete the plan become redundant.

- **DO_WHEN(cond, act):** Defines a conditional activity that is only carried out when (if) the corresponding condition is satisfied. This plan operator can be effectively combined with the OR plan operator to implement a choice mechanism (i.e either do A when X or B when Y).
- **DO_UNTIL(cond, act):** Defines an activity that is out repeatedly until the corresponding condition is satisfied.
- **DELAY(time):** Defines a delay of a prescribed time period. The specified time may be either a time in milliseconds or a number of iterations.
- **ATTEMPT(X, Y, Z):** Defines an plan where the agent should attempt X and, if X is successful, should do Y, otherwise the agent should do Z.
- **FOREACH(cond, act):** Defines a plan where the agent does the action for each variable binding that satisfies the condition.

While explicitly declared plans are sufficient for defining the behaviour of agents, however, there are number of cases where the same plan appears in a number of places. In order to facilitate the reuse of such plans, AFAPL supports the definition of partial plans via the PLAN construct. This construct is the counterpart (for plans) of the ACTION construct in that it associates each partial plan with a unique identifier and a set of pre- and post- conditions. Informally, the pre-conditions define the situation that must be satisfied for the plan to be realized, and the post conditions define the expected outcome of following the plan successfully.

Finally, the identifier can include a number of parameters that can be used to customise the underlying plan. For example, Figure 9 shows a plan for carrying out a health check on another agent whose name is "?name" and who is contactable by a set of addresses denoted by "?addr". The plan works by sending a "ping" request message to the specified agent that is being monitored, and waiting for a "pong" inform response message. If the agent received the pong message before a deadline of five seconds (5000 milliseconds), then the agent adopts a belief that the monitored agent is healthy otherwise it adopts a belief the monitored agent has failed.

Both actions and plans are realized through the adoption of commitments to the relevant action or plan and the subsequent realization of that commitment. Details of this are presented next.

## Commitments and Commitment Management

Commitments represent the outcome of the agent's decision-making process. Specifically, a commitment is defined to be the mental equivalent of a contract, which describes what the agent has decided to do, for whom it has decided to do it, when the chosen activity must be realized, and the terms and conditions under which the commitment may be dropped (known as the *maintenance condition*). For example, consider a Health Monitoring agent that has been charged with the task of monitoring the health of a number of application agents by "pinging" those agents on a regular basis via the use of the `healthCheck(...)` plan that was described in Section 4.2. At a certain point in time, the agent will decide to perform the health check on one of the agents that is is monitoring. In the case where this is the agent known as AnAgent, then the agent would adopt a commitment, shown in the upper part of Figure 10, to the `healthCheck(...)` plan.

The agent adopts this commitment through the firing of a special type of rule, known as a *commitment rule*. Commitment rules define the situations in which the agent should act. These situations are evaluated with respect to the current beliefs of the agent during each iteration of the interpreter cycule. Should the situation arise, then the agent will adopt a commitment the corresponding commitment. By way of illustration, the lower part of Figure 10 includes an example commitment rule, which states that, if the agent believes that it is time to check the health of a certain agent (whose name and contact addresses are known), then the agent should adopt a commitment to the `healthCheck(...)` plan. In addition to the variable defined in the situation, AFAPL also includes two additional system variables that may be used in a commitment rule: ?self is bound to the name of the agent, and ?now is bound to the current time. To illustrate how commitment rules work, consider the case where the agent believes: `BELIEF(timeToCheck(AnAgent ,addresses(http://localhost:4444/ acc)))`. Based on this belief, the commitment rule would be triggered, and the agent would adopt the commitment in the top part of Figure 10 (assuming the agent was called HealthMonitor and the current time was 11:32).

Our use of commitments is in contrast with the more popular BDI architecture (Rao & Georgeff, 1995) in which the future activities of an agent are driven by its desires, which represent the ideal state of the environment, and intentions, which are a chosen subset of those desires, which the agent considers to be achievable and which it is actively trying to achieve. An agent achieves its intentions by following a chosen course of action. How the agent selects that course of action is an implementation issue.

Typically, BDI-based AOP languages, such as AgentSpeak(L) and 3APL, adopt an approach in which intentions are represented by the plans that the agent decides to perform. Our use of commitments is somewhat similar. That is, commitments represent the courses of action that the agent has decided to try and achieve. The key difference between commitments and intentions is that commitments represent a decision to do something, they do not necessitate that the agent intends to bring about the resulting state of affairs, nor do they necessitate that the agent has a corresponding goal to bring about some future state of the world. The agent may simply be following a social convention that it must respond to requests from other agents, or may be reacting to some change in the environment.

Further, by adopting the concept of a commitment as the primary mechanism for representing the choices of the agent, we are able to associate a strength of commitment with that choice. The idea of commitment strength originates with the notion of a *commitment strategy* (Rao & Georgeff, 1991). In their work, a commitment strategy is a property of the underlying decision-making aparatus of an agent, and it defines how committed the agent is to its intentions. To illustrate this, they identified three potential commitment strategies that an agent could employ:

- *Blind Commitment:* The agent is committed to its intentions until it achieves them.
- *Single-Minded Commitment:* The agent is committed to its intentions only so long as it believes them to be achievable.

*Figure 10. An example of a Commitment and a Commitment Rule in AFAPL*

```
// Example Commitment
COMMIT(HealthMonitor, 11:32, BELIEF(true),
    healthCheck(AnAgent, addresses(http://localhost:4444/acc)));

// Example Commitment Rule
BELIEF(timeToCheck(?name, ?addr)) =>
COMMIT(?self, ?now, BELIEF(true), healthCheck(?name, ?addr));
```

- *Open-Minded Commitment:* The agent is committed to its intentions only so long as they still congruent with its goals.

In AFAPL, rather than forcing all commitments to adhere to one of the above strategies, we allow the developer to specify how committed the agent is to a given commitment through the inclusion of a *maintenance condition*. In cases where the agent should be blindly committed to some course of action, this mainteance condition is set to true. On the other hand, if the maintenance condition is set to define when the commitment is no longer achievable or (implicily) is no longer a goal of the agent, then the agent can become either single-minded or open-minded about that commitment.

The notion of a commmitment strategy, when applied directly to commitments, can be extended further to consider how more than just how a commitment is maintained. In general, such an expanded strategy can be termed a *commitment management strategy*. A commitment management strategy specifies a set of sub-strategies that define how an agent: (1) adopts new commitments, (2) maintains its existing commitments, (3) refines commitments to plans into additional commitments, (4) realizes commitments to primitive actions, and (5) handles failed commitments. These strategies realised through a meta-level process, known as *commitment management*.

The principal sub-strategy that underpins the behaviour of AF-APL agents is *commitment adoption*. Commitments are adopted either as a result of a decision to realise some activity, or through the refinement of an existing activity. The former type of commitment is known as a *primary commitment* and the latter as a *secondary* (or *tactical*) *commitment*. Primary commitments are adopted as a result of the triggering of a commitment rule,while secondary commitments, on the other hand, are adopted in response to the agents' efforts at realizing an existing commitment to a plan of action. That is, secondary commitments are adopted as part of the *commitment refinement* process that is enacted whenever an agent attempts to realize an activity to a plan. The secondary commitments correspond to each of the sub-activities that the plan operator has been applied to.

The commitment refinement process forces a hierarchical relationship between commitments that can be modelled as a tree-like structure as is presented in Figure 11. However, this structure is complicated by the additional constraints that are placed on secondary commitments as a consequence of the semantics of plan operators. For example, if an agent has committed itself to a plan that is a sequence of activities, then, when it refines this commitment into a set of secondary commitments, it can only consider the secondary commitment that corresponds to the first activity in the plan. That is, the sequential plan operator

*Figure 11. A view of the hierarchical relationships within a commitment structure*

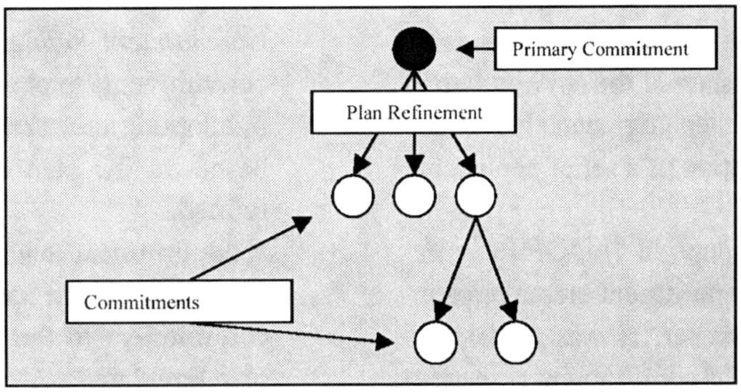

creates a dependency that limits when the agent may try to realize the secondary commitments. Similar dependencies are created for the other plan operators and are discussed in (Collier, 2001) along with further details of how failure of commitments are handled by the AFAPL interpreter.

## Interpreter Cycle

Agent Factory includes a built-in scheduler that organizes the execution of all of the agents residing on a given agent platform. Agents are allotted a time slice by the scheduler, and the interpreter cycle is started when that time slice begins. As can be seen in Figure 12, the interpreter cycle consists of five basic steps: belief update (UPDATE), commitment adoption (ADOPT), commitment management (MANAGE), action (ACT), and migration (MIGRATE). The interpreter is designed so that the first three steps are guaranteed to complete no matter how small the time slice is. The action step then uses as much of the remainder of the time slice as possible / necessary, sequentially executing actions that the agent has decided to perform. The final migration step simply checks whether certain conditions have arisen, and if so, contacts the migration service so that migration can start. We describe each of these steps in more detail below, mapping them onto relevant sections of the chapter.

As was discussed earlier, the belief update process employed in AFAPL has been decomposed into two sub-processes: perception (PERCEPT) and temporal update (TUPDATE). The first of these steps refers to the firing of the agent's preceptor units and the generation of a up-to-date snapshot of the current state of the environment. Conversely, the second step augments this snapshot through the evaluation of a set of persistent beliefs.

The commitment adoption (ADOPT) is in reality part of the commitment management (MANAGE) step. However, as was explained earlier, it is this part of the commitment management strategy that governs the overall behaviour of the agent as it is through the adoption of commitments, via the firing of commitment rules, that the agent decides what activities to perform. Consequently, it has been given first class status in our interpreter algorithm. In contrast, the commitment management (MANAGE) step of the algorithm can, if required, be further decomposed the following five phases:

- *Commitment Maintenance* (MAINTAIN): The maintenance condition of each commitment is checked, and those commitments whose condition is no longer true with respect to the beliefs of the agent are considered to have failed.
- *Commitment Failure Handling* (FAILURE): The failure of a commitment is propagated to its parents and children (if they exist). While all child commitments are set to a failed state, the impact of a child's failure on the state of the parent commitment depends upon the activity associated with that commitment. For example, the failure of a child whose parent is a sequence (SEQ) plan results in the failure of the parent, while the failure of a child whose parent is a optional (OR) plan only causes the parent to fail if it is the last option and all previous options have also failed.
- *Commitment Fulfilment* (FULFIL): The commitments are checked, and any leaf commitments (those commitments that have no children) that have either succeeded of failed are dropped.
- *Commitment Refinement* (REFINE): Any commitments to plans of action are refined by adoption a one or more sub-commitments based on the plan operator that is being refined.
- A commitment to a plan template *pt* (identified by the plan identifier) is refined to a commitment to the plan specified in that plan template.

*Figure 12. Overview of phases in the AFAPL Interpreter*

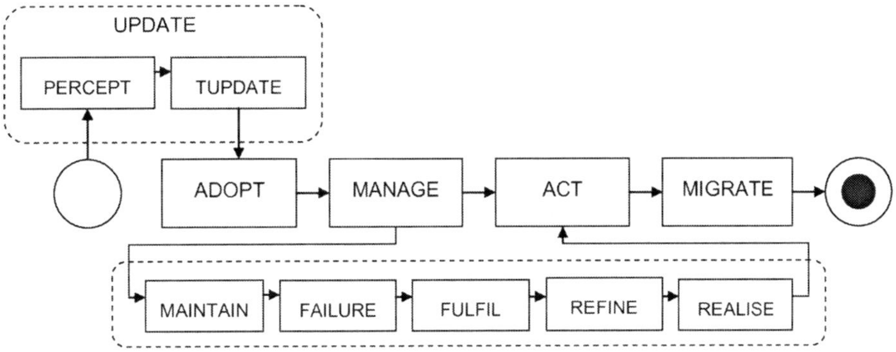


- *Commitment Realisation* (REALISE): Commitment realization applies only to commitments that are made to executing plan templates (specified by the plan identifier) or actions. Commitments made to these activities are realized when the time component of the commitment is satisfied and the preconditions of the corresponding activity are valid. For actions, the resulting behaviour is the scheduling of the commitment for execution, while, for plan templates it is the adoption of the underlying plan of action.

The action (ACT) step of the interpreter cycle deals with the actual execution of the agent's actions. During the commitment management phase, decisions are made as to which of the agents primitive activities should be realized. These activities are added to an internal schedule whose content is processed in this step. If enough time remains in the agents time slice, the action step removes items sequentially from the schedule in the order in which they were added and processes them one by one. This continues until either the schedule becomes empty or the agent's time slice finishes.

Finally, the migration (MIGRATE) step performs a check to see if the agent has decided to move to another agent platform. If this is the case, then the agent also checks to see if there are any commitments. If not, then the agent is removed from the scheduler and the migration platform service is notified, telling it to start the migration of the agent. In essence, migration in AFAPL starts by the agent performing the migrate(...) action, which sets an internal migration flag and stores details of the target platform that the agent will migrate to. Once the migration flag is set, then the interpreter suppresses the commitment adoption step so that no new commitments are adopted. Only when the outstanding (existing) commitments are completed (i.e. the commitment set becomes empty) is the agent is ready to migrate.

## Reuse of AFAPL Code

Support for the reuse of AFAPL code is provided in the form of an IMPORT statement that is based on the C #include statement. When building AFAPL programs, source code is written in a file that has a ".afapl" extension. These files are incorporated into the Java package structure, and can be referenced using standard Java naming conventions. For example, Figure 13 shows two example AFAPL programs where the second program reuses the code of the first program.

The first program builds on the health monitor example presented earlier, and implements the other half of the "ping" protocol (i.e. it defines how a monitored agent should respond to a health check). The second program is the AFAPL equivalent of the "Hello World" program. It defines a single action, called `alive`, whose implementation (actuator) will print out "Hello World!" to the console, and a single initial commitment to the `alive` action. The key point demonstrated here is the use of the IMPORT statement, which integrates the code for interacting with a HealthMonitor agent with the Alive agent program.

As is discussed later, the AFAPL compiler converts source code into a more machine-oriented target code, denoted by files with a ".agent" extension. As part of this process, the compiler combines all of the partial agent programs into a single block of code, which it then converts into the target code.

The ability to write reusable blocks of program code is an essential requirement of any practical programming language. While the reuse mechanism presented here is quite simplistic, it has had a significant impact on the usability of AFAPL as it facilitates the creation of code libraries that can easily be reused. This has led to the creation of a number of basic AFAPL programs that act as a starting point for developers. In this light, we have created the `com.agentfactory. core. fipa.agent.FIPACore` agent program, which contains the basic set of preceptors and actions required for agent communication. We have then built on this to create the `com.agentfactory. core.agent.BasicAgent` agent program which extends the core communication behaviours to include a number of core perceptors, actions and behaviours that we believe are core to all agents. This second program includes: actions for the adoption and retraction of beliefs; actions that support the binding to and unbinding from platform services; perceptors that generate beliefs about the current state of the agent platform; preceptors that generate beliefs about the current state of the agent (e.g. name, iteration); and so on.

## AFAPL TOOL SUPPORT

Support for the fabrication of agents using AFAPL is provided through a number of tools that are packaged as part of the AFAPL Development Kit (AFAPL-DK). These tools range from editors, compilers, and code libraries through to debuggers, test suites, and visual development tools. In the following sections, we highlight some of the more prominent tools in more detail.

*Figure 13. Two Example AFAPL Programs: Monitored.afapl and Alive.afapl*

```
// Program: Monitored.afapl (in agent package)
BELIEF(fipaMessage(request, sender(?name, ?addr), ping) =>
COMMIT(?self, ?now, BELIEF(true),
    inform(agentID(?name, ?addr), pong));

// Program 2: Alive.afapl
IMPORT agent.Monitored;

ACTION alive {
    PRECONDITION BELIEF(true);
    POSTCONDITION BELIEF(true);

    CLASS actuator.Alive;
}

COMMIT(?self, ?now, BELIEF(true), alive);
```

Two tools not discussed in detail here are the plugin for the Netbeans IDE and the VIsual Protocol EditoR (VIPER). A screenshot of the Netbeans plugin was presented earlier in Figure 4. In addition to the support provided for deploying agents using Agent Factory, this plugin also includes support for creating and compiling AFAPL programs. This includes templates for all components, syntax highlighting for the AFAPL files, and integration with the AFAPL Compiler.

VIPER (Rooney & Collier & O'Hare, 2004) is a visual development toolkit that is based on Agent UML (Bauer & Muller & Odell, 1999) Protocol Diagrams. It is comprised of two tools: a Protocol Editor that provides a visual tool for generating Agent UML Protocol Diagrams, and a Rule Editor that further supports the user by guiding them through the step of implementing the protocols in AF-APL. Two screenshots depiciting both the Protocol Editor and the Rule Editor are presented in Figure 14.

## The AFAPL Compiler

The primary responsibility of the compiler is to convert source code, denoted by a ".afapl" extension, into target code, denoted by a ".agent" extension that may be deployed. To achieve this, the compiler takes as input a source code file, and generates an AFAPL parse tree. Once the initial version of this tree has been generated, it is analysed, and any node relating to an IMPORT statement is expanded so that the resultant parse tree contains all of the source code associated with the target agent program. During this process, the detection of any syntactic errors results in the compilation process failing and an appropriate set of errors being displayed to the developer.

Following on from this, the compiler enters the first of two phases that, if completed successfully, ensure that the parse tree represents a syntactically, and as far as possible semantically, correct AFAPL program. In the first of these phases, it tries to collect various pieces of information about the program. Specifically, this phase of the compilation process involves two passes that:

- *Pass 1: Generate Ontology Database.* This pass searches the parse tree for nodes that contain ONTOLOGY constructs, and builds a temporary database of content language terms as defined in those ontologies.
- *Pass 2: Generate Activity Database.* This pass searches the parse tree for activity (i.e. action and plan) declarations and constructs a temporary database that contains the identifiers of each of the activities found.

In particular, the second pass extracts the identifiers of all activities (actions and plans) that are declared in the parse tree. A requirement of AFAPL is that each activity have a unique

*Figure 14. Screenshots of the VIPER Protocol Editor (left) and Rule Editor (right)*

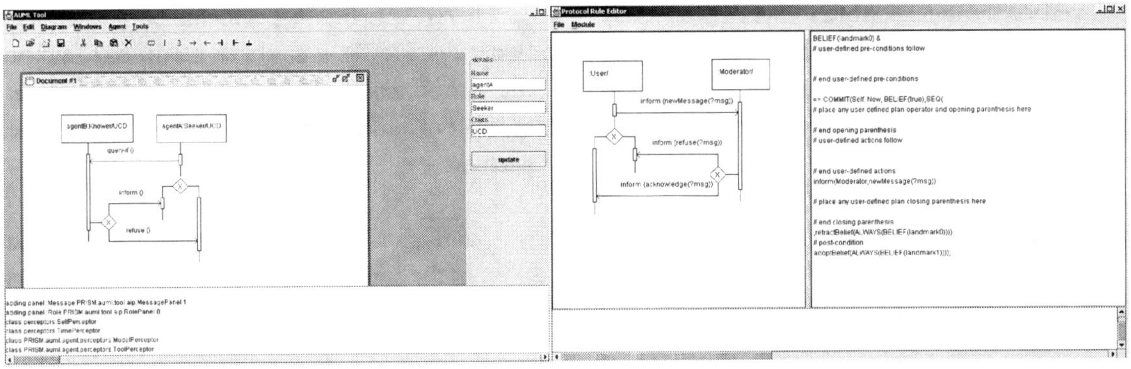

identifier. This requirement is enforced during this pass - if the compiler locates an activity identifier that is already in the activity database then the compiler halts and an appropriate error message is generated and displayed.

Once the compiler has collected this data, a second phase starts in which it carries out a series of checks that attempt to minimise the potential for semantic errors. Specifically, it performs two additional passes:

- *Pass 3: Check Beliefs.* This pass searches the parse tree for nodes that contain beliefs, and compares them against the terms that are stored within the ontology database.
- *Pass 4: Check Activities.* This pass searches the parse tree for plans and commitments and extracts all activity identifiers from them. It then compares the extracted identifiers against the database of activities that was compiled in pass 2.

Once these fours passes have been successfully completed, the compiler generates the resultant target code for the agent program, which can then be deployed using an appropriately configured instance of the Agent Factory Run-Time Environment.

## The AFAPL Inspector

Agent Factory provides support for the debugging of all types of agents via a generic debugging tool that is implemented as a run-time tool that can be deployed via the inclusion of a PLATFORM_GUI statement in the relevant platform configuration file. This tool is designed only for use with a single agent platform (other tools have been developed to provide multi-platform views), and provides support for inspecting the state of any agents or platform services that are resident on that platform. As a result, the tool currently provides two core views: an *agent view* and a *service view.*

The former view provides a list of agents together with a set of controls that can be used to start, stop and step individual or groups of agents. As expected, the start control resumes the execution of an agent, while the stop control suspends the execution of an agent. The step control is based on the step function that is offered by tradition debugging environments. Within AFAPL, a basic step is a single iteration of the AFAPL interpreter cycle. However, this has recently been extended to include support for breakpoints (Collier, 2006).

Individual agent architectures / interpreters are then able to extend this basic snapshot to include interpreter specific information, such as the current mental state of the agent. By default, the Agent Factory Debugger includes an inspection tool that can be used to inspect the state of any type of agent. This is achieved by extracting the features that are core to all agent implementations, and includes support for the inspection of: all messages sent and received, and the platform services to which the agent is bound. This information is retrieved from a snapshot object that can be generated by the agent architecture / interpreter on request. Specifically, on receipt of an end-of-time-slice message from an agent, the debugger requests a snapshot from that agent and stores the snapshot in a snapshot manager. This manager persists all previous snapshots to file and provides a mechanism for retrieving those snapshots on request. Visualisation of the snapshot is supported through the generic inspection tool (see Figure 15) that can be easily extended to include agent interpreter specific views.

For AFAPL, this inspector has been extended in two ways. Firstly, it has been modified to support the visualization of breakpoints, and secondly a number of additional views have been developed, such as:

- *Beliefs view*: Presents a view of the beliefs of the agent that can be customized either to show all beliefs or to show only the current

*Figure 15. A screenshot of the AF Debugger in which an AFAPL Inspector is being used to view the state of a health monitoring agent*

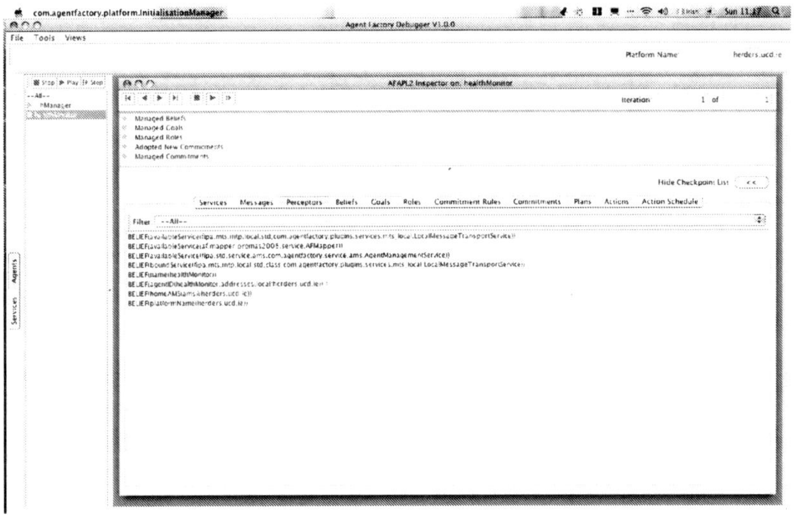


beliefs, belief rules or temporal beliefs of the agent.

- *Commitments view*: Presents a view of the agent's current commitments. Selecting a commitment, allows the developer to view the commitment structure that is rooted at that commitment.
- *Commitment rules view*: Presents a view of the agent's commitment rules. Selecting a rule allows the developer to see which of the currently active commitments were generated by that rule.
- *Action schedule view*: Presents a view of the current state of the action schedule and a list of the actions that were completed in the last iteration of the interpreter cycle. This second list includes information about the time taken and the success or failure of the action.

## FUTURE TRENDS

Multi-Agent Systems has become a large research area that encompasses a wide range of approaches, visions, and technologies. A recent Agent Technology Roadmap (Luck & McBurney & Shehory & Willmott, 2005), developed for AgentLink, the European Network of Excellence for Agent Research, outlined a number of issues and targets for the future of agent programming languages. These included a general lack of suitable tools, particularly for debugging and visualization; a lack of clarity in their semantics; missing concepts, such as support for organizational structures, social norms or models of cooperation; and a need for them to be evaluated through their use in the creation of large-scale agent systems.

As the roadmap document acknowledges, some of these issues have, at least partially, been solved already. However, there remain a number of outstanding issues. For example, little work has still been done on visualization of agent systems. Perhaps the most notable visualization tool currently available is Botia's (2005) *ACL Analyser*, which employs the ROCK clustering algorithm (Guha & Rastogi & Shim, 2000) to visualize agent interactions, and which can be easily configured to support multiple agent toolkits. However, this tool is an exception rather than the rule, and most other tools tend to work with only one specific platform or toolkit. Further, many of these tools only work well for small collections of agents, and do not scale to larger systems. The lack of such support is significant as it makes the development and debugging of large-scale agent systems more complex, if not impossible.

In addition to visualization, there is also a need to develop more comprehensive debugging tools. Traditionally, AOP researchers have focused on the development of inspection tools, such as the one presented earlier. While these tools are easy to use when debugging simple agent systems, they do not scale well. Again, multi-layered support is required that combines offline log analysis tools such as Lam and Barbers (2004) tracing tool; online monitoring and analysis tools; and the existing debugging tools. Additionally, there is a need to explore how other types of tool, such as profilers may be adapted to the agent domain.

From a program perspective, there is currently little consistency about the most appropriate set of features that underpin an agent system. This stretches from disparities in the choice of mental state to the way in which agents sense and effect their environment. These are further complicated by issues like the introduction of organizational concepts, such as roles and social norms; the need for features such as conversation management; and the incorporation of new work, such as argumentation. Some recent efforts in this area include the proposed extension of 3APL to include roles (Dastani & Dignum & Dignum, 2003) and the implementation of a role concept within AFAPL.

Finally, given the autonomous adaptive nature of agent technology, it is vital that AOP languages begin to provide guarantees regarding the operation of the implemented agents. This ranges from guarantees of behaviour in uncertain environments to the integration of autonomic (Kephart & Chess, 2003) capabilities so that agents may continue to operate in the event of partial failure. Delivering such guarantees is vital if AOP languages are to be a viable option for new problem domains, such as distributed sensor networks, or control architectures for constellations of satellites / exploratory space vehicles.

## CONCLUSION

In this chapter, we have presented an overview of a relatively new programming paradigm, known as Agent-Oriented Programming (AOP) (Shoham, 1993). AOP can be viewed as a specialization of Object-Oriented Programming in which objects are constrained to be agents, the state may only consist of mental attitudes, and message types are restricted to those provided by an underlying Agent Communication Language (ACL). In particular, the chapter has focused primarily on one AOP language, known as AFAPL, which supports the fabrication of agents that can be deployed through the Agent Factory framework.

AFAPL stands apart from many of the other current AOP languages by the fact that it has been extensively applied, over a prolonged period of time, to a wide range of non-trivial problem domains, such as mobile computing, robotics, virtual and mixed reality, and sensor networks. In fact, many of the enhancements made to both Agent Factory, and AFAPL have been as a direct result of experiences gained from work done in these areas. The primary exception to this is the JACK Agent Language (JAL) (Howden & Rönnquist & Hodgson & Lucas, 2001), which is part of JACK, a commercial agent development tool. Specifically, JACK has been used in the development of a number of commercial projects in a broad range of areas including defense, aerospace, telecommunications, manufacturing, and finance.

A common criticism of AOP languages has traditionally been the lack of tool support, and the custom nature of the tools that are provided. It is often argued that this leads to an increased learning curve for new developers; not only do they have to learn a new technology, but also they must learn how to use a new set of tools, and / or adapt existing tools to fill gaps in that tool set.

One approach to reducing this learning curve is to integrate such tool support into existing development tools. Unfortunately, this has not happened for many of the AOP languages. The two main exceptions to this are Jason (Bordini et al., 2005), the recent implementation of AgentSpeak(L) (Rao, 1996) and Agent Factory. As was described in the introduction to AOP, support for developing systems using Agent Factory has been provided in the form of a plugin for the Netbeans IDE. Similarly, Jason provides support for developing system via integration with the Eclipse IDE. JACK does provide an extensive toolset, however, this toolset is entirely custom built, and does not offer the range of support that is offered by more established IDEs such as Netbeans and Eclipse.

Another similar criticism of AOP languages is the lack of support for reuse. The ability to reuse code is often seen as vital to improving software development times. It allows developers to store generalized solutions to common problems that can be quickly adopted and adapted for different projects. An illustrative example of this is the Health Monitoring Service that is hinted at throughout the section on AFAPL. Such a service may be a common component of many implemented systems. Within AFAPL, a change in the design of this service requires updates in only two places: the HealthMonitor program, and the Monitored partial program. If no reuse mechanism had been provided then this change would need to be made to all agents that use the service. Currently, JACK and AFAPL are the only AOP languages that support code reuse.

Overall, the current state of AOP language research is promising. Many of the languages are continuing to develop and evolve as the state-of-the-art in Multi-Agent Systems research advances. Much ongoing work is being done on improving their robustness and expanding their feature sets to improve the expressiveness and simplify the task of creating more complex agent systems. This includes incorporating new concepts, such as roles and social norms, and expanding and improving the tool suites that are provided to support developers wishing to use an AOP language. That said, many of the languages now have a reasonably mature core set of constructs that are well defined and have been implemented robustly.

One of the main challenges facing AOP language developers over the next decade is to start to showcase the advantages that these languages offer over adopting a more traditional approach to agent development. This can only be achieved by using these languages to solve non-trivial problems and to highlight how adopting an AOP approach simplifies / improves the solution.

## REFERENCES

Bauer, B., Muller, J., Odell, J., (2001). Agent UML: A formalism for specifying multiagent interaction. In P. Ciancarini & M. Wooldridge (Eds.), *Agent-oriented software engineering* (pp. 91-103). Springer-Verlag.

Bjorner D. & Jones, C., (1978). The Vienna development method: The meta-language (LNCS 61).

Bordini, R. H., Hübner, J. F., & Vieira, R. (2005). Jason and the Golden Fleece of agent-oriented programming. In R. H. Bordini, M. Dastani, J. Dix, & A. El Fallah Seghrouchni (Eds), *Multi-agent programming: Languages, platforms and applications* (pp. 3-37). Springer-Verlag.

Botia, J. (2005). *Debugging huge multi-agent systems: group and social perspectives.* Paper presented at Programming Multi-Agent Systems (PROMAS), Agentlink Technical Forum 3, Budapest.

Clynch, N., & Collier, R. (2007). SADAAM: Software agent development – An agile methodology. In *Proceedings of the Workshop of Languages, Methodologies, and Development tools for multi-agent systems (LADS'007), Durham, U.K.*

Cohen, P., & Levesque, H. (1990). Intention is choice with commitment. *Artificial Intelligence 42,* 213-261.

Collier, R.W. (1996). *The realisation of agent factory: An environment for the rapid prototyping of intelligent agents.* Master of Philosophy Thesis, University of Manchester Institute of Science and Technology (UMIST), England.

Collier, R.W. (2001). *Agent factory: A framework for the engineering of agent-oriented applications.* Doctoral Thesis, University College Dublin, Ireland.

Collier, R.W., O'Hare G.M.P., Lowen, T., Rooney, C.F.B. (2003). Beyond prototyping in the valley of the agents. In *Multi-Agent Systems and Applications III: Proceedings of the 3rd Central and Eastern European Conference on Multi-Agent Systems (CEEMAS'03), Prague, Czech Republic* (LNCS 2691*).*

Collier, R., Rooney, C., & O'Hare, G.M.P. (2004). A UML-based software engineering methodology for agent factory. In *Proceedings of the 16th International Conference on Software Engineering and Knowledge Engineering (SEKE-2004), Banff, Alberta, Canada.*

Collier, R., Ross, R., & O'Hare, G.M.P. (2005). A role-based approach to reuse in agent-oriented programming. In *AAAI Fall Symposium on Roles, an interdisciplinary perspective (Roles 2005), November 3-6, Hyatt Crystal City, Arlington, Virginia, USA.*

Dastani, M., Dignum, V., & Dignum, F. (2003). Role-assignment in open agent societies. In *Proceedings of the Second International Conference on Autonomous Agents and Multiagent Systems (AAMAS'03), Melbourne.* ACM Press.

Dragone, M., Holz, T., & O'Hare, G.M.P. (2007). Using mixed reality agents as social interfaces for robots. In *Proceedings of the 16th IEEE International Symposium on Robot and Human Interactive Communication (RO-MAN 2007), Jeju Island, Korea.*

Duffy B.R., O'Hare G.M.P., Campbell A.G., Stafford J.W., & O'Grady M.J. (2005). NEXUS: fusing the real & the virtual. In *Proceedings of EUROCON 2005 - The International Conference on "Computer as a Tool", Belgrade, Serbia & Montenegro.*

Fikes, R.E., & Nilsson, N.J., (1971). STRIPS: A new approach to the application of theorem proving to problem solving. *Artificial Intelligence, 2,* 189-208.

Finin T., Labrou, Y., & Mayfield, J. (1997). KQML as an agent communication language. In J. Bradshaw (Ed.), *Software agents.* Cambridge, MA: MIT Press.

FIPA (2000). *The foundation for intelligent physical agents standards.* Retrieved May 1, 2008, from http://www.fipa.org

Guha, S., Rastogi, R., & Shim, K. (2000). ROCK: A robust clustering algorithm for categorical attributes. *Information Systems 25*(5), 345-366.

Hindriks, K.V., de Boer, F.S., van der Hoek, W., & Meyer, J.-J.Ch. (1999). Agent programming in 3APL. *Autonomous Agents and Multi-Agent Systems, 2*(4),357-401.

Howden, N., Rönnquist, R., Hodgson, A., & Lucas, A. (2001). JACK: Summary of an agent infrastructure. In *Proceedings of the 5th International Conference on Autonomous Agents, Montreal, Canada.*

Ingrand, F., Georgeff, M., & Rao, A. (1992). An architecture for real-time reasoning and system control. *IEEE Expert, 7*(6), 34-44.

Kephart, J. O., & Chess, D. M. (2003). The vision of autonomic computing. *Computer, 36*(1), 41-50.

Lam, D., & Barber, K. (2004). Debugging agent behaviour in an implemented agent system. In *Proceedings of 2nd International Workshop on*

*Programming Multiagent Systems Languages and tools (PROMAS 2004), New York, USA,*

Lillis, D., Collier, R., Toolan, F., & Dunnion, J. (2007). Evaluating communication strategies in a multi-agent information retrieval system. In *Proceedings of the 5th European Workshop on Multi-Agent Systems (EUMAS'07), Hammamet, Tunisia.*

Luck, M., McBurney, P., Shehory, O., & Willmott, S. (2005). Agent technology: Computing as interaction (A roadmap for agent based computing). *AgentLink.*

Marsh, D., Tynan, R., O'Kane, D., & O'Hare, G.M.P. (2004). Autonomic wireless sensor networks. *Engineering Applications of Artificial Intelligence Journal, 17*(7), 741-748.

McCarthy, J., (1979). Ascribing mental qualities to machines. In M. Ringle (Ed.), *Philosophical perspectives in artificial intelligence.* Harvester Press.

McCarthy, J. (1992). *Elephant 2000: A programming language based on speech acts.* Unpublished Manuscript. Retrieved May 1, 2008 from *http://www-formal.stanford.edu/jmc/*

McCarthy, J. (2007). *Elephant 2000: A programming language based on speech acts.* Invited talk at OOPSLA 2007, Montreal, Canada.

Muldoon, C., O'Hare, G.M.P., Phelan, D., Strahan, R., & Collier, R. (2003). ACCESS: An agent system for ubiquitous service delivery. In *Cooperative information agents VII* (LNCS 2782).

O'Hare G.M.P. (1996). Agent factory: An environment for the fabrication of distributed artificial systems. In *Foundations of Distributed Artificial Intelligence.* Sixth Generation Computer Series. Wiley Interscience Pub: New York.

O'Hare, G.M.P., Collier, R., Conlon, J., & Abbas, S. (1998). Agent factory: An environment for constructing and visualising agent communities. *9th AICS Conference, Irish Artificial Intelligence and Cognitive Science Conference, UCD, Dublin.*

O'Hare G.M.P., Duffy B.R., Collier R.W, Rooney C.F.B., & O'Donoghue R.P.S. (1999). Agent factory: Towards social robots. *First International Workshop of Central and Eastern Europe on Multi - agent Systems (CEEMAS'99), St. Petersburg, Russia.*

O'Hare, G.M.P., & O'Grady, M.J. (2003). Gulliver's genie: A multi-agent system for ubiquitous and intelligent content delivery. *Computer Communications, 26*(11), 1177-1187.

O'Hare G.M.P., O'Grady M.J., Marsh D., Ruzzelli A. G. & Tynan R. (2006). Autonomic wireless sensor networks: Intelligent ubiquitous sensing. In *Proceedings of ANIPLA 2006, International Congress on Methodologies for Emerging Technologies in Automation, University La Sapienza, Rome, Italy.*

Peng, L., Collier, R., Mur, A., Lillis, D., Toolan, F., & Dunnion, J., (2005). A self-configuring agent-based document indexing system. In *Proceedings of the 4th International Central and Eastern European Conference on Multi-Agent Systems, Budapest, Hungary*

Powell, M.S. (1989). A program development environment based on persistence and abstract data typesIn I. Rosenberg & D. Koch (Eds.), *Persistent object systems.* Springer Verlag.

Rao, A.S., & Georgeff, M.P. (1991). Modeling rational agents within a BDI architecture. In J. Allen, R. Fikes, & E. Sandwall (Eds.), *Proceedings of Second International Conference on Principles of Knowledge Representation and Reasoning* (pp. 473-484). San Mateo, CA: Morgan-Kaufmann.

Rao, A.S., & Georgeff, M.P. (1995). BDI agents: From theory to practice. In *Proceedings of the First International Conference on Multi-Agent Systems (ICMAS'95)* (pp. 312-319). San Francisco, CA.

Rao, A. (1996). AgentSpeak(L): BDI agents speak out in a logical computable language. In *Proceedings of the 7th European workshop on Modelling Autonomous Agents in a Multi-Agent World, Einhoven, Netherlands* (pp. 42-55).

Robinson, J. A. (1965). A machine-oriented logic based on the resolution principle. *Journal of the ACM, 12,* 23-41.

Rooney, C.F.B., Collier, R.W., & O'Hare, G.M.P. (2004). VIPER: VIsual Protocol EditoR. In *Proceedings of the 6th International Conference on Coordination Models and Language (COORDINATION 2004), Pisa, Italy, 24-27 February.*

Ross, R., Collier, R., & O'Hare, G.M.P. (2004). AF-APL – Bridging principles & practice in agent-oriented languages. In *Proceedings of the 2nd International Workshop on Programming Multi-Agent Systems Languages and Tools (PROMAS- 2004), New York.*

Searle, J.R. (1969). *Speech acts.* Cambridge, UK: Cambridge University Press.

Shoham, Y. (1993). Agent-oriented programming. *Artificial Intelligence, 60*(1), 51-92.

Thomas, S. R. (1995), The PLACA agent programming language. In M.J. Wooldridge & N.R. Jennings (Eds.), Intelligent agents. (LNCS 890, pp. 355-370).

Wooldridge, M. (2000). Reasoning about rational agents. Cambridge, MA: MIT Press.

Wooldridge, M., & Jennings, N.R. (1995). Intelligent agents: Theory and practice. *Knowledge Engineering Review, 10*(2).

## KEY TERMS AND DEFINITIONS

**Actuator:** An effecting unit that agents can use to manipulate their environment.

**AFAPL2:** An Agent-Oriented Programming language that is part of the Agent Factory framework.

**Agent:** An autonomous software and/or hardware entity that is situated in some environment, and which is capable of both reacting to changes in that environment and taking the initiative.

**Agent Communication Language:** A language that formalizes how agents may interact with one another and which is typically based on Speech Act Theory.

**Agent Platform:** The underlying apparatus that is provided to support the deployment and management of agents.

**Agent-Oriented Programming:** A programming paradigm in which agents are viewed as fundamental computational units, whose state consists of a set of mental attitudes, and which interact by passing messages using some Agent Communication Language.

**Belief:** A fact about the state of the agents' environment (this may include the agent itself).

**Belief-Desire-Intention (BDI):** A well-established mental state model in which beliefs represent the current state of environment, desires represent the ideal state of the environment, and intentions represent the subset of the agents' desires that it is committed to bringing about.

**Commitment:** An agreement that is made by an agent to do something at a given point in time, for a given agent (possibly itself), and which may in some cases only be maintained only so long as an associated set of conditions remains valid.

**Commitment Rule:** A core programming construct that defines situations (conjunctions of positive and negative beliefs) in which the agent should adopt a commitment.

**Intentional Agent:** An agent whose reasoning process is modeled as the interplay between a set of mental attitudes - such as beliefs, desires, intentions, obligations, or commitments - that together comprise the agents' mental state.

**Message Transport Service:** A type of platform service that provides support for inter-agent communication via some form of underlying network protocol (e.g. HTTP, IIOP, XMPP).

**Module:** A private resource that can only be utilized by the agent that it is associated with.

**Perceptor:** A sensing unit that agents can use to generate beliefs about the current state of their environment.

**Platform Service:** A shared resource that can be used by all agents residing on the given platform (assuming access to that resource has not been restricted).

# Chapter XVIII
# Modeling of Web Services using Reaction Rules

**Marko Ribarić**
*Mihailo Pupin Institute, Serbia*

**Shahin Sheidaei**
*Simon Fraser University, Canada*

**Milan Milanović**
*University of Belgrade, Serbia*

**Dragan Gašević**
*Athabasca University, Canada*

**Adrian Giurca**
*Brandenburgische Technische Universität Cottbus, Deutschland*

**Sergey Lukichev**
*Brandenburg University of Technology at Cottbus, Germany*

**Gerd Wagner**
*Brandenburg University of Technology, Germany*

## ABSTRACT

*The development process of Web services needs to focus on the modeling of business processes rather than on low-level implementation details of Web services, and yet it also needs to incorporate the support for frequent business changes. This chapter presents the UML-based Rule Language (URML) and REWERSE Rule Markup Language (R2ML), which use reaction rules (also known as Event-Condition-Action rules) for modeling Web services in terms of message exchange patterns. Web services that are being modeled in this way can easily be integrated in the wider context of modeling orchestration and choreography. In order to achieve proposed solution, we have developed a plug-in for the Fujaba UML tool (so called Strelka) and a number of model transformations for round-trip engineering between Web services and reaction rules. Also, the paper presents mappings of models of Web services with reaction rules into the Drools rule language, thus enabling the run time execution semantics for our rule-based models.*

## 1. INTRODUCTION

Web services provide a communication interface and a workflow management protocol for business systems. Their goal is set to change the Web from a static collection of information, to a dynamic place where different software components (business logics) can be easily integrated (Arroyo et al., 2004). Some of the factors that constrain web services from achieving that goal can be named as follow: when developers try to enrich the existing Web services with some new functionality, they have to implement necessary changes manually, which may lead to potential execution errors; furthermore, business rules, due to their declarative nature, are used for reflecting dynamic changes representing desired alteration in the business process. When business rules are used in a networked business, they are often large and complex, which makes them difficult to manage and change. There is a lack of automatic mechanism for updating Web services based on the business rules changes (Ribaric et al., 2008).

The need for a more straightforward way to express, and manage business rules as a separate part of Web services' composition has been recognized by researchers for the numerous benefits that it provides (Charfi & Mezini, 2004). However, integrating rule-based systems in a service-oriented environment is a complex task, due to the fact that both worlds have their own paradigms (Rosenberg & Dustdar, 2005). Many different factors involve in this integration. However, the most important one of these factors is the architecture of service oriented systems. In this architecture, layers play a great role. The layering architecture is needed to be represented and deployed in the world of rules which become challenging and could not be simple done. This integration approach should be loosely coupled, meaning that it is reasonable to expose business rules as services. The way of modeling Web services presented in this paper facilitates this integration process, and those services can be more easily integrated in the wider context of modeling orchestration and choreography.

A promising way to solve these problems is to use a high-level modeling approach combined with the use of rules. A modeling approach will allow developers to focus on a problem domain rather than on an implementation technology. This is why we propose using an approach based on Model Driving Engineering (MDE). Although, there have been several attempts to leverage MDE principles to model Web services, they are still very low level oriented, as they again focus on technical details covered either by WSDL (Bezivin et al., 2004) (Vara et al., 2005) or OWL-S (Timm & Gannod, 2005) (Gronmo et al., 2005). The use of rules implying that they are defined in a declarative way, they can dynamically reflect business logic changes at run-time without the need to redesign the whole system. Since Web services are used for integration of business processes of various stakeholders, it is important for them to reflect changes in business logic, or policies, as good as possible.

The solution that we propose is to use rules to represent and model business processes from the perspective of message exchange. That is, our modeling approach enables one to model Web services from the perspective of the underlying business logic regulating *how* Web services are used regardless of the context *where* they are used. To do so, our proposal is to leverage message-exchange patterns (MEPs) as an underlying perspective integrated into a Web service modeling language. This perspective has already been recognized by Web service standards (Chinnici et al., 2007), and our contribution is to raise this to the level of models.

As the leading Rule Modeling language we benefit from UML-based Rule Modeling Language (URML) (Lukichev & Wagner, 2006a) for our approach. URML having strong ties with R2ML (REWERSE Rule Markup Language) empowers us to model rules in various languages. URML supports modeling of domain vocabularies in addition to integrity, derivation, production and reaction rules. .Business rules, being similar to natural language rules, are proposed to be modeled by using reaction rules (also known as Event-Condition-Action, ECA, rules) in this paper. Reaction rules will perform a task (Action) under some circumstances (Conditions) if some events (Event) take place and can best describe a business rule.

We also have developed a plugin (called Strelka) for the well known UML tool Fujaba, which has a support for URML notation. Strelka also has the support transformation of URML models into Web Service Description Language (WSDL) and different types of rule languages such as Drools, Jess, and SWRL. In this paper, we discuss those transformations that are implemented by using a model transformation language entitled the ATLAS Transformation Language (ATL) (ATL, ver. 0.7).

The paper is structured as follows: in the next section we give a bit of the background, including, Web services, R2ML and URML languages with the emphasis on reaction rules. In section 3 we introduce our approach for modeling business rules and services following by section 4 in which we introduce Strelka (Lukichev & Wagner, 2006b), a tool we developed that is capable of presenting URML diagrams, and performing transformations. Section 5 describes mapping process between URML and WSDL that is a part of Strelka, and also covers a deployment of URML diagrams to the Drools rule engine. Section 6 summarized the related work done so far and in Section 7 we conclude this work by summarizing the major points.

## 2. BACKGROUND

In this section, we give a brief overview of the technologies and techniques relevant to the problem under study. This includes a short description of Web services, Drools rule engine, and rule language definitions based on MDE principles used in our modeling approach. A complete introduction into the ATLAS Transformation Language (ATL) can be found in the chapter entitled "Sharing ontologies and rules using model transformations".

### Web Services

A Web service is a loosely coupled component that exposes functionality to a client over the Internet (or an intranet) by using web standards such as HTTP, XML, SOAP, WSDL, and UDDI (Timm & Gannod, 2005).

SOAP is an XML-based protocol for exchanging information in a decentralized, distributed environment. SOAP builds on XML and common Web protocols (HTTP, FTP, and SMTP) (Coyle, 2002). A SOAP message is the basic unit of communication between SOAP nodes. The "envelope" element represents the root of a SOAP message structure. It contains a mandatory body construct and optional header construct (Gudgin et al., 2007). The header construct is where meta-information can be hosted. In a large number of service-oriented architectures, the header is an important part of the overall architecture, and although it is optional it is rarely omitted.

Web Service Description Language (WSDL) is a language for describing both the abstract functionality of a service and the concrete details of a Web service (Chinnici et al., 2007). At an abstract level, WSDL describes a Web service in terms of interfaces and the operations supported by the interfaces. An operation is an interaction with the service consisting of a set of (input, output, infault and outfault) messages exchanged between the service and other parties involved in the interaction (Chinnici et al., 2007). The messages are described independently of a specific wire format by using a type system, typically XML Schema. WSDL also describes the point of contact for a service provider, known as the endpoint – it provides a formal definition of the endpoint interface and also establishes the physical location (address) of the service.

Potential requestors need a way to discover Web services descriptors. It is necessary that these descriptors are collected and stored in a central registry. The key part of the Universal Description Discovery and Integration (UDDI) specification (Clement et al., 2004) presents standardizing information inside such a registry as well as specifying the way the information can be searched and updated.

Regardless of how complex tasks performed by a Web service are, almost all of them require the exchange of multiple messages (Erl, 2005). It is important to coordinate these messages in a particular sequence, so that the individual actions performed by the message are executed properly. Message exchange patterns (MEPs) are a set of templates that provide a group of already mapped out sequences for the exchange of messages (Erl, 2005). This basically means that MEPs define how services should be used, as they can coordinate input and output messages related to a certain operation.

The WSDL 2.0 specification defines three MEPs:

- in-only pattern – supports a standard fire-and-forget pattern (i.e., only one message is exchanged);
- robust in-only pattern – presents a variation of the in-only pattern that provides an option of sending a fault massage, as a result of possible errors generated while transmitting, or processing data;
- in-out pattern – presents a request-response pattern where two messages (input and output) must be exchanged.

However, WSDL 2.0 specification offers the possible use of five more patterns (Lewis, 2007):

- out-in pattern – also consists of exactly two messages (like in-out pattern), but now a service provider is the one that initiates the exchange of messages;
- out-only – consists of exactly one message (like in-only pattern), and is most often used for message notification;
- robust out-only - presents a variation of the out-only pattern that provides an option of sending a fault massage;
- in-optional-out – is similar to pattern in-out with one exception: sending an message that represents a response is optional, and because of this a requester that has started a communication should not expect this message;
- out-optional-in – is opposite to the previous pattern, where the input message is optional.

## Drools

Drools is a business rule management system (BRMS) with a forward chaining inference based rule engine, more correctly known as a production rule system, using an enhanced implementation of the Rete algorithm. BRMS is a production rule system with a focus on knowledge representation to express propositional and first order logic in a concise, non ambiguous and declarative manner. The core of a production rules system is an inference engine that is able to scale to a large number of rules and facts. The inference engine matches facts and data, against production rules, also called productions, to infer conclusions which result in actions. A production rule is a two-part structure using first order logic for knowledge representation (Proctor et al). Figure 1 presents a schema of a production rule. It executes the actions while the conditions are being satisfied. Drools representation contains the same structure, LHS (Left Hand Side) is referring to conditions while RHS (Right Hand Side) is referring to actions. Name of the rule which is presented by "name" is just for better understanding of rules and have no functional value. Attributes, marked with attributes in the rule are used for the rule internal functionality itself.

*Figure 1. First order logic representation (left) - Drools representation (right)*

```
when
    <conditions>
then
    <actions>
```

```
rule "name"
        attributes
        when
                LHS
        then
                RHS
end
```

The process of matching new or existing facts against production rules is called *pattern matching*, which is performed by the inference engine (part of Drools). Drools implements and extends the Rete algorithm The Drools Rete implementation is called ReteOO, signifying that Drools has an enhanced and optimized implementation of the Rete algorithm for object oriented systems (Schneier). For better understanding of rules and their structure please refer to the chapter "Object Oriented Rule Languages".

## R2ML

Existing Web rule languages (RuleML (Hirtle et al., 2006) and SWRL (Horrocks et al., 2004)) and rule modeling approaches (SBVR (SBVR, 2006) and PRR (Charfi & Mezini, 2004)) very nicely serve the purpose of rule interchange and rule representation independent of any specific platform, respectively. However, none of these languages offers a suitable modeling foundation that can be used for modeling Web services from a more abstract perspective of business processes and business rules in which they are used. That is, the language should be able to model reaction rules. Yet, such a modeling language should be closely related to the already existing software modeling languages such as UML and preferably defined by following MDE principles. Finally, the language should be closely related to the Web standards for defining ontologies (OWL) and rules (RIF).

We have decided to use REWERSE I1 Rule Markup Language (R2ML), as it fully satisfies the above requirements. The language is defined by a MOF-based metamodel which is refined by OCL constraints that precisely define relations between the language's constructs in addition to those defined by the metamodel; it has an XML schema defined concrete syntax; it has a UML-based graphical concrete syntax, so called UML-based Rule Modeling Language (URML), which we will introduce it in the section below; and it has a number of transformations with other rule languages (e.g., JBoss' Drools and OCL) allowing us to translate Web service models to the rule-based languages that can then regulate the use of Web services.

R2ML is a rule language that addresses all the requests defined by the W3C working group for the standard rule interchange format (Ginsberg et al., 2006). The R2ML language can represent different types of rule constructs, that is, it can represent different types of rules (Wagner et al., 2006) including: integrity rules, derivation rules, production rule, and reaction rules. Integrity rules in R2ML, also known as (integrity) constraints, consist of a constraint assertion, which is a sentence in a logical language such as first-order predicate logic or OCL. Derivation rules in R2ML are used to derive new knowledge (conclusion) if a condition holds. Production rules in R2ML produce actions if the conditions hold, while post-conditions must also hold after the execution of actions. A reaction rule is a statement of programming logic (Giurca et al., 2006) that specifies the execution of one or more actions in the case of a triggering event occurrence and if rule conditions are satisfied. Optionally, after the execution of the action(s), post-conditions may be made true. R2ML also allows one to define vocabularies by using the following constructs: basic content vocabulary, functional content vocabulary, and relational content vocabulary. Here we give short description of vocabulary constructs that we use in this paper. Vocabulary is a concept (class) that can have one or more VocabularyEntry concepts. VocabularyEntry is abstract concept (class) that is used for representing other concepts by its specialization. For example, one of the VocabularyEntry-s is an R2ML Class concept which represents the class element similar to the notion of the UML Class. An R2ML Class can have attributes (class Attribute), reference properties (class ReferenceProperty) and operations (class Operation).

Due to the importance for our Web service modeling approach, here we only describe the details of R2ML reaction rules. Reaction rules represent a flexible way for specifying control flows, as well as for integrating events/actions from a real life (Giurca et al., 2006). Reaction rules are represented in the R2ML metamodel as it is shown in Figure 2: triggeringEvent is an R2ML EventExpression (Fig. 4); conditions are represented as a collection of quantifier free logical formulas; producedAction is an R2ML EventExpression and represents a system state change; and (optional) postcondition must hold when the system state changes.

The R2ML event metamodel defines basic concepts that are needed for dynamic rule behavior (Figure 3). There are two types of events: atomic event (AtomicEventExpression) and composite events (AndNotEventExpression, SequenceEventExpression, ParallelEventExpression or ChoiceEventExpression). Characteristic of an atomic event is that it has no duration (duration = 0). For the sake of modeling Web services, we are using MessageEventExpression for both triggering events and produced actions. MessageEventExpression is used for modeling messages that are part of the definition of Web services, including, input, output, in-fault, and out-fault messages. Each MessageEventExpression has its own type – EventType. In terms of WSDL, message types are defined by XML Schema complex types, while in R2ML, EventType is a subclass of Class (already defined as a part of the R2ML Vocabulary). This means that each EventType

*Figure 2. Definition of reaction rules in the R2ML metamodel*

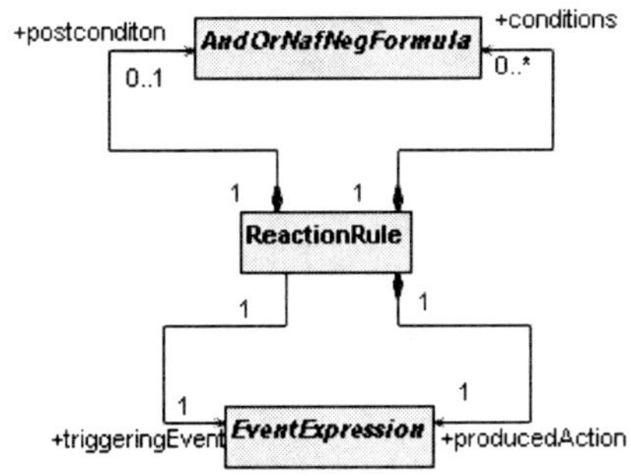


has its own attributes, associations, and all other features of R2ML classes.

## URML

UML-Based Rule Modeling Language (URML) is a graphical concrete syntax of R2ML. URML is developed as an extension of the UML metamodel to be used for rule modeling. In URML, modeling vocabularies is done by using UML class models. Rules are defined on top of such models. The URML reaction rules metamodel, which we use for modeling services, is shown in Figure 4a. The figure shows components of a reaction rule: Condition, Postcondition, RuleAction and EventCondition. The figure also shows that reaction rules are contained inside the UML package which represents Web services operation. This means, that such packages have a stereotype <<operation>> in UML diagrams. An instance of the EventCondition class is represented on the URML diagram as incoming arrow (e.g., see Figure 5), from a UML class that represents either an input message or an input fault message of the Web service operations, to the circle that represents the reaction rule. The UML class that represents the input message (input-Message in Figure 4b of the Web service operation is MessageEventType (a subclass of EventType from Figure 3) and it is represented by using the <<message event type>> stereotype on UML classes. The UML class that represents the input fault message (inFault in

*Figure 3. Event expressions in the R2ML metamodel*

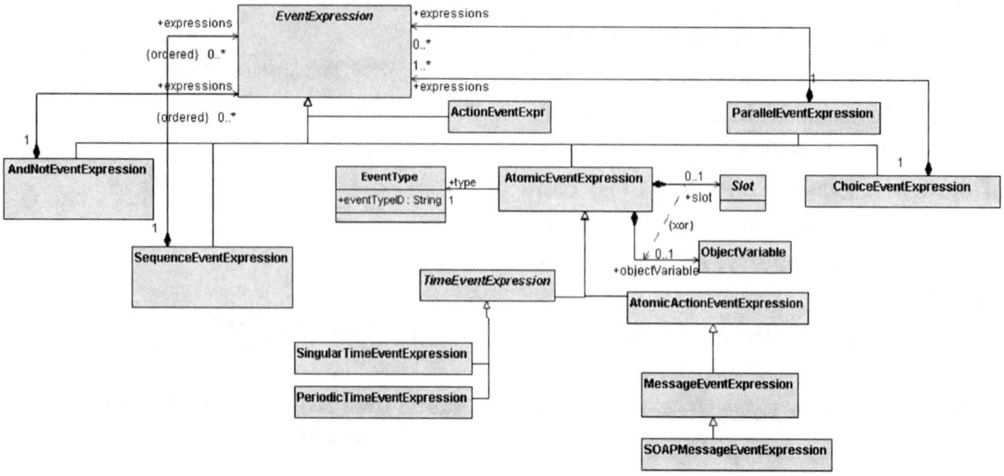


Figure 4b) of the Web service operation is FaultMessageEventType in the URML metamodel. In URML diagrams, FaultMessageEventType is represented by the <<fault message event type>> stereotype on UML classes. EventCondition contains an object variable (ObjectVariable in Figure 4c), which is a placeholder for an instance of the MessageEventType class. The object variable has a name that corresponds to the arrow annotation, which represents EventConditon.

An instance of the RuleAction class is represented as an outgoing arrow on the URML diagram, from the circle that represents the reaction rule to the class that represents either an output message or an output fault message of the Web service operation. The UML class that represents the output message (outputMessage in Figure 4c) of the Web service operation is MessageEventType and it is represented with the <<message event type>> stereotype on UML classes. The UML class that represents the output fault message (outFault) of the Web service operation is FaultMessageEventType in the URML metamodel and it is represented with the <<fault message event type>> stereotype on UML classes. RuleAction contains an object variable (ObjectVariable), which represents an instance of the MessageEventType class.

*Figure 4. a) Extension of the URML metamodel for reaction rules; b) Part of the URML meta-model for EventCondition; c) Extension of the URML metamodel for actions*

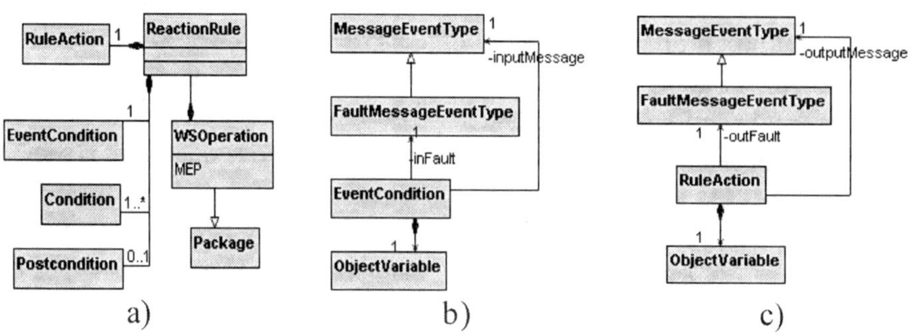

## 3. MODELING

In this section the idea behind modeling of rules and business processes will be exposed. Modeling will be performed with the help of URML and R2ML. In section 5 with the use of modeling, the transformation to the web services and automatic generation of codes will be described.

### Rule Modeling

As stated, in this paper we are concerned with reaction rules. Reaction rule formalizes event-condition-action behavioral model, where the action is executed on event with a condition satisfied. Let us analyze the following example: *On customer book item request, if the item is available, then approve order and decrease amount of items in stock.* In this rule *on customer book item request* is a triggering rule event, *if the item is available* is a rule condition, *approve order* is a rule action and *decrease amount of items in stock* is a postcondition, which can be expressed as a logical constraint, requiring less amount of items in stock, than before the rule execution.

There are several general advantages of using reaction rules for specifying web services:

- Requirements are often captured in the form of rules in a natural language, formulated by business people;
- Reaction rules are easier to maintain and integrate with other kinds of rules, used in business applications (integrity rules, which specify constraints the data must fulfill, derivation rules, which explain how a model element can be derived);
- The topic of rules validation and verification is well studied;
- Reaction rules emphasis on events gives a flexible way to specify control flow and web service interaction patterns.

Our approach is on the edge of the MDE, Web Services, and rules. It brings several large communities together and helps to facilitate the use of rules and MDE technologies in the growing area of Web Services.

### Business Rule Modeling

In this section we describe the concept of our approach talking about modeling of Business rules with the help of web services. In our approach, we look from the perspective of the potential patterns of the use of services. That is, we model services from the perspective of MEPs. It is important to point out that we **first start from the definition of a business rule that corresponds to a MEP under study**, but without considering the Web services that might be developed to support that rule. In this way, unlike other approaches to modeling of Web services, we are focused on the business rules describing how particular services are used, but without explicitly stating that we are talking about Web services. This approach enables us, not only to focus on the problems under study and the underlying business logic regulating the use of Web services, but also now we are able to translate such Web service modeling to both Web service languages and rule-based languages that can regulate how services are used at run-time.

Here, we describe how the in-out MEP is modeled in URML. The *in-out* MEP consists of exactly two messages: when a service receives an input message, it has to reply with an output message. This means that a fault message must be delivered to the same target node as the message it replaces, unless otherwise specified by an extension or binding extension. To show how this variation of the in-out MEP can be modeled by using reaction rules, we start our discussion from an example of the in-out MEP with an out-fault message. Let us consider the following business rule: *When a user sends a message, if the length of the message is less than five characters long return the message to the user, otherwise return an error message.*

The business rule is an in-out MEP and can be represented by two reaction rules, represented in semi-formal pseudo rule syntax:

```
ON UserMessageArrival (echoInParam)
IF echoInParam.length < 5 THEN
DO EchoStringResponse (echoInParam)

ON UserMessageArrival (echoInParam)
IF NOT echoInParam.length < 5 THEN
DO EchoStringError ("message too long")
```

In order to have these rules represented using our modeling notation, and also to be able to relate the rules with the elements of vocabularies, we model these rules by using URML. It is important to stress that all rules-based systems define business rules on top of business vocabularies. Thus, the URML graphical notation enables us to define business rules regulating the use of Web services by leveraging a human-comprehendible, and yet formally consistent representation, with the underlying business vocabularies (i.e., UML class models). These abovementioned two reaction rules represented in URML are shown in Figure 5.

In this particular case, we have the following situation: triggering event of either rule (i.e., UserMessageArrival) maps to the input message of the Web service operation. The action of the first reaction rule (i.e., EchoStringResponse), which is triggered when a condition is true, maps to the output message of the Web service operation. The action of the second reaction rule (i.e., EchoStringError), triggered on a false condition, maps to the out-fault message of the Web service operation.

As shown, a web service operation may be modeled by using more than one rule, where rules, that define an operation, are grouped in one package by the name of the operation. Otherwise, for modeling the basic in-out MEP, only one (first) reaction rule would be used. Besides the classes that model exchanged messages, the URML diagrams also model conditions (e.g., echoInParam. length < 5). Such condition constructs are modeled by using OCL filters. OCL filters are based on a part of OCL that models logical expressions, which can be later translated to R2ML logical formulas, as parts of reaction rules. However, these OCL filters cannot be later translated to Web service descriptions (e.g., WSDL), as those languages cannot support such constructs. But, we can translate our URML models into rule-based languages (e.g., Jess or Drools). This means that for each Web service, we can generate a complementary rule, which fully regulates how the service

*Figure 5. PIM level of the in-out message exchange pattern with out-fault, presented in URML*

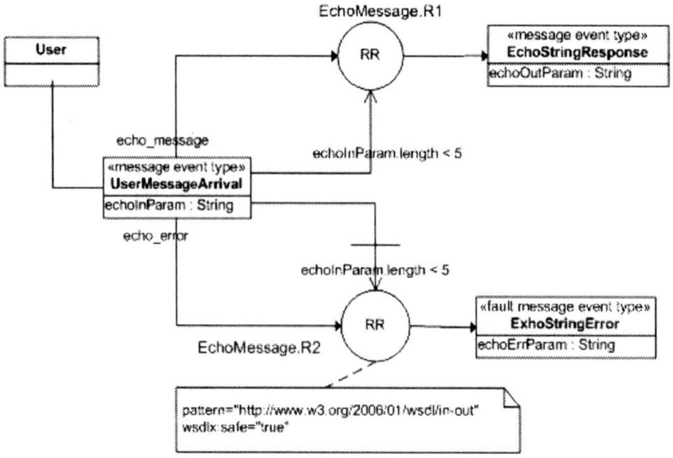

is used. Another benefit of this approach is that our generated Web services, and their regulating rules, are consistent, as they originate (i.e. they are generated from) the same rule-based Web service model.

## 4. STRELKA: OUR MODELING ENVIRONMENT

In this section, we describe Strelka – a tool that was implemented as a plugin for the Fujaba tool[1]. Fujaba is a UML modeling tool, developed at the University of Paderborn in Germany. Fujaba allows creation of different types of UML diagrams. It allows generation of Java code based on created model, and it offers a plugin interface that we used for Strelka implementation. Figure 6 shows, symbolically, the architecture of Strelka, as well as the functionality that it provides.

The first part of Figure 6 (leftmost dashed box) shows us two important features Fujaba has: it is capable of generating Java code from diagrams, and it offers easily extensible architecture suitable for plugin implementation. This box also symbolically shows that one of the steps that need to be undertaken when creating a plugin for Fujaba is to create a metamodel for the new elements Fujaba will be extended with. After this first box, in the figure, we have an OCL parser. The OCL parser, which we used in this version of Strelka, is developed at the Technical University of Dresden in Germany[2]. This parser is widely accepted open source tool – besides Fujaba, it can be also integrated into other tools: ArgoUML, Borland Together, Poseidon. The second dashed box on Figure 6 shows that we use JAXB in the process of mapping URML to R2ML. JAXB guaranties that the R2ML XML document we get as the result of serialization of URML diagram created in Strelka, is valid according to the R2ML XML schema. The reason we use JAXB, is that Fujaba does not have explicitly defined metamodel in some of the metamodeling languages (e.g. MOF, or Ecore). So, we could not use model transformations (and ATL), in the mapping process between URML and R2ML like we did in the process of mapping between R2ML and WSDL (this mapping we describe in the next section). And the last dashed box on Figure 6 shows that Strelka has an integrated support for calling model transformations between R2ML and WSDL.

As already mentioned, one of the steps in the process of creating plug-ins for Fujaba is defining a metamodel for all the new elements that we extend Fujaba with. This metamodel is created in Fujaba, and then Java code is generated from it. This Java code we use afterwards, in the process of generating plugin itself. As the process of defining metamodel, and Java code generation is complete, we can create class diagrams that can incorporate those newly created elements, as well as UML elements that were already available (e.g. classes and associations). Figure 7 presents fragment of the Rule metamodel created in Fujaba. Classes that are relevant to the modeling of reaction rules are:

*Figure 6. Strelka's functionality*

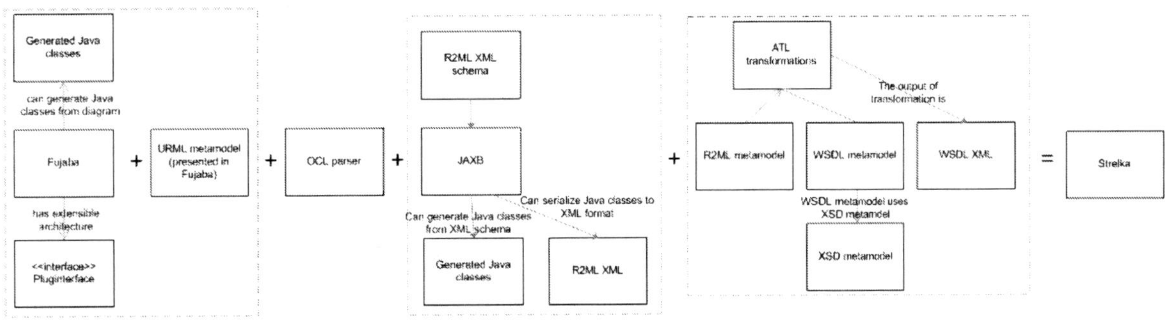

ReactionRule, MessageAction, EventCondition, MessageEventType, and FaultMessageEventType. Figure 7 shows relations between the ReactionRule class to the other classes in this metamodel. The ReactionRule class is a subclass of Rule class (just like IntegrityRule, ProductionRule, and DerivationRule classes, which are used for modeling integrity, production, and derivation rules respectively).

## 5. TRANSFORMING WEB SERVICE MODELS

In this section, we summarize the translation process from URML reaction rules to Web services, and then we describe how this URML model can be transformed to Drools rule language.

The process of getting WSDL documents from URML models is as follows: first, we create an URML diagram in Strelka, after this we serialize it in the R2ML XML concrete syntax (this functionality is built in Strelka, and it relies on the use of JAXB). Once URML reaction rules are serialized to R2ML XML, they are translated into WDSL documents (this translation is done at the level of metamodels as described below). It is important to say that this process allows for two way transformation between URML and WSDL, that is, it is possible to perform reverse engineering of existing services, and extract business rules already integrated to service implementation.

To support this approach, we have implemented a number of transformations between different languages and their representations (these transformations can be found on the ATL ZOO page (WSDL2R2ML, 2006):

- *URML and R2ML XML concrete syntax.* As previously stated, this is the only transformation that is not implemented by using ATL (ATL, ver. 0.7), but instead it is based on JAXB
- *R2ML XML-based concrete syntax and R2ML metamodel.* This transformation is important to bridge concrete (XML) and abstract (MOF) syntax of R2ML. This is done by using ATL and by leveraging ATL's XML injector and extractor.
- *R2ML metamodel and WSDL metamodel.* This transformation is the core of our solution and presents mappings between R2ML and WSDL at the level of their abstract syntax.

*Figure 7. Strelka's rule metamodel*

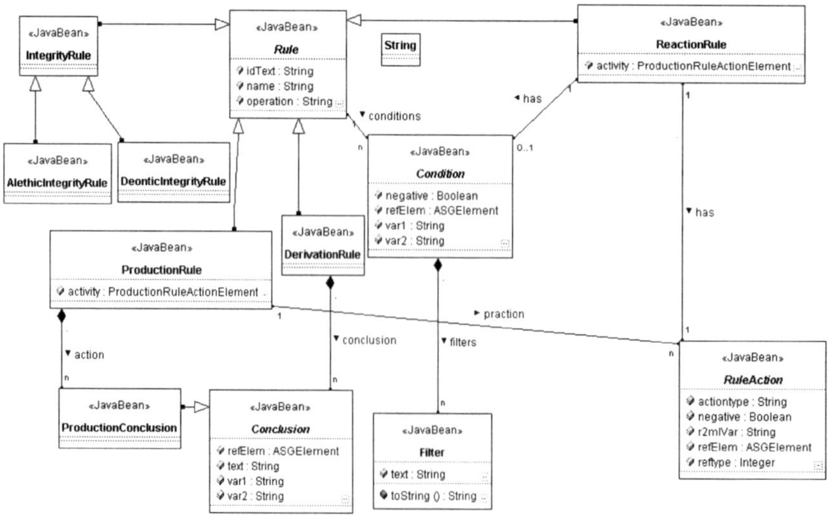

- *WSDL XML-based concrete syntax and WSDL metamodel.* This transformation is important to bridge concrete (XML) and abstract (MOF) syntax of WSDL. This is done by using ATL and by leveraging ATL's XML injector and extractor.

Here, we describe the third and the forth transformation: i.e. transformation between R2ML metamodel and WSDL metamodel, and transformation between WSDL XML schema and WSDL metamodel.

## Transformation Between WSDL Metamodel and R2ML Metamodel

Figure 8 shows the general principle of mappings between WSDL metamodel and R2ML metamodel (as can be seen on the figure, these mappings can be done in both directions).

Process of transforming WSDL metamodel elements to R2ML metamodel elements is conducted in the MOF technical space (TS) (Kurtev et al., 2002). Table 1 shows a conceptual mapping between these metamodels (this table does not show parts related to the mapping between the XML Schema language, which WSDL uses for defining message types and vocabularies, and the R2ML vocabulary).

Figure 9 shows an example of a matched rule from our (WSDL2R2ML.atl) transformation (for the Input class in the WSDL metamodel), and Figure 10 shows a graphical presentation of this mapping (for this particular class) between WSDL and R2ML metamodel. This matched rule initializes "sender", "objectVariable" and "type" features of the target R2ML MessageEventExpression element (we are putting a string "to_be_defined" in the "sender" feature, as the user will provide this information if required. This feature refers to the requester of the service).

As the result of the execution of WSDL2R2ML.atl transformation, we get an R2ML model. This model (which is an instance of the R2ML metamodel) can later be serialized into the R2ML XML concrete syntax, by our R2ML2XML.atl transformation and the use of the XMLExtractor class.

The WSDL2R2ML.atl transformation is defined as a sequence of ATL rules and ATL

*Figure 8. Mappings between models that conform to WSDL metamodel to models that conform to R2ML metamodel*

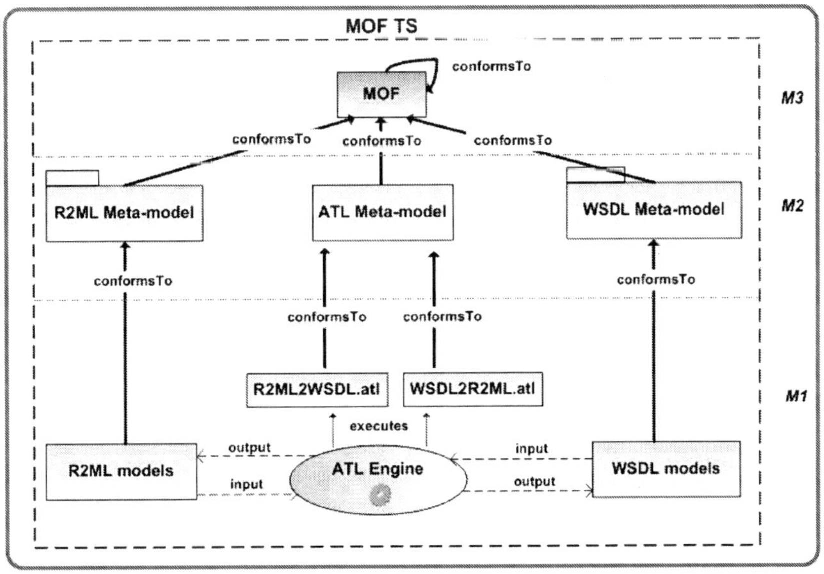

*Table 1. An excerpt of the mapping between the WSDL metamodel and the R2ML metamodel*

| WSDL metamodel | R2ML metamodel |
|---|---|
| Description | RuleBase |
| ElementType | Vocabulary |
| Interface | ReactionRuleSet |
| Operation | ReactionRule |
| Input | MessageEventExpression |
| Infault | MessageEventExpression |
| Output | MessageEventExpression |
| Outfault | MessageEventExpression |

*Figure 9. ATL matched rule that transforms WSDL Input element to R2ML MessageExpression element*

```
rule TriggeringEvent {
  from i : WSDL!Input (
     i.oclIsTypeOf(WSDL!Input)
  )
  to o : R2ML!MessageEventExpression (
     sender <- 'to_be_defined',
     objectVariable <- thisModule.ObjectVariable(i.element),
     type <- i.element
  )
}
```

helpers. It has 11 matched rules, 3 lazy rules and 3 helpers.

Transformation, in the opposite direction, from R2ML metamodel elements to WSDL metamodel elements (R2ML2WSDL.atl) is based on the same conceptual mappings presented in the Table 1. This transformation uses 9 matched rules, 4 lazy rules, 1 unique lazy rule and 5 helpers. Figure 11 shows one matched rule from R2ML2WSDL transformation. This rule is mapping R2ML ReactionRuleSet element into WSDL Service and Interface elements.

## Transformation of WSDL XML Schema to WSDL Metamodel

Because the concrete syntax of WSDL is defined in the XML technical space, and its abstract syntax (metamodel) is defined in the MOF TS, it is necessary to make a translation from concrete syntax elements to the abstract syntax elements. For this translation we have used XML metamodel (XML metamodel, 2006) as a bridge between XML and MOF TS.

To transform an XML document from XML TS to the XML model (instance of a XML metamodel) of the same rule in the MOF TS, we have used the XMLInjector class. By using a SAX parser, this class creates an XML model from an XML file. The next figure (Figure 12) shows general principle of mapping between WSDL XML concrete syntax and WSDL abstract syntax. Figure shows that the transformation can be done in both directions (i.e. we created bidirectional transformations).

Before we describe transformation process between WSDL XML schema and WSDL metamodel, we stress a fact that because WSDL does

*Figure 10. Graphical presentation o mapping shown on the Figure 9*

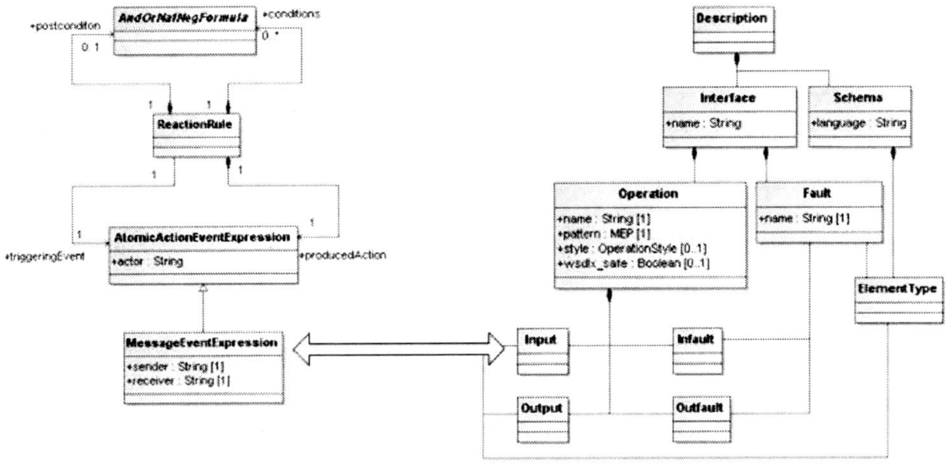


*Figure 11. ATL matched rule that transforms R2ML ReactionRuleSet element to WSDL Interface element*

```
rule Interface {
  from i : R2ML!ReactionRuleSet (
    i.oclIsTypeOf(R2ML!ReactionRuleSet)
  )

  to serv : WSDL!Service (
    name <- i.ruleSetID,
    interface <- inter,

    endpoint <- en
   ),
   inter : WSDL!Interface (
     name <- thisModule.setElemName(i.ruleSetID, 'Interface'),
     fault <- i.rules,
     operation <- let setOfGroupIDs : Set (String) = i.rules->iterate(p; s :
Set(String) = Set{} | s->including(p.groupID) ) in
     setOfGroupIDs->collect(e| thisModule.Operation(i.rules->asSequence()->
                    select(c| c.groupID = e)->first()) )
    ),
   en : WSDL!Endpoint (
     name <- thisModule.setElemName(i.ruleSetID, 'Endpoint'),
     address <- i.rules->select(c| c.oclIsTypeOf(R2ML!ReactionRule) and not

       c.producedAction.type.oclIsTypeOf(R2ML!FaultMessageType))-> first().pro-
                              ducedAction.sender
   )
}
```

not define its own language for defining data types, we are referring to the XSD metamodel that can be obtained from the Eclipse Model Development Tool project[3].

Transformation from WSDL XML schema to the WSDL metamodel has two primary steps:

- **Step 1:** This step is about transforming WSDL document from XML TS into MOF TS. In this step WSDL document (EchoService.xml on the Figure 12) is transformed into the WSDL metamodel that represents abstract syntax of the WSDL language in the MOF TS. By using XMLInjector class, WSDL XML document is transformed into model that conforms to the MOF based XML metamodel (step 1 on the Figure 12). As the result of this process we get XML model that can be presented in the XML XMI format (step 2 on the Figure 12).
- **Step 2:** This step is about transforming XML model (EchoService_XML on the Figure 12) into the WSDL based model (EchoService_WSDL on the Figure 12). ATL transformation XML2WSDL.atl (step 3 on the Figure 12) is performing this mapping. The output WSDL model (EchoService_WSDL) conforms to the WSDL metamodel. The XML2WSDL.atl transformation is executed on the model level (M1 level), by using information about elements from the (XML and WSDL) metamodel level (M2 level).

The next table (Table 2) shows fraction of mappings between WSDL XML schema and WSDL metamodel.

Figure 13 shows a matched rule from XML2WSDL.atl transformation for the "description" element of the XML metamodel. ATL rule Description transforms Element of the XML model (EchoService_XML) to Description (i.e. WSDL!Description) of the WSDL metamodel. This rule initializes "types", "interface", "binding" and "service" attributes of Description target model element by calling an ATL helper shown on the Figure 14.

The XML2WSDL transformation has 18 matched rules, one lazy rule, and 17 helpers (we have this number of helpers because we use helpers for mapping enumeration elements to their string representation – e.g. in-out MEP into the 'http://www.w3.org/2006/01/wsdl/in-out' string). The biggest problem in the XML2WSDL transformation was mapping of XML schema complex types. For solving this problem we found the (EclipseDoc, 2002) document very helpful.

*Figure 12. Transformation scenario: WSDL XML format into WSDL metamodel*

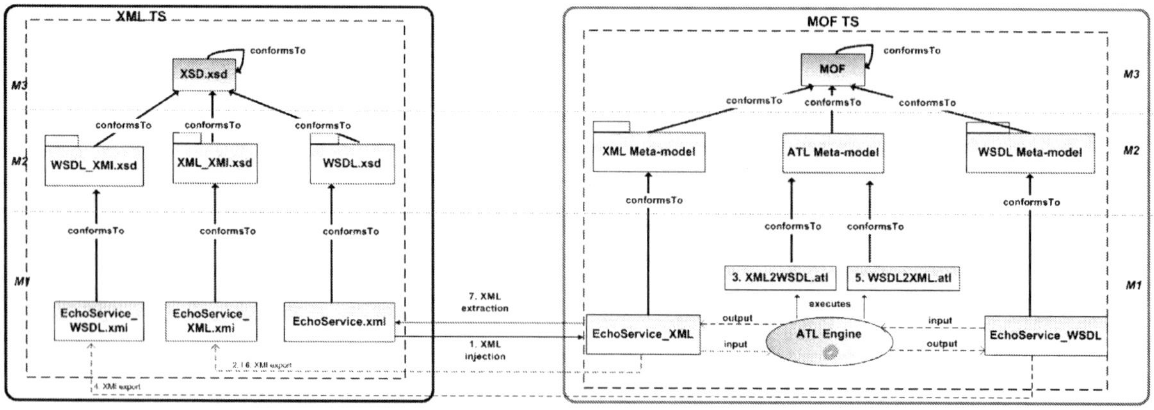

*Table 2. Fraction of transformation between WSDL XML schema and WSDL metamodel*

| WSDL XML schema | XML metamodel | WSDL metamodel | Description |
|---|---|---|---|
| description | Root name='description' | Description | The Root element that contains types, binding, service and interface elements |
| interface | Element name='interface' | Interface | This element presents WSDL interface, it contains operation and fault elements |
| operation | Element name='operation' | Operation | This element presents WSDL operation, it contains in/out-fault and in/output elements |

*Figure 13. Description matched rule*

```
rule Description {
from
  i : XML!Element (
    i.name = 'description'
  )
to
  o : WSDL!Description (
    types <- i.getElementsByName('types'),
    interface <- i.getElementsByName('interface'),
    binding <- i.getElementsByName('binding'),
    service <- i.getElementsByName('service')
  )
}
```

*Figure 14. Helpers we use in the rule in Figure 13*

```
helper context XML!Element def: getAttr(n : String) : Sequence(XML!Attribute) =
  self.getChildren(n)->select(e | e.oclIsKindOf(XML!Attribute))->first();

helper context XML!Element def: getAttrVal(n : String) : String =
  let d : XML!Attribute = self.getAttr(n) in
  if d.oclIsUndefined() then
     OclUndefined
  else
     d.value
  endif;
```

## Transformation of the WSDL Metamodel into the WSDL XML Schema

Transformation process from WSDL abstract syntax (i.e. WSDL metamodel) into the WSDL XML concrete syntax also contains two primary steps.

- **Step 1:** Transformation of WSDL model into XML model. Here a WSDL model (Echo-Service_WSDL on Figure 12) is transformed into XML model (EchoService_XML) by using the ATL transformation WSDL2XML.atl (step 5 on the Figure 12). After this transformation is executed on the input WSDL models, XML models (EchoService_XML) are placed in the model repository (EchoService_XML.xmi on the Figure 12).

- **Step 2:** Transformation from MOF TS into the XML TS, where an XML model (EchoService_XML on the Figure 12) that conforms to the MOF based XML metamodel (and that is created in the step 1), is transformed into the EchoService.xml document (step 7 on the Figure 12).

In the WSDL2XML transformation, we have 17 matched rules, 3 lazy rules, as well as 5 helpers. The WSDL2XML transformation (i.e. transformation from WSDL metamodel into the WSDL XML schema) is simpler then XML2WSDL transformation (WSDL elements are directly mapped to their corresponding representation from the WSDL XML schema, and the other way around is also true).

## Deploying URML Models on a Rule-Based Engine

Here, we describe the transformation process from a URML model to the Drools rule engine. For this purpose we are using an example of the in-out MEP, presented with the following business rule:

*When a user sends a request for Social Insurance Number (SIN), if the authentication is successful send the SIN to the user.*

This business rule is presented with one reaction rule that has a triggering event SINrequest, condition Person.Address=SINrequest.Address AND Person.Name=SINrequest.Name, and action SINresponse:

```
ON SINrequest (Address, Name)
IF Person.Address=SINrequest.Address AND
Person.Name=SINrequest.Name
DO SINresponse(SIN)
```

An URML diagram that models the corresponding Web service operation is presented in Figure 15, and Figure 16 illustrates the R2ML presentation of it that Strelka generated for us.

The first step in getting Drools rules, from the generated R2ML file is transformation of R2ML reaction rules to R2ML production rules. The second step is transformation of R2ML production rules to Drools rules, where we are using the translation provided by (Giurca & Werner, 2006).

In order to successfully realize the first step one must address the major difference between reaction rules and production rules: reaction rules have triggeringEvent element, while production rules do not have this element. So there is a need for representing triggeringEvent in production rules. If only atomic events are considered, it is legible to map the triggeringEvent element to the condition element in production rule. Figure 17 shows the corresponding R2ML production rule document that we get as the result of our transformation.

As Figure 17 shows, the producedAction part of the generated production rule is the same as in the reaction rule, while the conditions part is constituted from conditions of a reaction rule along with the triggeringEvent of reaction rule. Also, a vocabulary part of production rule is updated with the adequate class generated from triggeringEvent of reaction rule. (i.e. if no such class is already defined. This situation happens with the presentation of rules in older versions of R2ML). Table 3 provides a brief overview of the transformation from Reaction Rules to Production Rules.

Once we have generated a R2ML production rule from the R2ML reaction rule, we use the transformation provided by (Giurca & Werner, 2006) to transform R2ML production rule to a

*Figure 15. URML model of the in-out MEP*

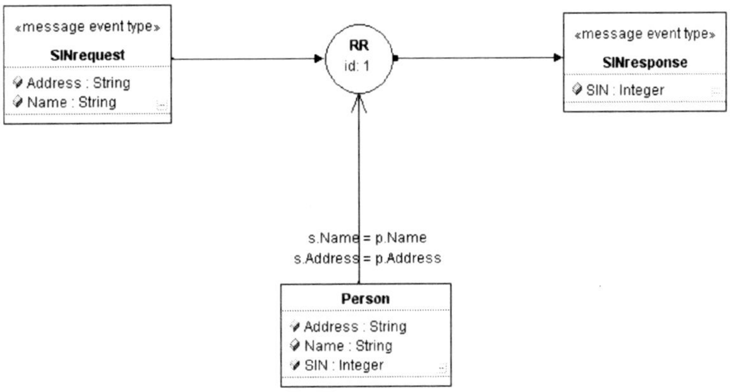


*Figure 16. R2ML serialization of URML model from Figure 15*

```
<r2ml:ReactionRuleSet r2ml:ruleSetID="ClassDiagram1">
    <r2ml:ReactionRule r2ml:ruleID="id469" r2ml:groupID="operation">
        <r2ml:triggeringEvent>
            <r2ml:MessageEventExpression r2ml:sender="not_defined" r2ml:eventType="SINrequest">
                <r2ml:ObjectVariable r2ml:name="s" r2ml:classID="SINrequest"/>
            </r2ml:MessageEventExpression>
        </r2ml:triggeringEvent>
        <r2ml:conditions>
            <r2ml:ObjectClassificationAtom r2ml:classID="Person" r2ml:isNegated="false">
                <r2ml:ObjectVariable r2ml:name="p"/>
            </r2ml:ObjectClassificationAtom>
            <r2ml:DatatypePredicateAtom r2ml:datatypePredicateID="swrlb:equal" r2ml:isNegated="false">
                <r2ml:dataArguments>
                    <r2ml:AttributeFunctionTerm r2ml:attributeID="Name">
                        <r2ml:contextArgument>
<r2ml:ObjectVariable r2ml:name="p" r2ml:classID="Person"/>
                        </r2ml:contextArgument>
                    </r2ml:AttributeFunctionTerm>
                    <r2ml:AttributeFunctionTerm r2ml:attributeID="Name">
                        <r2ml:contextArgument>
<r2ml:ObjectVariable r2ml:name="s" r2ml:classID="SINrequest"/>
                        </r2ml:contextArgument>
                    </r2ml:AttributeFunctionTerm>
                </r2ml:dataArguments>
            </r2ml:DatatypePredicateAtom>
 <r2ml:DatatypePredicateAtom r2ml:datatypePredicateID="swrlb:equal" r2ml:isNegated="false">
                <r2ml:dataArguments>
                    <r2ml:AttributeFunctionTerm r2ml:attributeID="Address">
                        <r2ml:contextArgument>
<r2ml:ObjectVariable r2ml:name="p" r2ml:classID="Person"/>
                        </r2ml:contextArgument>
                    </r2ml:AttributeFunctionTerm>
                    <r2ml:AttributeFunctionTerm r2ml:attributeID="Address">
                        <r2ml:contextArgument>
<r2ml:ObjectVariable r2ml:name="s" r2ml:classID="SINrequest"/>
                        </r2ml:contextArgument>
                    </r2ml:AttributeFunctionTerm>
                </r2ml:dataArguments>
            </r2ml:DatatypePredicateAtom>
        </r2ml:conditions>
        <r2ml:producedAction>
            <r2ml:MessageEventExpression r2ml:eventType="SINresponse"/>
           <r2ml:ObjectVariable r2ml:name="SIN" r2ml:classID="SINresponse"/>
        </r2ml:producedAction>
    </r2ml:ReactionRule>
  </r2ml:ReactionRuleSet>
</r2ml:RuleBase>
```

Drools rule. Figure 18 shows the generated Drools rule from the production rule presented in Figure 17. The left part of this figure shows the classes generated for Drools engine, and the right part of the figure shows the rule itself. Each rule fires an action when conditions are met - in the case below, SINrequest trigger this rule, while SINresponse is fired as the response to the information requested.

*Figure 17. Generated R2ML production rule document – part of rule containing conditions is shown on left – vocabulary part can be seen on the right*

```
<r2ml:ProductionRuleSet r2ml:ruleSetID="ClassDiagram1">
    <r2ml:ProductionRule r2ml:ruleID="id469" r2ml:groupID="operation">
        <r2ml:conditions>
        <r2ml:ObjectClassificationAtom r2ml:classID="SINrequest">
           <r2ml:ObjectVariable r2ml:name="s" r2ml:classID="SINrequest"/>
        </r2ml:ObjectClassificationAtom>
         <r2ml:ObjectClassificationAtom r2ml:classID="Person" r2ml:isNegated="false">
                <r2ml:ObjectVariable r2ml:name="p"/>
            </r2ml:ObjectClassificationAtom>
            <r2ml:DatatypePredicateAtom r2ml:datatypePredicateID="swrlb:equal" r2ml:isNegated="false">
                <r2ml:dataArguments>
                    <r2ml:AttributeFunctionTerm r2ml:attributeID="Name">
                        <r2ml:contextArgument>
                    <r2ml:ObjectVariable r2ml:name="p" r2ml:classID="Person"/>
                        </r2ml:contextArgument>
                    </r2ml:AttributeFunctionTerm>
                    <r2ml:AttributeFunctionTerm r2ml:attributeID="Name">
                        <r2ml:contextArgument>
                    <r2ml:ObjectVariable r2ml:name="s" r2ml:classID="SINrequest"/>
                        </r2ml:contextArgument>
                    </r2ml:AttributeFunctionTerm>
                </r2ml:dataArguments>
            </r2ml:DatatypePredicateAtom>
        <r2ml:DatatypePredicateAtom r2ml:datatypePredicateID="swrlb:equal" r2ml:isNegated="false">
                <r2ml:dataArguments>
                    <r2ml:AttributeFunctionTerm r2ml:attributeID="Address">
                        <r2ml:contextArgument>
                    <r2ml:ObjectVariable r2ml:name="p" r2ml:classID="Person"/>
                        </r2ml:contextArgument>
                    </r2ml:AttributeFunctionTerm>
                    <r2ml:AttributeFunctionTerm r2ml:attributeID="Address">
                        <r2ml:contextArgument>
                    <r2ml:ObjectVariable r2ml:name="s" r2ml:classID="SINrequest"/>
                        </r2ml:contextArgument>
                    </r2ml:AttributeFunctionTerm>
                </r2ml:dataArguments>
            </r2ml:DatatypePredicateAtom>
        </r2ml:conditions>
        <r2ml:producedAction>
            <r2ml:ObjectClassificationAtom r2ml:classID="SINresponse">
           <r2ml:ObjectVariable r2ml:name="SIN" r2ml:classID="SINresponse"/>
        </r2ml:ObjectClassificationAtom>
        </r2ml:producedAction>
    </r2ml:ProductionRule>
  </r2ml:ProductionRuleSet>
</r2ml:RuleBase>
```

```
<r2mlv:Vocabulary xmlns:r2mlv="http://www.reverse.net/I1/2006/R2ML">
    <r2mlv:Class r2mlv:ID="Person">
        <r2mlv:Attribute r2mlv:ID="Person.Name">
            <r2mlv:range>
                <r2mlv:Datatype r2mlv:ID="xs:string"/>
            </r2mlv:range>
        </r2mlv:Attribute>
        <r2mlv:Attribute r2mlv:ID="Person.Address">
            <r2mlv:range>
                <r2mlv:Datatype r2mlv:ID="xs:string"/>
            </r2mlv:range>
        </r2mlv:Attribute>
        <r2mlv:Attribute r2mlv:ID="Person.SIN">
            <r2mlv:range>
                <r2mlv:Datatype r2mlv:ID="xs:integer"/>
            </r2mlv:range>
        </r2mlv:Attribute>
    </r2mlv:Class>
    <r2mlv:Class r2mlv:ID="SINresponse">
        <r2mlv:Attribute r2mlv:ID="SINresponse.SIN">
            <r2mlv:range>
                <r2mlv:Datatype r2mlv:ID="xs:integer"/>
            </r2mlv:range>
        </r2mlv:Attribute>
    </r2mlv:Class>
    <r2mlv:Class r2mlv:ID="SINrequest">
        <r2mlv:Attribute r2mlv:ID="SINrequest.Address">
            <r2mlv:range>
                <r2mlv:Datatype r2mlv:ID="xs:string"/>
            </r2mlv:range>
        </r2mlv:Attribute>
        <r2mlv:Attribute r2mlv:ID="SINrequest.Name">
            <r2mlv:range>
                <r2mlv:Datatype r2mlv:ID="xs:string"/>
            </r2mlv:range>
        </r2mlv:Attribute>
    </r2mlv:Class>
</r2mlv:Vocabulary>
```

*Table 3. Presenting the basic of transformation from reastion rule to production rule*

| Reaction Rule | Production Rule |
|---|---|
| ReactionRuleSet | ProductionRuleSet |
| ReactionRule | ProductionRule |
| TrigerringEvent & Conditions | Conditions (There will be only one Conditions element generated although both TrigerringEvent and Conditions could be present in the Reaction Rule) |
| MessageEventExpression | ObjectClassificationAtom |

Table 4 presents the concept of transformation between R2ML production rule elements and the Drools generated code. The table is not presenting all steps of transformation as some of them are not a simple mapping between elements but gives a general idea of what is done.

## 6. RELATED WORK

In this section, we compare the proposed approach with some relevant solutions to the modeling of Web services that are based on the MDE principles.

(Bezivin et al., 2004) demonstrate how one can take advantage of MDE to develop e-business applications. In their approach, they start from UML and the UML profile for Enterprise Distributed Object Computing (EDOC) to define platform-independent models of e-business applications. In the next step, they translate such models into models that are based on metamodels of Java, WSDL, and JWSDP. Although this approach uses ATL, similar to what we do, it does not provide

*Figure 18. Generated Drools rule (right) and generated class of it (left)*

```
public class Person {
    public string Address;
    public string Name;
    public integer SIN;
 }
public class SINrequest {
    public string Address;
    public string Name;
 }
public class SINresponse {
    public integer SIN;
 }
```

```
rule "aRule"

  when
    #conditions
        p : Person()
        s : SINrequest()

        p : Person ( namePerson : Name )
        p : Person ( addressPerson : Address )
        s : SINrequest ( nameSINrequest : Name )
        s : SINrequest ( addressSINrequest : Address )

        eval (nameSINrequest.compareTo( namePerson ) == 0 )
        eval (addressSINrequest.compareTo( addressPerson ) == 0 )

  then
    #actions
        SIN : SINresponse()

end
```

*Table 4. Presenting transformation concept of R2ML production rule to Drools*

| R2ML | Drools |
|---|---|
| Class | Class |
| Attribute | Integer, String etc (Based on the value of r2mlv:ID) |
| Conditions | Conditions (and based on the object within this element the condition element will be generated) |
| producedAction | Actions (based on the elements residing within this element proper code with regards to the previously generated classes will be generated) |

two way transformations between models and service implementation. This is unlike our solution, because we support two-way transformations, and thus we enable reverse engineering of the existing Web services. Next, the translation of regular UML models, i.e., UML class related elements into WSDL is limited, as one does not have enough expressivity in UML diagrams to define all details of WSDL (i.e., one can not distinguish between input and output messages). This issue is addressed by using the UML profile for EDOC, but this approach is more focused on modeling distributed components rather on modeling business logics. In our approach, the use of reaction rules in URML models enables us to be closer to business processes, while OCL filter expressions can even further specify conditions under which a message can happen. This is very useful for potential integration of business process based on the use of Web services (Milanovic et al., 2006). Of course, such filter expressions overcome the potentials of the Web service technology, as they presume that there should also be a rule-based engine, which is able to interpret such conditions. However, the research on semantic Web services (Sheth et al., 2006) and Web rules (Nagl et al., 2006) demonstrates that this issue attracts more attention in the Web service research community.

(Vara et al., 2005) define a metamodel for WSDL and its corresponding UML profile that is used for modeling of Web services in the MIDAS-CASE MDA-based tool for Web information system development. They also support automatic generation of the respective WSDL description for the Web services that is being modeled. Although the MIDAS framework supports platform-independent models of services and service compositions, their definitions are very incomplete and one can hardly generate complete service models automatically. In our approach, we do not strictly base models of services on workflows in which services will be used, as we wanted to focus on how services are used (i.e., MEPs).

(Gronmo et al., 2005) and (Timm & Gannod, 2005) propose an approach to modeling semantic Web services by using UML profiles. Gronmo et al. have the goal to define platform-independent models of services, which can be translated to semantic Web services (OWL-S and WSMO). In their work, they abstract concepts from OWL-S and WSMO, and extend UML activity diagrams (i.e., they define a UML profile) accordingly, while for defining vocabularies they use the Ontology Definition Metamodel, and its corresponding UML profile defined in (Gasevic et al., 2005). Since both these approaches use XSLT, they can hardly translate service pre- and post-conditions from their UML definitions (i.e., OCL) to the languages of the potential target platforms (e.g. Jess). This approach does not consider modeling usage patterns or MEPs or error handling (i.e., business logic) like we do, but instead it focuses on the details specific for service platforms.

(Manolescu, et al., 2005) propose a model-driven approach to designing and deploying service-enabled web applications. This work is different from the ones above in the following aspects: i) it extends WebML, a well-known modeling language for developing Web applications; ii) it uses E-R models for defining data types; and iii) it focuses on the usage patters (MEPs) of Web services in order to define WebML extensions. Their modeling approach is based on the use of MEPs, but they are considering MEPs used in WSDL 1.1, and for each MEP they define corresponding WebML primitive. These new WebML primitives can be used in hypertext models of Web applications. This approach is the most similar to ours; both approaches fully focus on modeling business processes and potential usage patterns. However, the WebML approach does not explicitly define a Web service description, but it infers it from the definitions of hypertext models in which the service is used, which makes this process context dependent. That is, services are defined inside specific WebML workflows unlike our approach where we define workflow-independent services. In addition, they do not consider the use of preconditions of Web services, which is important for the reusability of services in different contexts. Although WebML has support for exception handling (Brambilla et al., 2005), this is also focused on the workflow level rather than on a service level. Finally, the WebML approach does not support reverse engineering of Web services.

The following table (Table 5) summarizes these approaches of modeling of Web services, and shows their relation to our approach we describe in this paper.

## 7. CONCLUSION

In this paper, we have demonstrated how the use of MDE principles can enable for rule-based modeling of Web services. By using the MDE principles, we have been able to develop a framework for modeling Web services from the perspective of how services are used in terms of message-exchange patterns (MEPs). Our approach enables developers to focus on the definition of business rules, which regulate MEPs, instead of focusing on low level Web service details or on contexts where services are used (i.e., workflows). Business rules are important assets of a business organization that embody valuable domain knowledge. They are stated in a way, which is very close to how users think and talk. Due to the declarative nature of rules, the business logic can easier be

*Table 5. Summary of solutions for modeling of Web services based on the MDE principles*

| Approach | Used metamodel | Model description | transformation | output |
|---|---|---|---|---|
| (Timm & Gannod, 2005) | UML profile | UML XMI | XSLT | OWL-S |
| (Bezivin et al., 2004) | MOF | MOF XMI | ATL | WSDL, Java, JWSDP |
| (Manolescu, et al., 2005) | Extended WebML | WebML model | XSLT (through visual editor) | WebRatio output |
| (Vara et al., 2005) | UML profile | UML XMI | MIDAS-CASE transformation unit | MIDAS output |
| Our approach | MOF | MOF XMI | ATL | R2ML, WSDL |

updated without the need to change to whole system. Furthermore, integrating business rules in process-oriented Web service composition can improve the quality and ease development. Once business rules are expressed explicitly as first-class entities, they can be reused across several compositions i.e., the same rule can be applied to many web service compositions.

By leveraging MDE, we have defined a rule-based modeling language that can be managed by universal MDE tools. The use of model transformations allows for transforming platform independent models of business logic to specific platforms such as Web services. Moreover, for each Web service, we can also generate a rule that will fully regulate the behavior of the service (i.e., how the service is used), and thus make sure that the business logic is fully followed. In this paper, we have not explained that part of the solution, but we are going to report on that in our future papers.

Our solution has much broader potentials that overcome the pure translation between rule-based business models and Web services. Unlike the WSDL definition of Web services, our models also have an option for defining pre- and post-conditions of services. However, WSDL can not express pre- and post-conditions in the descriptions of Web services. That is, Web service tools can not automatically support conditions under which some services can be used (as defined by OCL filters in Section 4). To address this problem, we expand our approach on W3C's Semantic Annotations for WSDL (SAWSDL) recommendation. In our current activities, we have developed transformations from our R2ML reaction rule-based models to several production rule languages (e.g., Drools, Jena2, and Jess). Our particular focus is on Drools, as the use of Drools allows us to directly enforce business rules to regulate the use of Web services deployed on JBoss' application server.

## REFERENCES

Arroyo, S., Toma, I., Roman, D., Drumm, C., Dimitrov, M., Spork, M., Nagypal, G., Domingue, J., & Henke, J. (2004). *Report on state of the art and requirements analysis (Deliverable 3.1)*. DIP EU FP6 Integrated Project.

ATLAS group, Nantes. (ver. 0.7) (n.d.). *Atlas transformation language - User manual*. Retrieved from http://www.eclipse.org/gmt/atl/doc/ATL_User_Manual[v0.7].pdf

Bézivin, J. (2005). On the unification power of models. *Software and System Modeling, 4*(2), 171-188.

Bezivin, J., Hammoudi, S., Lopes, D., & Jouault, F. (2004). Applying MDA approach for Web Service platform. In *Proc. of the 8th IEEE Int. Conf. on Enterprise Distributed Object Computing Conf.* (pp. 58–70).

Brahe, S., & Osterbye, K. (2006). Business process modeling: Defining domain specific modeling languages by use of UML profiles. In *ECMDA-FA 2006* (LNCS 4066, pp. 241-255). 2006.

Brambilla, M., Ceri, S., Comai, S., & Tziviskou, C. (2005). Exception handling in workflow-driven Web applications. In *Proc. of the 14th International WWW Conference* (pp. 170–179).

Charfi, A., & Mezini, M. (2004). Hybrid Web service composition: Business processes meet business rules. In *Proc. of the 2nd International Conference on Service Oriented Computing* (pp. 30-38).

Chinnici, R., Moreau, J., Ryman, A., & Weerawarana, S. (2007). Web Services Description Language (WSDL) Ver. 2.0 Part 1: Core Language. *W3C Recommendation*. Retrieved from http://www.w3.org/TR/2007/REC-wsdl20-20070626.

Clement, L., Hately, A., Riegen, C., & Rogers, T. (2004). *UDDI Ver. 3.0.2*. Retrieved from http://www.oasis-open.org/committees/uddi-spec/doc/spec/v3/uddi-v3.0.2-20041019.htm.

Coyle, F. (2002). *XML, Web services, and the data revolution*. Addison Wesley.

Czarnecki, K., Helsen, S. (2006). Feature-based survey of model transformation approaches. *IBM Syst. J. 45*(3), 621-645.

EclipseDoc (2002). *How abstract components are created when reading a schema – an animation in 9 slides (complexType)*. Retrieved from http://help.eclipse.org/help33/index.jsp?topic=/org.eclipse.xsd.doc/references/diagrams/images/xsd-diagram-ct02/img0.html

Erl, T. (2005). *Service-oriented architecture: Concepts, technology, and design*. Prentice Hall.

Falkovych, K., Sabou, M., & Stuckenschmidt, H. (2003). UML for the Semantic Web: Transformation-based approaches. In B. Omelayenko & M. Klein (Eds.), *Knowledge transformation for the Semantic Web* (pp. 92-106). Amsterdam: IOS Press.

Gasevic, D., Djuric, D., & Devedzic, V. (2005). Bridging MDA and OWL ontologies. *Journal of Web Engineering, 4*(2), 119-134.

Ginsberg, A., Hirtle, D., McCabe, F., & Patranjan, P. (2006). *RIF Use cases and requirements*. W3C Working Draft. Retrieved from http://www.w3.org/TR/rif-ucr.

Giurca, A., Lukichev, S., & Wagner., G. (2006). Modeling Web services with URML. In *Proc. of Workshop Semantics for Business Process Management*.

Giurca, A., & Werner, J. (2006). *Translator: R2ML to JBoss rules*. Retrieved from http://oxygen.informatik.tu-cottbus.de/translator/R2MLtoJBossRules/

Gronmo, R., Jaeger, M.C., & Hoff, H. (2005). Transformations between UML and OWL-S. In *Proc. of the 1st European Conf. on Model Driven Architecture - Foundations and Applications* (pp. 269–283).

Gudgin, M., Hadley, M., Mendelsohn, N., Moreau, J., Nielsen, H. F., Karmarkar, A., & Lafon, Y. (2007). *SOAP Ver. 1.2 Part 1: Messaging framework*. W3C Recommendation. Retrieved from http://www.w3.org/TR/soap12-part1/

Hirtle, D., Boley, H., Grosof, B., Kifer, M., Sintek, M., Tabet, S., Wagner, G (2006). *Schema specification of RuleML 0.91*. Retrieved from http://www.ruleml.org/spec/

Horrocks, I., Patel-Schneider, P. F., Boley, H., Tabet, S., Grosof, B., Dean, M. (2004). *SWRL: A Semantic Web rule language combining OWL and RuleML*. W3C Member Submission. Retrieved from http://www.w3.org/Submission/SWRL/

Jovanović, J., Gašević, D. (2005). XML/XSLT-based knowledge sharing. *Expert Systems with Applications, 29*(3), 535-553.

Kurtev, I., Bezivin, J., & Aksit, M. (2002). Technological spaces: An initial appraisal. In *Proceedings of the Confed. Int'l Conferences CoopIS, DOA, and ODBASE.*

Lewis, A. (2007). Web *Services Description Language (WSDL) Version 2.0: Additional MEPs.* W3C Working Group Note. Retrieved from http://www.w3.org/TR/2007/NOTE-wsdl20-additional-meps-20070626/

Lukichev, S., & Wagner, G. (2006a). Visual rules modeling. In *Proceedings of the 6th Inter-national Conference Perspectives of Systems Informatics* (pp. 467-673).

Lukichev, S., & Wagner, G. (2006). UML-based rule modeling with Fujaba. In *Proc. of the 4th International Fujaba Days* (pp. 31-35).

Manolescu, I., Brambilla, M., Ceri, S., Comai, S., & Fraternali, P. (2005). Model-driven design and deployment of service-enabled Web applications. *ACM Trans. Inter. Tech., 5*(3), 439-479.

Milanovic, M., Gasevic, D., Giurca, A., Wagner, G., & Devedzic, V. (2006). On interchanging between OWL/SWRL and UML/OCL. In *Proc. of the OCLApps Workshop* (pp. 81-95).

Miller, J., & Mukerji, J. (eds.) (2003). *MDA Guide Version 1.0.1.* OMG.

MOF (2005). Object Management Group, Meta Object Facility (MOF) Core, v2.0. *OMG Document formal/06-01-01.* Retrieved from http://www.omg.org/cgi-bin/doc?formal/2006-01-01.

Rupp, N.A. (n.d.). *The login of the bottom line: An introduction to The Drools Project.* Retrieved from http://www.theserverside.com/tt/articles/article.tss?l=Drools

Nagl, C., Rosenberg, F., & Dustdar, S. (2006). VIDRE– A distributed service oriented business rule engine based on RuleML. In *Proc. of the 10th IEEE International Enterprise Distributed Object Computing Conference* (pp. 35-44).

Object Management Group (2008). *Business Process Model and Notation (BPMN) specification 2.0.* Retrieved from http://www.omg.org/cgi-bin/doc?bmi/08-02-06

OCL (2006). Object constraint language. *OMG Available Specification, Version 2.0, formal/06-05-01.*

Proctor, M., Neale, M., Frandsen, M., Griffith Jr., S., Tirelli, E., Meyer, F., & Verlaenen, K. (n.d.). *Drools documentation.* Retrieved from http://downloads.jboss.com/drools/docs/4.0.7.19894.GA/html/index.html

Ribaric, M., Gasevic, D., & Milanovic, M. (n.d.). *R2ML: A rule-based approach to modeling of semantically-enriched Web services.*

Rosenberg, F., & Dustdar, S. (2005). Business Rules Integration in BPEL – A service-oriented approach. In *Proc. of the 7th International IEEE Conference on E-Commerce Technology* (pp. 476-479).

SBVR (2006). Object Management Group. *Semantics of business vocabulary and business rules (SBVR) – Interim specification.* Retrieved from http://www.omg.org/docs/bei/05-08-01.pdf

Schneier, B. (1992). *The Rete matching algorithm: A description of the pattern-matching algorithm underlying Drools.*

Seidewitz, E. (2003). What models mean. *IEEE Software, 20*(5), 26-32.

Sheth, A., Verma, K., & Gomadam, K. (2006). Semantics to energize the full services spectrum. *Communication of the ACM, 49*, 55-61.

Tabet, S., Wagner, G., Spreeuwenberg, S., Vincent, P., Jacques, G., de Sainte Marie, C., Pellant, J., Frank, J., & Durand, J. (2007). *Object Management Group: Production Rule Representation (PRR).* Retrieved from http://www.w3.org/2004/12/rules-ws/paper/53/

Timm, J., & Gannod, G. (2005). A model-driven approach for specifying Semantic Web services. In *Proc. of IEEE International Conference on Web Services* (pp. 313-320).

UML (2005). Object Management Group, Unified Modeling Language 2.0. Docs. *formal/05-07-04 & formal/05-07-05.*

Vara, J., Castro, V., & Marcos, E. (2005). WSDL Automatic Generation from UML Models in a MDA Framework. *International Journal of Web Services Practices, 1*(1-2), 1-12.

Wagner, G., Giurca, A., & Lukichev, S. (2006). R2ML: A general approach for marking up rules. In *Dagstuhl Seminar Proc. 05371.*

WSDL2R2ML (2006). *The WSDL to R2ML transformation scenario.* Retrieved from http://www.eclipse.org/m2m/atl/atlTransformations/#WSDL2R2ML

*XML metamodel* (2006). Retrieved from http://mdr.netbeans.org/mdrxml.html

## KEY TERMS AND DEFINITIONS

**Business Rules:** Business rules or business rule sets describe the operations, definitions and constraints that apply to an organization in achieving its goals.

**Drools:** A business rule management system (BRMS) with a forward chaining inference based rules engine, more correctly known as a production rule system, using an enhanced implementation of the Rete algorithm.

**Reaction Rules:** Also known as Event-Condition-Action Rules, are three components rules i.e. the *event* part specifies the signal that triggers the invocation of the rule, the *condition* part is a logical test that, if satisfied or evaluates to true, causes the action to be carried out and the *action* part consists of updates or invocations on the local data.

**R2ML:** REWERSE I1 Rule Markup Language, a rule markup language allowing rules interoperability.

**URML:** UML-based Modeling Language, a visual language for rules modeling.

**Web Service** A software system designed to support interoperable machine-to-machine interaction over a network.

**WSDL:** Web Service Description Language is a language for describing both the abstract functionality of a service and the concrete details of a Web Service.

## ENDNOTES

1 http://www.Fujaba.de
2 http://dresden-ocl.sourceforge.net/
3 http://www.eclipse.org/xsd/

# Section III
# Rule Applications

# Chapter XIX
# Distributed Business Rules within Service-Centric Systems

**Florian Rosenberg**
*Technical University Vienna, Austria*

**Anton Michlmayr**
*Technical University Vienna, Austria*

**Christoph Nagl**
*Technical University Vienna, Austria*

**Schahram Dustdar**
*Technical University Vienna, Austria*

## ABSTRACT

*Business rules enable a clear separation of concerns between the core business knowledge and the underlying application code. Service-oriented Computing, on the other side, enables flexible software systems and provides support for business processes based on software services with well-defined interface, descriptions and communication protocols. Yet, the alignment of software services and business rules has not been addressed in literature. In this chapter we present the ViDRE system that bridges the gap between these two paradigms by exposing business rules as Web services. In contrast to existing rule engines, our approach supports distributed rule execution using meta-rules which includes automatic transformation of rules on both client- and server-side.*

## INTRODUCTION

The design and development of adaptive business applications is a challenging task based on the fact that software applications are increasingly distributed and the requirements change frequently. Such changes involve a multitude of aspects, ranging from changes in the core business logic to changes in the external interfaces to other subsystems or even business partners. Such changes always imply costs and risks that come along while adapting a software system. It is a common goal to keep these costs minimal, therefore, two rapidly emerging trends receive a lot of attention in enterprise computing, namely *service-oriented architecture* (SOA) – or service-oriented computing (SOC) as a research area in general (Papazoglou, Traverso, Dustdar, & Leymann, 2007) – and *business rules*. SOA is an architectural paradigm that enables more flexible software systems and business processes by exposing some functionality as a dedicated software service. Such a service is described with a well-defined, document-centric interface based on well-known interface descriptions as for example the Web services technology using WSDL and SOAP (Weerawarana, Curbera, Leymann, Storey, & Ferguson, 2005). These services are typically distributed and can then be used to orchestrate different services in a business process or to deliver higher-level functionality in composed services (Dustdar & Schreiner, 2005).

Despite all advantages of the SOC paradigm, it does not elaborate on the internals of a service, i.e., how a service is implemented, its technology or paradigms for doing it. In this regard, business rules provide an elegant means to express complex business logic in terms of rules (for example "if-then-else" style production rules) and facts that can be effectively executed by a rule engine. A rule engine is a software component that is capable of executing such rules that are expressed in a given syntax. In the last years, various rule languages and engines have emerged, such as RuleML (The Rule Markup Initiative, 2001), JBoss Drools (JBoss, 2008), ILOG Rules (ILOG, 2008) or Prova (Kozlenkov, Paschke, & Schroeder, 2008). A main drawback of current rule engine implementations is the lack of alignment with service-oriented computing principles, that would allow an easier integration in existing applications by exposing rules as services, so-called *rule services*, and facilitate the creation of *decision services*, that is, a service that is implemented as a rule service with a special semantics of returning either true of false (the outcome of a decision). For example, consider a decision service for calculating insurance claims that accepts a claim as input and decides whether it is approved or denied based on complex calculations and rules.

The ViDRE (Vienna Distributed Rule Engine) project introduced in (Nagl, Rosenberg, & Dustdar, 2006) addresses some existing shortcomings of existing business rule solutions for the use within enterprise software systems that hamper an alignment and integration with the principles of SOAs. One of the main problems is the centralized way in which rules are managed in current business rules management systems (BRMS). Typically, the rules are stored in a central rule base, which poses some challenges when dealing with distributed services that need to access them, especially the increasing network communication overhead. Therefore, we advocate the use of ViDRE *rule service provider*s (VSP), i.e., a rule engine and its set of rules that are needed to execute a dedicated functionality (for example handling an insurance claim). A VSP enables the use of business rules technology in distributed environments by allowing a VSP to use different rule engines that perform the rules execution. This is achieved by a plug-in mechanism that uses the Java Rule Engine API, as developed by the JSR 94 (Java Specification Request) to unify the access to various rule engines (JSR 94, 2004).

Another important issue when using rules technologies is the lack of a standardized rule representation language. When using dedicated rule service providers, a unified rule representation format is necessary to enable communication among different providers. Within the scope of the ViDRE project, RuleML is used as an Interlingua and communication protocol for rules and facts; similar to what SOAP (Simple Object Access Protocol) is for Web services.

Finally, an important issue addressed by ViDRE concerns the execution of business rules. When dealing with distributed applications, rules cannot always be executed locally within an application; rules and facts may be required from other rule engines to facilitate a successful execution. Using a service-based design, each VSP and the unified communication protocol enable distributed rule execution.

The objective of this chapter is to provide a deeper understanding of a business rules framework that is aligned with SOC principles to facilitate and support the design and implementation of business applications. This is achieved by leveraging a distributed business rule infrastructure, consisting of multiple VSPs that communicate using a common rule representation format (RuleML). This infrastructure allows a flexible management of business rules and the ability to perform distributed rule execution. Due to its flexible way of exposing rules as services, it is well suited for creating rule and decision services.

This chapter starts by describing some background information on related scientific work in the area of business rules and rule engines. In addition, the basic concepts of the Java Rule Engine API and RuleML are depicted, as they are needed in the main part of the chapter to understand the distributed business rules execution. The main focus of this chapter is to describe the design and implementation aspects of a business rules infrastructure for executing distributed business rules. The emphasis is put on the novel features that address the alignment of rule engines with service-oriented principles and the use of an Interlingua for rules to hide the heterogeneity of various rule languages. This chapter concludes by discussing some of ViDRE's limitations and an outlook on future research in this area.

## BACKGROUND

In this section, we first present a discussion of related work and similar approaches from a scientific and industrial perspective. The second part will introduce the basic terminology and concepts that are later used and required for understanding the ViDRE approach.

### Related Work

Recently, business rules gain more attention from the scientific community as well as from industry, since they provide a means to more flexible and adaptable enterprise applications.

The work presented in (Cabitza, Sarini, & Seno, 2005) comes closest to ours with regard to distributed execution of business rules. It proposes an extension to the Jess rule engine (Jess, 2008) to enable distributed rule execution. Their distributed inference mechanism uses a shared working memory (SWM) to store facts and each rule engine uses this memory to access these facts. We do not use a shared memory; instead, we propose to use meta-rules to handle the distribution. Furthermore, we use RuleML as Interlingua and provide translators from RuleML to the corresponding rule engines, thus our approach is not limited to a concrete rule engine.

In (Rosenberg & Dustdar, 2005a), a loosely coupled integration of business rules into WS-BPEL (OASIS, 2007) – a standard for describing Web service based business processes – is proposed. The approach uses an enterprise service bus (ESB) and business rules by using an interceptor pattern to Web service invocations (e.g., WS-BPEL invoke activity) from a WS-BPEL engine. In contrast to the approach presented in this chapter, the authors focused on accessing different rule engines with the same API and generating Web services for it. In this chapter we focus on using a common rule representation language and enabling a distributed rule execution by using meta-rules.

The approach presented in (Rosenberg & Dustdar, 2005b) deals with deployment issues of business rules that occur when having various rule service providers that execute different rules. A coordinated deployment is necessary to guarantee that all providers receive their rules that are needed for execution. The ViDRE approach presented in this chapter can be seen complementary to that approach, as deployment also plays a major role in ViDRE. Currently, this is done manually by using an administration interface but that approach could be adopted to have an automated deployment for distributed rules.

In (Milanovic, et al., 2007), the authors present an approach that deals with problems that are involved in the business partner interchanges especially when heterogeneous business rules are used. The authors propose the use of a uniform rule representation language, called REWERSE II Rule Markup Language (R2ML), which is powerful enough to act as a rule interchange format. All business partners involved in a business process should still be able to leverage their own business rules technologies, thus, the KAoS policy language is used to transform to R2ML that allows processing in a uniform manner. In ViDRE, we follow a similar concept to use RuleML as an Interlingua between the partners in a distributed rule execution, which is not done in their work. Their focus is on the transformation step itself, whereas our goal is to enable distributed execution among rule service providers with heterogeneous business rule engines.

Patterns for applying business rules have been proposed by (Graml, Bracht, & Spies, 2007). The authors have analyzed a set of business processes with the goal to separate decision logic from the actual control flow in a business process. A rule engine handles the decision logic enabling business processes to be less vulnerable for changes. Our ViDRE approach could easily be integrated with their approach, as we provide a reasonable alignment with SOA technologies, thus, rule services as provided by ViDRE could act as decision services in their approach.

Schmidt (Schmidt, 2002) describes a distributed rule execution approach by encoding RuleML rules in the SOAP header that are processed by different SOAP intermediaries. Based on some conditions in one of the business rules a different execution path can be chosen. The focus of this work is on rule execution but it is not discussed how distributed rules can be specified and executed.

The Prova rule engine (Kozlenkov, Paschke, & Schroeder, 2008) is one of the few rule engines that supports the increasingly distributed nature of current business applications. It allows a distributed deployment of rule engines as agents and facilitates a message-based communication between the deployed agents to issue queries to local knowledge bases or submit tasks to other agents. In contrast to ViDRE, this system can only communicate with other Prova agents, other rule engines cannot be used without understanding the Prova language.

In contrast to Prova, the RuleML Responder system (Paschke, Boley, Kozlenkov, Craig, 2007) is a new type of business rule management system that allows a distributed deployment of arbitrary rule engines as rule inference services by leveraging the integration capabilities of ESB as a service broker and communication layer. The RuleML Responder system uses Reaction RuleML (Paschke, Kozlenkov, Boley, Tabet, Kifer, Dean, 2008) as a rule interchange format. Reaction RuleML is a sublanguage of RuleML specifically designed for encoding reaction rules. In contrast to ViDRE, the RuleML Responder uses a rule language that is able to capture reaction rules, which was not available at the time of implementing ViDRE. In practice, the support for arbitrary rule engines that can deal with reaction rules is limited, thus, it reduces the number of concrete rules engines which can be used in their framework.

## Business Rules

Using business rules implies some degree of declarative style of programming that only specifies what has to be done but omits much of the instructions on how to do it. It is the rule engine's task to decide on the best way to solve the given problem (KnowGravity Inc., 2002). Because declarative programs include only the important details of a solution, they can be easier to understand than procedural programs.

In literature the definition of a business rule is ambiguous. There is no industry standard definition for the term "business rule". The Business Rules Group (The Buiness Rules Group, 2000) describes a business rule as a "statement that defines or constrains some aspect of the business and it is intended to assert business structure or to control or influence the behavior of the business". We refer to three major types of business rules that have been identified in literature (Taveter & Wagner, 2001; The Buiness Rules Group, 2000). The rule engines that are used within ViDRE basically support the following rules types:

- **Derivation rules** (also called deduction rules or Horn clauses): A derivation rule is a statement of knowledge that is derived from other knowledge by inference or mathematical calculation. Derivation rules capture heuristic domain knowledge without explicitly storing information; hence it can be derived from existing or other derived information on demand. An example of a derivation rule is the following business rule: If the turnover of a customer is more than $5000 in the previous year, the customer is premium.
- **Integrity rules** (also called integrity constraints): An integrity rule is an assertion that must be satisfied in all evolving states. Integrity rules limit possible actions within well-defined boundaries. Crossing these boundaries will violate a constraint and cause some corrective actions. An example of an integrity constraint is: Online customers must be at least 18 years old. Integrity rules can also be used for validation, for example, to ensure that a form is filled out correctly.
- **Reaction rules** (also called "stimulus response rules", "action rules" or "event-condition-action (ECA) rules"): Reaction rules cause constructive actions when a certain event occurs and/or when a certain condition is met. Such an action could be the manipulation of one or more business objects, the triggering of another business activity, the creation of an entry in the "to do" list of a business actor (worfklow rule) or any combination of such actions. Depending on the event that triggers an action rule, the following two distinctions can be made: Object event rules are triggered when a business object manipulation occurs (e.g., a new account is created) whereas business event rules are triggered when certain business events occur (e.g., a money transfer is requested).

## Rule Markup Initiative (RuleML)

The RuleML Initiative was the first standardization effort to mark up rule-based systems. The participants of the RuleML Initiative constitute an open network of individuals and groups from both industry and academia. The initiative focuses on defining a standard rule language and an interoperability platform for integrating various business rule languages, inference systems, and knowledge representation paradigms. RuleML covers the entire rule spectrum, from derivation rules to transformation rules to reaction rules, and can thus specify queries and inferences in Web ontologies, mappings between Web ontologies, and dynamic behavior of workflows, services and agents. RuleML is a family of sublanguages whose root allows access to the language as a whole and whose members allow for the identification of customized subsets of the language.

ViDRE uses the object-oriented version of RuleML, to be able to reason about Java objects. The non object-oriented version is not suitable for ViDRE, because Java objects cannot be encoded without information loss. In a distributed environment we cannot assume that the encoded object is known on the server side, so we have to encode enough object information that the original object can be reconstructed based on the RuleML document. The object-oriented version augments RuleML slots with user level roles. These user level roles are used to encode the object attributes. We provide detailed examples of RuleML throughout the main parts of this chapter.

### Java Rule Engine API

The Java Specification Request (JSR) 94 defines an API for Java-based rule engines. In November 2003, the Java Community Process released the final version of the JSR 94 specification with the aim to standardize a set of fundamental rule engine operations such as parsing rule sets, adding objects to the inference process, firing rules and retrieving result objects from the engine. Although a generic API for parsing rule sets is defined, no rule language and semantic rule standards are specified. The main goal is an integration of different rule engines by using a simple adapter into client applications thus enabling interoperability and vendor independence. In order to achieve this vision a standard rule language and its semantics has to be defined. When JSR 94 first was initiated, there was a proposal to include a standard rule language, but it was rejected. There was also a suggestion to standardize Rete engines but no agreement could be reached. A number of vendors currently implement the JSR 94 specification, including JRules, Blaze Advisor, Jess and Drools.

The main components of JSR 94 are briefly described as follows: The `RuleAdministrator` interface provides methods to manipulate, register and unregister rule execution sets. A rule execution set is defined as a collection of rules belonging together, and can be initialized from external resources such as URIs, an XML element or various input streams.

The `RuntimeClient` API provides classes to be used by clients to perform the rule-execution cycle and to retrieve the results. It enables clients to acquire rule sessions and execute rules within that session, whereby only those rules that have been registered using the Rules Administrator API are accessible. In general, this API provides access to vendor specific implementations of the Rule Engine API, in a way similar to Java Database Connectivity (JDBC) software.

## VIENNA DISTRIBUTED RULE ENGINE

The aim of ViDRE is to provide a distributed infrastructure with the ability to use rule engines without settling with one specific rule engine vendor. A distinctive feature of the ViDRE approach is the distribution and collaboration between different heterogeneous business rule services, called *ViDRE service provider* (VSP). ViDRE's business rule service provides access to a rule engine in a service-oriented way. Hence, the reuse of business rules and facts in business applications can easily be established. Additional focus was paid to the ease of integration for client applications in heterogeneous environments and especially enterprise.

ViDRE leverages the Java Rule Engine API (JSR 94) standard in order to deal with the heterogeneity of various Java-based rule engines as described above. Unfortunately, as mentioned in section "Java Rule Engine API" it has several drawbacks. The lack of standardization regarding the rule representation and query language hampers interoperability with other applications or other business rule engines. Therefore, ViDRE combines the JSR 94 standard with RuleML, an emerging standardization proposal for knowledge representation. ViDRE uses the object-oriented RuleML version 0.9 because it is expressive enough to represent facts, queries and even Java objects. This combination enables to use different rule engines that support the JSR 94 standard without changing the knowledge or rule base.

ViDRE's distributed architecture is depicted in Figure 1 by outlining a full architecture that could be used in a real-world setting. A VCP provides three interfaces, a RuleML and a two WSDL interfaces. The RuleML interface accepts valid RuleML documents that can contain RuleML queries or facts that are used for query evaluation. Aligned with the principles of the Web services stack, RuleML documents can be sent via any arbitrary transport channel such as HTTP, JMS or even SMTP. The two WSDL interfaces provide access to the JSR94 client runtime and administration interface as Web service to facilitate a seamless management and execution of rule sets.

A client has two possibilities to access a VSP. The first one is to use the client runtime interface to trigger the rule execution cycle for a given business document or to issue RuleML encoded queries. The administration interface is primarily used for managing and administrating rule sets. The main reason for separating these two interfaces is security. It facilitates a seamless integration and separation of security policies and access control. On the client-side, the JSR94 API is used to access the Web service provided by a VSP. Therefore, ViDRE provides a pre-compiled client library to hide the implementation details from the programmer (such as the location of a VSP). Additionally, since the client library only deals with Java objects, the concrete rule representation language is hidden from the developer.

The second way of accessing a VSP is to use the SOAP-RuleML gateway. It allows provisioning a set of rules as a dedicated Web service using its own WSDL files. At runtime these SOAP requests are dynamically translated into a RuleML request and executed using the RuleML interface. From a client-side perspective, it enables the use of dedicated and fine-grained Web services that can be used and integrated into existing application and invoked as any other Web service (e.g., as a decision service).

ViDRE Client A and B use the VSP's WSDL services to send a RuleML query to be business rule engine (cp. Figure 1). This approach is suitable for integration with standalone clients because it

*Figure 1. ViDRE architecture. Reprinted from Rosenburg, Nagl, & Dustbar. "ViDRE - A Distributed Service-Oriented Business Rule Engine based on RuleML" In Proceedings of the 10^th^ International Enterprise Distributed Object Computing Conference (EDOC). ©2006, IEEE*

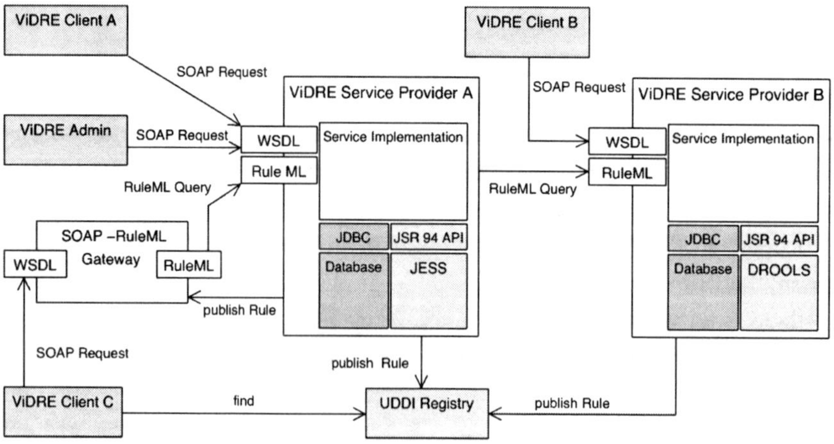

transparently handles the conversion from Java objects to RuleML documents on the client-side by using the client library.

ViDRE Client C uses the UDDI registry to look up a specific business rule published. Being successful, Client C uses the discovered WSDL file to access the rule engine through the SOAP-RuleML Gateway. This invocation method is reasonable for enterprise applications or in combination with service composition (Dustdar & Schreiner, 2005). A possible integration approach for BPEL is presented in (Rosenberg & Dustdar, 2005a). The ViDRE administration client uses the WSDL administration interfaces published by the business rule service provider.

ViDRE's service-oriented design approach enables a very flexible integration into existing application architectures. Moreover, it enables a distributed rule execution that is a highly desired functionality when implemented distributed applications. This distributes rule execution capabilities have been added to one existing rule engine that is encapsulated in a VSP. We describe our extension to the Jess rule engine (Friedman-Hill, 2003) in detail in section "Distributed Business Rules".

## ViDRE Service Providers

Each VSP has two distinctive features: a) providing a plugin functionality by leveraging JSR 94, b) implementing a transformation of RuleML requests to Java objects (and vice-versa) and c) implementing the distributed rule execution mechanism.

The plugin mechanism for connecting a rule engine in a VSP is based on the implementation of the `RuleMLTransformer` and the `ServiceProvider` interfaces. A concrete plugin implementation can be activated at runtime by specifying its class name in the VSP configuration file. ViDRE uses the `ServiceProviderFactory` to retrieve a concrete `RuleMLTransformer` as well as a concrete `ServiceProvider` for the registered rule engine.

The `RuleEngineService` class is the core of the service implementation. It holds a reference to the service provider API of the registered rule engine that is used to create rule runtimes and rule administration interfaces for the registered rule engine. Additionally, it provides the RuleML transformations by using a concrete `RuleMLTransformer`. One major task of the VSP implementation is to preprocess the incoming requests and transform them to a vendor-specific format understood by the registered rule engine, such as RuleML-to-Jess as used in our ViDRE implementation. The appropriate transformer can be retrieved from the `ServiceProviderFactory`. The transformation process is described in detail in the next section.

## Transforming Objects to RuleML

The communication with a ViDRE service provider is fully encoded in RuleML (incoming and outgoing traffic). The service implementation is responsible for translating between RuleML and the target language of the registered rule engine. ViDRE provides a set of pre-built transformers for Jess-to-RuleML and vice versa. A useful aspect when implementing a set of translators is their composability to achieve incremental translation. For example, a Jess-to-Drools translation can be implemented by composing a Jess-to-RuleML translator and a RuleML-to-Drools translator. A main disadvantage of this approach is the restriction to the least common denominator of the supported rule engine capabilities, which is induced by the different rule representation languages.

Figure 2 depicts the approach how a new Java fact can be added to a ViDRE service provider. Facts are added whenever a client rule session is executed, therefore, the objects contained in the session are transformed into a RuleML representation and then transmitted the service provider.

*Figure 2. Adding a Java Fact. Reprinted from Rosenburg, Nagl, & Dustbar. "ViDRE - A Distributed Service-Oriented Business Rule Engine based on RuleML" In Proceedings of the 10th International Enterprise Distributed Object Computing Conference (EDOC). ©2006, IEEE*

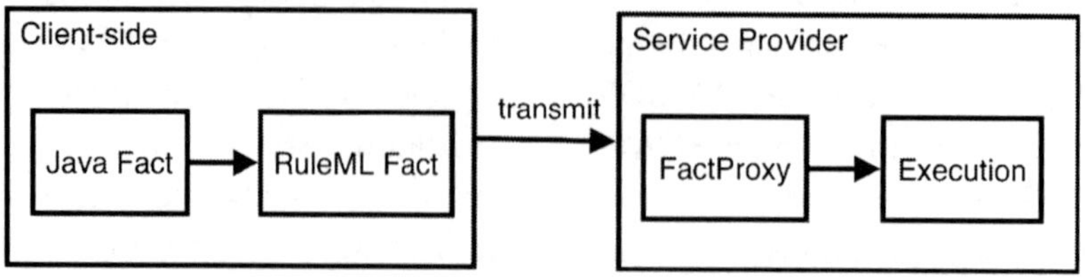


The first step is the transformation of a Java fact to a RuleML fact at the client-side. Listing 1 depicts an Order class representing an order from a simplified supply-chain example.

```
public class Order {
  private int id;
  private String customerId;
  private String product;
  private int quantity;
  // getters and setters
}
```

Listing 1 - Order Class

By adding the Java object to the client-side rule session, the fact is transmitted to a ViDRE service provider. The client library converts the Java object by using the `Object2RuleML` transformer. The resulting of this transformation is shown in Listing 2.

```
<Atom>
  <Rel>Order</Rel>
  <oid><Ind>1ca1a68</Ind></oid>
  <slot>
    <Ind>id</Ind>
    <Ind>785</Ind>
  </slot>
  <slot>
    <Ind>customerId</Ind>
    <Ind>154</Ind>
  </slot>
  <slot>
    <Ind>product</Ind>
    <Ind>Mainboard XY</Ind>
  </slot>
  <slot>
    <Ind>quantity</Ind>
    <Ind>1</Ind>
  </slot>
</Atom>
```

Listing 2 - RuleML Fact for the Order Class

The name of the fact is used as the relation name, e.g., `<Rel>Order</Rel>`. For each Java property an object attribute is created. ViDRE uses the slotted RuleML syntax to create a slot to associate the property name and the corresponding property value. Property names and values are represented as individual constants using the `<Ind>` element. ViDRE requires the slotted RuleML syntax to be able to reconstruct the internals of an object when doing a round-trip transformation.

In the right part of see right part of Figure 2 (server-side) the `Object2RuleML` transformer is used to transform the RuleML representation back into a Java object. The `<oid>` element contains a unique object identifier to preserve object identities. ViDRE uses a `FactProxy` implementation to emulate the Java objects on the server-side because they their lifecycle is restricted to the client-side rule session. Therefore, the `FactProxy` creates a transient in-memory representation a Java object that corresponds to the submitted RuleML in-memory. These objects are then added to the server-side rule session for executing the rules in a VSP.

## Service Provider Transformations

Server-side transformations require a more sophisticated translation because the RuleML syntax needs to be translated to a concrete rule language, such as the Jess language.

Listing 3 depicts a RuleML example that matches all `Order` objects from an exemplary built-to-order computer retailer. For this exemplary order, the `productId` represents a PC configuration (equals to 5) and places orders at the suppliers for the required components such as mainboard, memory, etc. We assume that the supplier is also running a VSP and an order is placed by asserting a new fact called `OrderParts`.

```
<Implies>
 <oid><Ind>If a computer is ordered, then order individual parts at supplier.</Ind></oid>
  <head>
   <Atom>
    <Rel>OrderParts</Rel>
    <slot>
     <Ind>quantity</Ind>
     <Var>quantityVar</Var>
    </slot>
    <slot>
     <Ind>orderId</Ind>
     <Var>orderIdVar</Var>
    </slot>
   </Atom>
  </head>
  <body>
   <And>
    <Atom>
     <Rel>Order</Rel>
     <slot>
      <Ind>orderId</Ind>
      <Var>orderIdVar</Var>
     </slot>
     <slot>
      <Ind>productId</Ind>
      <Ind>5</Ind>
     </slot>
     <slot>
      <Ind>customerId</Ind>
      <Var>customerIdVar</Var>
     </slot>
     <slot>
      <Ind>quantity</Ind>
      <Var>quantityVar</Var>
     </slot>
    </Atom>
   </And>
  </body>
</Implies>
```

Listing 3 - RuleML Rule Representation of an Order

A major challenge is the transformation of a RuleML document into a semantically equivalent representation in Jess by using the `RuleML2JessTransformer` component. Listing 4 depicts the result of the transformation in the LISP-like Jess syntax. A Jess rule consists of two parts. The first part of the rule consists of patterns to match facts (such as the `customerId ?customerIdVar` in the example). The second part consists of actions that are executed when the left-hand side of the rule is satisfied. Both parts are separated by the => symbol.

```
(defrule OrderParts
 "If a computer is ordered, then order individual part at supplier."
 (Order (customerId ?customerIdVar)
       (orderId ?orderIdVar)
       (productId 5)
       (quantity ?quantityVar))
 =>
 (bind ?orderParts (new OrderParts))
 (call ?orderParts setQuantity ?quantityVar)
 (call ?orderParts setOrderId ?orderIdVar)
 (definstance OrderParts ?OrderParts))
```

Listing 4 - Jess Rule of an Order

Several existing approaches (e.g., Grosof, Gandhe, & Finin, 2002) propose to use XSLT transformations to accomplish these conversions, however, this simple example illustrated that this can be very challenging. ViDRE uses a fast and efficient runtime translation of RuleML documents to Jess by leveraging the Jess Java API to add rules and facts directly to the rule engine.

### Encoding Reaction Rules

Reaction rules are an important means to express various action based on a number of event that occur in a system or business processes. Currently, RuleML focuses on derivation rules, however, a specific working group focuses specifically on reaction rules. One of the most recent efforts for implemented reaction rules is OO JDrew (OOj-Drew, 2008).

In ViDRE reaction rules are represented similar to derivation rules. From a client-side perspective there is no difference in handling reaction and derivation rules, however, a distinction is made on the server upon execution. In case of a derivation rule, the engine asserts a new fact whereas for a reaction rule it invokes a predefined function. Therefore, the rule engine has to know all user-defined functions, such as an `order-product` function. Internally, the rule engine looks up the function by its name and in case a function is available it gets a handle to that function. If no function can be found, the rule engine assumes that the rule is a derivation rule.

```
<Head>
 <Atom>
  <Rel>modify-fact</Rel>
  <slot>
   <Ind>factName</Ind>
   <Ind>Stock</Ind>
  </slot>
  <slot>
   <Ind>slotName</Ind>
   <Ind>inStock</Ind>
  </slot>
  <slot>
   <Ind>value</Ind>
   <Ind>(- inStockVar quantityVar)</Ind>
  </slot>
 </Atom>
</Head>
```

Listing 5 – Reaction Rule Example

Listing 5 illustrates the reaction part of a rule encoded in RuleML. This reaction performs an update to a slot of an existing fact. In this example the function modify-fact is called. Each slot in the RuleML Atom represents an input parameter. For the `modify-fact` function the fact name, the target slot and the new value is needed. Usually the new value depends on some variables matched by the condition part of the rule. In the example of Listing 5 the value is calculated using two variables, `inStockVar` and `quantityVar`, which must be bound in the condition part of the rule. The expression is written in prefix notation; hence the operator is written before its operands. The advantage of this notation is that the Jess rule engine can directly evaluate the expression without modification.

## Distributed Business Rules

One of the main contributions of ViDRE approach is the possibility to distribute rules and facts to align it with the increasing necessity of distributed processing and its applications. This section gives a detailed insight into the distributed inference process implemented in ViDRE.

### Distributed Rule Execution

The distributed rules execution does not really differ from the centralized rules execution. The activity diagram in Figure 3 illustrates both, the distributed, and the normal rules execution.

Upon activation of a distributed rule execution cycle, the pattern matcher determines which rules have to be activated. All previously activated rules combined with all newly activated rules form the conflict set. Depending on the conflict-resolution strategy, the rule engine selects the next rule to fire. If no rule can be fired the rule execution cycle terminates. In the other case, ViDRE's inference engine has to decide whether this rule is executed locally or remotely which is described in detail below. For a normal location-bound rule

*Figure 3. Distributed rule execution. Reprinted from Rosenburg, Nagl, & Dustbar. "ViDRE - A Distributed Service-Oriented Business Rule Engine based on RuleML" In Proceedings of the 10th International Enterprise Distributed Object Computing Conference (EDOC). ©2006, IEEE*

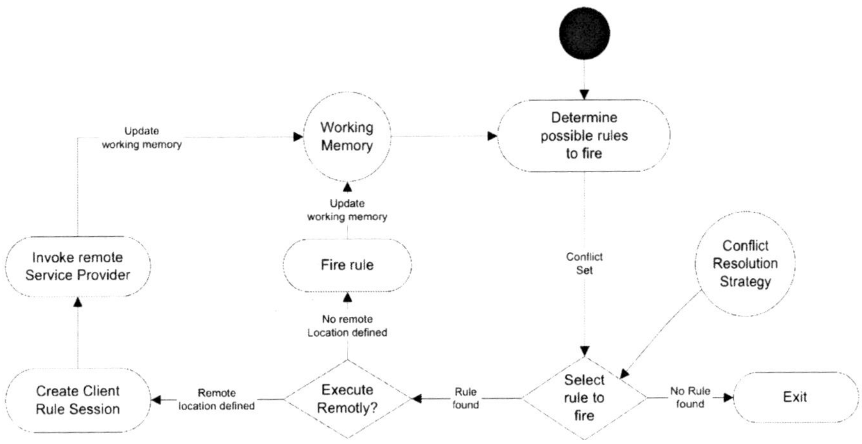


the consequence part of the rule is executed and newly asserted facts are added to the working memory. These newly derived facts potentially activate or block rules in the conflict set, thus the execution cycle starts all over. If the activated rule is a distributed rule and the consequence of the rule should be executed at a remote service provider, ViDRE creates a client session and invokes the remote service provider. The result of the remote service invocation is added to the local working memory and the rule execution cycle starts again.

## Distribution with Meta-Rules

Meta-rules are typically used in rule-based systems to specify preferences concerning the behavior of the inference process or to influence the conflict resolution strategy. They are distinguished from ordinary rules by their role that is used to instrument the reasoning. This is required to solve the problem, rather than to actually perform the reasoning (Jackson, 1999).

In the following, we provide some examples based on our prototype implementation which currently only supports Jess, therefore, the examples are show using the Jess rule language. However, the described methods can be applied to arbitrary rule engines. Since Jess does not offer mechanisms for explicitly defining meta-rules, ViDRE uses the salience property of Jess to emulate this behavior. In general, Jess favors activated rules with a higher salience during the conflict-resolution thus it ensures that meta-rules are always evaluated before business rules.

The question remains to be answered, how to leverage meta-rules for a distributed rule execution? Therefore, we distinguish between two types of rules. The first type has a remote consequence part implying that the assertion of a derived fact or the enabled action should be performed on a remote service provider. In this case a meta-rules play is used to encode the fact that the consequence of the activated rule is a remote rule and define the location where the consequence should be executed.

For each remote rule, ViDRE adds a meta-rule and a meta-fact. In Listing 6 a meta-rule describing remote mainboard orders at a supplier for each newly placed PC order is shown. The meta-fact `META _ LOCATION` associates to each `Order-Mainboard` object a URI. This URI encodes the target location where the consequence should be executed. The prototype implementation uses the prefix META for that kind of meta-information.

```
(defrule OrderMainboard
  "On reception of a new computer order, order a mainboard remotely at the supplier"
  (OrderParts (orderId ?orderIdVar)
              (quantity ?quantityVar))
  =>
  (bind ?OrderMainboard (new OrderMainboard))
  (call ?OrderMainboard setQuantity ?quantityVar)
  (call ?OrderMainboard setOrderId ?orderIdVar)
  (definstance OrderMainboard ?OrderMainboard))

(assert (META_LOCATION (name "OrderMainboard")
  (uri "rules://madrid.vitalab.tuwien.ac.at/supplier")))

(defrule META_REMOTE_OrderMainboard
  "if a remote location is defined call
   defRemoteInstance"
  ?obj <- (OrderMainboard)
  (META_LOCATION (name "OrderMainboard") (uri ?uri))
  =>
  (defRemoteInstance ?obj ?uri))
```

Listing 6 - Remote Jess Rule

The meta-rule REMOTE _ OrderMainboard matches all newly instantiated OrderMainboard objects if the object is a remote fact and, therefore, a META _ LOCATION fact is defined. The function defRemoteInstance is a user defined Jess function, which is responsible to define the fact at the remote location.

The second remote rule type is has premises that contain remote-facts. A remote premise is only reasonable in combination with backward chaining. In order to perform backward chaining in Jess the remote fact has to be declared as backward chaining reactive. With backward chaining it is possible to incorporate data from outside the knowledge base, e.g., a relational database. Jess reasoning engine is strictly a forward chaining engine; backward chaining is simulated in terms of forward chaining rules.

ViDRE uses backward chaining reactive facts in order to pull required data needed for the inference process from other ViDRE service providers. In Listing 7, all customers having an order quantity of more than 10 are updated to premium customers. The required data can be pulled from a remote service provider (as defined by the META _ LOCATION fact). Therefore, ViDRE hooks into the backward chaining mechanism of Jess. For each backward chaining reactive pattern on the left-hand side of the rule, Jess automatically asserts a fact named after the desired fact with prefix "need-", e.g., (need-Order nil nil). This fact can be used to trigger a forward chaining rule, with the premise (need-Order nil nil).

```
(defrule PremiumCustomer
 (Order (customerId ?customerIdVar)
        (orderId ?orderIdVar)
        (productId ?productIdVar)
        (quantity ?quantity))
 (test (> ?quantity 10))
 =>
 (bind ?PremiumCustomer (new PremiumCustomer))
 (call ?PremiumCustomer setCustomerId ?customerIdVar)
 (definstance PremiumCustomer ?PremiumCustomer))

(do-backward-chaining Order)

(assert (META_LOCATION (name "Order")
 (uri "rules://rome.vitalab.tuwien.ac.at/retailer")))

(defrule META_REMOTE_Order
 "pull facts from remote location"
 ?obj <- (need-Order (customerId ?customerId)
                     (orderId ?orderId)
                     (productId ?productId)
                     (quantity ?quantity))
 (META_LOCATION (name "Order") (uri ?uri))
 =>
 (runRemoteQuery ?obj ?uri))
```

Listing 7 - Remote Jess Fact

Listing 7 shows a remote backward chaining fact. The function do-backward-chaining declares the Order fact as backward chaining reactive fact. Similar to the remote-rule example, the location of the remote service provider is encoded in the META _ LOCATION fact. The meta-rule META _ REMOTE _ Order is activated as soon as the do-backward-chaining function was called and a remote location for the remote fact was specified. The function `runRemoteQuery` is another user defined Jess function, which issues a query to ViDRE's service provider with the given URI.

Writing distributed rules by hand is a cumbersome and error-prone task, therefore, the ViDRE administration client provides support for creating distributed rule-sets. Section "Managing the Rules" gives a brief introduction about how to administer remote rules with ViDRE's administration client.

### Remote Rule Set URI

There is no convention for local rule set URIs, but for remote rule execution the host address and the target rule set name have to be encoded in the rule set URI. ViDRE's rule set URIs have to follow the convention: `rules://<hostname>:<port>/<rule-set-name>`

The first part indicates that this URI is a rule set URI. This part is not mandatory and also not validated, but it is useful to differentiate a rule set URI from other URIs. Subsequently the hostname together with the listening port is specified. Like in any client application the hostname and port is needed to establish a connection to the remote host. Thereafter the name of the target rule set has to be encoded.

## Accessing the Knowledge

As mentioned above, a ViDRE's service provider can be accessed transparently by the provided client library or can use the exposed Web services directly. Alternatively, a generic RuleML interface is available. This section focuses on different client invocation methods offered by ViDRE and discusses different application areas. The client library implements the JSR 94 API, which we do not discuss in detail in this chapter. For details on using JSR 94, we refer the interested reader to the JSR 94 specification (JSR 94, 2004).

### Generic RuleML Interface

ViDRE business rule service providers offer a generic RuleML interface. This generic RuleML interface is mainly intended for collaboration between different heterogeneous business rule engines. The generic interface can be used by any other application to directly send RuleML documents, which can contain RuleML queries and RuleML facts. The RuleML interface response is also encoded in a RuleML document. Similar to SOAP requests the RuleML documents can be sent via any arbitrary transport channel (HTTP, SMTP, JMS, etc).

The prototype implementation deploys a servlet that accepts HTTP requests and forwards the `Inputstream` to ViDRE's RuleML parser. The rule set URI is encoded in the query string of the HTTP request. The servlet uses JNDI (Java Naming and Directory Interface) to lookup the `RuleEngineService` implementation.

### Publish Rules and Queries

Another way to access ViDRE's service provider is to access the business rules via the ViDRE SOAP-RuleML gateway. The aim of the SOAP-RuleML gateway is to provide the ability to expose business rules and queries as Web services in form of WSDL documents. These rules can be easily accessed by client applications using the corresponding WSDL file of the SOAP-RuleML gateway. The main goal of this approach is to provide a set of rules as rule service or decision service as stated in the introduction of this chapter.

Consider the following business rule in Listing 8. This rule is responsible to calculate the discount of a customer for a specific product. Exposing this rule as Web service requires the identification of a corresponding operation that maps this business rule's input and output behavior. Obviously, this rule has two input parameters, a customer, a product and the calculated discount as an output parameter.

```
<Implies>
 <oid>
   <Ind>The discount for a customer buying a
product is 7.5 percent if the customer is premium
and the product is luxury.</Ind>
 </oid>
 <head>
  <Atom>
   <Rel>Discount</Rel>
   <slot>
    <Ind>customerName</Ind>
    <Var>x</Var>
   </slot>
   <slot>
    <Ind>productId</Ind>
    <Var>y</Var>
   </slot>
   <slot>
    <Ind>amount</Ind>
    <Ind>7.5 percent</Ind>
   </slot>
  </Atom>
 </head>
 <body>
  <And>
   <Atom>
    <Rel>Customer</Rel>
    <slot>
     <Ind>name</Ind>
     <Var>x</Var>
    </slot>
    <slot>
     <Ind>name</Ind>
     <Ind>premium</Ind>
    </slot>
   </Atom>
   <Atom>
    <Rel>Product</Rel>
    <slot>
     <Ind>product_id</Ind>
     <Var>y</Var>
    </slot>
    <slot>
     <Ind>type</Ind>
     <Ind>luxury</Ind>
    </slot>
   </Atom>
  </And>
 </body>
</Implies>
```

Listing 8 - RuleML Rule for Customer Discount

In general, the number of input and output parameters cannot be determined so easily. Due to the fact that a client does not supply all input parameters, a fully automated approach does not make sense. For example, consider a shopping cart in a Web shop, where not only the current product but also previous transactions have to be taken into account for discount calculation. Furthermore, rules can produce more than one output, for example a list of result objects, or the rule can produce no appropriate result object, which is the case for reaction rules. Therefore, ViDRE implements a semi-automatic approach. The ViDRE administration client supports the publishing process of rules and queries. Rules of a specific rule set can be listed and can be selected as Web service candidates. ViDRE administration client proposes input and output parameters that can further be augmented and modified. Thereafter, ViDRE administration client generates a WSDL file representing the business rule. A short snippet of the resulting WSDL file is illustrated in Listing 9. This WSDL file can be saved to the file system for later application; for example, a deployment to a public UDDI directory or can be deployed directly to the SOAP-RuleML gateway.

## SOAP-RuleML Gateway

In the previous section we have generated a Web service description file (WSDL) for business rules, omitting the service location and implementation. Obviously, the description of a service is not enough. We also have to provide a service endpoint for executing the requests. ViDRE does not generate Web service implementations for business rules, but rather provides a generic Web service that uses a VSP to fulfill the requests.

The SOAP-RuleML Gateway acts as a generic Web service that intercepts the SOAP message, translates the message into an equivalent RuleML

*Listing 9. WSDL file for the customer discount rule*

```
<types>
 <schema targetNamespace=...>
  <import namespace="http://schemas.xmlsoap.org/soap/encoding/"/>

  <complexType name="Customer">
   <sequence>
    <element name="name" type="string"/>
    <element name="type" type="string"/>
   </sequence>
  </complexType>

  <complexType name="Product">
   <sequence>
    <element name="product_id" type="string"/>
    <element name="type" type="string"/>
   </sequence>
  </complexType>

  <complexType name="Discount">
   <sequence>
    <element name="customerName" type="string"/>
    <element name="productId" type="string"/>
    <element name="amount" type="string"/>
   </sequence>
  </complexType>

  <complexType name="TechException">
   <sequence>
    <element name="message" type="string"/>
   </sequence>
  </complexType>
 </schema>
</types>

<message name="TechException">
 <part name="TechException" type="ns2:TechException"/>
</message>

<message name="DiscountRules_calculateDiscount">
 <part name="cust" type="ns2:Customer"/>
 <part name="prod" type="ns2:Product"/>
</message>

<message name="DiscountRules_calculateDiscountResponse">
 <part name="result" type="ns2:Discount"/>
</message>

<portType name="DiscountRules">
 <operation name="calculateDiscount" parameterOrder="cust prod">
  <input message="tns:DiscountRules_calculateDiscount"/>
  <output message="tns:DiscountRules_calculateDiscountResponse"/>
  <fault name="TechException" message="tns:TechException"/>
 </operation>
</portType>
```

representation and forwards it to the generic RuleML interface of the service. Hence, the SOAP-RuleML Gateway acts as a client to a VSP. The response can then be directly encoded in a SOAP document.

The implementation of the SOAP-RuleML Gateway is based on the JAX-RPC handler mechanism. A JAX-RPC handler can be used to modify SOAP request or response messages on the client and the server side. Figure 4 illustrates ViDRE's SOAP-RuleML Gateway that is implemented as a Web service. The request handler transforms the SOAP document into a semantically equivalent RuleML document that is forwarded to the VSP. The resulting response document is transformed back into a SOAP message and sent back to the callee.

## Managing the Rules

Writing (distributed) facts and rules by hand is a cumbersome and error-prone task. It is even worse if they have to be written in RuleML on the client-side. Therefore, we implemented the ViDRE administration client, a Java Swing GUI, which provides support for authoring distributed rule sets. The client supports all management tasks, and allows a rapid development of rule-based applications. The administration client allows connecting to each service provider via its Web service interface. Although business rules should be editable and auditable by business users, ViDRE's administration tasks are designed specifically for people with little programming skills rather than pure business analysts. The administration of a distributed business rules environment not only requires business knowledge but also fundamental IT knowledge. The following section focuses on ViDRE's administration and maintenance tasks.

## Administration Console

ViDRE's administration client is a Java Swing user interface, designed for handling various administrative tasks, such as managing rule sets. Figure 5 illustrates a screenshot of ViDRE's administration client. It enables the user to perform the following tasks:

- *Create Rule Sets:* a rule set represents a valid RuleML document that can be created from scratch or based on an existing RuleML document. ViDRE's administration client supports the creation of RuleML facts and rules. In general, facts are represented by Java objects that can be loaded facts from serialized Java objects or by importing Java class files that can be edited within the application. Rules can be constructed nearly the same way. Rule premises consist of patterns also represented by Java objects augmented with search constraints. Remote rules and facts can be specified by adding a URL to the consequence of the rule. The URL causes ViDRE to execute the consequence on a remote rule engine.

*Figure 4. SOAP - RuleML Gateway*

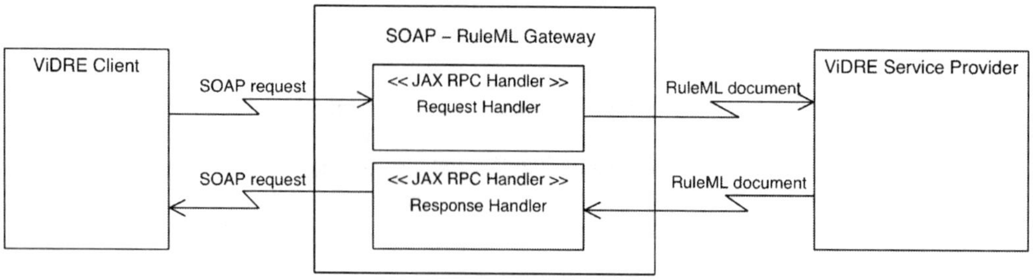

*Figure 5. ViDRE administration console*

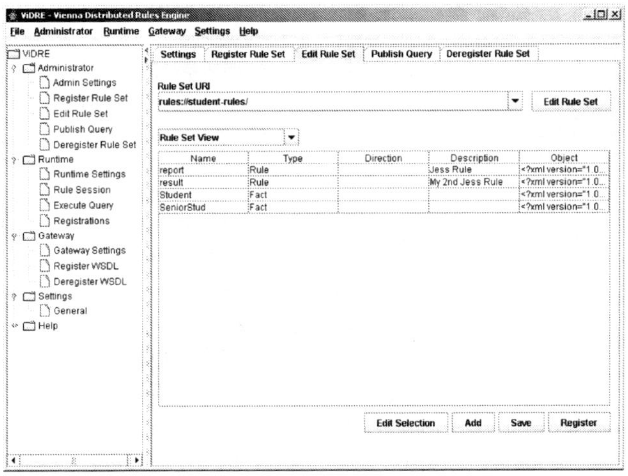


- *Register Rule Sets:* registers a RuleML document containing rules and facts against a specified URI. The rule set can be used and edited after registering the rule set at ViDRE's service provider.
- *Edit Rule Sets:* rule sets are not static. ViDRE's administration client supports the editing of registered rule sets. Rule sets are loaded from the service provider and registered against the same URI after editing.
- *Deregister Rule Sets:* deregisters a rule set at the service provider.
- *Publish Rules and Queries:* in order to expose rules and knowledge as Web services, ViDRE's administration client provides support to automatically publish rules and queries associated with rule sets.
- *List Registrations*: lists all registrations of a specified ViDRE endpoint.
- *Gateway Configuration:* the SOAP-RuleML Gateway can be easily configured and administrated. Therefore, rules exposed as Web services can be inspected.
- *Execute Rules and Queries:* for debugging rules and queries, the administration client offers an interface for creating rule sessions and executing rules and queries.

## Defining Local and Remote Rules

As afore mentioned, defining distributed rules hand-crafted is a cumbersome task. Therefore, ViDRE's administration client supports the creation and deployment of local and remote rules.

Figure 6 illustrates ViDRE's rule editor for the RuleML rule presented in Listing 10. The rule editor is a straightforward implementation that helps to create and edit RuleML rules. A rule can be created in a constructive way, without writing a line of RuleML. Normally a production rule consists of two major parts, the premises (also called preconditions) and the consequences. ViDRE's rule editor tries to reflect this rule configuration by providing the ability to add conditions to the body of the rule and to specify the consequence part of the rule.

Conditions apply to Java facts, hence ViDRE builds conditional patterns out of serialized Java objects or class files similar to the creation of facts. Each field specified in a conditional pattern can be a constant (individual) or a variable. In order to specify relations between different patterns or to specify constraints on variable fields constraints can be added to the condition part of the rule.

In our example rule the conditional pattern is the student object with the two variables name

*Figure 6. ViDRE Rule Editor*

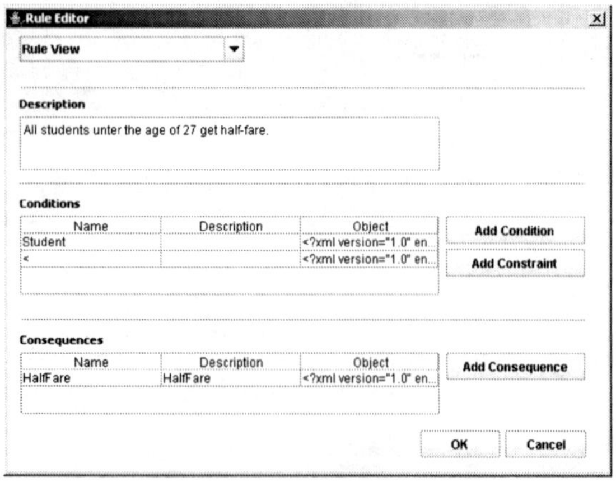


and age. Thus this pattern would match all student objects. In order to match only students younger than 27, we have to add a constraint to the age variable. ViDRE's rule editor only supports simple constraints to pattern variables. Complex relations and constraints have to be encoded in the RuleML document. Therefore, the rule editor provides a RuleML view that can be used to write more complex rule patterns in RuleML directly.

ViDRE's rule editor supports two different consequence types. The derivation rule is characterized by the fact that the consequence consists of a newly derived fact and secondly, that the consequence represents an action. ViDRE's RuleML assembler does not distinguish between these two types, thus on the server-side all facts named after a function are traded as functions.

In section "Distributed Business Rules" we introduced the concept of distributed inference with meta-rules. In order to create these distributed rules each pattern or consequence can be augmented with a URI. The user does not have to write any meta-rules or facts. Behind the scenes each URI is automatically encoded as RuleML fact like in Listing 10. The creation of meta-rules is automatically done by the RuleML transformer on the server-side.

```
<Atom>
  <Rel>URI</Rel>
  <slot>
    <Ind>name</Ind>
    <Ind>Student</Ind>
  </slot>
  <slot>
    <Ind>uri</Ind>
    <Ind>rules://rome.vitalab.tuwien.ac.at:8080/
    Students</Ind>
  </slot>
  <slot>
    <Ind>type</Ind>
    <Ind>forward</Ind>
  </slot>
</Atom>
```

Listing 10 - URI encoded in RuleML

## Discussion and Limitations

ViDRE was designed with the purpose of providing business rules as "rule services" by hosting rules using a ViDRE service provider allowing a simple interchange of rule engines that execute the rules. Even though, ViDRE is implement in Java and uses the JSR94 interface to comply with industry-related interfaces, the fact that rule services can be accessed using SOAP or plain RuleML makes it easy to use for non-Java clients. When using a rule engine within a VSP, the main constraint is that the rule engine has to be Java-based as our approach uses the JSR94.

One of the main limitations of our approach is its scalability: On the one hand, the use of meta-rules to specify the location of other rules during a distributed rule execution cycle might be cumbersome as the number of distributed rules evolves but on the other hand, they provide the necessary decoupling of business rules and meta-rules to specify the location of other rules using the same syntax. When used in large-scale environments, we envision an automated generation of these meta-rules based on a deployment module that knows where the rules are deployed.

Another important aspect is the use of RuleML as an Interlingua to exchange rules between different ViDRE service providers that may use different rule engines to execute the rules. There is an ongoing standardization effort at the World Wide Web Consortium (W3C) to standardize a Rule Interchange Format (RIF). The current RIF draft (W3C – Rule Interchange Format Working Group, 2008) seems to be very promising, as they can encode different rule types and are also capable of encoding meta-rules. Thus, it would only be an additional effort to use RIF as Interlingua and provide transformers from RIF to the concrete rule engine syntax (e.g., the Jess language).

## Alignment of Business Rules with the SOC Approach

Service-orientation has some distinctive attributes that can, when applied to business rules, foster an easier integration between enterprise applications and business rules. As stated in (Papazoglou, Traverso, Dustdar, & Leymann, 2007, p. 1), services have the following attributes: autonomous, distributed, platform-independent, describable, publishable and loosely coupled.

After showing the details of ViDRE, which combines service-oriented principles with traditional business rules approaches, both approaches are now compared according to the main architectural attributes of SOC. Table 1 summarizes the main points and methods of both approaches.

This comparison shows methods that are used to realize the architectural properties within the SOC approach. In particular we focus on what the business rules approach offers and how it compares or complements the service-oriented paradigm. This shows that ViDRE is feasible and is aligned with the SOC approaches making it a suitable candidate for use within enterprise systems that increasingly build upon these paradigms.

## FUTURE TRENDS

Distributed business rules enable a number of possibilities and application scenarios that can be efficiently implemented using ViDRE. Nevertheless, there are a lot of further research possibilities regarding distributed business rules. First and foremost, the creation and specification of distributed business rules can still be improved. Currently, meta-rules are used to specify the location of another rule that is used for the distributed inference process. This has the advantage that meta-rules are well integrated into the rules schema because they are represented as rules. Nevertheless, specifying the location informa-

*Table 1. Comparing SOC and the business rules approach*

| Attribute \ Approach | Service Oriented Computing | Business Rules |
|---|---|---|
| **distributed** | Web services, RESTful services | Rule services, Decision services |
| **platform-independent** | Programming language independent; support by all major programming languages | Pluggable rule engines in a VSP |
| **describable** | WSDL, WADL (for RESTful Services) | RuleML and WSDL as description languages |
| **publishable** | UDDI, ebXML, custom repository | UDDI, ebXML, custom repository |
| **autonomous & loosely coupled** | Registry support to decouple service provider and consumer; allow to build composite service by assembling services | SOAP-RuleML Gateway for creating loosely coupled rule services; Registry support for decoupling; Meta-rules manage dependencies between distributed rules |

tion of the other VSPs in the meta-rules may lead to some problems in dynamically changing environments.

Secondly, architecting enterprise applications that use business rules with complex logic is still in very early stage. Similar to design patterns in software engineering, such patterns for the use of business rules and especially distributed business rules need to be defined. This will help users to understand the objectives and implications of using it within their applications. The work done by (Graml, Bracht, & Spies, 2007) can be seen as a starting point in the area of business rules in combination with business processes that use rules to outsource decision logic.

## CONCLUSION

The SOC paradigm aims at building flexible software systems and business processes by using well-defined interfaces and communication protocols. However, the actual implementation and especially the business logic of the different services are not defined. This is where the business rules paradigm comes into play, following the concept of rules as a mechanism for separating implementation and business logic which is often desired in software systems.

The ViDRE approach described in this chapter brings together the advantages of both paradigms by providing business rules as services using RuleML as communication protocol, and JSR 94 as API for accessing different rule engines. In contrast to existing approaches which mainly build on centralized rule bases, ViDRE supports distributed rule execution by federating various rule engines. This enables executing business rules on, as well as retrieving facts from remote rule engines by using meta-rules. Furthermore, ViDRE is responsible for transforming rules from Java objects to RuleML on the client side, and from RuleML to the different rule languages (such as Jess) on the service provider side. Our approach performs an efficient runtime transformation of rules using the Jess Java API. The rules are finally exposed as Web services which may be accessed via SOAP-RuleML gateways. Additionally, ViDRE provides an administration console for creating, editing, publishing, and executing rules and queries, as well as registering and deregistering of rule sets at the service provider.

The samples used throughout the chapter have illustrated the usefulness of this approach, while the alignment of business rules with the SOC paradigm is also addressed. The combination of the two paradigms proposed by the ViDRE approach solves some difficulties of existing business rule solutions by enabling the integration with enterprise software systems following the principles of SOA.

## REFERENCES

Cabitza, F., Sarini, M., & Seno, B. D. (2005). Djess - A context-sharing middleware to deploy distributed inference systems in pervasive computing environments. In *Proceedings of the International Conference on Pervasive Services (ICPS'05)* (pp. 229-238). IEEE Computer Society.

Dustdar, S., & Schreiner, W. (2005). A survey on Web Services composition. *International Journal of Web and Grid Services, 1*(1), 1-30.

Forgy, C. L. (1982). Rete: A fast algorithm for the many pattern/many object pattern match problem. *Artificial Intelligence, 19*(1), 17-37.

Friedman-Hill, E. (2003). *Jess in Action - Java Rule-based Systems.* Manning Publications Co.

Graml, T., Bracht, R., & Spies, M. (2007). Patterns of Business rules to enable agile business processes. In *Proceedings of the 11th Enterprise Distributed Object Computing Conference* (pp. 365-378). IEEE Computer Society.

Grosof, B. N., Gandhe, M. D., & Finin, T. W. (2002). SweetJess: Translating DAMLRuleML to JESS. In *Proceedings of the International Workshop on Rule Markup Languages for Business Rules on the Semantic Web.* Italy: Sardinia.

ILOG. (2008). *ILOG Rules*. Retrieved April 19, 2008, from http://www.ilog.com/products/rules/

Jackson, P. (1999). *Introduction to Expert Systems.* Addison Wesely Longman (3rd ed).

JBoss. (2008). *JBoss Drools*. Retrieved April 18, 2008, from http://www.jboss.org/drools/

Jess. (2008). *Jess, the rule engine for the Java platform*. Retrieved November 05, 2008, form http://herzberg.ca.sandia.gov/

JSR 94. (2004, August 04). *JSR94: Java Rule Engine API.* Retrieved April 18, 2008, from http://www.jcp.org/en/jsr/detail?id=94

KnowGravity Inc. (2002, July 26). Retrieved April 2008, from http://www.knowgravity.com/pdf-e/BRA - Concepts English.pdf

Kozlenkov, A., Paschke, A., & Schroeder, M. (2008). Retrieved April 18, 2008, from http://www.prova.ws/

Milanovic, M., Kaviani, N., Gasevic, D., Giurca, A., Wagner, G., Devedzic, V., et al. (2007). Business process integration by using general markup language. In *Proceedings of the 11th IEEE International Enterprise Distributed Object Computing Conference* (pp. 353-364). IEEE Computer Society.

Nagl, C., Rosenberg, F., & Dustdar, S. (2006). ViDRE – A distributed service-

oriented business rule engine based on RuleML. In *Proceedings of the 10th IEEE International Enterprise Distributed Object Computing Conference* (pp. 35-44). IEEE Computer Society.

OASIS. (2007, April 11). *Web services business process execution language version 2.0.* Retrieved April 18, 2008, from http://docs.oasis-open.org/wsbpel/2.0/OS/wsbpel-v2.0-OS.html

OOjDrew. (2008). Retrieved April 18, 2008, from http://www.jdrew.org/oojdrew/

Paschke, A., Boley, H., Kozlenkov, A., Craig, B. (2007), *Rule responder: A RuleML-based pragmatic agent web for collaborative teams and virtual organizations.* Retrieved from http://responder.ruleml.org

Paschke, A., Kozlenkov, A., Boley, H., Tabet, S., Kifer, M., Dean M. (2008). *Reaction RuleML Version 0.3. Retrieved* from http://ibis.in.tum.de/research/ReactionRuleML/index.htm

Papazoglou, M., Traverso, P., Dustdar, S., & Leymann, F. (2007, November). Service-oriented computing: State of the art and research challenges. *IEEE Internet Computing*, 64-71.

Rosenberg, F., & Dustdar, S. (2005a). Business rules integration in BPEL - A service-oriented approach. In *Proceedings of the 7th International IEEE Conference on E-Commerce Technology (CEC'05)* (pp. 476-479). Munich: IEEE Computer Society.

Rosenberg, F., & Dustdar, S. (2005b). Design and implementation of a service-oriented business rules broker. In *Proceedings of the 1st IEEE International Workshop on Service-oriented Solutions for Cooperative Organizations (SoS4CO'05), co-located with the 7th International IEEE Conference on E-Commerce Technology (CEC'05)* (pp. 55-63). Munich: IEEE Computer Society.

Schmidt, R. (2002). Web services based execution of business rules. *International Workshop on Rule Markup Languages for Business Rules on the Semantic Web.*

Taveter, K., & Wagner, G. (2001). Agent-oriented enterprise modeling based on business rules. In *Proceedings of the 20th International Conference on Conceptual Modeling (ER'01)* (pp. 527-540). London: Springer.

The Buiness Rules Group. (2000, July). *Defining business rules - What are they really?* Retrieved April 10, 2008, from Business Rules Group: http://www.businessrulesgroup.org/first_paper/BRG-whatisBR_3ed.pdf

*The Rule Markup Initiative.* (2001). Retrieved April 18, 2008, from http://www.ruleml.org

*W3C – Rule Interchange Format Working Group.* (2008). Retrieved November 2, 2008 from http://www.w3.org/2005/rules/wiki/RIF_Working_Group

Weerawarana, S., Curbera, F., Leymann, F., Storey, T., & Ferguson, F. D. (2005). *Web services platform architecture.* Prentice Hall.

## KEY TERMS AND DEFINITIONS

**Business Rule:** A statement that defines or constrains some aspect of the business and is intended to assert business structure or to control or influence the behavior of the business.

**Distributed Rule:** A distributed rule is a business rule that is not physically located at one central rule base but spans over multiple rule bases at different rule engines. The execution of a distributed rule involves more that one rule engine.

**Fact:** A fact asserts an association between two or more terms by expressing it as a relationship between them. A term is a word or phrase that has a specific meaning for a given business domain.

**Inference:** The act of reasoning from facts in the knowledge base or evidence.

**Meta-Rule:** It describes the location of other rules that are needed to fully specify a distributed rule. The meta-rule encodes the location of other rules that are needed for the execution of a distributed rule.

**Rule Engine:** A rule engine is a software component that manages business rules and facts, and is further responsible for executing and enforcing the rules.

**RuleML:** The goal of the Rule Markup Initiative is to develop RuleML as the canonical Web language for rules using XML markup, formal semantics, and efficient implementations. RuleML covers the entire rule spectrum, from derivation rules to transformation rules and reaction rules.

**Service:** Services are autonomous, platform-independent entities that can be described, published, discovered, and loosely coupled in novel ways. They perform functions that range from answering simple requests to executing sophisticated business processes requiring peer-to-peer relationships among multiple layers of service consumers and providers. Any piece of code and any application component deployed on a system can be reused and transformed into a network-available service.

# Chapter XX
# Sharing Ontologies and Rules Using Model Transformations

**Milan Milanović**
*University of Belgrade, Serbia*

**Dragan Djurić**
*University of Belgrade, Serbia*

**Dragan Gašević**
*Athabasca University, Canada*

**Vladan Devedžić**
*University of Belgrade, Serbia*

## ABSTRACT

*Web Ontology Language (OWL), Semantic Web Rule Language (SWRL) and Model-Driven Engineering (MDE) are technologies being developed in parallel, but by different communities. They have common points and issues and can be brought closer together. Many authors have so far stressed this problem and have proposed several solutions. The result of these efforts is the recent OMG's initiative for defining an ontology development platform. However, the problem of transformation between Semantic Web ontology and rule languages and MDE-based languages has been solved using rather partial and ad hoc solutions, most often by XSLT. In this paper, we relations between the Semantic Web languages and MDE-compliant languages as separate technical spaces. In order to achieve a synergy between these technical spaces, we present ontology and rule languages in terms of MDE standards, recognize relations between the OWL and SWRL langauges and MDE-based ontology languages, and propose mapping techniques. In order to illustrate the approach, we use an MDE-defined architecture that includes the ontology and rule metamodels and ontology UML Profile. We also show how MDE techniques, such as model transformations, can be used to enable sharing rules and ontologies by using REWERSE Rule Markup Language (R2ML), a proposal for a general rule language. The main benefit of this approach is that it keeps the focus on the language concepts (i.e., languages' abstract syntax - metamodels) rather than on technical issues caused by different concrete syntax. Yet, we also provide transformations that bridge between both languages' concrete (XML) and abstract (MOF) syntax.*

## INTRODUCTION AND MOTIVATION

The Semantic Web is based on the use of ontologies that should provide an explicit definition of the domain conceptualization. Employing the rich AI research experience and being driven by practical needs for the use on the Web, the W3C has adopted the Web Ontology Language (OWL) as a standard ontology language (Bechhofer et al., 2004). Although the adoption of OWL means that Semantic Web applications can exchange their ontologies and tool vendors can develop reasoners and query languages over OWL, there is also a need to have some other mechanisms for defining knowledge. This is mainly manifested through advanced mechanisms for enriching ontologies by using rules. Thus, we should also define a standardized Semantic Web rule language that will be based on OWL to provide an additional reasoning layer on top of OWL. On the other hand, there are many Semantic Web applications that might use (OWL) ontologies whose business logic is implemented by using various rule languages (e.g., F-Logic, Jess, and Prolog) (Sheth et al., 2006). In this case, the primary goal is to have a rule exchange language for sharing rules, and hence enabling reusability of their business logics.

The above arguments motivated the research in the (Semantic) Web community to look at their different aspects. The most important proposal for the first group of rule language is Semantic Web Rule Language (SWRL) (Horrocks et al., 2004) that tends to be a standardized reasoning layer built on top of OWL. However, this is just one submission to such a language, while the research in Semantic Web services (e.g., WSMO and SWSL) introduces/relies on other rule languages besides SWRL such as SWSL-Rules or F-Logic (Sheth et al., 2006). In fact, this can be addressed by the second group of research efforts for Semantic Web rules manifested in the Rule Interchange Format (RIF) initiative (Ginsberg, 2006), which tries to define a standard for sharing rules. That is, RIF should be expressive enough, so that it can represent concepts of various rule languages. Besides RIF, one should also develop a (two-way) transformation between RIF and any rule language that should be shared by using RIF. Currently, there is no official submission to RIF, but RuleML (Hirtle et al., 2006) and the REWERSE I1 Rule Markup Language (R2ML) (Wagner et al., 2006) are two well-known RIF proposals.

Researchers have also been trying to integrate the ongoing software engineering efforts with the concept of the Semantic Web for a while (Kogut, 2002). The main question they want to answer is how to develop the Semantic Web ontologies enriched by rules using well-accepted software engineering languages and techniques in order to have a large number of practitioners developing and using ontologies in real world applications. Many researchers have previously suggested using UML in order to solve this problem. However, UML is based on object-oriented paradigm, and has some limitation regarding ontology development (e.g., properties in ontology languages are first-class concepts, while UML properties (i.e. attributes and associations) are defined in the scope of a class they belong to (Baclawski, 2002). Furthermore, UML classes and their inheritance cover behavioral characteristics of abstractions[1]. For a detail overview of differences between UML and ontology language, see (Baclawski, 2002). Hence, we can only use UML in initial phases of ontology and rule development. We believe that these limitations can be overcome using Model Driven Engineering (MDE) techniques (Kent, 2002). In addition, if we want to offer a solution consistent with MDE, we should also support automatic generation of completely operational ontology definitions with rules (e.g. in OWL/SWRL languages) that are model driven (Selic, 2003). Currently, the most important direction toward this goal is the one pursued by a dedicated research group within OMG that tries to converge many different proposals of solutions to this problem (ODMrfp, 2003). The result of this effort should be a standard language (i.e., a metamodel) based on the MDE and the W3C Web Ontology Language (OWL) recommendation (Bechhofer et al., 2004).

In this chapter, we analyze how MDE principles can be leveraged to bridge between SWRL/OWL abstract and concrete syntax. We also show how model transformations between R2ML and SWRL/OWL can be used to enable interchanging OWL ontologies enriched with SWRL rules with various other rule languages (e.g., Object Constraint Language - OCL) via R2ML. However, the presented solution is completely based on the abstract syntax of both languages, unlike other similar approaches that mainly focus on a concrete syntax (mainly XML-Schema-based) without efficient mechanisms to check whether the implemented transformations are valid w.r.t. the abstract syntax. The problem of sharing ontologies and rules is further hampered by the lack of a transformation language for the Semantic Web. Motivated by our positive experience with the recent OMG standard of the Ontology Definition Metamodel (ODM) (Gašević et al., 2006; ODM, 2006), we propose using MDE principles and model transformations to address this issue. Consequently, we define the abstract syntax of R2ML and OWL/SWRL by means of metamodels. While R2ML is actually fully built by using metamodeling principles, SWRL and OWL are not. Thus, we leverage the OMG's standard meta model for OWL (i.e., ODM) and a comprehensive metamodel for SWRL, the Rule Definition Metamodel (RDM) proposed in (Brockmans & Haase, 2006). As SWRL is based on OWL, we adapted RDM to be based (i.e., to inherit concepts) on the standard ODM. Our solution also covers mappings between the abstract syntax of both languages and their XML-based concrete syntax, but this is completely decoupled from the transformations between R2ML and RDM (ODM).

## BACKGROUND

In this section, we describe the basic principles of MDE (e.g., metamodeling, MDA, and model transformations), how MDE standards are used for defining Semantic Web rule and ontology languages.

### Model Driven Engineering: A Brief Introduction

Model-Driven Engineering (MDE) is a new software engineering discipline in which the process heavily relies on the use of models (Bézivin, 2005). Models are the central MDE concepts and they are specified by using modeling languages (e.g., UML or ODM), while modeling languages are defined by metamodels. A metamodel is a model of a modeling language. That is, a metamodel makes statements about what can be expressed in the valid models of a certain modeling language (Seidewitz, 2003). The OMG's Model Driven Architecture (MDA) is a possible architecture for MDE (Miller & Mukerji, 2003). MDA consists of three layers, namely: M1 (model) layer for defining models of systems under study; M2 (metamodel) layer for defining model languages (e.g., ODM is defined on this layer); and M3 (metametamodel) layer where only one metamodeling language is defined (i.e. MOF) (MOF, 2006). The relations between different MDA layers can be considered as instance-of or conformant-to, which means that a model is an instance of a metamodel, and a metamodel is an instance of a metametamodel. The MDA architecture uses XML Metadata Interchange (XMI), the OMG's standard that defines mappings of MDA-based metametamodels, metamodels, and models onto XML documents and XML Schemas (XMI, 2005). Another MDE architecture is Eclipse Modeling Framework (EMF), which is different from MDA just in using Ecore on the M3 layer instead of MOF.

Model transformations play an important role and represent the central operation for handling models in the MDA (Miller & Mukerji, 2003). Model transformations are the process of producing one model from another model of the same system (Miller & Mukerji, 2003). The OMG adopted the MOF2 Query View Transformation (QVT) specification (QVT, 2005) to address this need. In our research, we have decided to use ATLAS Transformation Language (ATL, 2008) as the primary language and tool for model transformations, as it is one of the most important contributions to QVT, is the official Eclipse recommendation for model-to-model transformations, and yet is an open-source solution. However, the actual ATL implementation is different from the QVT standard. ATL integrates an important concept – technical spaces (Kurtev et al., 2002), where a technical space (TS) is a working context with a set of associated concepts, body of knowledge, tools, required skills, and possibilities. Although some technical spaces are difficult to define, they can easily be recognized (e.g., XML, MDA, databases, and Semantic Web). In our case, this is important when bridging between abstract and concrete syntax of the same rule language (e.g., between MOF-based based SWRL metamodel and SWRL XML-Schema). This practically means that ATL has tools for automatic injection/extraction of XML rules into/from the MOF representation. Therefore, we can use the same transformation language for bridging between a language's concrete and abstract syntax and between abstract syntax of different languages, which reduces the learning curve of technologies needed to provide a complete solution.

The declarative part of ATL is based on *matched rules* notation. Such a rule consists of an input pattern that is connected to source models and an output pattern that is created in target models. Figure 1 shows an ATL rule, with the XML metamodel *Root* element in its input pattern, and with the RDM metamodel *OWLOntology* element in its output pattern.

ATL has two imperative constructs: *called rule* and *action block*. A called rule is invoked as a procedure, whereas action blocks are sequences of imperative constructs that can be used in both matched and called rules. An example of a called rule and an action block is shown in Figure 2. The rule in this figure defines the input point for transformation, creates a Metamodel element, and assigns it to the *metamodel* variable, which is located in the actual transformation (module).

ATL also support *unique lazy* rules. They are "lazy", since they invoke explicitly from some other rule. They are "unique", since they only create an element of the target model once for a specific element of the source model, while all other executions of the same rule for that specific input element create references to the only target definition. This type of rules substitutes called templates in XSLT with the main distinction that XSLT does not posses any mechanism for unique element definitions, and is very hard to implement in XSLT (Jovanović & Gašević, 2005). The unique lazy ATL rule that transforms *DataVariables* from the XML metamodel into the RDM metamodel is shown in Figure 3. It transforms an Element of the XML metamodel (XML!Element) into the *DataVariable* element of the RDM metamodel (RDM!DataVariable) following all conditions given above.

*Figure 1. An example of ATL matched rule*

```
rule Ontology {
       from    i : XML!Root ( i.name <> 'ruleml:imp' )
       to      o : RDM!OWLOntology (
                     elements <- i.children->select(c | c.oclIsKindOf(XML!Element))
               )
}
```

*Figure 2. A called rule and an action block in ATL*

```
entrypoint rule Metamodel() {
        to
                t : KM3!Metamodel
        do {
                thisModule.metamodel <- t;
        }
}
```

*Figure 3. An example of ATL unique lazy rule*

```
unique lazy rule DataVariable {
        from i : XML!Element
        to
                o : RDM!DataVariable (
                        name <- i.children->first().value
                )
}
```

ATL language also include notion of *helpers* which can be viewed as the ATL equivalent to Java methods. They make it possible to define factorized ATL code that can be called from different points of an ATL transformation. In Figure 4 we show an example of ATL helper, which returns a sequence of all XML elements which name is 'ruleml:imp'.

## Ontologies

There is an ongoing effort by OMG that tries to define a suitable language for modeling Semantic Web ontology languages in the context of MDA, which is currently near the completion of the first version (ODM, 2006). This approach of ontology modeling in the scope of MDA, which is shown in Figure 5, defines several specifications:

- Ontology Definition Metamodel (ODM).
- Ontology UML Profile – a UML Profile that supports UML notation for ontology definition.
- Two-way mappings between: OWL and ODM, ODM and other metamodels, ODM and Ontology UML Profile and from Ontology UML Profile to other UML profiles.

Ontology Definition Metamodel (ODM) is designed to comprehend common ontology concepts. ODM is a metamodel of the OWL ontology language. OWL is at the Logical layer of the Semantic Web, on top of RDF Schema (Schema layer). In order to make use of graphical modeling capabilities of UML, an ODM should have a corresponding UML Profile. This profile enables graphical editing of ontologies using UML diagrams as well as other benefits of using mature UML CASE tools.

Both UML models and ODM models are serialized in the XMI format, so that the two-way transformation between them can be done by using XSLT or model transformation languages such as ATL. OWL also has representation in the XML format, so another pair of XSL Transformations or model transformations should be provided for two-way mapping between ODM

*Figure 4. An example of ATL helper*

```
helper def: getAllRules() : Sequence(XML!Element) =
        XML!Element.allInstances()->select(e | e.name = 'ruleml:imp')->asSequence();
```

Figure 5. The Semantic Web and MDA TS

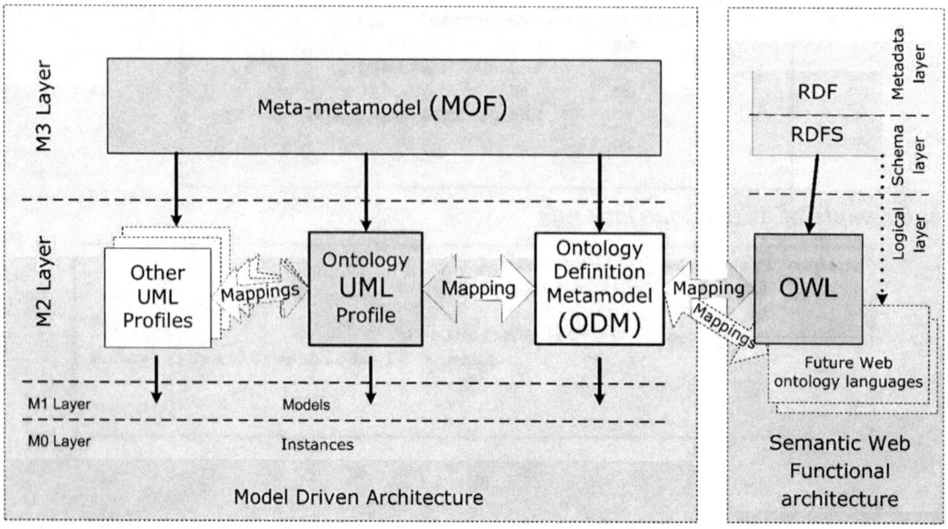


and OWL. For mapping from the ODM into other metamodels or from Ontology UML Profile into another, technology-specific UML Profiles, additional transformations can be added to support usage of ontologies in design of other domains and vice versa.

In its finally adopted version, ODM is organized as a composition of several metamodels. We show the architecture of the current ODM submission in Figure 6. In order to support well-known Semantic Web ontology languages, ODM has two separate metamodels, namely OWL and RDF. These languages are W3C standards which form a basis for the Semantic Web, thus are the central part of ODM. Other metamodels have 2-way mappings to and from RDFS/OWL. That means that they are also used as a mediator between other metamodels.

Since ODM needs an expressive logics language that can be used on the Semantic Web to describe string-based expressions, the ODM has the Common Language (CL) metamodel. Apart from W3C ontology languages, ODM supports other ontology language standards (ISO) by having defined the Topic Maps (TM) metamodel. Topic Maps is a metalanguage designed to express the "aboutness" of an information structure with key model elements *topic* and *association*. Common Logic represents a family of knowledge representation languages. Common Logic, is a first order logic, analogous to predicate calculus, and is the successor to KIF (Knowledge Interchange Format). Both Topic Maps and CL have XML serializations, and were designed to express semantics for knowledge exchanged over the World Wide Web. These languages overlap with some parts of OWL as might be expected, but are used for different purposes and have different or no requirements for automated reasoning. CL is more expressive than OWL, and is better suited to applications involving declarative representation of rules and processes, for example. The Description Logics (DL) metamodel, which was planned to be a central metamodel in the early versions of the ODM draft, has been moved a bit on the side, being left as a non-normative metamodel of the ODM specification, while his central part in ODM is now filled by the RDFS and OWL metamodels.

Taking into account the importance of UML ODM developers want to create mappings between the standard UML metamodel and the RDFS and

*Figure 6. The organization of ontology definition metamodel*

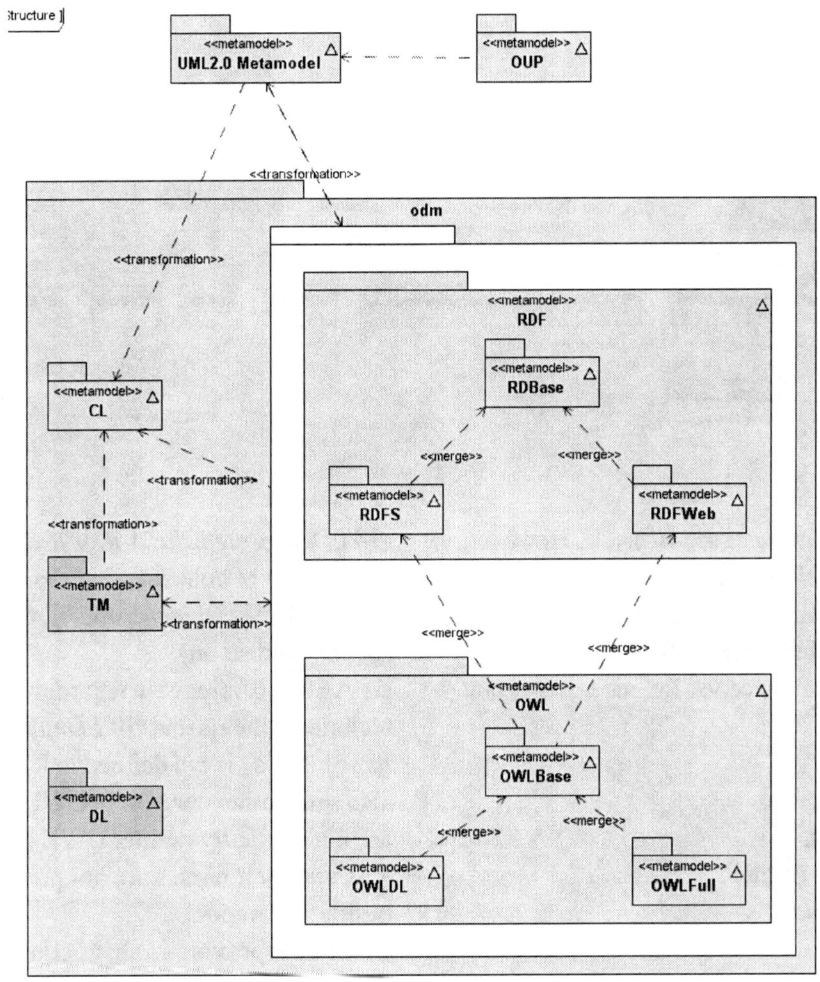


OWL metamodel, and hence establish a way for employing present UML models in ontology development. Furthermore, they define Ontology UML Profile (OUP) using standard UML extension mechanisms. The purpose of OUP is to enable the use of the standard UML graphical notation for developing ontologies. *Resource* is one of the basic RDFS concepts; it represents all things described by RDF and OWL. It may represent anything on the Web: a Web site, a Web page, a part of a Web page, or some other object named by URI. Compared to ontology concepts, it could be viewed as a root concept, the Thing. In RDFS TS, *rdfs:Resource* is defined as an instance of *rdfs:Class*. Since we use Meta-Object Facility (MOF) as a metamodeling language, this concept will be defined as an instance of MOF Class named *RDFSResource*. It is the root class of most other concepts from RDFS and OWL metamodels that will be described: *RDFSClass*, *RDFSProperty*, *RDFStatement* etc. The RDFS metamodel concepts hierarchy of is shown on class diagram in Figure 7. More details about these concepts can be found in (Gašević et al., 2006). Note that, since RDFS is self-defined and *rdfs:Class* inherits *rdfs:Resource*, in RDFS TS

*Figure 7. The hierarchy of concepts in RDF metamodel*

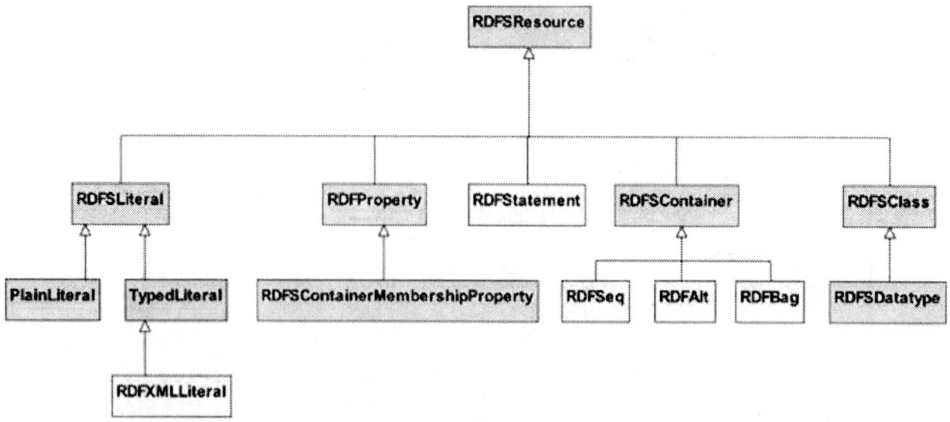


*rdfs:Resource* is an instance of itself. However, in MOF TS, *RDFSResource* is an instance of MOF *Class*, not RDFS *Resource*, because MOF is meta-metamodel, not RDFS.

Among shown concepts, the most important ones are:

- *RDFSResource*;
- *RDFSClass*;
- *RDFProperty*, and
- *RDFStatement*.

They are the base for forming Subject-Predicate-Object triples, like "this section – describes – RDFS metamodel".

OWL is built on top of RDF(S), using RDF(S) as both metametamodel (M3) and a metamodel that is a base for extension (M2). In MDE, the first dependency (RDFS as a meta-metamodel) is replaced by using MOF as a metametamodel. The second dependency means that OWL metamodel concepts extend RDFS metamodel concepts. Figure 8 shows the hierarchy or OWL concepts – we can see that most OWL concepts inherit RDFS concepts *RDFSResource*, *RDFProperty* and *RDFClass*. *RDFSClass* is a base concept of OWL concepts that represents classes (*OntologyElement, OWLClass*, *OWLRestriction* and *OWLDeprecatedClass*), *RDFProperty* is inherited by a plenty of concepts that represent properties in OWL (*OWLObjectProperty*, *OWLDatatypeProperty* and so on).

A little difference in regards to OWL, in RDFS technical space is that *OWLOntology* inherits *Ontology*, which is not defined in RDF(S). There are also some other concept in OWL metamodel that are not explicitly definer OWL language (RDFS TS), and their names are not prefixed with OWL in this metamodel.

Classes provide an abstraction mechanism for grouping resources with similar characteristics. Like RDF classes, every OWL class is associated with a set of individuals, called the class extension. As OWL in many cases refines the concept of a class, it needs to model it separately, e.g. to inherit *RDFSClass* with a new concept, *OWLClass*.

The OWL metamodel together with the RDFS metamodel makes the base for ontology development in MDE. The main advantage is that it is compatible with OWL, which is one of the key technologies of the Semantic Web.

IBM's tool Integrated Ontology Development Toolkit (IODT) is the most complete implementation of ODM (Pan et al., 2006).

*Figure 8. The hierarchy of concepts in OWL metamodel*

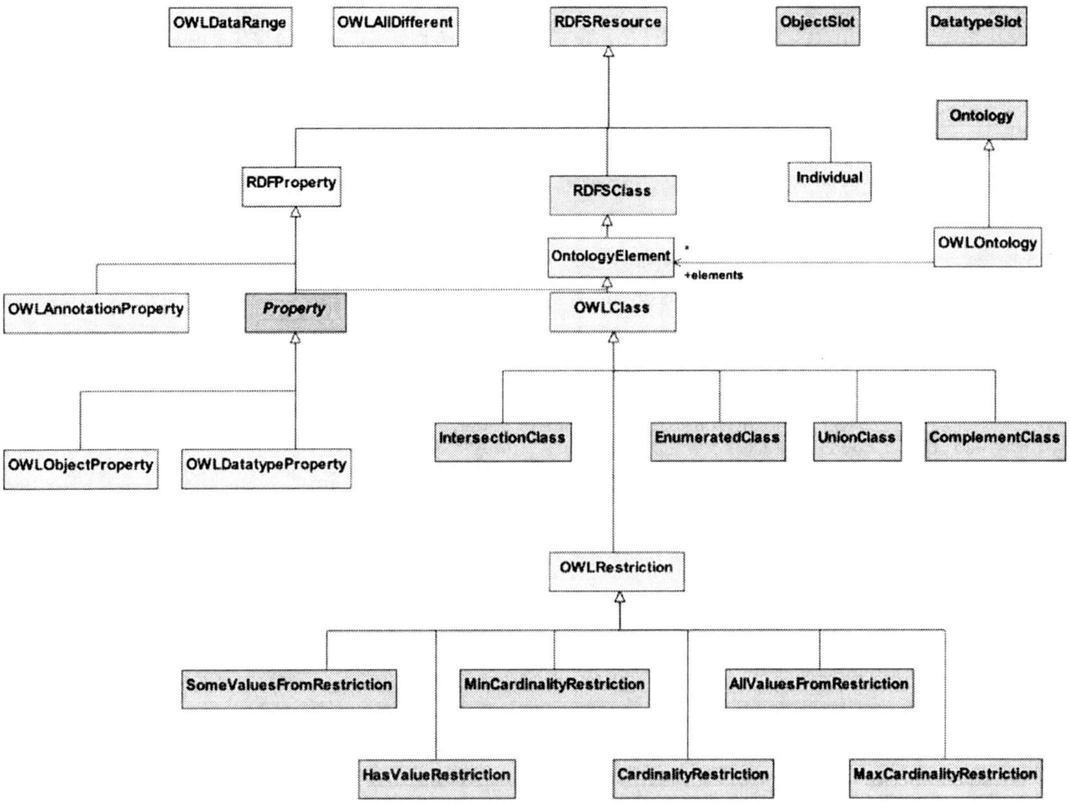


## Rules

In this section we describe rule languages, namely, SWRL, its metamodel RDM, and R2ML.

### Semantic Web Rule Language (SWRL)

The most recent initiative in the W3C consortium for rule languages is Semantic Web Rule Language (SWRL), which actually represents a combination of the OWL and RuleML languages (Horrocks et al., 2004). Like RDF and OWL, SWRL is based on classical first-order logic. SWRL is restricted to OWL DL expressions, so it does not have support for high-level expressions that are allowed in RDF and OWL Full. This language has rules in the form of an implication between an antecedent (body) and a consequent (head). The intended meaning can be read as "whenever the conditions specified in the antecedent hold, then the conditions specified in the consequent must also hold". Both the antecedent (body) and consequent (head) consist of zero or more atoms. Multiple atoms are connected with the conjunction operator. Atoms in these rules can be of the form C(x), P(x,y), sameAs(x,y) or differentFrom(x,y), where C is an OWL description, P is an OWL property, and x,y are either variables, OWL individuals or OWL data values. SWRL has two concrete syntax, namely, the RDF/XML concrete syntax (Becket, 2004) and the OWL XML presentation syntax (Hori et al., 2003).

## Rule Definition Metamodel (RDM)

We have already mentioned that we use RDM as an abstract syntax for the SWRL language, but instead of using it with the ODM proposed in (Brockmans & Haase, 2006), we decided to use the standard ODM metamodel. In this chapter, we do not explain RDM here in detail, but refer readers to the complete reference given in (Brockmans & Haase, 2006). We just explain how RDM defines rules. The RDM *Rule* concept is a subclass of *OntologyElement* (see Figure 9), while *OntologyElement* is defined in the ODM metamodel as an element of the ODM *OWLOntology* class. An RDM (as well as SWRL) rule consists of an antecedent and a consequent, also referred to as the body and head of the rule, respectively. Both the RDM antecedent and consequent consist of a set of atoms which can possibly be empty.

## REWERSE I1 Rule Markup Language

In difference with OWL and SWRL, which are not originally defined by a metamodel, the R2ML is defined by means of a metamodel. The R2ML metamodel is defined by using the MOF metamodeling language. R2ML supports four kinds of rules, namely, integrity rules, derivation rules, production rules, and reaction rules. Since SWRL rules can represent integrity rules, we describe here R2ML integrity rules. An *integrity rule*, also known as (integrity) constraint, consists of a constraint assertion, which is a sentence in a logical language such as first-order predicate logic or OCL (see Figure 10). The R2ML framework supports two kinds of integrity rules: the *alethic* and *deontic* ones. An alethic integrity rule can be expressed by a phrase, such as "*it is necessarily the case that*" and a deontic one can be expressed

*Figure 9. Rule definition metamodel (adapted from (Brockmans & Haase, 2006))*

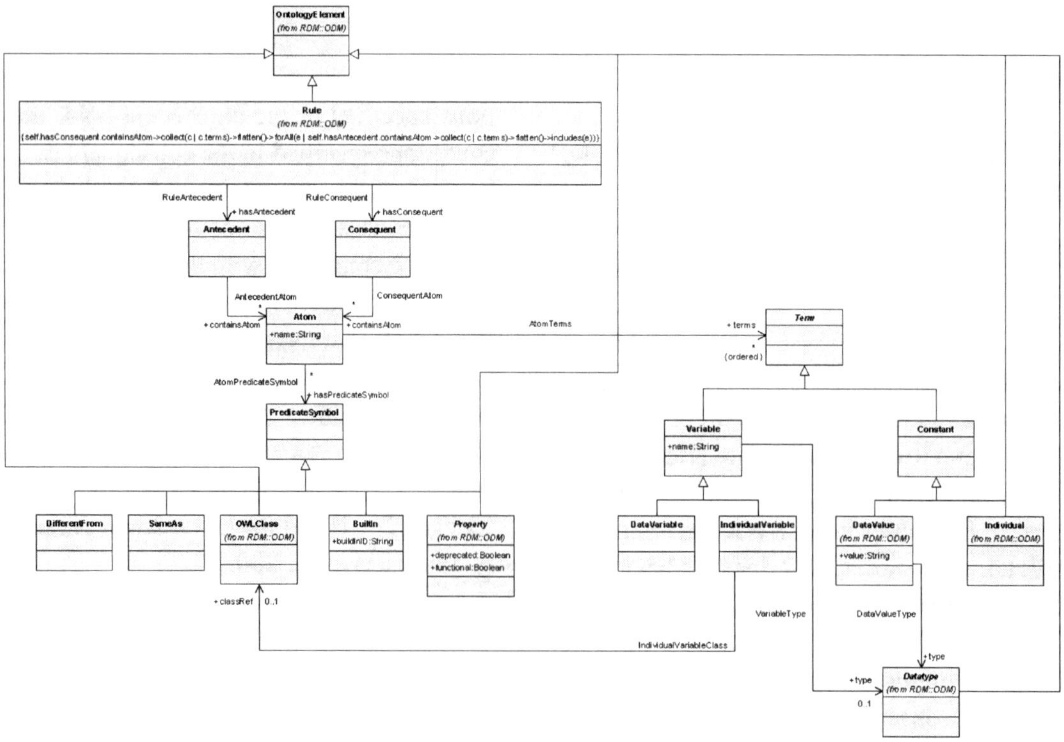

*Figure 10. The metamodel of integrity rules*

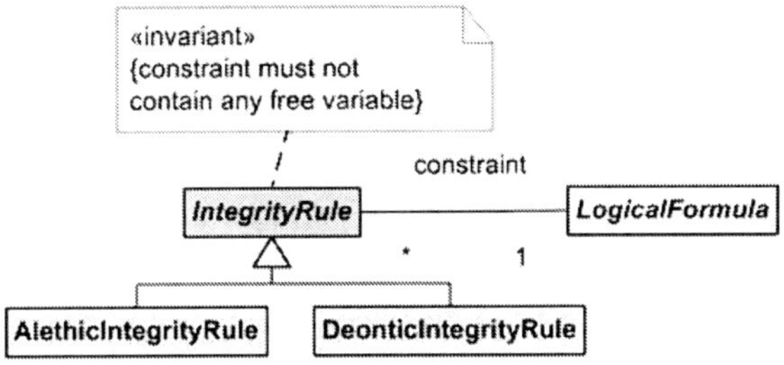


by phrases, such as "*it is obligatory that*" or "*it should be the case that.*"

The corresponding *LogicalFormula* must have no free variables, that is, all the variables from this formula must be quantified. R2ML defines the general concept of *LogicalFormula* (see Figure 11) that can be *Conjunction*, *Disjunction*, *NegationAsFailure*, *StrongNegation*, and *Implication*. The concept of a *QuantifiedFormula* is essential for R2ML integrity rules, and it subsumes existentially quantified formulas and universally quantified formulas. Figure 11 also contains elements such as *AtLeastQuantifiedFormula*, *AtMostQuantifiedFormula*, and *AtLeastAndAtMostQuantifiedFormula* for defining cardinality constraints with R2ML rules.

*Atoms* are basic constituents of formulas in R2ML. Atoms are compatible with all important concepts of OWL/SWRL. R2ML distinguishes object atoms, data atoms, and generic atoms. Here we present just R2ML atoms necessary for our goal, that is, atoms used in SWRL – object and data (see (R2ML, 2008) for a complete description and use of all supported atoms). An R2ML *ObjectClassificationAtom* refers to a class and consists of an object term. Its role is for object classification, i.e., an *ObjectTerm* is an instance of the referred class. An R2ML *AttributionAtom* consists of an object term as "subject", and a data term as "value". Example: startDate(r1, sd), where startDate is an attribute, r1 is an object term and sd is a data term.

## MODEL TRANSFORMATIONS TO SHARE RULES AND ONTOLOGIES

In this chapter, we describe our idea of bridging OWL ontologies enriched with SWRL rules into the Software Engineering world by using MDE techniques.

### Proposed MDE-Based Solution

In order to develop Semantic Web ontologies enriched by rules using well-accepted software engineering languages and techniques and to

*Figure 11. The concept of a logical formula in R2ML*

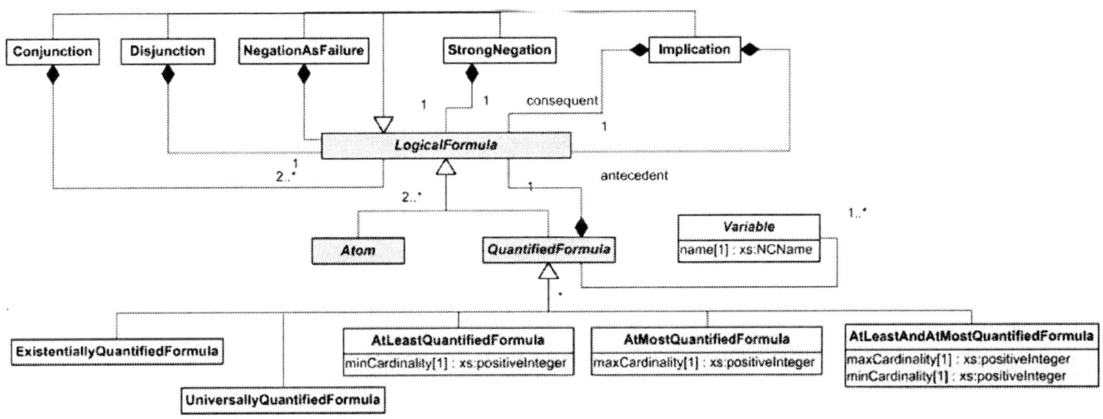

share them with other software environments and at the same time to use existing ontologies and rules in software engineering language, we propose a model transformation-based approach. Our solution consists from the following steps. Firstly SWRL rules represented in the SWRL XML format (SWRL.xsd and OWL.xsd) (Hori et al., 2003) are transformed to models compliant with RDM (Brockmans & Hasse, 2006) (see Figure 12). Since SWRL is actually based on OWL, we include transforming OWL ontologies along with SWRL rules, that is, transforming the OWL XML-based concrete syntax into RDM. This is done by using the XML injection (ATL feature) of OWL/SWRL documents by instantiating the MOF-based metamodel of XML. Then, such MOF-based SWRL rules (XML models) are transformed to models compliant with RDM. This transformation between the XML metamodel and RDM is implemented in both ways, and it is a bridge between SWRL concrete and abstract syntax.

## Transformation between SWRL (OWL) XML Schema and the RDM Model

Transformation of the SWRL rules and OWL ontologies from the SWRL (OWL) XML concrete syntax into the SWRL abstract syntax (i.e., meta-model) consists of two primary steps as follows.

***Step 1.*** This step consists of transformation of the SWRL/OWL rules from the XML technical space into the MOF technical space. SWRL is represented in the XML concrete syntax that represents combination of the OWL XML concrete syntax (Hori et al., 2003) with SWRL XML concrete syntax (Horrocks et al., 2004). In this step, SWRL (OWL) documents (Rules.xml in Figure 12) are transformed in form that conforms to MOF. SWRL/OWL rules are transformed to the RDM metamodels (Brockmans & Haase, 2006), which represent the SWRL/OWL language abstract syntax in the MOF TS. To do this, we first use the XML injector, (see Figure 12, step 1: XML injection), a part of ATL that automatically transforms SWRL XML documents like the one given in Figure 13 (without any manually written transformation) into the models conforming to the MOF-based XML metamodel that defines

*Figure 12. The transformation scenario: the OWL/SWRL XML format into the instances of the RDM metamodel*

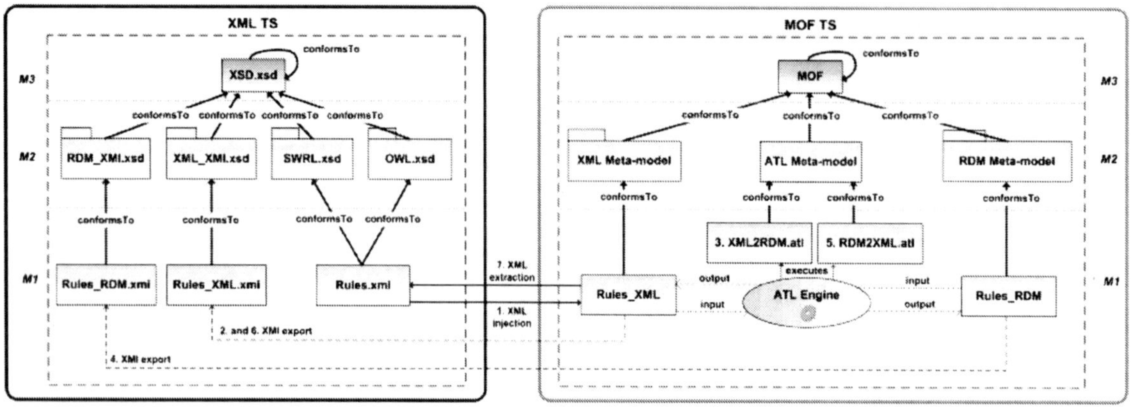

*Figure 13. A SWRL rule in XML-based concrete syntax*

```
<swrlx:OWLOntology swrlx:name = "Ontology">
  <ruleml:Implies>
      <owlx:Annotation>
          <owlx:Documentation>Gold customers get a 10% discount
                              on purchases of $500 or more
          </owlx:Documentation>
      </owlx:Annotation>
      <ruleml:body>
         <swrlx:individualPropertyAtom swrlx:property="hasStatus">
            <ruleml:var>customer</ruleml:var>
            <owlx:Individual owlx:name="gold"/>
         </swrlx:individualPropertyAtom>
         <swrlx:datavaluedPropertyAtom swrlx:property="hasTotalPurchase">
            <ruleml:var>customer</ruleml:var>
            <ruleml:var>total</ruleml:var>
         </swrlx:datavaluedPropertyAtom>
         <swrlx:builtinAtom swrlx:builtin="greaterThanOrEqual">
            <ruleml:var>total</ruleml:var>
            <owlx:DataValue owlx:datatype="int">500</owlx:DataValue>
         </swrlx:builtinAtom>
      </ruleml:body>
      <ruleml:head>
         <swrlx:datavaluedPropertyAtom swrlx:property="hasDiscount">
            <ruleml:var>customer</ruleml:var>
            <owlx:DataValue owlx:datatype="int">10</owlx:DataValue>
         </swrlx:datavaluedPropertyAtom>
      </ruleml:head>
   </ruleml:Implies>

   <owlx:ObjectProperty owlx:name="hasStatus">
       <owlx:domain>
            <owlx:Class owlx:name="Customer"/>
       </owlx:domain>
       <owlx:range>
            <owlx:Individual owlx:name="gold"/>
       </owlx:range>
  </owlx:ObjectProperty>
</swrlx:OWLOntology>
```

XML (e.g., Node, Element, and Attribute). Once we inject SWRL XML rules into a MOF-based representation (Rules_XML in Figure 12), we can manipulate with them like with any other type of MOF-based models. Thus, such XML models can be represented in the XMI. XMI format (in Figure 12, step 2: XMI export). This is again an integrated ATL feature that requires no manual work

For OWL ontology enriched by SWRL rule defined in Figure 13, after XML injection into the MOF technical space, we attained XML model as an instance of the XML metamodel, which is represented in the XML XMI format in Figure 14. This rule is contained in ontology (with the name "Ontology") and it is defined as: *Gold customers get a 10% discount on purchases of $500 or more.*

***Step 2.*** In this step, we transform between XML models (Rules_XML from Figure 12) and RDM-compliant models (Rules_RDM from Figure 12). This actually requires developing two ATL transformations (Figure 12, step 3: XML2RDM.alt and step 5: RDM2XML.atl), and hence this is the bridge between the SWRL XML-based concrete syntax and SWRL abstract syntax. Both transformations are executed on the M1 level, but they require the input and output models to be compliant to the input and output metamodels (i.e., XML and RDM), respectively. This way we check validity of all input SWRL XML-based rules w.r.t. the RDM metamodel. Since we have implemented transformations in both directions, we can transform RDM rules into the XML models, that can be later exported into SWRL XML concrete syntax (Figure 12, step 6: XML export) and obtain

*Figure 14. XML XMI representation of the SWRL rule shown in Figure 13*

```
<XML.Root xmi.id = 'a1' name = 'ruleml:imp' value = ''>
    <XML.Element.children>
      <XML.Element xmi.id = 'a2' name = 'ruleml:_body' value = ''>
        <XML.Element.children>
          <XML.Element xmi.id = 'a3' name = 'swrlx:builtinAtom' value = ''>
            <XML.Element.children>
              <XML.Attribute xmi.id = 'a4' name = 'swrlx:builtin' value = 'greaterThanOrEqual'/>
              <XML.Element xmi.id = 'a5' name = 'owlx:DataValue' value = ''>
                <XML.Element.children>
                  <XML.Attribute xmi.id = 'a6' name = 'owlx:datatype' value = 'int'/>
                  <XML.Text xmi.id = 'a7' name = '#text' value = '500'/>
                </XML.Element.children>
              </XML.Element>
              <XML.Element xmi.id = 'a8' name = 'ruleml:var' value = ''>
                <XML.Element.children>
                  <XML.Text xmi.id = 'a9' name = '#text' value = 'total'/>
                </XML.Element.children>
              </XML.Element>
            </XML.Element.children>
          </XML.Element>
          <!-- ... -->
        </XML.Element.children>
      </XML.Element>
    </XML.Element.children>
</XML.Root>
```

the rules in the form given in Figure 13, which is important when transforming SWRL rules into the SWRL XML concrete syntax. Note also that once we transform SWRL rules into the RDM representation, we can also export SWRL rules in the RDM XMI format (Figure 12, step 7: XMI export), and thus we can share SWRL rules with any MOF-compliant repository. This is another important contribution to the RDM metamodel itself (Brockmans & Hasse, 2006) that improves its practical value to be used by other MOF-based tools. Finally, say that we have also changed the RDM metamodel (Brockmans & Hasse, 2006), so that it is now based on our proposal for ODM metamodel instead of being based on the one defined in (Brockmans & Hasse, 2006).

The SWRL rule, shown in Figure 13 is transformed from the XML TS into the MOF TS by using ATL's XML injector (1. XML injection is represented in Figure 12). Subsequently, this rule can be exported into the XML XMI format by using ATL tools, as it is shown in Figure 14. This transformation step is shown in Figure 12 (1. XML injection and 2. XMI Export). In the XML2RDM.atl transformation, source elements from the XML model are transformed to the target elements of the RDM model. XML2RDM.atl transformation is done on the M1 level (i.e., the model level) of the MDA, and it uses the information about elements from the M2 (metamodel) level (i.e., the XML and RDM metamodels).

The XML model shown in Figure 14 (i.e., the model is shown in the XMI-based concrete syntax of the XML metamodel) can be transformed to the corresponding RDM model, by using the XML2RDM.atl transformation shown in Figure 12 (step 3. XML2RDM.atl). The result of executing this transformation on the XML model shown in Figure 14 is the RDM model shown in Figure 15.

In this transformation, the XML model elements shown in Figure 14 are transformed to the equivalent RDM model elements (it also includes ODM model elements). This transformation is shown in Figure 15. Table 1 shows an excerpt of mappings between the SWRL XML Schema (with OWL XML Schema), XML metamodel and RDM metamodel.

Mappings between elements of the XML metamodel and elements of the RDM metamodel are

*Figure 15. RDM model and RDM XMI representation of the transformed XML model from Figure 14*

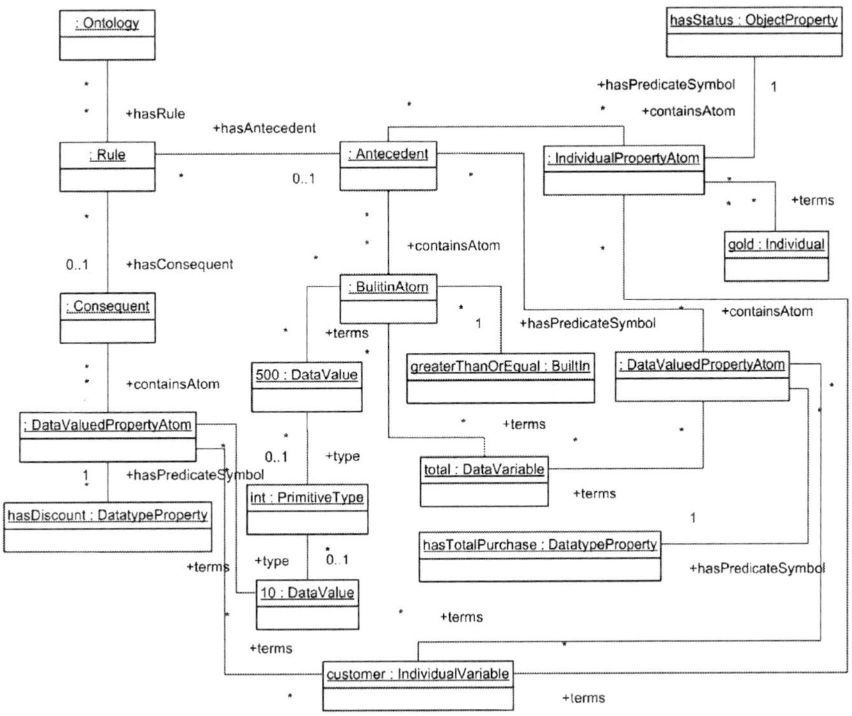


*a) UML object diagram of RDM model transformed from the XML model which is shown in Figure 14*

```
<RDM.ODM.Ontology>
  <RDM.ODM.Ontology.hasRule>
    <RDM.Rule xmi.id = 'a2'>
      <RDM.Rule.hasAntecedent>
        <RDM.Antecedent xmi.idref = 'a3'/>
      </RDM.Rule.hasAntecedent>
      <RDM.Rule.hasConsequent>
        <RDM.Consequent xmi.idref = 'a4'/>
      </RDM.Rule.hasConsequent>
    </RDM.Rule>
  </RDM.ODM.Ontology.hasRule>
</RDM.ODM.Ontology>
<RDM.Antecedent xmi.id = 'a3'>
  <RDM.Antecedent.containsAtom>
     <RDM.Atom xmi.idref = 'a7'/>
     <RDM.Atom xmi.idref = 'a8'/>
     <RDM.Atom xmi.idref = 'a9'/>
  </RDM.Antecedent.containsAtom>
</RDM.Antecedent>
<!--...-->
<RDM.Atom xmi.id = 'a9' name = 'IndividualPropertyAtom'>
  <RDM.Atom.terms>
     <RDM.IndividualVariable xmi.idref = 'a1'/>
     <RDM.ODM.Individual xmi.idref = 'a10'/>
  </RDM.Atom.terms>
  <RDM.Atom.hasPredicateSymbol>
     <RDM.ODM.ObjectProperty xmi.idref = 'a11'/>
  </RDM.Atom.hasPredicateSymbol>
</RDM.Atom>
```

*b) RDM XMI representation of the transformed XML model which is shown in Figure 14*

*Table 1. An excerpt of mappings between the OWL/SWRL XML schema, XML meta-model and the RDM meta-model*

| SWRL XML Schema | XML meta-model | RDM meta-model | Description |
|---|---|---|---|
| individualProperty Atom | Element name ='swrlx:individualProperty Atom' | Atom | Atom which contain property name and two elements that can be individual, variable or value. |
| OneOf | Element name = 'owlx:OneOf' | EnumeratedClass | Contain individual enumeration. |
| var | Element name = 'ruleml:var' | IndividualVariable | Represents a variable. |
| sameIndividualAtom | Element name = 'swrlx:sameIndividualAtom' | Atom | Denotes equality between set of individuals or variables. |
| maxcardinality | Element name = 'owlx:maxcardinality' | MaxCardinality Restriction | Declares restriction on maximal cardinality value. |

defined as a sequence of rules in the ATL language. These rules use additional helpers in defining mappings (similar to functions in programming languages). Each rule in the ATL has one input element (i.e., an instance of a meta-class from a MOF based metamodel) and one or more output elements. In fact, the ATL transformation takes an input XML model from a model repository and creates a new model compliant to the RDM metamodel. In the XML2RDM.atl transformation, we mainly use ATL matched rules. A matched rule matches a given type of source model element, and generates one or more kinds of target model elements. These rules are activated by the ATL rule engine for each element of a given type of the source model.

During implementation of these transformations, a certain number of different ATL rules and helper operations have been created. Table 2 shows a different number of ATL rules for every transformation.

A number of matched rules are the same for both transformations. For every element which must be defined as unique in the MOF TS (i.e., RDM metamodel), we used unique lazy rules for their creation, which when called create a new element with the same content from the same input element. In the opposite direction we used only lazy rules, because we had to create more than one output element with the same contents from the same input element. The number of helper operations is larger in the XML2RDM.atl transformation, because we need to search (recursively) through the input XML model to find instances of the same IndividualVariable elements.

Figure 16 gives an example of a matched rule which is in fact an excerpt of the XML2RDM.atl transformation for the Root class (XML!Root) of the XML metamodel. The ATL rule Ontology transforms the Root of the XML model (RuleBase_XML) into the Ontology element (i.e., RDM!Ontology in Figure 16) of the RDM metamodel. The Ontology element (the Root of an XML tree) captures a collection of different ontology elements and SWRL rule.

Along with above-mentioned transformations, we have additionally defined transformations between RDM and R2ML language, that is RD-

*Table 2. Number of different ATL rules in the XML2RDM.atl and RDM2XML.atl transformations*

| Transformation/rules | Matched rules | Lazy rules | Unique lazy rules | Helpers |
|---|---|---|---|---|
| SWRL XML Schema to RDM meta-model | 35 | 0 | 3 | 23 |
| RDM meta-model to SWRL XML Schema | 35 | 3 | 0 | 1 |

*Figure 16. ATL matched rule for ontology RDM element*

```
module XML2RDM;
create OUT : RDM from IN : XML;


rule Ontology {
        from i : XML!Root (
                        i.name <> 'swrlx:OWLOntology'
                )
        to
                o : RDM!Ontology (
                                elements <- i.children->select(c | c.oclIsKindOf(XML!Element))
                        )
}
```

M2R2ML.atl and R2ML2RDM.atl (Milanović et al. 2007), and as we also have defined transformations between R2ML language and other rules languages (R2ML, 2008), such as UML/OCL, JBoss, Jess, F-Logic, so it is possible to share OWL ontologies and SWRL rules with these languages.

Figure 17 shows a transformation scenario between the RDM model and the R2ML model by using ATL. From this figure we can see that it is possible to execute transformations in both directions.

The RDM model shown in Figure 15 is transformed into the analogous R2ML model by using the RDM2R2ML.atl transformation. The resulting R2ML model is shown in Figure 13.

## FUTURE TRENDS

By using the approach presented in this chapter, ontology and rule development can be done in MDE or XML technical spaces, but ontologies and rules can be also used in different phases of the Software Engineering development process, such as Analysis & Design, Implementation and Deployment, Run-time, Maintenance and Testing (Happel & Seedorf, 2006) (Gašević et al., 2008). Cranefield was the first to explore the synergy of software modeling languages and ontologies (Cranefield, 2001). He started from the assumption that there are similarities between the standard concepts of UML and those of ontologies (e.g., classes, relations, and inheritance). Having this in mind, he proposed the use of UML for modeling ontologies due to the wide-acceptance of UML by software engineers and many already-developed UML models, which would facilitate adoption of ontologies by software practitioners. Moreover, software engineers can also benefit from the use of ontology reasoning services (e.g., consistency checking) to reason over UML models. In this way, one can connect conceptual modeling and ontology development. This motivated several other researchers to look into the problem of similarities and differences between ontology and software modeling languages (mainly UML).

*Figure 17. Transformation of the RDM model elements into the R2ML model elements and vice versa*

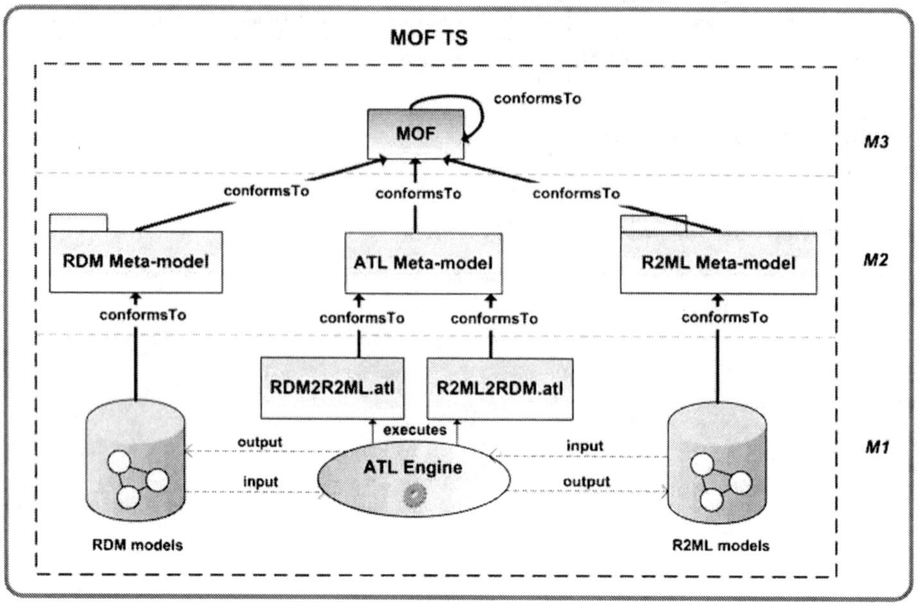


Note also that application (and domain) ontologies and rules can also be used in the design of software architectures. (Grønmo et al., 2005) demonstrated how MDE principles can be used to model Semantic Web services (i.e., OWL-S) by extending UML activity diagrams. This demonstrates the importance of further exploration of how Semantic Web ontologies and rules can be integrated into custom modeling languages (e.g., Business Process Modeling Notation - BPMN). This effort can have several contributions to software engineering such as (i) improved tractability of software models when maintaining software and (ii) improved software integration capacity, especially, in the context of service-oriented architectures.

There also have been several approaches that propose the use of ontology alignment techniques to attack the problem of model transformation. The ModelCVS system addresses this problem by transforming (i.e., lifting) metamodels into ontologies (i.e., transforming Ecore to ODM) (Kappel, 2006)). Then, such obtained ontologies are further refactored to represent explicitly some hidden concepts that are usually not precisely represented in metamodels, but should be placed in ontologies. Finally, ontology matching algorithms are executed over such ontologies (e.g., COMA++), and discovered mapping relations are encoded into the ATL transformations. The ontology-based model transformation (ontMT) approach in an attempt that semantically annotates metamodels with the concepts from a reference ontology of a domain (Roser & Bauer, 2006). ontMT makes use of such semantic annotations to reason over concepts of the metamodels being mapped and generates model transformations (i.e., ATL) from inferred mapping relations.

Both of these applications of ontologies to model transformations have been recognized as valuable contributions to the MDE area. However, there are many important research questions that should be solved such as: combinations of both approaches to make the process of model transformation more effective; applying ontologies and ontology matching at the model (M1) level of MDA, for example, to improve software refactoring; and applications of ontology matching to contribute round-trip engineering (i.e., code generation and reverse engineering) by complementing the efforts for model-to-text and text-to-model transformations (Jouault et al., 2006).

## CONCLUSION

To the best of our knowledge, there is no available solution to transforming ontology and rule languages on the level of their abstract syntax and by using model transformation languages. The main benefit of our solution is that the mappings between the abstract syntax of ontology and rule languages (e.g., OWL and SWRL in our case) are completely independent of their concrete syntax. Thus, we do not have to reconsider the mappings between two different languages when supporting various concrete syntax of the languages under study. Since the mappings between the same language's abstract and concrete syntax are straightforward, the effort and the price of support for concrete syntax are lower. For example, in the case of SWRL, we may use two concrete syntax, namely, the RDF/XML concrete syntax (Becket, 2004) and the OWL XML presentation syntax (Hori et al., 2003). Currently, we only support the first one, but when we implement the support for the second one, we will not reconsider the mappings between R2ML and SWRL, but only between SWRL RDF concrete syntax and RDM, that is, the SWRL abstract syntax.

A similar approach to ours is applied in the ODM specification (ODM, 2006) where the (model) transformations between OWL and the languages such as UML, Topic Maps, and ER models are defined at the level of their abstract syntax (i.e., metamodels). Our solution goes one step further and demonstrates how to bridge between concrete and abstract syntax of Semantic Web languages. Besides the obvious benefit of developing transformations between rule languages on the level of abstract syntax, the use of model transformations and languages such as ATL is more suitable than XSLT. Although, in principle, we could use XSLT to map between abstract syntax thanks to XMI in which all MOF-based metamodels can be stored, the available analysis of the use of XSLT for sharing knowledge indicates that XSLT is hard to maintain where modifications of input and output formats can completely invalidate previous versions of XSLTs (Jovanović & Gašević, 2005). Even some recent experiences in transforming rule languages (e.g., SWRLp) report on constraints of XSLT (e.g., to transform unique symbols) that can only be overcome by XSLT extensions implemented in other languages such as Java and Jess (Matheus, 2004). In the case of using model transformations and engines such as ATL, we also provide a convenient support for transforming semantic Web rules in the XMI format, and thus interoperability with MOF-based tools and integration with the current trends in software engineering.

In future research we plan to use the proposed approach to provide mappings between R2ML and OMG's initiatives for Semantics of Business Vocabulary and Business Rules (SBVR) and Production rules. This also nicely fits into the OMG's initiative for Business Rule Management (http://www.omg.org/busrulesmgmtrfi) by enabling the use of Semantic Web rules, which will be another important aspect of our future research.

## REFERENCES

ATLAS Transformation Language (ATL). (2008). Retrieved from http://www.sciences.univ-nantes.fr/lina/atl.

Baclawski, K., Kokar, M.K., Kogut, P., Hart, L., Smith, J.E., Letkowski, J., Emery, P. (2002). Extending the Unified Modeling Language for ontology development. *International Journal Software and Systems Modeling (SoSyM), 1*(2), 142-156.

Bechhofer, S., van Harmelen, F., Hendler, J., Horrocks, I., McGuinness, L., D., Patel-Schneider, F., P., & Stein, A., L. (2004). OWL Web Ontology Language Reference. *W3C Recommendation*. Retrieved from http://www.w3.org/TR/owl-ref/.

Beckett, D. (Ed). (2004). RDF/XML Syntax Specification (Revised). *W3C Recommendation*, http://www.w3.org/TR/rdf-syntax-grammar/.

Bézivin, J. (2005). On the unification power of models. *Software and System Modeling,4*(2), 171-188.

Brockmans, S., & Haase, P. (2006). *A Metamodel and UML Profile for Rule-extended OWL DL Ontologies - A Complete Reference* (Tech. Rep.). Universität Karlsruhe (TH).

Cranefield, S. (2001). UML and the Semantic Web. In *Proceedings of the Semantic Web Working Symposium,* (pp. 113-130).

Gašević, D., Djurić, D., & Devedžić, V. (2006). *Model driven architecture for Ontology development.* Berlin-Heidelberg: Springer.

Gašević, D., Kaviani, N., & Milanović, M. (2008). Ontologies and software engineering. In S. Staab & R. Studer (Eds.), *Handbook on Ontologies.* Germany: Springer.

Ginsberg, A. (2006). RIF Use cases and requirements. *W3C Working Draft.* Retrieved from http://www.w3.org/TR/rif-ucr/.

Grønmo, R., Jaeger, C., M., & Hoff, H. (2005). Transformations between UML and OWL-S. In *Proc. of the 1st European Conference Model Driven Architecture - Foundations and Applications* (pp. 269-283).

Happel, H., & Seedorf, S. (2006). Applications of Ontologies in software engineering. In *Proc. of the Int'l WSh. on Semantic Web Enabled Software Engineering.*

Hirtle, D., Boley, H., Grosof, B., Kifer, M., Sintek, M., Tabet, S., & Wagner, G. (2006). *Schema specification of RuleML 0.91.* Retrieved from http://www.ruleml.org/spec/

Hori, M., Euzenat, J., & Patel-Schneider, F. P. (2003). OWL Web ontology language, XML presentation syntax. *W3C Note*. Retrieved from, http://www.w3.org/TR/owl-xmlsyntax

Horrocks, I., Patel-Schneider F.P., Boley, H., Tabet, S., Grosof, B., & Dean, M. (2004). SWRL: A Semantic Web rule language combining OWL and RuleML. *W3C Member Submission*. Retrieved from http://www.w3.org/Submission/SWRL.

Jovanović, J., & Gašević, D. (2005). XML/XSLT-based knowledge sharing. *Expert Systems with Applications, 29*(3), 535-553.

Jouault, F., Bézivin, J., Kurtev, I. (2006). TCS: A DSL for the specification of textual concrete syntaxes in model engineering. In *Proc. of the 5th Int'l Conf. on Generative Prog. and Component Eng.* (pp. 249-254).

Kappel, G., Kapsammer, E., Kargl, H., Kramler, G., Reiter, T., Retschitzegger, W., Schwinger, W., Wimmer, M. (2006). Lifting metamodels to ontologies: A step to the semantic integration of modeling languages. In *Proc. of the ACM/IEEE 9th Int'l Conf. on Model Driven Eng. Languages and Sys.* (pp. 528-542).

Kent, S. (2002). Model driven engineering. In *Proceedings of IFM2002* (LNCS 2335).

Kogut, P., Cranefield, S., Hart, L., Dutra, M., Baclawski, K., Kokar, M., & Smith, J. (2002). UML for Ontology development. *The Knowledge Engineering Review, 17*(1), 61-64.

Kurtev, I., Bézivin, J., & Aksit, M. (2002). Technological spaces: An initial appraisal. In *Proceedings of the CoopIS, DOA'2002 Federated Conferences, Industrial track, Irvine, USA.*

Matheus, C.J. (2004). SWRLp: An XML-Based SWRL presentation syntax. In *Proceedings of the 3rd International Workshop on Rules and*

*Rule Markup Languages for the Semantic Web*, Hiroshima, Japan (pp. 194-199).

Milanović, M., Gašević, D., Giurca, A., Wagner, G., & Devedžić, V. (2007). *Model transformations to share rules between SWRL and R2ML*. Paper presented at the 3rd International Workshop on Semantic Web Enabled Software Engineering (SWESE 2007), Innsbruck, Austria.

Miller, J. & Mukerji, J. (Eds.) (2003). MDA Guide Version 1.0.1. *OMG Doc. omg/03-06-01*. Retrieved from http://www.omg.org/cgi-bin/doc?omg/03-06-01.

OMG, Meta Object Facility (MOF) Core, v2.0. (2006). *OMG Document formal/06-01-01*. Retrieved from http://www.omg.org/cgi-bin/doc?formal/2006-01-01

OMG, MOF QVT Final Adopted Specification. (2005). *OMG document 05-11-01*. Retrieved from http://www.omg.org/docs/ptc/05-11-01.pdf

OMG, Meta Object Facility (MOF) 2.0 XMI Mapping Specification, v2.1. (2005). *OMG Document formal/2005-09-01*. Retrieved from http://www.omg.org/cgi-bin/doc?formal/2005-09-01

OMG, Ontology Definition Metamodel Request for Proposal. OMG Document: ad/2003-03-40, (2003). Retrieved from http://www.omg.org/cgi-bin/doc?ad/2003-03-40

OMG Ontology Definition Metamodel (ODM) (2006). Sixth revised submission. *OMG Document ad/2006-05-01*. Retrieved from http://www.omg.org/docs/ad/06-05-01.pdf

Pan, Y., Xie, G., Ma, L., Yang, Y., Qiu, Z., & Lee, J. (2006). Model-driven ontology engineering. *Journal of Data Semantics VII*, 57-78.

*REWERSE I1 Rule Markup Language (R2ML)* (2008). Retrieved from http://oxygen.informatik.tu-cottbus.de/rewerse-i1/?q=node/6.

Roser, S. & Bauer, B. (2006). An approach to automatically generated model transformations using ontology engineering space. In *Proc. of the 2nd Int'l WSh. on Semantic Web Enabled Software Eng.*

Seidewitz, E. (2003). What models mean. *IEEE Software, 20*(5), 26-32.

Selic, B. (2003). The pragmatics of model-driven development. *IEEE Software, 20*(5), 19-25.

Sheth, A., Verma, K., & Gomadam, K. (2006). Semantics to energize the full services spectrum. *Comm. of the ACM, 49*(7), 55-61.

Wagner, G., Giurca, A., & Lukichev, S. (2006). A usable interchange format for rich syntax rules integrating OCL, RuleML and SWRL. In *Proceedings of the Workshop of Reasoning on the Web*. Edinburgh, UK.

## KEY TERMS AND DEFINITIONS

**ATL:** A hybrid (i.e., declarative and imperative) transformation language, based on the OMG

**MDA:** An approach to application design and implementation that encourages efficient use of system models in the software development process, and it supports reuse of best practices when creating families of systems.

**MOF 2:** A metadata management framework, and a set of metadata services that enable the development and interoperability of model- and metadata-driven systems.

**OCL:** Norm for both its data types and its declarative expressions. ATL and QVT share some common features, as they initially shared the same set of requirements defined in QVT Request for proposals.

**OCL:** A formal language used to describe expressions on UML models. These expressions typically specify invariant conditions that must hold for the system being modeled or queries over objects described in a model.

**ODM:** A family of MOF metamodels, mappings between those metamodels as well as mappings to and from UML, and a set of profiles that enable ontology modelling through the use of UML-based tools.

**OWL:** Web Ontology Language designed for use by applications that need to process the content of information instead of just presenting information to humans. OWL can be used to explicitly represent the meaning of terms in vocabularies and the relationships between those terms. This representation of terms and their interrelationships is called an ontology.

**QVT:** A standard for model transformation languages

**R2ML:** A MOF-defined general rule language that captures integrity, derivation, production, and reaction rules. It is a well-known proposal for RIF.

**RIF:** Rule Interchange Format is W3C initiative that should define an intermediary language between various rule languages, but it should not provide a formally defined semantic foundation for reasoning on the Web such as OWL for ontologies.

**SWRL:** A Semantic Web Rule Language based on a combination of the OWL DL and OWL Lite sublanguages of the OWL Web Ontology Language with the Unary/Binary Datalog RuleML sublanguages of the Rule Markup Language. It thus enables Horn-like rules to be combined with an OWL knowledge base.

# Chapter XXI
# Using Rule-Based Concepts as Foundation for Higher-Level Agent Architectures

**Lars Braubach**
*University of Hamburg, Germany*

**Alexander Pokahr**
*University of Hamburg, Germany*

**Adrian Paschke**
*Freie Universität Berlin, Germany*

## ABSTRACT

*Declarative programming using rules has advantages in certain application domains and has been successfully applied in many real world software projects. Besides building rule-based applications, rule concepts also provide a proven basis for the development of higher-level architectures, which enrich the existing production rule metaphor with further abstractions. One especially interesting application domain for this technology is the behavior specification of autonomous software agents, because rule bases help fulfilling key characteristics of agents such as reactivity and proactivity. This chapter details which motivations promote the usage of rule bases for agent behavior control and what kinds of approaches exist. Concretely, these approaches are in the context of four existing agent architectures (pure rule-based, AOP, Soar, BDI) and their implementations (Rule Responder, Agent-0 and successors, Soar, and Jadex). In particular, this chapter emphasizes in which respect these agent architectures make use of rules and with what mechanisms they extend the base functionality. Finally, the approaches are generalized by summarizing their core assumptions and extension mechanisms and possible further application domains besides agent architectures are presented.*

## INTRODUCTION AND MOTIVATION

Software agents are computational entities that can act autonomously without user intervention and interact with each other in order to fulfill given tasks. In the recent years, agent technology has evolved in to a large and highly active research field. Therefore, nowadays many different forms of agent applications exist, such as information agents, interface agents, mobile agents, and agents for problem solving (Nwana 1995), many of which also have been put successfully into practice (Jennings and Wooldridge 1998). Besides the application domain, implemented agent systems can also be distinguished by the employed 'agent architecture', i.e. the control structures that facilitate the specification and execution of agent behavior. Often these agent architectures build upon or have been influenced by rule-based technology.

In this chapter, rule-based technology is seen as a declarative approach to programming. On a conceptual level, a rule-based approach enforces a separation of the state of a system (i.e. the working memory) from the behavior (i.e. state transition rules). Different types of rules exist for implementing different types of systems. E.g. in production systems so called forward chaining rules are used, which are composed of a condition and an action part, such that the system will evaluate the condition of a rule and execute the corresponding action, when the condition holds. On the other hand, backward chaining rules (containing antecedence and consequence parts) are used in expert systems and allow deriving new knowledge (consequence) from existing facts (antecedence). For implementing such systems, sophisticated mechanisms have been developed for e.g. representing knowledge in the state, efficiently evaluating rules, deciding which rule to execute when multiple rules match at the same time (conflict resolution) and dealing with knowledge created by rules, which are no longer activated (truth maintenance).

The general approach of this chapter is to present and discuss agent architectures as one very interesting application domain for rule concepts. Therefore, the chapter shows the current state of the art with respect to rules for describing behavior of intelligent agents. One specific objective of this chapter is to show why rule concepts are important for the specification of agent architectures and in what different ways they can be used as basis for such architectures. On a more generic level, the objective of this chapter is to explain how higher-level concepts can be built on rule concepts and what advantages can be drawn from such developments.

The outline of the chapter is as follows. In the next section a short overview of the field of agents and multi-agent systems will be given, with a special focus on the role of agent architectures. Section 3 will provide a description scheme for and a categorization of existing agent architectures with regard to their incorporation of rule-based technology. An in-depth discussion of each of the four categories, including a detailed analysis of one representative in each case, follows in Sections 4-7. Section 8 will provide an outlook on interesting areas of future research before Section 9 closes the chapter with a summary and conclusion.

## BACKGROUND ON AGENTS AND MULTI-AGENT SYSTEMS

The field of agent technology emerged during the Nineties of the last century and has its roots in different areas of computer science such as artificial intelligence (AI), software engineering (SE), and distributed computing (Luck et al. 2005). In agent technology, an agent is seen as an independent software entity situated in an environment that is capable of controlling its own behavior (i.e. an agent can act without user intervention). Although agent technology is a very diverse field with many sometimes quite unrelated sub areas, general consensus exists, that agents can be ascribed the following set of properties (Wooldridge 2001):

- *Autonomy.* An agent decides on its own, how to accomplish given tasks. This is also the case for agents that act on behalf of a user.
- *Reactivity.* An agent continually monitors its environment and automatically reacts to changes in a timely manner, if necessary.

- *Proactivity.* An agent does not only react to stimuli from the outside, but also starts new actions on its own in order to pursue its own (given) objectives.
- *Social ability.* Agents are commonly situated in an environment composed of other (software or human) agents. For accomplishing their tasks, agents can engage in dialogs with other agents and interact in cooperative or competitive ways.

Despite the diversity of the field, two broad strands can be identified, focusing on individual agents and multi-agent issues, respectively. With regard to individual agents, different approaches have been developed for allowing agent programmers to control the practical reasoning processes inside the agents. Hereby, *practical reasoning* means the process of deciding about the action(s) to perform, given current knowledge and (e.g. sensory) input. Multi-agent systems on the other hand, highlight the aspect of *interaction*, that is, communication between agents to coordinate their actions.

Among the prime advantages of the agent metaphor over other approaches are the natural and intuitively understandable concepts. In this respect, agent technology was and continues to be influenced by numerous other disciplines like philosophy or biology (for models of decision processes) as well as sociology or economics (for models of agent interaction). The concepts and technologies that have been developed in the agent field have already successfully been deployed in many different application areas, such as electronic negotiations, shop-floor management, transportation logistics, business process management, computer games and military decision support (Jennings and Wooldridge 1998).

## Agent Architectures, Frameworks and Platforms

One important question of multi-agent system construction concerns the mechanism that agents apply for conducting their practical reasoning. For this purpose generic *internal agent architectures* have been devised, which on the one hand describe the building blocks agents consist of and on the other hand an interpreter that works on these building blocks.

As a broad range of different internal agent architectures have been developed, several classification schemes have been proposed in the literature. In the well-known scheme of Wooldridge and Jennings (1995) a distinction is made between, *reactive*, *deliberative* and *hybrid* agent architectures. Reactive agent architectures emphasize the importance of fast reactions to changes within dynamic environments. In the strictest interpretation reactive agents are not allowed to posses a symbolic knowledge representation and act in direct consequence to the perceived percepts. In contrast to reactive agent architectures, deliberative architectures rely on a symbol-based reasoning process, which requires an agent to posses local knowledge and are therefore typically processing-intensive and time consuming. In order to alleviate the weaknesses, that each of the aforementioned architecture types exhibits, *hybrid architectures* have been devised. They try to unify aspects from both approaches and therefore combine timely reactions with well-planned behavior hybrid architectures have gained high attention in practice and nearly all internal architectures, which are supported by agent frameworks, build on the balanced reactive as well as deliberative actions.

On a technical level, the individual and multi-agent concepts have been concretized into various software frameworks and agent platforms. A detailed overview can be found in (Braubach et al. 2006). In the following special attention will be paid to agent architectures and platforms that incorporate or are based upon some kind of rule mechanism.

## RULE-BASED AGENT ARCHITECTURES

In this section the foundations for the following architecture and system descriptions will be laid. The selection of architectures is based on a rule-based categorization, which aims at a distinction of agent architecture with regard to the way they employ rule-based mechanisms. Moreover, a general scheme is introduced, which will be used for the explanation of the architectures. This scheme consists of different steps and as it is applied to all architecture in the same way, the evaluations will be directly comparable.

### Architecture and System Categorization

In the literature a multitude of different agent architectures and systems have been proposed. In order to describe these artifacts with respect to rule properties a new classification scheme has been devised (see Figure 1), which allows assigning the representatives to the different categories. The categorization is done along two axes termed *deliberation cycle* and *programming concepts.* The deliberation cycle dimension describes the way the agent interpreter is constructed. In this respect it is distinguished between a controlled rule engine (1), which extends a rule engine with agent concepts, a normal rule engine (2) and an agenda-based interpreter (3), which does not rely on rules at all. Regarding the programming concepts dimension it is differentiated between procedural behavior description (A) based on mainstream programming languages such as Java, rule approaches without (B) and with mentalistic

*Figure 1. Agent architecture classification*

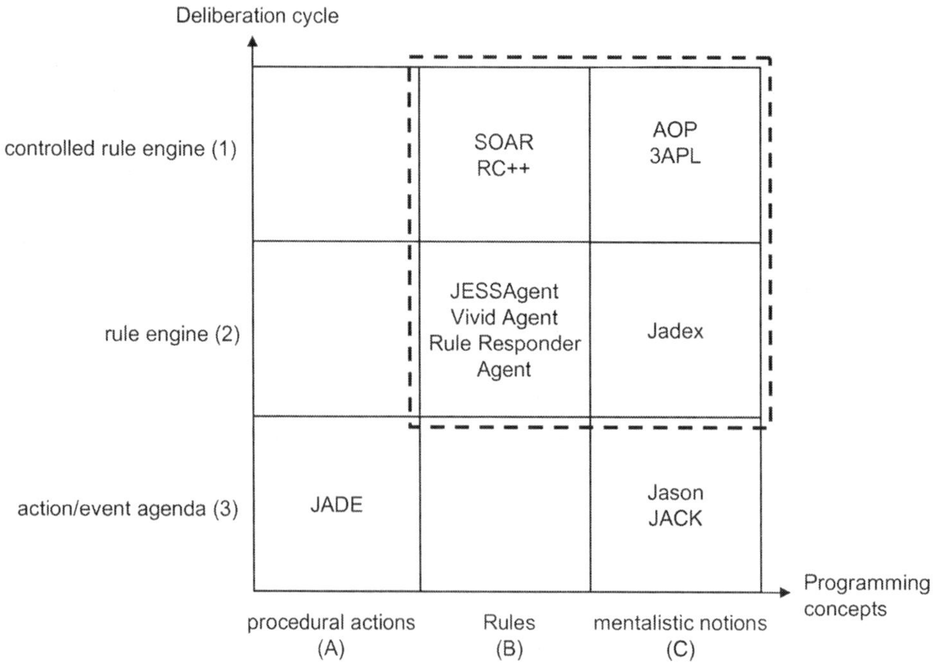

notions (C). Several well-known agent platforms have been assigned to one of the resulting nine categories. From these nine categories only those four will be presented in greater detail, which rely on rule concepts (framed categories). Therefore, agent platforms such as JADE (Bellifemine et al. 2005), Jason (Bordini et al. 2005), and JACK (Winikoff 2005) are not considered in this chapter.

## The Description Scheme

Each of the architecture descriptions follows the coarse scheme that first the category is introduced, then a specific representative of that category is described in detail and finally a category generalization and conclusion are given. The explanation of the concrete architecture representative is done according the following sections:

- **Architecture and system description.** This section provides detailed information about the selected architecture and system. Therefore, the underlying agent architecture is explained with respect to its components and their interplay. Furthermore, a closer look will be taken at the agent's deliberation cycle, which determines its basic functioning.
- **Agent properties.** In this section, an agent-based perspective will be used for discussing how essential agent properties are supported by the selected architecture. The essential agent properties that are used in this chapter are those defined in the strong notion of agency (Wooldridge 2001). In this definition it is assumed that an agent, in addition of being autonomous, should be capable of *reactive*, *proactive* and *communicative* behavior. Additionally, it is important if *mentalistic notions* can be used for describing the agent behavior in an abstract and intuitive way. For each of these properties it is especially described, which architecture concepts are used for their realization and which aspect rules play in this context.
- **Rule-based mechanisms.** In contrast to the preceding section, this section takes up a rule-based perspective and elaborates on specific rule mechanisms employed within the architecture. The following topics will be discussed: *rule types*, *state representation*, *rule evaluation mechanism*, *conflict resolution* and *truth maintenance*. With regard to rule types it is discussed, which kinds of rules are used (e.g. production, ECA or deduction rules). The state representation considers the working memory assembly including aspects like fact representation, type system and garbage collection. The rule evaluation mechanism determines in which way the rules are processed, i.e. which algorithm is used and conflict resolution refers to the way how the system decides which activation to fire if more than one is present. Finally, truth maintenance relates to the usage of logical facts and rules.
- **Rule incorporation.** In this section it will be discussed how the rule mechanisms have been integrated into the architecture. This is done on two different levels. On the programming language level it will be discussed, how rules are incorporated into the programming concepts, i.e. in which way a developer gets in contact with rule specifications. On the system level it is considered, which part rules play for the interpreter architecture, i.e. in what way the execution depends on rule mechanisms.

After setting the scene and scheme for discussion, the following sections will deal with the identified categories, respectively.

## RULE ENGINES AND RULES (B2)

The rule-based approach can directly be used as basis for agent behavior control. Using this kind of architecture basically requires that the rule base is properly connected with the agent's sensors and effectors in order to allow an agent to receive percepts and execute actions. To be able to exhibit reactive behavior and process incoming messages it is necessary that 1) incoming messages will trigger the execution of processing rules and 2) the external knowledge representation fits to the internal one or is mapped accordingly. Examples of this domain are e.g. JADE/Jess agent (Cardoso 2007), Vivid Agents (Schroeder and Wagner 1998) and Rule Responder (Paschke et al. 2007).

In JADE/Jess an agent is constructed in a way that allows access to the production rule engine (JESS). JADE/Jess agents have e.g. been used in the application domain of agent based scheduling in hospitals (Krempels and Panchenko 2003). Vivid agents are controlled systems whose state comprises the mental components of knowledge, perceptions, tasks, and intentions. Their behavior is represented by means of action and reaction rules, which follow the event-condition-action (ECA) paradigm and are underpinned by formal transition semantics for concurrent action planning. Applications are e.g. the distributed system diagnosis and the monitoring of communication protocols (Schroeder and Wagner 1998).

Rule Responder (Paschke et al. 2007) builds upon existing technologies in multi-agent systems, in particular the Prova Agent Architecture (Dietrich et al. 2003) which evolves from the earlier works in the field, e.g. Vivid Agents (Schroeder and Wagner 1998). It blends and tightly combines them, with modern distributed Web rule and Semantic Web technologies, and service oriented and event driven architectures. Application domains of Rule Responder (http://responder.ruleml.org) include, e.g., virtual organizations (rule-based collaborative agents), Event-driven BPM, Rule Based Service Level Agreements (RBSLA), Health Care and Life Science eScience Service Infrastructure.

In the following Rule Responder architecture will be detailed as representative of this category "B2". Its architecture has evolved from earlier agent architectures, integrates existing agent vocabularies (e.g. FIPA ACL) and uses RuleML / Reaction RuleML for rule interchange and serialization.

### Rule Responder

The Rule Responder middleware extends the Semantic Web towards a Pragmatic Web infrastructure for collaborative agent networks where independent agents engage in conversations by exchanging messages and cooperating to achieve collaborative goals (Paschke et al. 2007). Rule Responder utilizes modern enterprise service technologies and the Semantic Web with intelligent agent services that access data and ontologies, receive and detect events (complex event processing), and make rule-based inferences and autonomous pro-active decisions and reactions based on these representations. For a description of the syntax, semantics and implementation of the underlying logical formalisms see (Paschke 2007).

The core of a Rule Responder agent is a Prova rule engine (http://sourceforge.net/project/showfiles.php?group_id=50817) which has a modular knowledge base (module = set of facts and rules) to implement several different roles an agent might play. Each role has its own set of reaction rules to react on detected situations (complex events) and its own set of decision rules to interpret goals and derive decisions according to conditional proofs. To sense the environment and trigger actions, query data from external sources such as databases, call external procedural code such as Enterprise Java Beans, and receive / send messages from / to other agents or external services the Prova engine provides a set of built-in functions and additionally can dynamically instantiate any Java objects and

*Figure 2. Rule responder agent architecture*

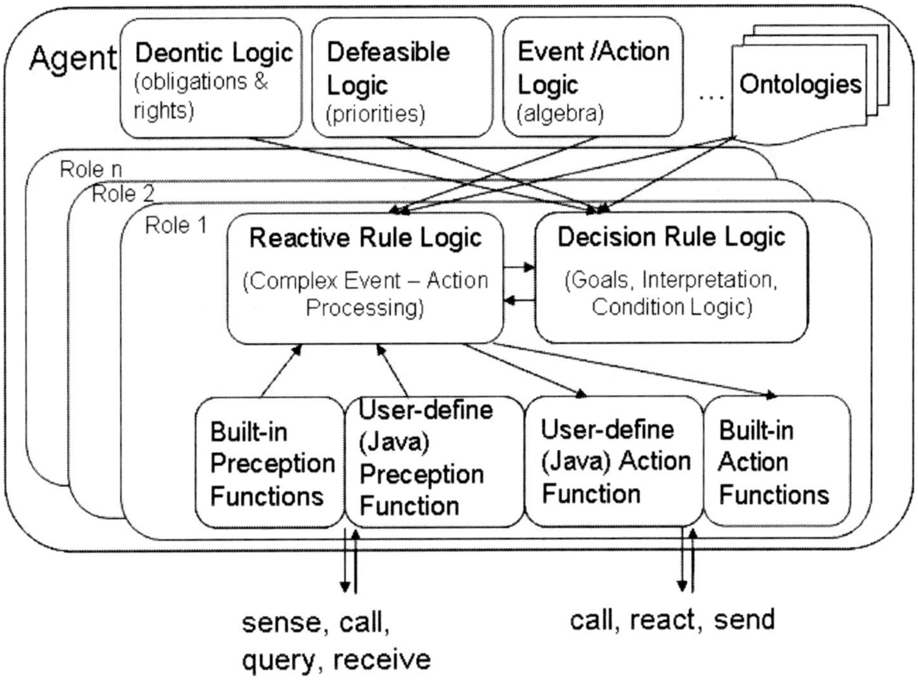


call their API methods. Additional libraries can be imported, e.g., to represent rights and obligations of agents, implement conflict handling rules, or describe complex events and actions. The agent might use vocabularies defined as Semantic Web ontologies (e.g. RDFS or OWL based) or Java class hierarchies to give its' rules a domain-specific meaning. The vocabularies can be used within the conversation with other agents to enable a semantic and pragmatic interpretation.

For the deployment of agents on the Web and for the communication in agent networks the Rule Responder enterprise service middleware is used. The three core parts of the Rule Responder middleware are (1) Reaction RuleML (http://ibis.in.tum.de/research/ReactionRuleML/) as a common platform-independent rule interchange format to interchange rules, events and data between agent services and other Semantic Web tools, (2) a highly scalable and efficient enterprise service bus (ESB) as agent/service-broker and communication middleware (Paschke et al. 2007a), and (3) Prova rule engines as platform-specific (Java) rule-based agent execution environments.

The agents are deployed as distributed web-based services on an ESB. They have local autonomy and are decoupled via exchanging Reaction RuleML messages either asynchronously or synchronously over the ESB, or in an ad-hoc conversation directly between two agents using agent protocols such as Jade-HTTP. Arbitrary transport protocols can be used to transport rule sets, queries and answers as payload of Reaction RuleML event messages between the agent services. Each agent dynamically imports or pre-compiles and loads a possibly distributed and modular rule base (represented as one or more Reaction RuleML scripts published on the web or stored locally, e.g., in a (XML) database or file), which implements the decision and (behavioral) reaction logic of the agent.

## Deliberation Cycle

Basically, a Rule Responder agent follows the sense-reason-act pattern. However, Rule Responder does not define one particular deliberation cycle, but allows configuring a Rule Responder agent with user-defined conversation-based negotiation and coordination protocols or workflow patterns. Via constructs for asynchronously sending and receiving event messages an agent interacts with the environment. The payload of incoming event messages is interpreted with respect to the local conversation state, denoted by the conversation id, and the pragmatic context which is defined by a possibly complex pragmatic performative(s). Depending on the pragmatic context the message payload is used e.g., to update the internal knowledge of the agent (e.g. adds new facts or rule sets), add new tasks (goals), or detect a complex event pattern (from the internal event instance sequence). Several reasoning processes might run in parallel local to their conversation flows. Inactive reactions (conversation partitions) are removed from the system. Self activations by sending a message to the receiver "self" are possible. Several expressive logic formalisms are provided for updating the knowledge base (transactional update logic), defining and detecting complex events (complex event algebra) (Paschke et al. 2007a), handling situations (states) (event calculus), and for reasoning (e.g. deontic logic for normative reasoning on permissions, prohibitions, obligations) (Paschke 2007) and planning (abductive reasoning on plans and goals (Paschke and Bichler 2005).

## Agent Properties

**Reactive behavior.** External events (changes in the environment) and data from data sources such as Web resources, relational databases and external tools, web services or (Java) object representations can be dynamically accessed and integrated during runtime via the expressive homogenous and heterogeneous integration interfaces of Prova, or via publish-subscribe or reactive monitoring processes using the ESB, event / action messages and reaction rules (Paschke et al. 2007a). The logic for reactivity of the agent is implemented in terms of reaction rules and conversation-based messaging reaction rules. Complex event patterns can be defined in terms of an event algebra (based on an interval-based

*Figure 3. Rule responder middleware*

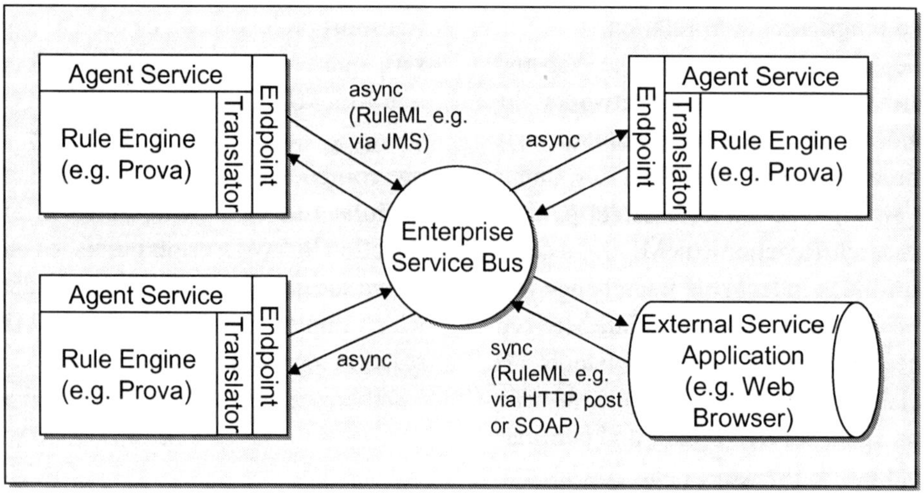

event calculus variant) and in terms of messaging workflow-like conversations

**Proactive behavior.** The decision and planning logic for (semi-) autonomous and pro-active decisions and actions, e.g. starting a new sub-conversation, is implemented in terms of derivation rules and messaging reaction rules. An interval-based event calculus axiomatisation (Paschke et al. 2007a) allows representing the agent's states as changeable fluents which are initiated or terminated by occurred (complex) events. For a set of believed (planned) goal states a set of hypotheses (~a plan represented as a sequence of planned events that need to occur) can be determined by an abductive extension of the event calculus meta program. (Paschke and Bichler 2005). Contradictions are detected with respect to predefined (application-specific) integrity constraints and test suites (set of test cases) from which violation free (hard integrity constraints) and covered (level of coverage according to the covered test cases) plans are selected (using a priority based defeasible logic) (Paschke 2007).

**Communication.** Typically, Rule Responder agents form collaborative networks where the agents interact in conversation dialogs which might follow a pre-defined negotiation or coordination protocol (see Paschke et al. 2006). This might also include negotiation and discussion about the meaning of ontological concepts, since agents might use their own micro-ontologies / vocabularies and must agree on relevant shared concepts to enable an efficient communication and knowledge interchange between the agent nodes.

**Mentalistic notions.** Rule Responder does not define a specific set of mentalistic notions as first-class programming constructs. Instead, interchanged messages beside the conversation's meta data and payload also carry the pragmatic context of the conversation such as communicative situations / acts, mentalistic notions, organizational and individual norms, purposes or individual goals and values. For instance, a standard nomenclature of pragmatic performatives which can be integrated as external (Semantic) vocabulary/ontology is defined by the Knowledge Query Manipulation Language (KQML) (Finin et al. 1993), the FIPA Agent Communication Language (ACL) which gives several speech act theory-based communicative acts or the normative concepts on obligations, permissions and prohibitions of Standard Deontic Logic (SDL).

## Rule-Based Mechanisms

**Rule types.** Rule Responder on the basis of the RuleML language family and the Prova rule engine supports different rule types:

- *Derivation rules* to describe the agent's decision logic
- *Integrity rules* to describe constraints and potential conflicts
- *Normative rules* to represent the agent's permissions, prohibitions and obligation policies
- *Global reaction rules* to define global reaction logic which are triggered on the basis of detected complex events (event patterns defined by an event algebra)
- *Messaging reaction rules* to define conversation-based workflow reaction and behavioral logic based on complex event processing

**State representation.** States, denoting changeable agent's properties, such as its obligations, are represented as explicit event calculus fluents which are initiated and terminated by events. Updates to the internal knowledge base of the agent are treated as modules which transit the knowledge base to an extended knowledge state, where a set of rules or facts are added or removed to the previous knowledge state. Each module in the knowledge base (add rule/fact set) is identified by a unique module object id (oid). A complete sequence of state transitions, e.g. a sequence of updates in an execution path of firing reaction rules, might be rolled-back by backtracking the update operators on the remembered sequence module oids (Paschke et al. 2007a).

**Conflict resolution.** To overcome arising conflicts, preserve integrity of the knowledge base in each state and ensure an unique declarative outcome of update sequences (active rules) or complex updates (complex update actions) Rule Responder supports integrity constraints (ICs) and test cases (TCs). ICs and TCs are used to verify and validate the actual or any hypothetically future knowledge state (Paschke 2007).

**Truth maintenance**. Truth is defined locally with respect to an explicitly closed scope (constructive view) on the agent's knowledge base (KB). Scoped queries apply on the constructive views which are defined over module(s) or rule sets, e.g. all rules, which are written by a particular author (defined in the rule meta data). By default, un-scoped queries (goals) apply globally on the complete knowledge base of the agent, i.e. the current knowledge state. Hypothetical updates are treated as uncommitted modules in the KB which are tested against the defined integrity constraints and might be rolled-back if they violate the integrity of the KB (Paschke 2007).

### Rule Incorporation

In Rule Responder rules of various types (derivation, integrity, reaction, messaging) are first class citizens to program agent behavior and agent decision logic. The core Prova rule language and engine is not a specific agent language but a highly expressive, hybrid, declarative and compact rule programming language which combines the declarative programming paradigm with object-oriented programming (in particular Java) and the Semantic Web approach.

Due to the natural integration of Prova with Java, it offers a syntactically economic and compact way of specifying agents' behavior while allowing for efficient Java-based extensions to improve performance of critical operations. Via its capabilities to integrate external vocabularies such as Semantic Web ontologies and Java type systems the rule language can be custom-tailored to a specific agent vocabulary giving the used rule constructs a specific meaning with a clearly defined semantics, e.g., based on the FIPA vocabulary. In addition to backward-reasoning derivation rules, which for instance are used to implement decision or planning logic, Prova supports messaging reaction rules specifying the sequences or branches of actions that an agent responds with upon detection of a matching pattern in the inbound messages which the agent receives from other agents or the environment (other tools, services). Several meta-programs written as Prova scripts exist (Paschke 2007), e.g., for deontic reasoning about obligations, prohibitions and permissions of (agent) actions. These scripts can be added selectively as modules to the rule based of an agent hence providing the needed (meta) reasoning, planning and reaction capabilities to formalize the agent behavior in terms of rules.

On the platform-independent and web-based rule interchange level, Reaction RuleML is used. Via translator services the interchanged RuleML messages are translated into the platform-specific execution syntaxes of the agents' rule execution environments such as Prova.

## Conclusion

In this section rules are directly employed to implement the agents' decision and reaction logic. Expressive declarative rule languages in combination with external data models and vocabularies are used to formalize coordination mechanisms and agent behavior. The main design philosophy behind these approaches is the minimalism and simplicity of the additionally introduced agent-specific syntax extensions to the core rule language.

Below, advantages and disadvantages of the B2 category are presented:

+ Highly-expressive rule languages allow declarative programming of complex agent coordination mechanisms and agent behavior; representation restrictions are only due the well-known trade-off between expressiveness and computational complexity
+ Rules can be serialized in common rule interchange formats such as RuleML / Reaction RuleML, published on the Web and interchanged between different platform-specific rule engines (given these execution environments, such as Prova, provide adequate expressiveness)
- Long learning curve for users / agent engineers, due to the highly flexible and expressive rule languages
- High-level conceptual abstractions which should be provided as specific agent programming constructs in the language need to be explicitly introduced either by meta programming techniques or via integration of external functionalities and vocabularies.

## Generalization

The approaches in this category directly employ general rule languages (e.g. Reaction RuleML) to implement agent functionalities. Extensibility of the general rule language is achieved by integrating external vocabularies and external data and functionalities, e.g. object-oriented procedure calls. Rule languages are primary used in the expert system field. In this context various application domains such configuration, diagnosis and interpretation of data have been considered. For an extensive survey see Giarratano and Riley (2005).

## CONTROLLED RULE ENGINES AND MENTALISTIC NOTIONS (C1)

This category encompasses approaches that aim at introducing abstract mentalistic notions as programming language constructs. Each of these approaches proposes a specific concerted set of mental state components and introduces a language with specific types of rules to operate on these components (e.g. commitment rules, which operate on commitments). As a natural way of controlling the relation between different types of mental components and rules respectively, these approaches are not implemented as a simple rule base, but use the specific rules only as part of an extended interpreter architecture.

All approaches in this category are intended to be general purpose agent programming languages. Nevertheless, due to the rules operating directly on the mental state of the agent, these approaches are best suited for agents operating in dynamic environments, where quick reactions to environmental changes are advantageous. Such environments are common, e.g. for agents operating on mobile devices such as PDAs or cell phones and for controlling autonomous robots.

Prominent representatives of this category are the AOP and the 3APL/2APL language families. 3APL ("An Abstract Agent Programming Language") and its successor 2APL ("A Practical Agent Programming Language") are developed at the University of Utrecht (Hindriks et al. 1999, Dastani et al. 2007). In addition to implementing agent applications, these languages are also used for teaching purposes. In contrast to 3APL/2APL, AOP languages are developed by different groups for different purposes including academic and commercial projects. Because of the greater diversity, in the following, the AOP approach will be presented in more detail.

## AOP

Agent Oriented Programming (AOP, cf. Shoham 1993) is one of the first approaches being motivated by the desire of using agent-oriented concepts as part of a high-level programming language. In the AOP paradigm an agent is cast as a special type of object (in the object-oriented sense). The state of an agent (unlike an arbitrary object) has a fixed and predefined structure, which only incorporates mental attitudes, such as beliefs and obligations. The approach is backed by the assumption that taking an *intentional stance* (Dennett 1971) towards a software system under development is advantageous in certain scenarios (McCarthy 1979). Therefore, the representation of mental attitudes as first-class programming constructs is seen as a desirable extension of traditional software engineering practices.

Three basic attitudes are proposed by Shoham for representing an agent's mental state: beliefs, obligations and capabilities. These are represented using a formal temporal language that contains the modal operators $BEL^t_a\varphi$ ("agent $a$ currently believes that sentence $\varphi$ is or will be true at time $t$"), $OBL^t_{a,b}\varphi$ ("agent $a$ is obliged to agent $b$ to do or bring about $\varphi$ at time $t$"), and $CAN^t_a\varphi$ ("agent $a$ is able to do or bring about $\varphi$ at time $t$"). Sentences ($\varphi$) are defined in a recursive way, such that e.g. beliefs about obligations of other agents can be expressed. Although some other common attitudes can be represented by combinations of the basic attitudes (e.g. commitments can be seen as obligations of the agent to itself), Shoham sees this basic model only as a starting point and therefore does not claim completeness.

Besides such a *formal language* for representing the mental state of an agent as stated above a concrete *programming language* is needed, which follows the semantics of the formal language and allows to specify the agent behavior using additional constructs. Shoham concretizes his initial AOP approach by introducing a simple programming language called Agent-0 (Shoham 1993). Based on Agent-0 other AOP languages (cf. Figure 4) have been conceived that introduce additional constructs for making programming more convenient. In the second generation of AOP, PLACA (Thomas 1995) extends Agent-0 with planning features, while Agent-K (Davies and Edwards 1994) supports KQML (Finin et al. 1993) as a standardized agent communication language. Agent-K has been further extended to GOAL (Byrne and Edwards 1996), which incorporates planning similar to PLACA, but also on a multi-agent level (e.g. group goals, which comprise commitments of several agents). Successors of PLACA are RADL, which is the language of the commercial AgentBuilder toolkit (http://www.agentbuilder.com/) and AF-APL, belonging to the open source AgentFactory framework (Ross et al. 2005). Both languages try to advance the practical usability of the language, e.g., by providing convenient mechanisms for integration with Java. Due to its simplicity, we will use Agent-0 to discuss the basic principles of the AOP architecture and refer to the advanced features of the newer languages as necessary.

### Architecture Components

The separation between a (more or less formal) state representation and programming constructs acting upon that state quite nicely fits to the idea of a rule-based system. Not surprisingly, all proposed AOP languages employ a rule-based specification of agent behavior. In Agent-0, only one rule type is supported: so called *commitment rules*. These rules are a special form of event-condition-action (ECA) rules. The event part basically describes patterns of messages that may be received from other agents. The condition part refers to the current mental state (e.g. the beliefs) of the agent. The action part is quite restricted compared to traditional ECA-rules: instead of directly executing actions, a rule may only create new commitments towards actions, associated to some specific time point. As commitments are just added to the

*Figure 4. AOP language family*

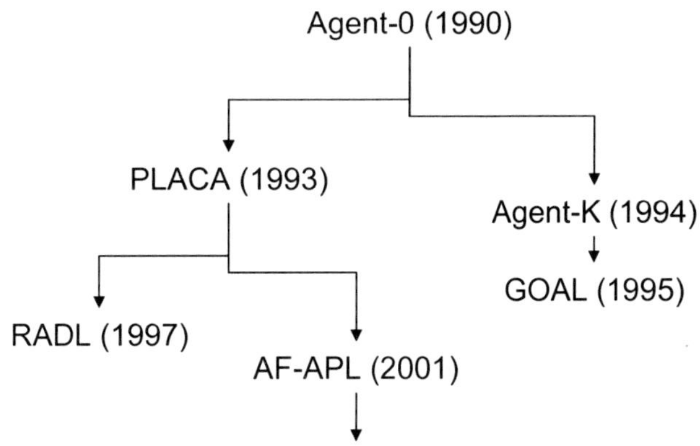


mental state of the agent, the triggering of a rule does not lead directly to the execution of actions. Instead actions are queued as commitments in the mental state of the agent and automatically executed, when their time is due.

As an example, a simple library agent is shown in figure 5 (adopted from Shoham 1990 and Thomas 1993). It has the abilities to shelve books (line 2) and Xerox documents (line 3). Initially (at time 1), it knows the customers Alice (line 6) and Bob (line 7). Finally, it has one commitment rule to do whatever requested from customers. Whenever the agent receives a request to perform some action at a given time (event pattern in line 10) and the sending agent is known to be a customer (mental condition in line 11) the agent will adopt the commitment to execute the action for the sender as requested (line 12).

The architecture of an AOP agent is illustrated in figure 6. Most important part of an agent program is the set of commitment rules (bottom right). Agent-0 provides only a few basic actions like sending messages to other agents. Domain specific actions (e.g. a move operation for a robot) can be defined using so called capabilities (bottom left). Capabilities specify a guard (i.e. a mental condition), which must be true for executing the associated action. The actions itself are specified by mechanisms outside the agent language such as a procedure in a traditional programming language. Both commitment rules and capabilities are specific at design time. At runtime, the agent continuously updates it's initially given beliefs (top left) and creates/executes/drops commitments (top right) according to the commitment rules and the current situation. At the heart of the agent, there is a clocked interpreter that is responsible for incrementing the current time and managing the other components.

## Deliberation Cycle

The deliberation cycle in AOP basically consists of three steps:

1. Updating beliefs
2. Updating commitments
3. Executing due commitments

In the first step, the interpreter takes care of updating the agent's beliefs when receiving INFORM messages from other agents. While the original Agent-0 interpreter just adds any information into beliefs, newer languages provide advanced mechanisms like considering authenticity and authority of agents (AgentBuilder) or leave the interpretation of message contents up to the agent developer (AgentFactory). In the second step, the interpreter checks all commit-

*Figure 5. Example of a library agent*

```
CAPABILITIES :=
  ((shelve ?x) (book ?x))
  ((xerox ?x) (document ?x))

INITIAL BELIEFS :=
  (1 (customer alice))
  (1 (customer bob))

COMMITMENT RULES :=
  (COMMIT       (?sender REQUEST (DO ?time ?action))
                (B (now (customer ?sender)))
                (?sender (DO ?time ?action)))
```

*Figure 6. AOP agent architecture*

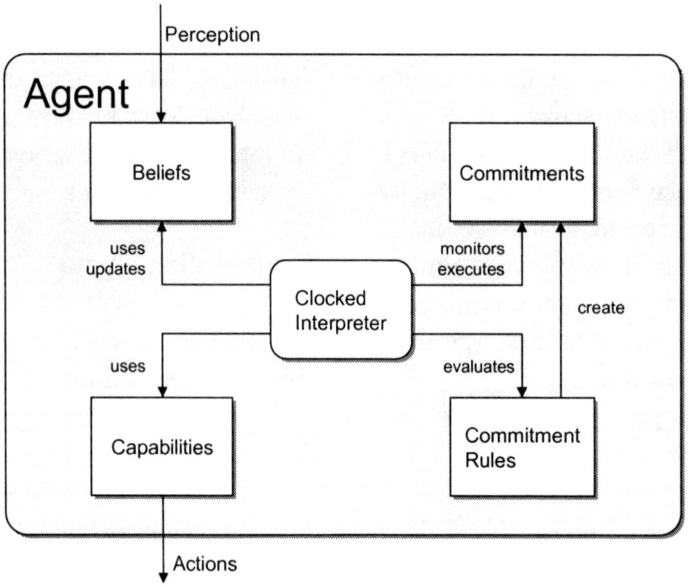


ment rules and creates new commitments when required. Moreover, existing commitments are dropped, when the agent no longer believes to be able to perform the desired action or when it is "unrequested" to perform the action (i.e. the original requestor revokes the request). In the third step, the interpreter checks, whether some of the current commitments are due at the current time point and executes the associated actions.

## Agent Properties

**Reactive behavior.** Reactive behavior can be expressed quite naturally in AOP. Although rules do not directly lead to actions but only to future-directed commitments instead, the keywords *self* and *now* allow to specify commitments to actions, which should be executed instantaneously. Therefore a direct ECA-style of reactive agent programming can be done in AOP quite easily.

**Proactive behavior.** The basic AOP model has only very limited support for pro-active behavior, in the sense that the agent will not adopt a commitment to an action when it already has a commitment to refrain from that action. Newer AOP languages support commitments to complex objectives in the sense of procedural goals that an agent pursues by executing predefined sequences of actions (called plans), but these goals usually lack a declarative aspect (i.e. a condition that can be used to check, if the goal is achieved).

**Communication.** Communication is an important aspect of AOP agents. AOP languages provide primitives for communicative actions as well as message patterns that can be used in the event part of a commitment rules. Moreover, the interpreter usually provides mechanisms that automatically deal with incorporating information received via messages into the beliefs of an agent.

**Mentalistic notions.** The semantics of the mentalistic notions in AOP (e.g. beliefs or commitments) follows an intuitive interpretation, which is therefore quite easy to understand. Newer languages extend the initial basic model by introducing further notions such as goals or plans. Although a wide range of complex behavioral patterns can be developed with most AOP languages, these approaches remain on a quite technical level and do not integrate the mentalistic notions in an abstract, self-contained model of practical reasoning or problem solving.

## Rule-Based Mechanisms

**Rule types.** The core concept of AOP are the commitment rules, which are a specialized form of ECA-rules. In the basic model, these rules are restricted in the sense that only the creation of commitments is allowed in the action part. Newer AOP languages partially relax this restriction and allow also sequences of actions, where actions may include sending messages, updating beliefs or creating new commitments.

**State representation.** Conceptually, the state of an AOP agent is represented using a first order logical language, extended with modal operators for the mental attitudes. Such a representation, in principle, allows reasoning over arbitrary complex logical statements. Nevertheless, because such a representation is usually computationally intractable, substantial simplifications are applied in the implemented interpreters, e.g. allowing only atomic statements for facts.

**Rule evaluation mechanism / Conflict resolution.** Most AOP implementations follow a naive approach, i.e. in each execution pass of the deliberation cycle, all rules are tested and the activated rules are executed in the order they are specified in the agent program. Due to the moderate number of rules in most agent programs, the performance of such an approach has been found to be sufficient for many practical settings.

**Truth maintenance.** No AOP system to date employs or provides truth maintenance functionality.

## Rule Incorporation

In AOP, rules are the primary programming construct for representing the behavior of agents. Unlike pure rule-based programs, AOP languages introduce a set of agent-oriented concepts and constructs with a predefined semantics. In this respect, the mentalistic notions of the architectural model help structuring the rule based program into intuitive agent-oriented concepts. I.e. the working memory is structured into flat factual knowledge (beliefs) as well as knowledge about the agent's skills (capabilities) and intended actions (commitments). Additionally, capabilities are used as an access point to functionality that is implemented in imperative programming languages.

Another difference to pure rule-based programs is that the form of rules is restricted. This means that the mentalistic notions that may be used in the event, condition and action parts are fixed in a way to support the intended semantics of the agent model (e.g. in actions, only commitments may be created). Finally, the interpreter supports the developer by automatically performing additional operations related to the semantics of the constructs (e.g. updating beliefs and executing or dropping commitments automatically).

## Conclusion

This section has introduced approaches, which strive to provide high-level abstractions as first-class programming constructs. To implement the resulting languages, interpreters are developed, which provide additional functionality besides the simple execution of rules. These approaches can be seen as a restriction and specialization of rule-based systems for certain application domains, where the abstractions and constructs provided by an approach will be beneficial.

Below, advantages and disadvantages of the C1 category are presented:

- \+ The main motivation of the approach is to provide high-level conceptual abstractions, which are made available to the programmer directly as first-class programming language constructs with an intuitive semantics. The resulting languages are therefore quite easy to learn and the built-in semantics allows quickly developing complex applications with only a few statements.
- \+ Due to still using rules as the main programming construct, approaches from this category are especially well suited for encoding reactive behavior. Different reactions to varying situations can be intuitively described and easily maintained by adding or removing behavioral rules.
- \- The rule-based style of behavior representation has led to architectures, which have difficulties representing long term activities. Although this is not an inherent problem of the category per se, all currently available approaches lack high-level concepts for such long term activities.
- \- The restrictions imposed by the architecture prevent the developer from using arbitrary rules. Therefore, the expressive power of the language is reduced in favor of better supporting a certain application domain.

## Generalization

As a generalization, we consider approaches that introduce higher-level concepts (not necessarily mentalistic notions) and provide an extended or controlled rule base for interpreting programs using these contructs. One emerging research/ application field, where this kind of approach is quite common is the area of Complex Event Processing (CEP, see e.g. Luckham 2002, Fiege et al. 2006). Thereby, a complex event is a composition of a specific pattern of many simpler events. The purpose of CEP is to filter and extract meaningful domain-level events from low-level occurrences (e.g. a missed deadline might be common while three missed deadlines in a row indicate a serious problem). Complex events are defined by event patterns, which are described in en event algebra with operators to relate events top each other (e.g. *between*, *before*, *after*, ...). CEP languages extend traditional ECA-rules by providing the operators of the event algebra as first-class programming constructs. For implementing CEP languages, the rule-based execution needs to be extended, because the system has to keep a repository of past events for all incomplete matches. Moreover, the system has to make sure that past events are automatically removed from the memory, when they are no longer needed.

## CONTROLLED RULE ENGINES AND RULES (B1)

This category contains rule engine based architectures, which have changed the typical way of iteratively executing activations via an agenda. Nonetheless, the offered programming concepts have not been changed, i.e. the agent behavior specification is mainly done by programming traditional rules.

As Figure 1 shows, two typical representatives for this category are RC++ (Wright and Marshall 2003) and SOAR (Lehman et al. 1996). The motivations behind those approaches are quite different. RC++ is an extension to the C++ programming language, which incorporates rules directly into the base language. It therefore realizes a conservative extension approach, which aims at a tight integration with the underlying procedural core language. The RC++ language is a general purpose language, but has been designed with a clear application focus in mind. It should facilitate the programming of game AI for the game console. In contrast, SOAR has been developed to be a general problem solver for arbitrary complex problems. Hence, the SOAR architecture has primarily been used for knowledge-rich intelligent agent applications. Examples include intelligent control, natural language understanding, human behavior experiments and simulation (Wray and Jones 2006). As SOAR has advanced the ordinary rule interpretation in many directions, it will be explained in more details in the following.

## SOAR

The official project start of SOAR (originally for State, Operator And Result) can be dated back to 1983. Initially, the work on SOAR was influenced strongly by Alan Newell and his former students John Laird and Paul Rosenbloom. In the course of time an international research community emerged that jointly works on extensions and enhancements for SOAR.

One main motivation for the conception of SOAR was Newell's (1990) idea of a Unified Theories of Cognition (UTC). Such an UTC should encompass all relevant areas of human abilities into one major psychological theory, which is able to explain e.g. problem solving, learning, perception, language understanding, emotion and dreaming aspects. In order to advance research in this direction and to overcome the multitude of existing psychological micro-theories, each applicable for one very specific phenomenon, he proposed SOAR as a promising UTC candidate. The SOAR architecture has been devised to be able to cope with all tasks that could be principally posed to an intelligent agent, i.e. it should be able to handle simple routine jobs as well as complex problem solving tasks using different solution strategies. Furthermore, interaction with the environment and learning from past experiences should be facilitated by the architecture. Despite these challenging requirements it has been a primary design principle to keep the architecture complexity as small as possible (*parsimony principle*). Therefore, the number of architecture mechanisms has been reduced to a minimum, i.e. the architecture comprises only one representation for long-term and short-term memory as well as one mechanism for goal construction and learning.

Being a multi-disciplinary effort, different viewpoints exist regarding SOAR as a theory of general intelligence or human cognition, and also as an agent architecture or a programming language. Discussing SOAR as an agent architecture is useful, because SOAR demonstrates, how algorithms from artificial intelligence can be integrated into an architecture that allows building software systems exhibiting general intelligent behavior (Wray and Jones 2006). This viewpoint will therefore allow comparing SOAR to the other agent approaches described in this chapter.

## Architecture Components

The SOAR architecture (Lehman et al. 1996, Laird et al. 2006) is based on the hypothesis that behavior determination should be done by manipulating knowledge only (*physical symbol system hypothesis*). SOAR proposes state space searching as core mechanism for problem solving. Primary concepts are therefore problem spaces and operators, which are applied to transfer one state to the follow-up state. Declarative goal achievement can be realized by specifying a target goal state and apply an operator search until the desired state has been reached.

In Figure 7 the SOAR architecture, its main components and their interplay are depicted. The interpreter performs the problem space transformations by operating on its memories. Generally, two different kinds of memories are distinguished. The working memory contains information about the current situation of the agent including sensor data, results of inferencing as well as active operators. SOAR's working memory is hierarchically structured and contains substates for each subgoal, which are connected via state objects. The long-term memory of a SOAR agent contains its problem solving capabilities in form of production rules. In SOAR three different kinds of rules are supported. *Operator proposal rules* are used for finding applicable operators. Given that an operator was selected, *operator application rules* help to realize its behavior. The third kind of rules is *state elaboration rules*, which are used for deriving new knowledge from the existing knowledge in the working memory. In the following a more detailed review of the interpreter's deliberation cycle will be given.

## Deliberation Cycle

SOAR's deliberation is performed in a typical perceive-decide-act cycle. It consists of the following five phases, which are continuously executed:

**Input phase:** In the input phase external data from the environment is transferred to the SOAR working memory. The external data can be in arbitrary format and specific input mapping

*Figure 7. SOAR agent architecture*

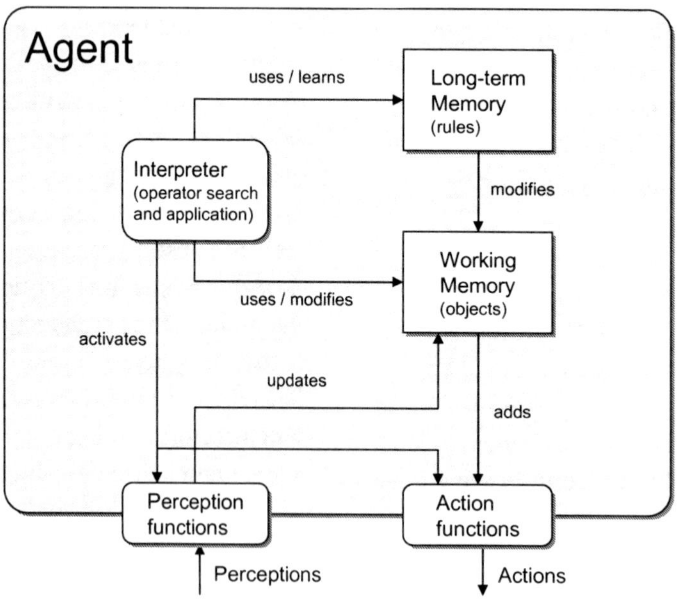

functions need to be provided, which transform the external representation to the internal triple representation. This design separates the perception from the cognitive processes and allows a concurrent execution of both.

**Proposal phase:** In the proposal phase two different activities are performed. On the one hand all activated state elaboration rules are fired in parallel. Given that these rules lead to new activations, these will also be applied. The process stops when no activations are available any more and *quiescence* has been reached. In addition operator proposal rules are fired. These rules have the one and only purpose to suggest applicable operators.

**Decision phase:** The aim of this phase is to select one of the proposed operators for execution. For this purpose a built-in evaluation mechanism is applied, which has the task to weigh operator preferences and select the best one. If the decision yields a single winning operator, this operator will be marked for application. Otherwise, it is assumed that the knowledge for operator selection is insufficient and additional knowledge needs to be obtained. This situation is called an *impasse* and can basically have three different reasons:

- No operator can be selected, because no operator has been proposed.
- No operator can be selected, because multiple equivalent operators have been proposed.
- An operator can be selected but the agent is missing information for operator execution.

In case of an impasse the interpreter automatically tries to resolve it by using the built-in subgoaling mechanism (*automatic subgoaling hypothesis*). For the subgoal a new problem space will be created, which is a substate of the original state. As the creation of subgoals can be applied in each problem space this may lead to arbitrary deep goal hierarchies. When the subgoal processing generates information suitable for resolving the tie, the results can be merged into the higher-level state leading to a removal of the substate. In order to avoid the possibly expensive search for resolving the same tie more than once, SOAR employs a learning mechanism called *chunking*. This mechanism enables a SOAR agent to summarize and generalize a problem solving strategy used in the substate to a new rule. The rule will be added to the long-term memory and therefore can be utilized in future problem solving activities.

**Application phase:** In this phase the execution of the selected operator will be performed. As the operator is itself part of the working memory it can be used in the condition part of operator application rules. Similar to the proposal phase, all activated application rules are fired in parallel until quiescence has been reached.

**Output phase:** The output phase serves for propagating information about actions to the environment. Similar to the input phase user-defined functions are used to transform the internal knowledge representation to an external one.

## Agent Properties

**Reactive behavior.** In SOAR reactive behavior can be encoded via operator proposal and application rules. New environmental information is consumed via user-specific input functions, which add the new information to the input-links of the state. Operator rules can be specified with condition parts that are sensitive to those changes. They produce operators and provoke the invocation of application rules. These application rules can trigger external actions by adding data to the output-links of the state, which will be interpreted by the also user-specified action functions once in every deliberation cycle.

**Proactive behavior.** Proactive behavior can be achieved by goal directed search and planning strategies. For this purpose the target state to achieve can be observed by a dedicated monitoring rule. In addition, the goal can be represented e.g. via state attributes that guide the operator proposal process. Whenever a goal is to be achieved, dedicated search rules and possibly heuristic rules for controlling the search directions can be employed. In this respect, automatic subgoaling in impasse situations and the distinction between proposal, selection and application rules facilitates the implementation of proactive behavior.

**Communication.** In SOAR, interaction and communication facilities have to be provided on the user level by exploiting the in- and output functions. On top of these, simple environment-based agent interaction, e.g. in the TanksSoar Game, as well as sophisticated social models, such as the STEAM teamwork model based on the Joint Intention theory, can be implemented.

**Mentalistic notions.** Mentalistic notions are not directly used in SOAR. Designing SOAR solutions requires thinking in terms of problem spaces and operators, whereas the programming of SOAR agents is done with (different kinds of) production rules only. Wray and Jones (2006) state that this minimalist design of the programming concepts leads to an "'assembly language' for intelligent systems".

## Rule-Based Mechanisms

**Rule types.** SOAR is based on forward-chaining and production rules are used only. In contrast to normal rule-based systems, a further distinction of generic rule types has been introduced. In this respect it is distinguished between operator proposal and application rules as well as state elaboration rules. These different kinds of rules help in realizing complex search-based behavior and establish a meaningful separation of concerns between what can be done (operator proposal) and how it can be accomplished (operator application).

**State representation.** The basic representation of facts is the same as in the OPS5 expert system (Brownston et al. 1985) SOAR is based on. The working memory consists of facts, which are represented by object-attribute-value (OAV) triples. A value of a triple can either be a constant or an identifier of another object. SOAR uses a weak type system, which does not require types to be defined before usage, i.e. one can add new attributes to an object as needed. On the contrary this also means that the system cannot apply type checks and typing errors cannot be automatically detected. Despite these similarities to OPS5, SOAR also exhibits important distinctions. One important aspect is the hierarchical organization of substates according to the subgoaling mechanism. A substate represents a new problem space, which can contain arbitrary new information and is additionally linked to the superstate.

**Rule evaluation mechanism.** The rule evaluation is based on the RETE algorithm (Forgy 1982). In the context of the SOAR project also alternatives (e.g. TREAT, Miranker 1989) have been evaluated, but the RETE algorithm proved to be the most effective one. In the course of the development several improvements to the basic RETE algorithm have been found and applied (e.g. Doorenbos 1995).

**Conflict resolution.** Due to the specific execution mode of SOAR, conflict resolution is handled completely different from standard rule-based systems. In the proposal and application phases the rules are fired in parallel until quiescence is reached. This renders conflict resolution within these phases unnecessary. On the contrary, within the operator selection phase an elaborated conflict resolution mechanism is applied for choosing among the proposed operators, i.e. conflict resolution is not applied with respect to activated rules but to operators. The resolution strategy is predefined and is based on operator preferences, which are assigned to the operators within the proposal phase. The resolution has then the task to determine the best operator. If this is impossible due to an impasse situation, subgoaling will be applied for resolving the tie.

**Truth maintenance.** In SOAR between permanent and temporary working memory elements is distinguished. State changes that are effect of operator execution in the operator application phase are considered as permanent and are said to be o-supported (operator-supported). However, all state changes that occur within the proposal phase are volatile and rely on the production that produced them to continue to match. SOAR supports this by logical facts that exist only as long as the productions condition is valid. If it gets invalid the fact is retracted automatically.

### Rule Incorporation

At the programming language level SOAR incorporates rules as primary programming concept. In contrast to traditional rule-based approaches, rules are used in SOAR for realizing higher-level architecture concepts. SOAR does not strive for an integration of the SOAR language with common mainstream programming languages, i.e. integration with external systems is possible only by using the input and output functions. This also means that SOAR actions, i.e. the right hand side of a rule, can only affect the working memory of the agent.

At the interpreter level, SOAR builds on an enhanced rule engine, which also makes up its only core component. The interpreter deliberation cycle is largely different from the well-known recognize-select-act execution cycle employed in standard rule systems. The reason for this significant change is mainly caused by the introduction of operator-based problem space search. Hence, the interpreter uses rules for the realization of problem space exploration via rule application.

## Conclusion

In this category proven rule-based programming concepts have been made usable on top of custom agent interpreters. These agent interpreters extend rule-kernels in different directions and especially try to adapt the deliberation cycle according to the sense-think-act loop of situated agents. This means that the extended interpreters take into account that an agent regularly senses its environment and exert actions on it.

In the following a short overview about the pros and cons of the B1 category is given:

- \+ Rules are a well-known programming concept and exploiting only rules leads to a flat learning curve for people having already a background in this area.
- \+ Reactive agent behavior can be encoded easily in terms of rules and therefore the rule metaphor fits for specifying simple behavior patterns.
- \- The representation of proactive behavior is rather unintuitive and difficult using rule descriptions only. One important reason for this is that proactive behavior needs to describe what is to be achieved and in addition how it can be accomplished. If no explicit concept for the representation of the agent's goals are available, they have to be encoded indirectly, which renders the descriptions hard to understand.
- \- Procedural knowledge is difficult to capture using rule descriptions. Especially, if a plan consists of multiple logical steps (performing an action, sending a message, waiting for the response, performing the next action), this cannot be easily captured in one rule, but needs to be scattered among several rules.

### Generalization

The approach of using rules for programming in a system environment that extends the basic rule engine behavior has not been considered to a great extent outside the agent community. Despite this fact, at least one other application area exists, in which a similar approach has been pursued. It is widely recognized that the object-oriented programming paradigm fits well for implementing many application domains. Nevertheless, other approaches such as logic programming have emerged and can partly complement the weaknesses of purely procedural representations. In (Lodha et al. 2005) an architecture for the combination of rule and object-oriented programming has been proposed. Despite these efforts, the mainstream software development does not focus on extending rule interpreters, but instead puts effort in a better integration of rule engines as one part of a business infrastructure. In the advent of service-oriented approaches e.g. rule services and repositories have been proposed, which have the task to centrally bundle rule functionalities and make them available enterprise-wide.

## RULE ENGINES AND MENTALISTIC NOTIONS (C2)

In this category approaches are subsumed that introduce mentalistic notions as programming constructs and rely on a rule engine as execution machinery for the agent interpreter. In contrast to category C1 the agent architecture is built on top of a rule kernel using standard rule programming. This means that the rule engine directly becomes an agent interpreter and no changes at the execution level need to be applied. Nonetheless, from the agent programmer's point of view the difference between both categories is blurred, because in both cases high-level mentalistic programming concepts are offered for application development.

Figure 1 highlights that currently Jadex is the only system which falls into this category. Jadex offers belief-desire-intention (BDI) abstractions for programming agents. It has been designed as general purpose agent architecture, which aims at a wide applicability within various kinds of application domains. Some of the most interesting applications have tackled agile business process management (Burmeister et al. 2006) and dynamic appointment scheduling in hospitals (Paulussen et al. 2006). In the following the details of the underlying Jadex architecture will be presented.

### Jadex

Foundation of the Jadex architecture is the BDI model, which has been originally proposed by Bratman (1987) as a framework for explaining rational behavior of human agents. BDI assumes a two-staged practical reasoning process which consists of a goal deliberation and a means-end reasoning phase (Wooldridge, 2000). Goal deliberation aims at deciding what to do by selecting a coherent set of goals which should be pursued and the downstream means-end reasoning phase has the purpose to find means for the realization of a specific goal. The BDI model has been picked-up by Rao and Georgeff (1995) who conceived the PRS (Procedural Reasoning System) agent architecture on basis of the BDI model. This architecture focuses mainly on the means-end reasoning process and was subsequently extended by Pokahr et al. (2005a) supporting also the goal deliberation phase. The resulting architecture is based on a rule-kernel but exhibits only the BDI concepts as programming constructs to the developer.

#### Architecture Components

In Figure 8 a high-level view of the Jadex architecture is depicted. It basically shows that the agent interpreter operates on the three attitudes: beliefs, desires and intentions. These basic notions represent, what the agent knows about the world (beliefs), which things its tries to achieve (desires), and the already committed courses of actions it is currently using (intentions).

In order to decide which goals to pursue the interpreter evaluates its current set of desires and selects a coherent subset, which does not contain any conflicting goals. The Jadex interpreter uses the "easy deliberation strategy" (Pokahr et al. 2005b) for performing the selection process, which requires that meta information for goals has been defined.

*Figure 8. Abstract PRS agent architecture (cf. Georgeff and Lansky 1987)*

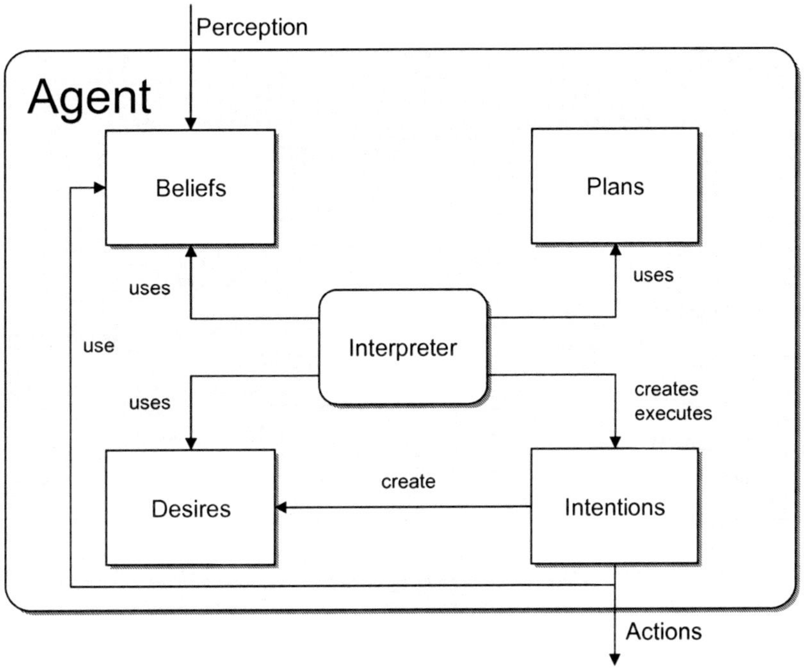


The means-end reasoning is done for each of the active goals separately. Foundation of this process is that a library of predefined plans is available. A plan hereby represents a procedural piece of knowledge, which encodes how a certain objective can be handled. In contrast to a single action, a plan can be arbitrary complex including sophisticated control structures and may consist of many subsequent steps. In addition to the plan code, also meta information about the plans is stored in the plan library. This allows the means-end reasoning process to search the library for relevant plans, i.e. plan that have declared to be usable for achieving the current goal. From this plan set only those will be kept, whose context condition holds, i.e. which are applicable with respect to the situation the agent is in. If still more than one plan is available, further reasoning for choosing among them can be applied (meta-level reasoning), or a single one can be picked according to a simple heuristics (e.g. random or first one). This plan will then be instantiated and will be added to the active intentions of the agent. During plan execution the plan can access the agent's beliefs and also can issue subgoals for further task delegation. If a plan cannot achieve its associated goal, the goal is not automatically considered as failed. Instead the means-end reasoning mechanism will try to accomplish the goal using other applicable plans until no plan is left or the goal could finally be attained.

## Deliberation Cycle

In contrast to classical PRS agents (Rao und Georgeff 1995) and also many other agent architectures, Jadex does not employ a specific deliberation cycle. Instead, Jadex has a built-in rule program, which controls the interpreters operation. This rule program encodes the complete BDI-logic as explained above and realizes goal-deliberation as well as means-end reasoning.

In Figure 9 the detailed Jadex agent architecture is presented. It can be inferred that the

*Figure 9. Detailed Jadex agent architecture*

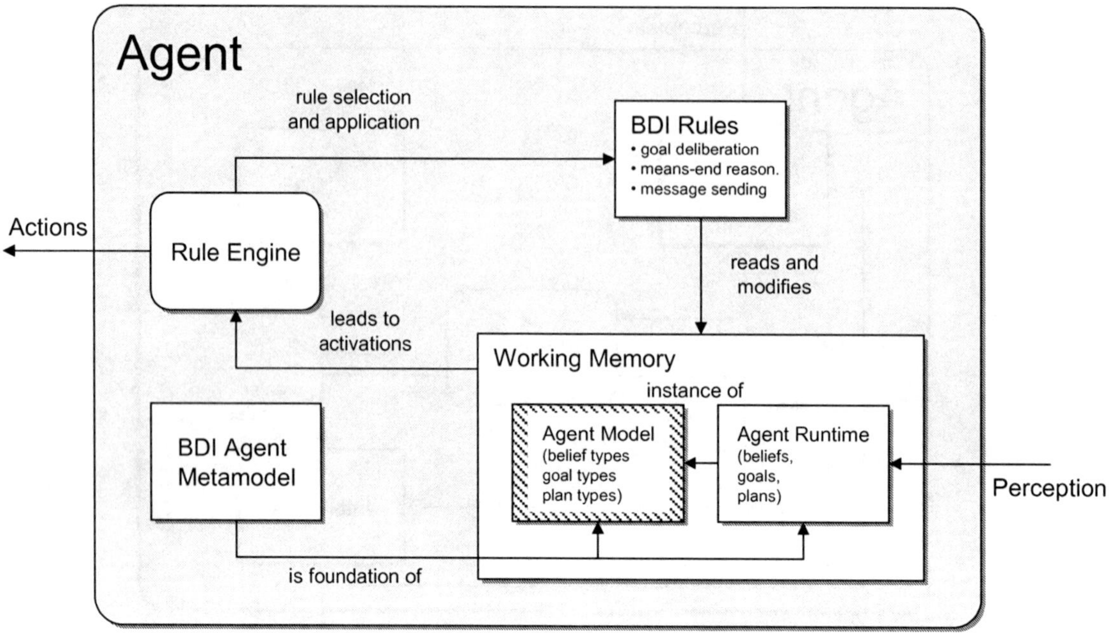


only active component is the rule engine, which operates on a predefined set of BDI rules. The execution cycle is therefore exactly the same as in a normal rule engine and consists of the recognize-select-act loop that is repeated until no new activations can be found. In contrast to normal rule programs Jadex agents are equipped with a fixed set of BDI rules. These rules are responsible for realizing all agent as well as BDI-specific behavior, i.e. they model e.g. message receiving as well as goal and plan handling. They rely on a predefined agent memory structure, which is encoded in the BDI agent metamodel. This model specifies exactly which kinds of attributes an agent type and an agent instance may have, i.e. the agent type metamodel e.g. constitutes that the agent has specific slots for storing beliefs, plans and goals. At runtime the working memory therefore contains two information sources. The agent model represents the agent class, which is defined by the developer and mainly consists of the definition of the structural elements (beliefs, goals and plan meta-information) as well as the procedural plan code. Similar to object-oriented systems an instance of this class is held in the memory, which e.g. contains runtime information about the currently active goals and plans.

## Agent Properties

**Reactive behavior.** A Jadex agent can receive new information via incoming messages as well as through direct senor update via its beliefs. Immediate reactions are possible to both kinds of events. On the one hand plans can be triggered by incoming message events and on the other hand belief changes can be tracked for defining data-driven plans. Within agent plans arbitrary actions can be performed. This includes changes to the agent state, initiating environmental actions as well as sending and waiting for messages.

**Proactive behavior.** Jadex has been conceived with specific support for proactive behavior. In Jadex, goals are a first-class concept and are explicitly represented within the agent. In this respect it is distinguished between different kinds of goals, which express different motivations towards their accomplishment. Currently, the platform supports perform, achieve, query and maintain goals. The simplest goal kind is the perform goal, which directly leads to action execution and does not contain a declarative part. In contrast achieve goals are associated with a declarative description of a desired world state. Therefore an achieve goal is finished only once this condition is satisfied or it has been determined that it cannot be attained in the future. Query goals can be used to retrieve information. These goals only lead to plan execution, if the requested information is not immediately available from the existing agent knowledge. The most complex goal types are maintain goals, which have the purpose to continuously monitor a specified world state. Whenever the goal detects that the world state is not valid any longer it will get active and lead to plan execution for the re-establishment of the invalid state. Jadex handles all these different goal kinds within the same execution machinery, which is only possible, because a generic goal lifecycle has been defined for all kinds of goals. This goal lifecycle is used for the goal deliberation as well as for the means-end reasoning. It basically partitions the adopted agent goals into active and inactive sets. The goal deliberation can then shift goals between these states as needed and the means-end reasoning processes the goals of the active set as needed. Details about the goal representation and execution can be found in (Braubach et al. 2005).

**Communication.** The Jadex architecture includes message-based communication facilities. Even though the architecture does not prescribe a specific message format, the FIPA standard (http://www.fipa.org/) has been adopted for representing messages. As in Jadex the agent architecture is decoupled from the platform architecture, messages can be sent using any transport layer the platform offers. The standard implementation uses simple TCP/IP based communication channels.

**Mentalistic notions.** Mentalistic notions play a crucial role in Jadex. The programming model is built on the intentional stance and considers mentalistic notions as an adequate vehicle for constructing agents. As all important concepts are represented explicitly, the behavior of agents can be introspected at runtime. This introspection reveals, which goals and plans are currently pursued and also which belief values are stored. This allows programmers to easily grasp the agent's state and understand immediately the motivations for the externally perceived agent actions. In addition, further explanation mechanisms have been developed (Lam and Barber 2005) and also made available in Jadex, which visualize mentalistic event traces and interconnections.

## Rule-Based Mechanisms

**Rule types.** The Jadex agent interpreter uses a forward-chaining rule engine and employs production rules for rule specification. Within Jadex two different kinds of rules can be distinguished. The system relies on a set of fixed BDI rules for the realization of the BDI base functionality. On the other hand, conditions are part of many higher-level concepts, such as an achievement condition of a goal. This means that a programmer can define the condition part of a goal, which is internally handled as rule. The action part of such an implicit rule is determined by its context and cannot be altered by the developer.

**State representation.** The state representation of Jadex has two facets. Internally, the working memory is organized in an object-attribute-value style similar to other rule engines. The state is strongly typed and relies on a type system for verifying that objects are correctly accessed. This also means that types have to be defined in beforehand of their usage. Additionally, the state provides automatic garbage collection facilities that simplify the memory management, as unreachable state objects are cleaned up automatically. For the agent programmer an object-oriented perspective is offered as access point to the working memory. Hence, externally the state is perceived as a structure of Java objects and can be accessed and modified by using normal method calls.

**Rule evaluation mechanism.** In Jadex, the RETE algorithm is used for rule evaluation (Forgy 1982). The original mechanisms has been adapted to the strong type system and is similar to the functionality of JESS (Friedman-Hill 2003).

**Conflict resolution.** In general, the importance of rule ordering is kept low in Jadex, i.e. the BDI rules are formulated in a way that tries to avoid ambiguities and the overall functionality of the agent should not depend on the rule order. Conflict resolution has to be applied in Jadex when multiple activations of BDI rules are present in the agenda. In this case a simple FIFO strategy is used for fetching the next activation to execute.

**Truth maintenance.** Truth maintenance is currently not used in Jadex.

### Rule Incorporation

Jadex offers a hybrid programming language approach and separates structural from procedural specifications, i.e. an agent is programmed using an XML dialect for defining the agent type and ordinary Java for the plan bodies. Within the structural part conditions are currently specified in a CLIPS-like language (Giarratano and Riley 2005). Probably this will change in the future and a more Java-like syntax will be adopted for rules, too (possibly similar to Drools, http://www.jboss.org/drools/). As the plans are written in Java, an integration with external libraries and systems is possible using the standard Java mechanisms. Hence, in Jadex agent programming is done using intuitive mentalistic notions and rules appear only as conditions within predefined higher-level constructs such as goals.

The Jadex agent interpreter is a pure rule engine, which operates on a fixed set of BDI rules. These rules encode the agent and BDI knowledge especially for realizing the goal deliberation and means-end reasoning mechanisms.

## Conclusion

This category has highlighted that rule engines can be used as backbone for realizing higher-level architectures with more abstract concepts than rules itself. The extensions performed in this category hide the internal rule execution mechanism to a great extent and instead present other conceptual notions for the programming of agents. The idea of this category is therefore twofold: using established rule engines for the efficient agent execution and provide intuitive mentalistic notions to the agent developer.

A short list of the main advantages and disadvantages summarizes the main properties of Category C2:

+ Mentalistic notions for programming agents realize the intentional stance and therefore exhibit advantages with respect to the specification and understanding of its behavior. Moreover, it can help in debugging or improving the agent.
+ Using rule engines for agent implementation has the big advantage that algorithms such as RETE, TREAT or LEAPS exist, which ensure an efficient and scalable execution.
- Rules may not be applicable for implementing all kind of higher-level system behavior. This means that an agent interpreter should introduce further extension (cf. category C1) if a pure rule implementation is not adequate for the problem at hand.
- The usage of higher-level architectures for programming does not pay off for all applications. For specific problems it will be simpler and more convenient to use a rule-based solution, e.g. if an agent is only required to exhibit reactive abilities.

### Generalization

The proposed approach of using rule engines as execution kernel for higher-level has been used outside the agent research area as well. One reason for this is that rule engines have a long track of research and are proven reliable technology today. In the area of workflow management systems, questions of how to describe and implement a process have to be tackled. For the design of processes notations such as EPCs (event process chains) and BPMN (business process modeling notation) have emerged. The execution of workflows is in most cases performed within specialized workflow engines, which can handle workflow specifications. As EPCs do not posses an executable semantics per se, in Knolmayer et al. (2000) an approach for a transformation of EPC descriptions into rules has been proposed. For each of the different EPC connector types rules patterns have been conceived so that an automatic translation can be achieved. The integration of rule and workflow concepts has also been recently discussed in the context of the Drools rule engine and the JBoss workflow management system. The idea here is also to use the Drools engine as core of a workflow system.

## FUTURE TRENDS

Rule bases have been used for building intelligent agents at least since the eighties of the last century. This chapter has shown that rules are nowadays an important part of many prevalent agent architectures. The continued integration of agent concepts and rule-based technology opens up many interesting areas for future developments. Some opportunities for future research are illustrated in the following.

From the multi-agent perspective distributed systems are one important application area. Despite this fact, in many commercial projects not multi-agent but service-oriented architectures are employed. This has advantages with respect to the proven and standards-driven technology base but also leads to problems, when the passive service character is insufficient and e.g. dynamic and proactive behavior is needed. In order to cope with this dilemma the event-driven architectures (EDA) have been proposed as an extension for SOA. Core concept of EDA is that the environment is observed via a network of agents, which can detect business-relevant situations using complex event processing (CEP). For building EDA-enabled infrastructures a combination of all three aforementioned technology strands is necessary.

Another upcoming trend is the enrichment of the Internet with semantics and its transformation to the SemanticWeb (Antoniou and van Harmelen 2004). This shall enable not only humans but also machines to understand web content and use it for fulfilling tasks. In a typical application case a human user could just delegate a task to a dedicated agent and let the agent perform this task in the background. The agent then possibly would gather information from the web and interact with other agents. For exploiting the semantics and deducing information from the given ontology knowledge, often a rule-based machinery is used. Hence, the bringing together of rule-based knowledge representation and reasoning with agent technology could be considered as enabling research area for the semantic web.

## CONCLUSION

In this chapter agent architectures have been described as one application area for rule bases. It has been shown in what different ways agent architectures make use of rule concepts and in what directions they extend existing rule-based approaches.

To identify the differences and commonalities of the approaches, a categorization has been introduced. This categorization is made up of two dimensions: The first dimension considers the execution and distinguishes between pure rule engines and approaches that introduce additional execution logic in an extended or controlled rule engine. The second dimension takes a programmers viewpoint and differentiates approaches that introduce higher level constructs such as beliefs or goals from those that do not.

Each of the resulting four categories has been investigated and a prominent representative of each category has been presented in detail. Besides a general description of the features of a representative, the focus has been put on the incorporation of rule mechanisms into the overall architecture. Discussions of the advantages and disadvantages of the approaches reveal that all approaches legitimately coexist, as all have their strength and weaknesses with respect to certain application requirements.

## REFERENCES


Antoniou, G, & van Harmelen, F. (2004). *A Semantic Web primer.* MIT Press.

Bass, L., Clements, P., & Kazman, R. (2005). *Software architecture in practice.* Addison-Wesley.

Bellifemine, F., Bergenti, F., Caire, G., & Poggi, A. (2005). JADE - A Java agent development framework. In R. Bordini, M. Dastani, J. Dix, A. El Fallah Seghrouchni (Eds.), *Multi-agent programming: Languages, platforms and applications* (pp. 125-147). Springer.

Bordini, R., Hübner, J. F., & Vieira, R. (2005). Jason and the golden fleece of agent-oriented programming. In R. Bordini, M. Dastani, J. Dix, A. El Fallah Seghrouchni (Eds.), *Multi-agent programming: Languages, platforms and applications* (pp. 3-37). Springer.

Bratman, M. (1987). *Intention, plans, and practical reason.* Harvard University Press.

Bratman, M., Israel, D., & Pollack, M. (1988). Plans and resource-bounded practical reasoning. *Computational Intelligence*, *4*(4), 349-355.

Braubach, L., Pokahr, A., Lamersdorf, W., & Moldt, D. (2005). Goal representation for BDI agent systems. In R. Bordini, M. Dastani, J. Dix, A. El Fallah Seghrouchni (Eds.), *Proceedings of the 2nd International Workshop on Programming Multiagent Systems, Languages and Tools (PROMAS 2004)*, 3rd International Joint Conference on Autonomous Agents & Multi-Agent Systems (AAMAS'04) (pp. 46-67). New York: Springer.

Braubach, L., Pokahr, A., & Lamersdorf, W. (2006). Tools and standards. In S. Kirn, O. Herzog, P. Lockemann, O. Spaniol (Eds.), *Multiagent engineering - Theory and Applications in enterprises* (pp. 503-530). Springer.

Burmeister, B., Steiert, H.-P., Bauer; T., & Baumgärtel H. (2006). Agile processes through goal- and context-oriented business process modeling. In *Proceedings of Business Process Management Workshops*, Vol. 4103, (pp. 217-228). Springer.

Brownston, L ., Farrell, R., Kant, E., & Martin, N. (1985). *Programming expert systems in OPS5.* Addison-Wesley.

Brooks, R. A. (1989). How to build complete creatures rather than isolated cognitive simulators. In K. VanLehn (Ed.), *Architectures for intelligence* (pp. 225-239).

Byrne, C., & Edwards, P. (1996). Refinement in agent groups. In G. Weiß (Ed.), *Proceedings of the IJCAI'95 Workshop on Adaption and Learning in Multi-Agent Systems* (pp. 22-39). Springer.

Cardoso, H. L. (2007). *Integrating JADE and Jess.* Retrieved from http://jade.tilab.com/doc/tutorials/jade-jess/jade_jess.html

Dastani, M., Hobo, D., Meyer, J.-J. (2007). Practical extensions in agent programming languages. In *Proceedings of the Sixth International Joint Conference on Autonomous Agents and Multi-*

*agent Systems (AAMAS'07)* (pp. 923-925). ACM Press.

Davies, W., & Edwards, P. (1994). Agent-K: An integration of AOP and KQML. In *Proceedings of the Third International Conference on Information and Knowledge Management (CIKM'94).* ACM Press.

Dennett, D. (1971). Intentional systems. *Journal of Philosophy, 68*, 87-106.

Dietrich, J., Kozlenkov, A., Schroeder, M., & Wagner, G. (2003). Rule-based agents for the Semantic Web. *Electronic Commerce Research Applications, 2*(4), 323-338.

Doorenbos, R. (1995). *Production matching for large learning systems.* PhD Thesis, Carnegie Mellon University, UMI Order No. GAX95-22942.

Finin, T., Weber, J., Wiederhold, G., Genesereth, M., McKay, D., & Fritzson, R. et al. (1993). *Specification of the KQML agent-communication language – plus example agent policies and architectures.* (EIT TR 92-04).

Fiege, L., Mühl, G., Pietzuch, P. (2006). *Distributed event-based systems.* Springer.

Forgy, C. (1982) Rete: A fast algorithm for the many patterns/many objects match problem. *Artificial Intelligence,* 19(1), 17-37.

Franklin, S., Graesser, A. (1997). Is it an agent, or just a program? A taxonomy for autonomous agents. In J. Müller, M. Wooldridge, N. Jennings (Eds.), *Proceedings of the 3rd Workshop on Intelligent Agents III, Agent Theories, Architectures, and Languages (ATAL 1996)* (pp. 21-35). Springer.

Friedman-Hill, E. (2003). *Jess in action, Java rule-based systems.* Manning Publishers.

Georgeff, M., Lansky, A. (1987). Reactive reasoning and planning: An experiment with a mobile robot. In *Proceedings of the 6th National Conference on Artificial Intelligence (AAAI 1987)* (pp. 677-682).

Giarratano, J. C., & Riley, G. D. (2005). *Expert systems. Principles and programming.* Thomson Corse Technology.

Hindriks, K., Boer, F. de, Hoek, W., & van d., Meyer, J.-J. (1999). Agent programming in 3APL. *Autonomous Agents and Multi-Agent Systems, 2*(4), 357-401.

Jennings, N. R., & Wooldridge, M. J. (1998). *Agent technology - Foundations, applications and markets.* Springer.

Knolmayer, G., Endl, R., Pfahrer, M. (2000). Modeling processes and workflows by business rules. In W. M. Aalst, J. Desel und A. Oberweis (Eds.), *Business process management, models, techniques, and empirical studies* (pp. 16-29). Springer.

Krempels, K.H., & Panchenko, A. (2003). An approach for automated surgery scheduling. In E.K. Burke & H. Rudov (Eds.), *Proceedings of the 6th International Conference on the Practice and Theory of Automated Timetabling (PATAT '06).*

Laird, J., Congdon, C.B., & Coulter, K. (2006). *The Soar user's manual. Version 8.6.3.* University of Michigan. Retrieved from http://sitemaker.umich.edu/soar

Lam, D. N., & Barber, K. S. (2005). Comprehending agent software. In F. Dignum, V. Dignum, S. Koenig, S. Kraus, M. P. Singh, & M. Wooldridge (Eds.), *Proceedings of the 4th International Joint Conference on Autonomous Agents and Multiagent Systems (AAMAS 2005)* (pp. 586-593). ACM Press.

Lehman, J. F., Laird, J., & Rosenbloom, P. (1996). A gentle introduction to Soar, an architecture for human cognition. In S. Sternberg, D. Scarborough (Eds.), *Invitation to cognitive science*, 4, 212-249. MIT Press.

Lodha, B., Dinesha, K. V., & Kumar, P. (2005). Need for incorporating a rule-based component in conventional object oriented systems. In

*Proceedings of the International Conference on Information Technology: Coding and Computing (ITCC'05).*

Luck, M., McBurney, P., Shehory, O., & Willmott, S. (2005). *Agent technology: computing as interaction (A roadmap for agent based computing).* AgentLink.

Luckham, D. (2002). *The power of events.* Addison Wesley.

McCarthy, J. (1979). Ascribing mental qualities to machines. In M. Ringle (Ed.), *Philosophical perspectives in artificial intelligence* (pp. 161-195). Humanities Press.

Miranker, D. (1989). TREAT: A new and efficient match algorithm for AI production systems. In *Proceedings of National Conference on Artificial Intelligence* (pp. 42-47), Pittman/Morgan Kaufman.

Newell, A. (1990). *Unified theories of cognition.* Harvard University Press.

Newell, A., & Simon, H. A. (1976). Computer science as empirical enquiry. *Communications of the ACM, 19,* 113-126.

Nwana, H. (1995) Software agents: An overview. *Knowledge Engineering Review, 11*(2), 205-244.

Paulussen, T.O., Zöller, A., Heinzl, A., Braubach, L., Pokahr, A, & Lamersdorf, W. (2006). Agent-based patient scheduling in hospitals. In S. Kirn, O. Herzog, P. Lockemann, & O. Spaniol (Eds.), *Multiagent engineering - Theory and applications in enterprises* (pp. 255-275). Springer.

Paschke, A., & Bichler, M. (2005). SLA representation, management and enforcement. In *Proceedings IEEE International Conference on e-Technology, e-Commerce, e-Service (EEE'05).*

Paschke, A., Kiss, C., & Al-Hunaty, S. (2006). NPL: Negotiation Pattern Language - A design pattern language for decentralized (agent) coordination and negotiation protocols. In R. Banda (Ed.), *E-negotiation - An introduction.* ICFAI University Press.

Paschke, A. (2007). *Rule-based service level agreements - Knowledge representation for automated e-contract, SLA and policy management.* München: IDEA Verlag GmbH. Paschke, A., Boley, H., Kozlenkov, A., & Craig, B. (2007). Rule responder: RuleML-based agents for distributed collaboration on the pragmatic Web. In *Proceedings of the 2nd international Conference on Pragmatic Web* (pp. 17-28). ACM Press.

Paschke, A., Kozlenkov, A., & Boley, H. (2007a) A homogenous reaction rules language for complex event processing. In *Proceedings of International Workshop on Event Drive Architecture for Complex Event Process (EDA-PS 2007).*

Pokahr, A., Braubach, L., & Lamersdorf, W. (2005a) A BDI architecture for goal deliberation. In F. Dignum, V. Dignum, S. Koenig, S. Kraus, M. Singh, & M. Wooldridge (Eds.), *Proceedings of 4th International Joint Conference on Autonomous Agents and Multiagent Systems (AAMAS 2005)* (pp. 1295-1296). ACM.

Pokahr, A., Braubach, L., & Lamersdorf, W. (2005b). A goal deliberation strategy for BDI agent systems. In T. Eymann, F. Klügl, F, W. Lamersdorf, M. Klusch, & M. Huhns (Eds.), *Proceedings of the Third German Conference on Multi-Agent System Technologies (MATES-2005)* (pp. 82-94). Springer.

Rao, A., Georgeff, M. (1995). BDI Agents: From theory to practice. In V. Lesser, L. Gasser (Eds.), *Proceedings of the First International Conference on Multiagent Systems* (pp. 312-319). MIT Press.

Ross, R., Collier, R., O'Hare, G. (2005). AF-APL - Bridging Principles and practice in agent oriented languages. In R. Bordini, M. Dastani, J. Dix, & A. El Fallah Seghrouchni, (Eds.), *Proceedings of Second International Workshop on Program-*

*ming Multi-Agent Systems (ProMAS 2004)* (pp. 66-88). Springer.

Schroeder, M., & Wagner, G. (1998) Vivid agents: theory, architecture, and applications. *Applied Artificial Intelligence, 14*, 645-675.

Shoham, Y. (1993) Agent-oriented programming. *Artificial Intelligence, 60*(1), 51-92.

Thomas, S. R. (1995). The PLACA agent programming language. In M. Wooldridge, N. Jennings (Eds.), *Proceedings of the 1st International Workshop Intelligent Agents I, Agent Theories, Architectures and Languages (ATAL 1994)* (pp. 355-370). Springer.

Winikoff, M. (2005). JACK intelligent agents: An industrial strength platform. In R. Bordini, M. Dastani, J. Dix, & A. El Fallah Seghrouchni (Eds.), *Multi-agent programming: Languages, platforms and applications* (pp. 175-193). Springer.

Wooldridge, M., & Jennings, N. (1995) Agent theories, architectures, and languages: A survey. In M. Wooldridge, N. Jennings (Eds.), *Proceedings of International Workshop on Intelligent Agents I, Agent Theories, Architectures, and Languages (ATAL'94)* (pp. 1-39). Springer.

Wooldridge, M. (2000). *Reasoning about rational agents*. MIT Press.

Wooldridge, M. (2001). *An introduction to multiagent systems*. John Wiley & Sons.

Wray, R., Jones, R. (2006) Considering Soar as an agent architecture. In R. Sun (Ed.), *Cognition and multi-agent interaction - From cognitive modeling to social simulation*. Cambridge University Press.

Wright, I., & Marshall, J. (2003). The execution kernel of RC++: RETE*, A faster RETE with treat as a special case. *International Journal of Intelligent Games and Simulation, 2*(1), 36-48.

## KEY TERMS AND DEFINITIONS

**Agent Architecture:** Following the definition of 'software architecture' from Bass et al. (2005) as "the structure or structures of the system, which comprise software elements, the externally visible properties of those elements, and the relationships among them", we regard an 'agent architecture' as the control structures that facilitate the specification and execution of agent behavior.

**BDI Model:** The Belief-Desire-Intention (BDI) model of human practical reasoning, developed by the philosopher Michael Bratman (1987), is a model for assessing the rationality of human actions. Unlike earlier models, which only consider desires and beliefs, the BDI model introduces future-directed intentions, which are composed to plans, as an important and irreducible concept. In the agent research community, the philosophical model has been slightly adapted to specify the behavior of software agents in terms of beliefs, goals and plans.

**Deliberation Cycle:** The term 'deliberation cycle' refers to a conceptual "main loop" of an agent interpreter to illustrate the basic mode of operation (e.g. "sense-reason-act").

**Intentional Stance:** The term 'intentional stance' was coined by philosopher Daniel C. Dennett (1971) and refers to a viewpoint in which we use (human) mental properties for explaining the behavior also of animals or inanimate things. McCarthy (1979) has argued that the intentional stance is also useful for developing software systems.

**Mentalistic Notions / Mental Attitudes:** The terms mentalistic notions resp. mental attitudes refer to human properties like beliefs and goals when used for describing software agents.

**Multi-agent System:** (Wooldridge 2001, p. 3) defines: "A multi-agent system is on that consists of a number of agents, which interact with one another [...]."

**Physical Symbol System Hypothesis:** The Physical Symbol System Hypothesis has been formulated by Newell and Simon (1976) and states that: "A physical symbol system has the necessary and sufficient means of general intelligent action."

**Software Agent:** Although a great variety of definitions of the term 'software agent' exists (Franklin and Graesser 1997), a widely accepted definition is given by Jennings and Wooldridge (1998, p. 4): "an agent is a computer system situated in some environment, and that is capable of *autonomous action* in this environment in order to meet its design objectives".

**UTC:** (Newell 1990) introduced the term Unified Theories of Cognition (UTC) as an objective for the psychology research field. A Unified Theories of Cognition should overcome the existing variety of psychological theories (over 3000 in 1990) and offer a unified explanation framework for human cognition.

# Chapter XXII
# Using the Semantic Web Rule Language in the Development of Ontology-Driven Applications

**Martin O'Connor**
*Stanford University, USA*

**Mark Musen**
*Stanford University, USA*

**Amar Das**
*Stanford University, USA*

## ABSTRACT

*The Semantic Web Rule Language (SWRL) is an expressive OWL-based rule language. SWRL allows users to write Horn-like rules that can be expressed in terms of OWL concepts to provide more powerful deductive reasoning capabilities than OWL alone. Semantically, SWRL is built on the same description logic foundation as OWL and provides similar strong formal guarantees when performing inference. Due to its description logics foundation, rule-based applications developed using SWRL have a number of relatively novel characteristics. For example, SWRL shares OWL's open world assumption so certain types of rules that assume a closed world may be difficult or impossible to write in SWRL. In addition, all inference in SWRL is monotonic so deductions cannot be updated or retracted. These formal characteristic have a strong influence on the development and use of SWRL rules in ontology-driven applications. In this chapter, we describe the primary features of SWRL and outline how, despite some limitations, SWRL can be used to dramatically increase amount of knowledge that be represented in OWL ontologies.*

## INTRODUCTION

The Semantic Web project is a shared research plan that aims to provide explicit semantic meaning to data and knowledge on the World Wide Web (Berners-Lee et al., 2001). One of the goals of the Semantic Web is to enable applications to integrate data and knowledge automatically through the use of standardized languages that describe the content of Web-accessible resources. The Web Ontology Language (OWL; OWL, 2004) is emerging as a core Semantic Web language. OWL provides a language for constructing ontologies that provide high-level descriptions of Web content. This language is built on a description logic foundation and provides strong formal consistency guarantees when checking ontologies for inconsistencies and when deducing new knowledge. OWL has three sublanguages—OWL Lite, OWL-DL, OWL Full—that provide varying levels of expressive power. An important characteristic of OWL-DL is that it provides strong *decidablilty* guarantees – that is, the consistency checking and inference processes are guaranteed to terminate with definite conclusions no matter how large or complex the underlying ontologies.

OWL-DL's formal underpinnings also limit its expressive power. While it provides a rich set of modeling constructs, it restricts those constructs to the set that meet its decidability guarantees. As a result, certain concepts can not be expressed in OWL-DL. In particular, OWL-DL has poor support for reasoning with data values and certain types of interrelationships between multiple entities in an ontology can not be represented. OWL Full, which is slightly more expressive than OWL DL, shares these limitations. The proposed OWL 2 standard (OWL 2, 2008) addresses some, though not all, of the limitations.

Some researchers have proposed adding rules to OWL to expand its expressiveness. The Semantic Web Rule Language (SWRL; Horrocks et al., 2004) is one of the primary results of these activities. Semantically, SWRL is built on the same description logic foundation as OWL and provides strong formal guarantees when performing inference. SWRL provides considerably more expressive power than OWL alone, particularly when dealing with complex interrelationships between OWL individuals, or when reasoning with data values. It is rapidly become the *de facto* OWL rule language.

## Background

Rule-bases developed using SWRL have a number of relatively novel characteristics. Instead being a standalone rule language, SWRL can be considered as a formal extension of the OWL language. Each SWRL rule is a sort of OWL axiom that is added to and interacts with existing OWL axioms in an ontology. That is, SWRL rules are formal logical statements about entities in an OWL ontology. This logical underpinning has significant consequences for rule development. A major consequence is that SWRL rules cannot be considered independently from OWL axioms during inference: SWRL rules and standard OWL axioms must be reasoned with together, which can make the reasoning process computationally expensive. This interdependence is such that a serial reasoning with OWL axioms followed by inference with SWRL rules does not guarantee completeness (Motik at al., 2005). As a result, OWL reasoners have to be extended to support SWRL, which is not a trivial undertaking.

As a practical consequence, SWRL rule development is deeply intertwined with the development of the associated OWL ontology. Since a SWRL rule may potentially interact with any OWL axiom in an ontology care must be taken to ensure that they do not conflict. Fortunately, these conflicts can be identified by a SWRL-aware OWL reasoner. An important further consequence is that SWRL rules bases cannot be considered independently of their associated OWL ontology.

The formal underpinnings of SWRL also results in some unexpected expressive limitations when writing rules, which has important consequence for rule development. SWRL rules are not as free form as many rule languages and many common rule constructs are not allowed. Techniques to work around these limitations may be required for many applications.

In this chapter, we explore some of SWRL's idiosyncrasies and show that despite them SWRL is a very expressive language that dramatically increase the amount of knowledge that can be represented in OWL ontologies. We describe how SWRL can be used to address a variety of tasks in the development of knowledge-driven applications.

## Basic Language Features of SWRL

SWRL allows users to write Horn-like rules that can be expressed in terms of OWL concepts. It provides deductive reasoning capabilities that can use those rules to infer new knowledge from an OWL ontology. SWRL's design is strongly influenced by the Rule Markup Language (RuleML; RuleML, 2008). In particular, it is based on the Unary/Binary Datalog sublanguage of RuleML and is thus very close to Datalog in terms of its language features. RuleML is a central part of the Rule Interchange Format (RIF; RIF, 2008), which was developed to provide an interchange format for rules on the Web. SWRL's position within the RIF family also helps ensure that its evolution will align with future standards in Web-centric rule development.

A set of simple examples will serve to illustrate some of SWRL's basic features. Consider, for example, a rule expressing that a person with a male sibling has a brother Since all SWRL rules are expressed in terms of OWL entities, this rule would require OWL representations of the concepts of 'person', 'male', 'sibling' and 'brother'. Intuitively, the concept of person and male can be captured using an OWL class called Person with a subclass Man; the sibling and brother relationships can be expressed using OWL properties hasSibling and hasBrother, which are attached to Person. The rule in SWRL would be[1]:

```
Person(?x) ^ hasSibling(?x,?y) ^ Man(?y) ->
hasBrother(?x,?y)
```

Executing this rule would have the effect of setting the hasBrother property of all persons with a male sibling to contain that sibling. Similarly, a rule that asserts that all persons who own a car should be classified as drivers can be written:

```
Person(?p) ^ hasCar(?p, true) ->
Driver(?p)
```

Again, this rule would require that the property hasCar and the class Driver exist in an OWL ontology. Executing this rule would have the effect of classifying all car-owner individuals of type Person to also be members of the class Driver.

SWRL also supports the use of OWL class descriptions in rules. For example, a rule that classifies an individual as a parent if it is a member of a class with a hasChild property with a minimum cardinality of one can be written:

```
(hasChild >= 1)(?x) -> Parent(?x)
```

One of SWRL's most powerful features is its ability to support user-defined methods or *built-ins* (SWRL Built-ins, 2004). Built-ins are user-define predicates that can be used in rules. A number of core built-ins for common mathematical and string operations are defined in the SWRL Submission. For example, the built-in greaterThan can be used to determine if one number is greater than another. A sample SWRL rule using this built-in to help classify persons aged greater than 17 as adults can then be written:

```
Person(?p)^ hasAge(?p,?age) ^
swrlb:greaterThan(?age,17) -> Adult(?p)
```

When executed, this rule would classify individuals of class Person with a hasAge property value greater than 17 as members of the class Adult.

Built-ins can also assign (or bind) values to arguments. For example, the SWRL atom swrlb:add(?x, 2, 3) uses the core SWRL built-in add method to add two integer literals. If x is unbound when this built-in is invoked, it will be assigned the value 5 when the built-in returns. If x is already bound when the built-in is invoked, it will simply determine if its value is 5. This ability to effectively assign values to variables in rules through built-ins provided a significant increase in expressivity.

In OWL-DL, built-ins can take any number or combination of OWL datatype property values. Datatypes can be XML Schema Datatype or arbitrary user-defined types. Built-ins can also take object, class and property values, in which case the ontology will become OWL Full. SWRL allows new libraries of built-ins to be defined and used in rules. Users can define built-in libraries to perform a wide range of tasks. Such tasks could, for example, include currency conversion, temporal manipulations, and taxonomy searches. This extension mechanism provides a means of continuously increasing SWRL's expressive power.

## SWRL's Semantics and Rule Development Consequences

SWRL's formal semantics are outlined in the SWRL Submission (SWRL, 2004). These semantics are based on description logic, specifically the SHOIN(D) variant that underlies OWL-DL. Formally, SWRL extends OWL by adding an *if-then* conditional. OWL already has several types of conditionals but they are variable-free and highly constrained. In contrast, SWRL rules can have arbitrary mixing of variables and OWL expressions. A SWRL rule's *if* part is referred to as the *body* and the *then* part is referred to as the *head*. Both the body and head consist of positive conjunctions of atoms. SWRL does not support negated atoms or disjunction. Informally, a SWRL rule may be read as meaning that if all the atoms in the antecedent are true, then the consequent must also be true.

A major characteristic of SWRL is that it shares OWL's open world assumption. An OWL ontology by default is assumed to be incomplete: unless something is stated to be false it is assumed to be true. Because of the open world assumption, rules that attempt to make inferences based on, for example, the number of individuals of a particular class in an ontology are not usually expressible. SWRL's adoption of open world semantics distinguished it from most rule languages and has significant consequences for rule development.

### Monotonicity

The most significant consequence of SWRL's description logic foundation is monotonicity. Since SWRL rules are logical statements, ontology information cannot be modified or retracted by rules: new assertions must accumulate monotonically and must not contradict previous assertions. If a rule reaches a conclusion that differs from previously asserted knowledge, then the ontology is inconsistent. These previous assertions may be ground facts or may have have been reached by another rule, by a previous invocation of the same rule, or from a conclusion deduced from OWL axioms.

Hence, SWRL rules cannot be used to modify information in an ontology. If SWRL rules allowed ontology modifications nonmonotonicity would ensue. For this reason, it is also not possible to retract or remove information from an ontology using SWRL. For example, assume we have a rule that indicates that a driver older than 25 is insurable by assigning a boolean property to true:

```
Driver(?d) ^ hasAge(?d, ?age) ^
swrlb:greaterThan(?age, 25) -> isInsurable(?d,
true)
```

This rule will add the value of true to the isInsurable property for all drivers that satisfy the antecedent. It does not change the existing value for that property. If, for example, a driver has a previous assignment of false for that property, a successful firing of this rule for that driver will result in the property having two values. If the isInsurable value is functional (which it most likely should be), an OWL reasoner will indicate an inconsistency when applied to the resulting ontology.

Similarly, one may be tempted to write the following rule to increment a driver's age:

```
Driver(?d) ^ hasAge(?d, ?age) ^ swrlb:add(?newage, ?age, 1) -> hasAge(?d, ?newage)
```

As with the previous rule, a successful invocation of this rule will result in a driver having two values for their hasAge property, which is probably not the intention. Worse still, the rule will be invoked for each new value and continue to assign incremented age values indefinitely.

A further consequence of SWRL's monotonicity is that negation as failure is not supported. The following rule, for example, is not valid:

```
Person(?p) ^ ¬ hasCar(?p, ?c) -> CarlessPerson(?p)
```

It is easy to see that the addition of a car to the ontology could invalidate this rule's conclusion. Only individuals with an explicit OWL axiom stating that they have no car can be safely be concluded to be without a car:

```
Person(?p) ^ (hasCar = 0)(?p) -> CarlessPerson(?p)
```

While SWRL does not support negated atoms or negation as failure, classical negation is possible in SWRL with the use of OWL class descriptions in rules. For example, we can write the following rule that asserts that of an individual is not a member of class Person then it should be classified as a member of the class NonHuman:

```
(not Person)(?x) -> NonHuman(?x)
```

Of course, with OWL's (and SWRL's) open world assumption this conclusion can only be reached for individuals for which it can definitely be concluded that they can not be members of the class Person. For example, if an OWL axiom states that members of the classses Person and Dog are disjoint then this rule can safely conclude that all dogs are not humans.

An addition consequence of SWRL's logical underpinnings is that SWRL rules cannot directly reason with non-ontology entities: rules can refer to only to OWL entities such as classes, properties, individuals or data values. However, this restriction has the positive consequence of ensuring portability. A SWRL rule that uses only OWL concepts is guaranteed to have the same meaning irrespective of the environment in which it is executed. In addition to making rules more easily sharable, this formal guarantee allows automatic integrity checking when adding rules to a rule base or when merging rule bases. The ability to automatically detect inconsistencies between an arbitrary set of SWRL rules can greatly assist in the management of rule bases as they grow larger. Built-ins do allow a limited form of workaround for the restriction that only OWL entities may be referred to in rules. Users can define custom built-in libraries that refer to and manipulate non-OWL entities. However, such operations would be beyond the semantics of OWL and SWRL; any deductions made by rules using those built-ins would not be formally safe.

## Open World Semantics

SWRL's adoption of OWL's open world semantics has unexpected consequences when writing even very simple rules. For example, one might expect that a rule that infers that if two OWL individuals of type Author cooperate on the same publication that they are collaborators could be written as:

```
Publication(?p) ^ hasAuthor(?p, ?y) ^
hasAuthor(?p, ?z) -> collaboratesWith(?y,
?z)
```

However, OWL's open world semantics do not allow one to assume that two individuals are automatically distinct if they have different names - that is, OWL does not have a unique name assumption. (Additionally, due to normal rule pattern matching, variables y and z can also match the same individual in this rule.)

SWRL supports sameAs and differentFrom clauses to determine if individuals refer to the same underlying individual or are distinct. For example, a rule that indicates that individuals are cooperators if they are co-authors on a paper can use the differentFrom clause in association with an owl:allDifferents axiom declaration over all individuals of class Author. It can be written:

```
Publication(?a) ^ hasAuthor(?x,?y) ^
hasAuthor(?x,?z) ^ differentFrom(?y,?z) ->
cooperatedWith(?y,?z)
```

Similarly, SWRL will only deduce that two individuals are the same as each other if there is an explicit owl:sameAs OWL axiom defined for the individuals in an ontology.

Because of the open world assumption, rules that attempt to enumerate individuals or properties are not always possible. One can not write a rule that make an inference based on, say, the number of individuals or property values in an ontology unless OWL statements explicitly state those numbers. So, the following rule to deduce that a publication is a single author publication based on the number of its authors is not expressible in SWRL:

```
Publication(?p) ^ hasAuthor(?p,?a) ^
<has one hasAuthor value> ->
SingleAuthorPublication(?p)
```

A publication may have only one known author but other authors may also exist - unless it has been explicitly stated using OWL axioms that they do not. That is, an OWL cardinality restriction must have been declared to restrict the number of authors, e.g.,

```
Publication(?p) ^ (hasAuthor = 1)(?p) ->
SingleAuthorPublication(?p)
```

This rule does not match publications in an ontology with a single known author; instead, it matches publications for which *it can be proven* (directly or indirectly) that they are members of a class that has the specified cardinality restriction on a hasAuthor property.

## Decidability

SWRL's additional expressivity over OWL comes at the expense of *decidability*. That is, while OWL reasoners are guaranteed to terminate when reasoning over an OWL ontology, reasoning with SWRL rules is not. However, researchers have defined a means of limiting the interpretation of SWRL rules to ensure decidability (Motik et al., 2005). Rules interpreted in this way are referred to a *DL-Safe* rules. Decidability is ensured by restricting inference with rules to known individuals. More precisely, all variables in a DL-Safe SWRL rule bind only to known individuals. For complex reasons (Motik et al, 2005), the ability to bind to individuals that are not known causes undecidability. Restricting rules to only bind known individuals has been proven to ensure decidability.

It may not be immediately obvious why variables in a SWRL rule would ever bind to anything other than known individuals. However, SWRL rules are not 'standalone' – as mentioned above, they are a sort of OWL axiom and interact with other OWL axioms in an ontology. Consider, for example, the following rule:

```
Vehicle(?v) ^ Motor(?m) ^ hasMotor(?v, ?m)
-> MotorizedVehicle(?v)
```

which classifies a vehicle as a motorized vehicle if it has a motor. Clearly, this rule will classify any individual of class Vehicle as a MotorizedVehicle if it has a hasMotor property with an individual of class Motor as a value.

Assume we now define a subclass of Vehicle called Car with the associated restriction (hasMotor some Motor) and define a single individual of this class in our ontology. Since we have stated that the car has a motor, we would expect that the car individual would be classified as a motorized vehicle. However, since there is no specific motor declared for the hasMotor property in the ontology (only the statement that it has some motor), a DL-safe implementation of SWRL would not infer that the car is a MotorizedVehicle. Doing so would mean that the variable m in the rule be bound to an individual that is not explicitly known. There are many other situations in which variables in SWRL rules could bind to individuals that are not known.

Clearly, DL-safety restricts the expressive power of SWRL somewhat. DL-Safe rules may produce incomplete inferences - that is, they may not generate all the deduction that are entailed by a particular ontology. However, any deductions that are reached are formally sound.

## Extending SWRL with Built-Ins

SWRL's built-in extension mechanism can be used to dramatically extend the expressive power of the language. We describe two general purpose built-in libraries that have been developed that expand the ways in which SWRL can be used. We then describe a how these built-ins can be used to turn SWRL into a powerful data integration language.

## Querying OWL Ontologies with SQWRL

SWRL is a rule language, not a query language. However, many ontology-based applications require the ability to extract information from ontologies in addition to reasoning with the information in those ontologies. To support this knowledge extraction, a query language called SQWRL (Semantic Query-Enhanced Web Rule Language; SQWRL, 2008) was developed to extend SWRL to support querying of OWL ontologies.

SQWRL is implemented as a built-in library using the standard SWRL built-in mechanism. It is syntactically and semantically compatible with standard SWRL. The SQWRL built-in library contains SQL-influenced built-ins that can be used in a rule to construct retrieval specifications for information stored in an OWL ontology. For example, the following SQWRL query retrieves all persons in an ontology whose age is less than 25, together with their ages:

```
Person(?p) ^ hasAge(?p,?a)
^ swrlb:lessThan(?a,25) ->
sqwrl:select(?p,?a)
```

This query will return pairs of persons and ages. The sqwrl prefix is used to identify SQWRL operators.

To list all cars owned by each person, we can write:

```
Person(?p) ^ hasCar(?p,?car) ->
sqwrl:select(?p,?car)
```

This query will return pairs of individuals and their cars. Assuming a person can have more than one car, multiple pairs would be displayed for each person.

Basic counting is also supported by SQWRL, provided by an operator called count. If we wished to, for example, get a count of the number of cars owned by individuals in an ontology, we could write:

```
Person(?p) ^ hasCar(?p, ?c) -> sqwrl:select(?p)
^ sqwrl:count(?c)
```

This query would return a list of individuals and counts, with one row for each individual together with a count of the number of cars that they own. Individuals that have no cars would not be matched by this query.

Basic aggregation is also supported. Four operators called min, max, sum, and avg provide this functionality. For example, a query to return the average age of persons in an ontology (for which an age is known) can be written:

```
Person(?p) ^ hasAge(?p, ?age) ->
sqwrl:avg(?age)
```

SQWRL has access to all available SWRL built-in libraries. For example, to retrieve the name of all males in an ontology and prepend the title "Mr." to each name, we can use core SWRL stringConcat built-in:

```
Person(?p) ^ Male(?p) ^ hasName(?p, ?name) ^
swrlb:stringConcat(?fullname, "Mr. ", ?name)
->sqwrl:select(?fullname)
```

The ability to freely use built-ins in a query provides a means of continuously expanding the power of the query language. Crucially, it provides an interoperation bridge to deal with other knowledge formats. SQWRL itself is formally based on OWL-DL and SQWRL queries can be expressed only in terms of core OWL concepts. However, the built-in mechanism provides the ability to refer to non OWL DL concepts in queries. A TBox built-in library, for example, allows direct querying of OWL classes and properties, something that would not be possible in OWL DL. Similarly, an ABox built-in library could support the direct querying of ABox entities.

## Modeling Dynamic Information with SWRL

Because of SWRL's description logic foundation, SWRL cannot easily be used to model dynamic information: SWRL rules, like OWL axioms, are logical statements so deductions must remain true forever. However, many real-world processed are dynamic. To capture this changing information a representation of the temporal dimension of those processes is required. OWL itself has no temporal model and adding a temporal dimension to description logic based formalism is not trivial. However, time can be modeled at the user level in OWL.

There are a variety of common temporal models. One of the most widely adopted in the temporal relational database world is the *valid-time model* (Snodgrass, 1995). In this model, a piece of information—which is often referred to as a *fact*—can be associated with instants or intervals denoting the times that they are held to be true. When this model is used in a relational system, temporal information is typically attached to all rows in a temporal table. This approach effectively adds a third dimension to two-dimensional relational tables. In these tables, every tuple holds temporal information denoting the information's valid-time. Conceptually, this representation means that every tuple is held to be true or valid during the time or times associated with this tuple. In Shahar's (Shahar, 1999) encoding of this model, for example, the core concept in the model is an *extended proposition* class that represents information that extends over time. These extended propositions can be used to consistently represent temporal information in ontologies. For example, a set of participant visits in a clinical trial data can be represented by defining a visit record class that inherits the valid time property from the extended proposition class. The valid time property will then hold a visit's actual occurrence time. This consistent representation of temporal information allows standard temporal operators top be applied consistently and greatly simplifies the temporal reasoning task.

This model can be easily represented in OWL (Temporal Ontology, 2008). The core class modeling an entity that can extend over time in the OWL ontology can be represented by an OWL class called ExtendedProposition. This class can be associated with a property called hasValidTimes that holds the time or times during which the associated information is held to be true. This property can be modeled by an abstract class called Time, which has subclasses Instant and Period. These classes represent instants and intervals, respectively. The class Instant is associated with the property hasTime, and the Period class, is associated with the properties hasBeginning and hasFinish. Periods and instances also have granularities associated with them.

This classes and properties in this ontology can be used to consistently represent temporal information in OWL ontologies. For example, a set of visits in a protocol tracking application can be represented by defining a class called Visit that subclasses the extended proposition class. It can contain the hasValidTime property from that class, which holds its visit times. Similarly, an extended proposition can be used to represent a drug regimen, with a value of type string to hold the drug name and a set of periods in the valid time property to hold drug delivery times. These extended propositions can then be associated with a class using OWL properties.

Once all temporal information is represented consistently using this ontology, SWRL rules can be written in terms of this ontology. Since the core SWRL language has limited temporal reasoning capabilities, a temporal built-in library is necessary to represent complex temporal operators (Temporal Built-ins, 2008). Using these built-in operators in conjunction with the temporal ontology permits expression of complex temporal rules that can represent information about dynamic processes. For example, in modeling visits in a protocol as extended propositions, a new SWRL rule can indicate that a second visit in a particular protocol must occur within two weeks of the first visit, as follows:

```
Participant(?p) ^ hasVisit(?p, ?v1)
^ hasVisit(?p, ?v2) ^ Visit1(?v1) ^
Visit2(?v2) ^ temporal:hasFinish(?v1,
?finishV1) ^ temporal:hasStart(?v2,
?startV2) ^ durationGreaterThan(?finishV1,
?startV2, 2, temporal:weeks) ->
NonConformingPatient(?p)
```

In this rule, the temporal prefix identifies concepts and built-ins from the temporal library.

SWRL thus allows the knowledge-level encoding of complex temporal rules using concepts from the temporal ontology. These rules can freely mix temporal concepts with underlying ontology concepts representing domain entities. More importantly, dynamic processed can be modeled by adopting the valid-time temporal ontology and writing SWRL rules with temporal built-ins

## Data Integration with SWRL

Data integration is central challenge of the Semantic Web. Semantic Web applications must typically deal with a variety of information formats. While there are many dimensions to the data integration challenge, two functionalities are crucial: (1) the ability to *query* diverse data formats; and (2) the ability to *map* between different formats. The capability to perform this querying and mapping at the knowledge level is highly desirable. The types of mapping required range from low level structural or syntactic mappings to transformations that require extensive domain knowledge. Similarly, queries must be able to deal with information at varying levels of complexity.

In general, a first step to integrate a particular data format is to build a model describing that format. For example, integrating information contained in an XML document in to an OWL ontology could involve describing the possibly structure and content of an XML document in an OWL ontology. This OWL representation would have classes to represent standard XML constructs, such as document, element, attributes, and namespaces. Instances of this ontology would be used to represent XML document that is to be mapped or queried. Custom software would be required to populate that ontology either in batch mode or on demand. The reverse transformation of ontology-level information to XML may also be supported. Once the integrated information is represented in OWL, it can be queried directly or can be transformed into high-level concepts in a domain ontology. We show that with the added power of custom built-in libraries SWRL provides an expressive language to deal with these mapping and querying tasks.

We describe several tools that support a variety of mappings to support data integration with SWRL. These tools are: (1) XMLMapper, a library that supports the mapping of information described in XML documents to and from OWL ontologies; and (2) DataMaster, a tool that supports mapping of relational data to OWL. While these tools presently only support relational and XML data formats, SWRL's built-in mechanism provides an extension point for dealing with additional formats.

## XMLMapper

The XMLMapper tool (XMLMapper, 2008) can be used to read and write XML streams. This tool can be used to read an XML document and automatically transform it to an OWL ontology that represents the document. This OWL model is described in terms on an OWL XML ontology. This ontology has classes that represent standard XML constructs, such as document, element, attributes, and namespaces. Instances of this ontology are used to represent both the structure and content of an imported XML document. The reverse transformation from this ontology to XML is also supported.

Once it is populated by the XMLMapper tool SWRL rules and SQWRL queries can be written against the OWL XML ontology to support both querying and mapping. Suppose, for example, that we have the following simple XML document containing a list of patients and some of their details:

```
<Patients>
  <Patient>
    <PID>3454-34</PID>
    <Age>43</Age>
    ...
    <ZIP>94402</ZIP>
  </Patient>
...
<Patents>
```

and are asked to return the average age of patients per ZIP code. If we map this document to OWL using the OWL XML Ontology we can write the following SQWRL query:

```
swrlxml:XMLElement(?ep) ^ swrlxml:hasName(?ep,"Patent") ^ swrlxml:hasSubElement(?ep, ?eAge) ^ swrlxml:hasName(?eAge,"Age") ^ swrlxml:hasContent(?eAge,?cAge) ^ swrlxml:convert(?age, ?cAge, xs:Integer) ^ swrlxml:hasSubElement (?ep,?eZIP) ^ swrlxml:hasName(?eZIP,"ZIP") ^ swrlxml:hasContent(?eZIP, ?ZIP) -> sqwrl:select(?ZIP) ^ sqwrl:avg(?age)
```

The swrlxml prefix here identifies the OWL XML Ontology. Of particular interest here is the convert built-in, which is based on the XPath 2.0 (XPath, 2007) language. This built-in is used to convert the content of an XML element to a particular XML Schema type. The OWL XML Ontology includes a variety of XPath-based built-ins for processing XML documents. The Core SWRL built-in library (SWRL Built-ins, 2004) also includes dozens of built-ins that are based on

XPath and XQuery (XQuery, 2007) functions. In association with the OWL XML ontology, these built-ins allow SQWRL to support XQuery-like queries on XML documents.

This ontology can also be used to map to and from XML documents to domain ontologies. Consider, for example, a simple mapping of patients from the above XML document to a domain ontology with prefix dom containing a Patient class with various subproperties. We can write a SWRL mapping rule as follows:

```
swrlxml:XMLElement(?ep) ^
swrlxml:hasName(?ep,"Patent") ^
swrlxml:hasSubElement(?ep,
?eAge) ^ swrlxml:hasName(?eAge,"Age")
^ swrlxml:hasContent(?eAge,?cAge)
^ swrlxml:convert(?age, ?cAge,
xs:Integer) ^ swrlxml:hasSubElement
(?ep,?eZIP) ^ swrlxml:hasName(?eZIP,"ZIP")
^ swrlxml:hasContent(?eZIP, ?ZIP)
-> dom:Patient(?ep) ^ dom:hasAge(?ep, ?age)
^ dom:hasZIP(?ep, ?ZIP)
```

This rule maps individual patient elements to OWL individuals of type Patient in the ontology and also copies the age and ZIP code value from the elements and assigns them as property values of the corresponding individuals. Reverse mappings to support the generation of XML documents from OWL ontologies are also possible.

## DataMaster

DataMaster is a relational-to-OWL mapping tool (DataMaster, 2008). It can serve as a relational data access mechanism in ontology-driven applications. It supports the mapping of data stored in relational databases to concepts described in OWL. DataMaster has two primary components: (1) schema and mapping ontologies that describe both the schema of an arbitrary relational database and the mapping of data stored in these schemas to triples in an OWL ontology; and (2) software to produce these ontologies, either automatically or semi-automatically with user guidance.

The OWL schema ontology provides a knowledge-level description of a relational schema. It describes schemas in a database and associated tables together with the columns and column data types contained in each table. It also describes primary and foreign key relationships for tables in a schema. This schema ontology is used by a *mapping ontology* to describe how relational tables are to be mapped to OWL concepts. The fundamental goal is to specify the mapping of rows in a relational table to triples in an RDF model, which will then be mapped to OWL classes, properties and individuals. This process cannot normally be performed automatically, and additional user markup is usually required.

DataMaster can be used with any relational database with JDBC/ODBC drivers. A user interface is provided that supports user-driven configuration of the importation process. A user can use this interface to connect to a database and select the portions of the database that they wish to import. If the user decides on a schema-only import the database schema is read and represented in OWL using the Relational-OWL (de Laborda and Conrad, 2005) ontology. If a content import is requested, the user can select a number of mapping options, such as, for example, how table and column names are mapped, and how relational column types are mapped to XSD Schema types.

Once data have been mapped to OWL, SWRL rules or SQWRL queries can be written against the corresponding OWL concepts. In general, the structure of an imported database will not be in a form that corresponds to domain level entities in an ontology so SWRL mapping rules will be required to map instances from the imported ontology to an ontology containing domain-level entities.

DataMaster also has a dynamic mapping layer that uses the schema source and mapping ontologies to map data on demand (O'Connor et al., 2007a,b). Instead of importing all database content, it provides a data access layer to dynamically translate knowledge-driven data requests to queries on a relational database. This layer works with an ontology level query engine to allow retrieval from a relational database. At run-time, the software uses the source and mapping ontologies to transform the data in a relational database to OWL entities. A SWRL rule engine or a SQWRL query engine can interact with this software to retrieve these mapped OWL entities. The engines take rules or queries written in terms of OWL classes, properties, and individuals and generates requests to the mapping software for the OWL entities in a relational database. The mapping software then generates SQL queries to retrieve that appropriate data from the database identified by the schema and mapping ontologies.

A variety of optimization techniques can be used to improve the performance of this mapping. Consider, for example, a query that retrieves all patients older than 17 years:

```
Person(?p) ^ hasAge(?p,?age) ^
swrlb:greaterThan(?age,17) ->
sqwrl:select(?p)
```

Assuming that persons and their age properties are mapped from a database, a naïve implementation would retrieve all person individuals and their ages and then apply the greaterThan built-in to the retrieved data. However, knowledge about the operations implemented by built-ins can be used to optimize data access. For example, the parameters to built-ins can be processed to generate SQL queries that exclude unnecessary data. In this case, an additional SQL clause can exclude patients not aged over 17. Given the greater performance of modern relational database systems, these optimizations can lead to significant performance improvements for some queries. DataMaster uses a built-in annotation ontology to define the types of operations built-ins perform. The mapping software then reads these annotations and used them to generate optimized SQL requests for data.

Optimizations can also be achieved by rewriting SWRL queries. For example, a SWRL rule's meaning is independent of the order of its constituent atoms so these atoms can be reordered without changing meaning. This reordering can reap considerable benefits because normal atom ordering can conspire to reduce performance. Typically, users will place more general atoms near the start of a SWRL rule or SQWRL query and then use progressively more specific atoms. By reordering atoms to place more specific-and thus less likely to be satisfied-atoms near the beginning of a rule or query, many unnecessary evaluations can be avoided. Clearly, atoms that use mapped relational data should be reordered as late as possible in a query.

Standard database optimization techniques can also be used. For example, an artifact of the triple-to-relational mapping approach is that some queries can generate expensive multi-way relational joins. Adding intermediate views in addition to auto-generated primary key columns to tables and then creating indexes with them can improve performance dramatically. We have seen significant improvements to some queries using this technique. This approach has been in use for many years in object-relational mapping software and can be used without breaking legacy database applications that worked with the unmodified schemas.

## FUTURE TRENDS

A key goal of Semantic Web researchers is to increase the expressivity of ontology languages while maintaining a strong semantic foundation for those languages. Tractable reasoning is also a major goal. The OWL language meets these goals. It has a strong formal basis in description logic research and the OWL Lite and OWL-DL variants of the language are decidable. Current efforts center on the development of the OWL 2 (OWL 2, 2008) evolution of the language, which addresses some expressive limitations of OWL 1.0. SWRL is not undergoing an equivalent version evolution; instead, because of its relatively novel features, the current language is being evaluated in use, with no immediate plans for extensions.

Research in the use of rules in the Semantic Web is driven primarily by the RIF Working Group (RIF, 2005) whose goal is to produce W3C recommendations for rule interchange. The primary goal is to produce a core rule language and a set of extensions which together allow rules to be translated between rule languages and rule systems. Meeting the needs of both the Semantic Web and business rules communities is a central requirement. The ability to leverage the decades of experience with rules systems in academia and industry is also an aim. The RuleML (RuleML, 2008) initiative is also close aligned with the RIF. The final specification of the core RIF Basic Logic Dialect (BLD, 2008) is close to completion and is similar to SWRL in terms of its features and expressivity.

The future of rules on the Semantic Web will be decided by users. In normal software practice, the vast majority application logic is encoded in software and rules are used only in specialized cases. The assumption for the Semantic Web, however, is that rules will be more central because they can be layered naturally on ontology languages. As can be seen in this chapter, it is relatively easy to add rules to OWL and the layering is very natural. In addition, this layering can maintain a semantic compatibility with the underlying formalism. Hence, the intuition that rules are more likely to be seen as a natural extension of the underlying ontology languages, which will lead to natural adoption. Expressivity is the key to the success of this adoption process. If a rule language is constrained and inexpressive, it is unlikely to gain widespread adoption.

Of course, Semantic Web applications themselves are unlikely to be built from rules and most will likely be built using standard applications languages such as Java. However, encoding the central logic of these applications using rules would have tremendous advantages on the Semantic Web. The ability to encode software logic at the knowledge level would encourage the development of sharable rule bases. For example, high level policies, procedures, and constraints about a particular ontology or Semantic Web application can be published as rules. These rules can be 'understood' by client applications. Similarly, rules can be used to encode business processes in the same as business rules languages are used at the moment. This ability to encode and share high level knowledge is central to the success of the Semantic Web.

## CONCLUSION

One of the goals of the Semantic Web is to provide rich knowledge-level descriptions of information on the Web. The development and adoption of standardized ontology languages such as RDF and OWL is key to its success. While OWL in particular provides a rich language, it does have expressive limitations. Higher levels of expressivity are required to represent the very complex types of interrelationships that can exist between web resources. SWRL is one of the first steps in providing this type of increased expressivity. The attractiveness of SWRL is that it provides this increase by building on the formal semantic of OWL. SWRL rules are fully compatible with OWL ontologies and can be formally verified against an ontology. The cost of this expressivity is lack of decidability of inference with SWRL, but, as show in the chapter, even this shortcoming can be overcome by adopting minor restrictions in how rules are interpreted. These restrictions do not typically have significant practical consequences. In addition to its formal underpinnings, one of SWRL's most useful features is its ability to incorporate user-defined libraries. This extension mechanism provides a very powerful means of expanding SWRL's expressiveness and of increasing the types of information that can be reasoned with using rules. As a result of these features, SWRL provides a concise, simple, formally sounds, and powerful language for representing complex knowledge-level information.

## REFERENCES


Berners-Lee, T., Lassila, O., & Hendler, J. (2001, May). The Semantic Web. *Scientfic American.*

Datamaster (2008). *DataMaster.* Retrieved November, 10th, 2008 from http://protegewiki.stanford.edu/index.php/DataMaster

de Laborda, C.P., & Conrad, S. (2005). RelationalOWL - A data and schema representation format based on OWL. *Conceptual Modeling, 43*, 89-96.

*FIC Basic Logic Dialect* (2008). Retrieved November 10, 2008 from http://www.w3.org/TR/rif-bld/

Horrocks, I., Patel-Schneider, P.F., Boley, H., Tabet, S., Grosof, B., Dean, M. (2004). *SWRL: A Semantic Web rule language combining OWL and RuleML.* W3C.

Motik, B., Sattler, U., Studer, R. (2005). Query answering for OWL-DL with rules. *Journal of Web Semantics, 3*(1), 41-60.

O'Connor, M.J., Tu, S.W., Das, A.K., Musen, M.A. (2007a). Querying the Semantic Web with SWRL. *International RuleML Symposium on Rule Interchange and Applications (RuleML2007), Orlando, FL.* Springer Verlag.

O'Connor, M.J., Shankar, R.D., Tu, S.W., Nyulas, C., Parrish, D.B., Musen, M.A., & Das, A.K. (2007b). *Using semantic web technologies for knowledge-driven querying of biomedical data.* Paper presented at the 11th Conference on Artificial Intelligence in Medicine (AIME 07), Amsterdam, Netherlands.

OWL, (2004). *OWL Web ontology language reference.* Retrieved November, 10th, 2008 from www.w3.org/TR/owl-ref, 2004.

OWL 2, (2008). *OWL Web ontology language 1.1.* Retrieved November, 10th, 2008 from http://www.w3.org/TR/owl2-profiles/

*RIF Working Group* (2005). Retrieved November 10, 2008 from http://www.w3.org/2005/rules/wiki/RIF_Working_Group

RuleML (2008). *The rule markup initiative.* Retrieved November 10, 2008 from http://www.ruleml.org/

Shahar, Y. (1999). *Timing is everything: Temporal reasoning and temporal data maintenance in medicine.* Paper presented at the Seventh Joint European Conference on Artificial Intelligence in Medicine and Medical Decision Making (AIMDM'99), Aalborg, Denmark.

Snodgrass, R.T. (1995). *The TSQL2 temporal query language.* Boston: Kluwer Academic Publishers.

SQWRL (2008). *SQWRL.* Retrieved November 10, 2008 from http://protege.cim3.net/cgi-bin/wiki.pl?SQWRL

SWRL (2004). *SWRL: A Semantic Web rule language cCombining OWL and RuleML.* Retrieved November 10, 2008 from http://www.w3.org/Submission/SWRL/

SWRL Built-ins (2004). *Built-ins for SWRL.* Retrieved November 10, 2008 from http://www.daml.org/2004/04/swrl/builtins.html

Temporal Ontology (2008). *SWRL Temporal Ontology.* Retrieved November 10, 2008 from http://swrl.stanford.edu/ontologies/built-ins/3.3/temporal.owl

Temporal Built-ins (2008). *SWRL Temporal Ontology*. Retrieved November 10, 2008 from http://protege.cim3.net/cgi-bin/wiki.pl?SWRLTemporalBuiltIns

XMLMapper (2008). *SWRLTab XMLMapper Tool*. Retrieved November 10, 2008 from http://protege.cim3.net/cgi-bin/wiki.pl?SWRLTabXMLBuiltIns

XPath (2007). *XML Path Language (XPath) 2.0*. Retrieved November 10, 2008 from http://www.w3.org/TR/xpath20/

XQuery (2007). *XQuery 1.0: An XML Query Language*. Retrieved November 10, 2008 from http://www.w3.org/TR/xquery/

## KEY TERMS AND DEFINITIONS

**Datalog:** Datalog is a declarative, formal logic based query language based on Prolog.

**Description Logic:** Description logic (DL) is a knowledge representation formalism for representing and reasoning about knowledge. DL languages have formal logic-based semantics that allow them to deduce new knowledge from explicitly asserted knowledge.

**Open World Assumption:** The open world assumption is describes the assumption in formal logic that one cannot conclude that some statement is false unless that it is explicitly stated to be is false. The assumption is that the statement has not yet been added to the knowledge base, not the more common closed world assumption that a statement that is not known to be true is false.

**OWL:** The Web Ontology Language (OWL) is emerging as a core Semantic Web language. OWL provides a language for constructing ontologies that provide high-level descriptions of Web content. This language is built on a description logic foundation and provides strong formal consistency guarantees when checking ontologies for inconsistencies and when deducing new knowledge.

**Semantic Web:** The Semantic Web project is a shared research plan that aims to provide explicit semantic meaning to data and knowledge on the World Wide Web. One of the goals of the Semantic Web is to enable applications to integrate data and knowledge automatically through the use of standardized languages that describe the content of Web-accessible resources.

**SQWRL:** SQWRL (**S**emantic **Q**uery-Enhanced **W**eb **R**ule **L**anguage; pronounced *squirrel*) is a SWRL-based query language that can be used to query OWL ontologies. SQWRL provides SQL-like operations to format knowledge retrieved fro

**SWRL:** The Semantic Web Rule Language (SWRL) is an expressive OWL-based rule language. SWRL allows users to write rules that can be expressed in terms of OWL concepts to provide more powerful deductive reasoning capabilities than OWL alone. Semantically, SWRL is built on the same description logic foundation as OWL and provides similar strong formal guarantees when performing inference.

## ENDNOTE

1 The SWRL Submission (SWRL, 2004) does not detail a standard presentation syntax for the language; the examples presented in this paper reflect the presentation syntax adopted by the Protégé-OWL SWRL Editor.

# Chapter XXIII
# Business Rules Management for Business Processes:
## From Modeling to Deployment

**Marwane El Kharbili**
*IDS Scheer AG, Germany*

## ABSTRACT

*The power of rule-based solutions has been demonstrated over a wide range of domains and a number of industrial-scale solutions and business rules have now proven their usability in complex real world scenarios. But the use of business rules in conjunction with business process management is still a young research field. Business process management (BPM) is a new paradigm for companies to carry out their value-creating activities. Bringing agility and flexibility to business process management is one of the most pressing challenges we are facing today. In this chapter, we make the case for rule-enabled BPM by motivating the need for introducing business rules in BPM and studying the possible advantages of combining business rule management (BRM) and BPM techniques. A discussion of possible uses of business rules (BRs) in business processes (BPs) is made. Furthermore, we also propose a lifecycle for BPM-oriented business rule management, and illustrate this using a business scenario. Hence, the aim of this chapter is to provide readers with insights into issues conceptual BRM applied to BPM in a business context, not from a formal, but from a methodological point of view.*

## INTRODUCTION AND MOTIVATION

When speaking about artificial intelligence, most of us would think about some intelligent dog-robot playing with their children or an autonomous shop assistant giving you advice to find the best shoe for you. Needless to say such visions are more likely to be found in science fiction book than on reports of enterprise management specialists. However, business people are still talking about artificial intelligence in the enterprise. Well how then? This is what we will further discuss in this chapter by motivating the use of a business rules-oriented approach to business process management.

BPM is a discipline encompassing among others, the organization, modeling and implementation of business processes. Business processes capture the logic behind the business activities of companies. These companies have to react to changes in the market because each change requires changes to the business, in order to keep it aligned with enterprise goals. Moreover, this has to be done in a cost-effective way. Another aspect is the ability of quickly switching strategies to adapt to new enterprise goals. Companies grow with their business and business processes have to be extended while kept efficient. This agility is a considerable advantage in competitive markets. As corporations turn towards a business process-oriented vision of their activities, agility and flexibility requirements towards the business rely on the degree to which these requirements are fulfilled by the design of business processes.

Imagine a shoe reseller that has a series of special offers and reductions on the price of the shoes it sells, depending on the time of the year, on seasons and on special celebrations. This shoe reseller wants to introduce new reduced prices for a certain set of shoe models for customers purchasing these shoes in the next 3 days, and with an additional bonus reduction for purchases of more than a pair of shoes. If this reseller implements this offer by "hard-modeling" it directly in his business processes, it generates additional implementation costs for an action that will only last some days. It may even be that the costs of this implementation surpass the benefits generated by this action. One can only imagine what the situation would be for a regional or nationwide reseller, introducing such complexity in its business with a huge number of offers. This small example shows us what complexity can reside in business processes. For this kind of business rules embedded in the business processes, a business rules approach seems most promising.

Another aspect impacting business processes are regulatory compliance and governance issues. Companies are now obliged to comply with laws and regulations, emulating from a number of governmental bodies and standardization bodies, or simply have to ensure their business is running with respect to guidelines they have defined in other to implement a certain strategy. For example, in the financial sector, the Sarbanes-Oxley Act, SOX (2002), requires companies to define controls on their financial reporting. There are also numerous examples of domain-dependent regulations coming from the banking such as Basel (2004), health HIPAA (2006), or security management (ISO17799, 2008); ISO27002, 2008) sectors. There are several ways of achieving this, such as using internal controls or policies El Kharbili (2008). Business rule themselves are one very intuitive way of implementing policies.

We understand that requirements defined on business processes in order to keep up with change and regulations let us observe BPM from a new angle. One important asset companies have are corporate business rules. Business rules exist either as formally modeled or as volatile rules present in business artifacts such as manuals, guidelines or best practices, are the result of years of experience and constitute very important proprietary knowledge for companies. By extracting business rules which are hard coded in the business processes and in the know-how of business specialists, the knowledge on which the business relies could be made much easier to manage. As explained in Figure 1, the direct implementation of compliance measures and business change management logic (represented by the red connections) are bypassed by being implemented as business rules. The control of business processes is done by the business rules.

Currently, business process management does not separate the pure business logic of the business from the control flow organizing and orchestrating activities. The constraints placed on the business have to be outsourced from the business processes in order to be manageable. Through this separation between business processes and business rules, changes on the rules governing the business can be implemented without having to change the processes that are controlled by them. Reactivity to change is made easier. Agility, as far as the business rules management methodology used allows it, can thus be achieved. In this chapter, an introduction to the topic of business process

Figure 1. Achieving agility through a business rules oriented approach to BPM

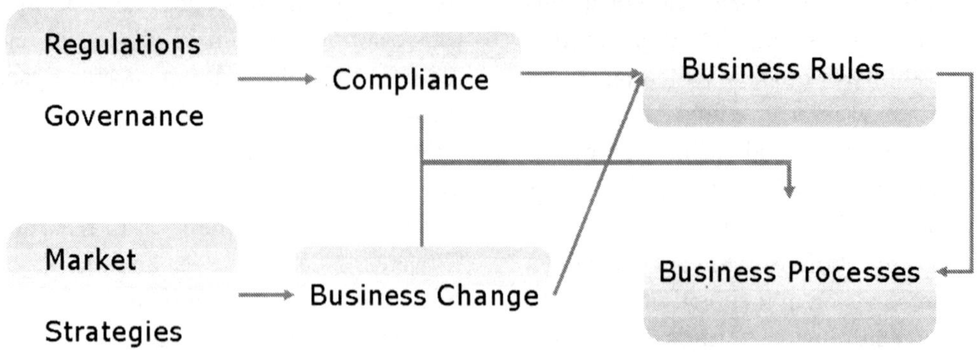


management will be given, where we will study aspects of BPM that require the use of business rules. Our goal is then to introduce a business rules management methodology tailored to the discipline and practitioners of business process management. We will also illustrate how this methodology works in a business scenario.

## BACKGROUND

Business processes are flow-oriented representations of how a company's business is carried out. As said in Weske (2007), "while at an organizational level, business processes are essential to understand how companies operate, business processes also play an important role in the design and realization of flexible information systems". Business Process Management (BPM) is the discipline of information systems that studies techniques and methods for managing business processes. In Aalst (2003), a more workflow management oriented definition of BPM states: "Business Process Management (BPM) includes methods, techniques, and tools to support the design, enactment, management, and analysis of operational business processes".

BPM can also be viewed as taking a model-driven approach towards information systems. BPM enables designing and implementing information systems and applications that support the business. This is achieved by first of all creating models representing different aspects of the business to be implemented, and using these models for application generation (Stein, 2007; Ouyang (2006)). Languages such as BPEL (see Alves (2007)) are one example how such a model driven approach is achieved. The definition of BPM does not explicitly encompass the definition of rules. BP notations are many and research has been concentrating on defining formal notations and formalisms for representing and managing business processes. For instance the extended Event-driven process chains (eEPC, see Keller (1992)) as part of the ARIS framework for BPM (see Scheer (1999)) and the Business Process Modeling Notation (BPMN, see BPMN (2006)) are widely used nowadays.

Nevertheless, as processes represent the business, it can be expected that business rules are partly present in one form or another in business processes, which is confirmed by experience in BPM. Business rules are appealing as a mean to achieving BPM agility. The term business rules is used here to refer to rules that constrain or guide aspects of the business. Rules have been extensively used in research on expert systems (production systems). Production systems contain learning algorithms capable of managing bases of facts and rules and creating new facts out of the latter through inference. Production systems require human assistance in order to fulfill their goal, by requesting human users to validate results delivered by the production system and learning from this validation. The most widespread algorithm invented for production systems is certainly the RETE algorithm (Forgy, 1982; Forgy, 1987). An enhanced version of RETE (RETE/UL) has also been published in Doorenbos (1995), and one known concurrent of RETE is LEAPS (see endnote 1, Batory (1994)). The wide use of such algorithms is due to their performance when faced with a huge number of rules, which was the main criterion of adoption at the beginnings of expert systems. Examples of expert systems very known in research on artificial intelligence are CLIPS (see Riley (2004), endnote 2) and OPS5 (OPS5).

Current business rule management systems (BRMS) have their roots in expert systems and are partly based on the latter. Many of the current industrial BRMS (see endnote 3) have their business rules engines (BRE) built on enhanced versions of the RETE algorithm. Even open-source rule engines such as JESS (see endnote 4, Jess (2008)) are based on the RETE algorithm. But expert systems have been designed for scenarios fundamentally different from the way BPM systems approach business. For example, modern BRMS cannot expect from their users to be able to check or complete the knowledge of the system. BRMS are used in scenarios where the knowledge is distributed over the whole company and must be common to all users in order to keep consistency. Moreover, BRMS are not expected to deliver non-deterministic answers because they are deployed in business production environment where decisions must always be the same every time the rules are provided with the same information and apply in the same conditions. Corticon (2004) Provides further explanations of these differences. In Huntington (1999) a retrospective of rule-based systems from the perspective of knowledge management is given and the history of expert systems and uses of the latter are explained. The author recognizes the pitfalls of using expert systems in the industry but concludes his work by sustaining the promise of industrial level use of the latter, particularly in conjunction with the web.

BRM is such a complex topic due to the pervasiveness of BRs and their use in various sub-disciplines of information systems, computer science and artificial intelligence. In Bonatti (2004), a very complete review of all the languages related to BRM is made. The document covers modeling policies as rules. Policies, are the main initial content driving rule modeling, and are specified in regulations, standards, norms, laws and governance guidelines. Rules are one very intuitive way of implementing policies, because humans just think naturally in rules. In Bonatti (2004), the authoring group studies usage scenarios for rules –based policies, from security policies to BRs. One particular type of Brs the authors identify are production rules in the IF CONDITION THEN ACTION. These BRs are the ones we concentrate on for BPM.

In this report of the REWERSE network of excellence , Antoniou (2004), the necessity to visualize and verbalize business rules in order to make their use accessible to business users is motivated. It is necessary to express the syntactical structure of business rules, to express dependencies between rules and their elements, to include rules in other models, and to allow testing rules. The authors proceed to a survey of existing visual languages and also propose elements of a new visual language for rules. In Kardasis (2004), a repository schema for managing BRs is presented, seeking to facilitate the rule collection process, assist the transition from analysis to design and implementation of the information system, and to support change. The main flagship preaching the use of BRs in BPM, is the ability to handle change. In Vasilecas (2006), in order to react to changes in and around the business and also to respond to the necessity to assist in change-driven decision-making, a method to extract BRs out of information systems and transform them into software artifacts is proposed. In Endl (1998) the constructs and fundamental concepts of a rule-based method for modelling business processes and the specification of workflows are described. The authors show how control flow types van be modelled by rules and classical event-condition-action (ECA) approaches to business rules are extended to include data and organizational aspects. The ideas presented are illustrated by a proposed architecture. In Cysneiros (2000), the use of business rules in requirements engineering is proposed. The idea explained here is because ISO 9000 standards require companies to do a bookkeeping of their policies and procedures to ensure conformance with the standards' requirements, these assets should be used to elicit business rules in order to govern the business. A conceptual model to achieve this is detailed in the paper. This is a more business-near use of BRs because the authors went from business-level requirements, in this case quality standard requirements, and inmplemented them using BRs.

In the context of dynamic workflow systems where the workflow models are not deterministic, because they are composed on the fly according the desired use to be made of them, the authors of Zeng (2002) use business rule inference to generate dynamic workflows. Here business rules are used in order to cope with the exploding space of possible workflows. As the applications or BRM are numerous, there is a variety of proposed methodologies and lifecycles for BRM. In Ross (1997), a comprehensive methodology for managing rules is presented. But as noticed in Taveter (2001), the introduced notation called Ross notation has strong deficiencies for modeling BP related aspects because it largely concentrates on databases and lacks dynamic representational elements such as events or triggers. In Bajec (2004), a methodology is proposed for keeping business rules aligned with their implementation. The methodology's scope is the whole organizational information system. Requirements for an implementation are also elicited. In Rosca (2002), a methodology regarding BRs as high-level business assets and addressing BR acquisition, deployment and evolution aspects of the BR lifecycle is introduced. The approach considers the business under three aspects: enterprise, decision and rule aspects. BPs with all their specificities are not considered. In Rosca (2004), a lifecycle is presented that covers BR management from acquisition through deployment and evolution. In Bajec (2001), a business rule management scenario that integrates activities for managing business rules in organizations, and activities that support business rule management across the IS development life cycle. Here, the authors remind the reader that numerous researches have been done on business rules for databases systems as a mean to contain and enforce rules (through triggers and procedures).

We have seen that interest in BRs is not recent and that BRM build on very solid theoretical foundations in artificial intelligence (AI). We have also seen that there is a lack for a common methodology for BRM tailored to the needs of BPM. There are numerous propositions of lifecycles and corresponding frameworks. Again, the many different uses of BRs in IS and the different approaches to incorporating BRM into BPM we have seen signal the complexity and size of BRM. We argue that an efficient use of BRM technologies and tools can only be achieved by precisely scoping the use scenarios taken into consideration. This is what we will show in the next pages, by showing what is intended by the use of BRs in BP models and what requirements can be made on a BR modeling framework for that purpose. We also elaborate on an adapted business rule management lifecycle (BRLC).

## MAIN FOCUS

BRs are present in all aspects of IS, in enterprise architectures, in databases, in governance guidelines and business strategies, and in BPs. As we have seen in the previous section, lifecycles and management techniques for BRs have held great importance for making the use of BRs efficient from the beginning. The research works cited in the previous section attest that. But none of these approaches has tackled the problem from the perspective of BPM. It is our aim in this section to introduce the advantages of combining the two disciplines and showing why BPM profits from BRs and how. We will also present a lifecycle for BP-oriented BRM that takes various requirements that we will define into account. This lifecycle is used in a widely used industry framework for BPM. A small example will be used to support our arguments.

## Business Rules and Change Management in Enterprise Decision-Making

Some of the biggest challenges for BPM are reactivity to change and support or automatic generation of applications. Figure 2 shows a layered vision of the enterprise and gives an idea about the relationships between concepts such as goals, strategy, policies, BPs and BRs. Examples of these external changes are changes to the market situation such as the introduction of a concurrent product that has better performances or better quality. Or the availability of recent technologies that create new market opportunities which the company cannot allow itself to miss, because that would constitute a danger to its competitiveness. Another important reason for change are regulations, laws and governance guidelines. Any change in a law or the introduction of a new regulation lead to a corresponding repercussion on the organization's BPs. Change management is not a trivial task in corporations, and keeping consistency of the business while conducting changes, as well as correctly propagating changes throughout the BPs and information system layers of the enterprise is not an optional requirement.

As we can understand it, change can generate not only costs in BP, system maintenance and additional testing, but also cause BPs to be redesigned and then redeployed. This is a particularity of BP-oriented enterprises which should not be omitted. BPs are not only used for documentation purposes, they are the fundament of the underlying information systems that implement them. This is particularly emphasized in model-driven approaches to BPM applications, as discussed in the beginning of the previous section of this

*Figure 2. Enterprise layers*

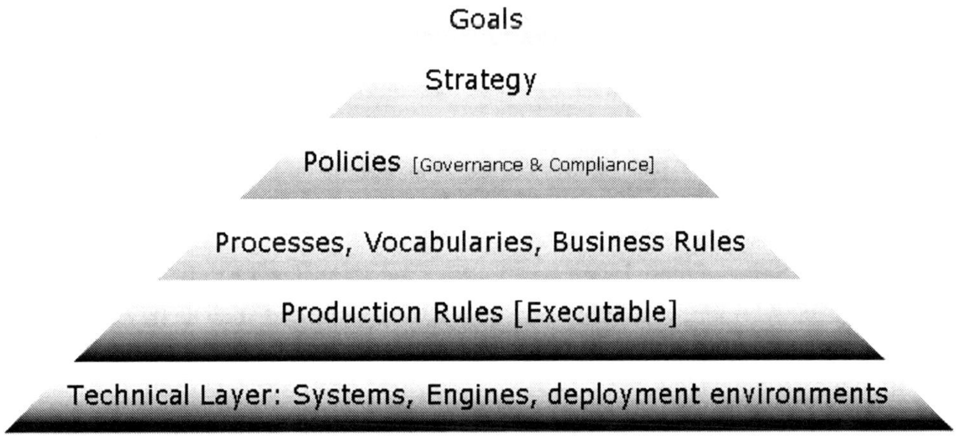

chapter. The main reason for this is that after conducting change on BPs, instances of BP most likely have to be stopped, and restarted, otherwise consistency problems can arouse by having BP instances based on different models running concurrently. It is these observations that make use believe that investing in the overhead of creating BRM components that outsource the hard-coded BRs from the BPs is a more efficient alternative to change implementation directly in processes. Decision-making has to be put into the hands of business experts, not IT developers. Decision-making processes have to be documented and loosely coupled from BPs. Decision-making assets have to be manageable, auditable and traceable. By grouping a set of decisions from the business decisions space into decision clusters and formalizing these decisions using rules, decision management is made scalable, traceable, consistent and automatable. The grouped decisions build on the same set of facts and use the same set of BRs. To model these we need an adapted model, which we call Rulesheet. Rulesheets will be introduced later. Another aspect of using business rules is related to run-time aspects of BPM. Enhanced business process controlling is made possible by auditing rules execution and decision-making helps adjusting and correcting strategies, discovering weak or wrong spots in the business processes and can help monitor compliance to governance and regulatory policies. Such capabilities should be delivered by the BRMS.

Like business processes, business rules are also model driven by nature. The authors of Wagner (2006) defend a similar idea as well. The way business rules are formalized doesn't have an influence of the knowledge carried by rules. Business rules can be modeled using a variety of formalisms and be implemented or enforced using different languages. Some of these languages are proper to rule execution engines, other are openly accessible programming (OPSJ or IRL) or constraint modeling languages (OCL). Business rules contain the decision-making logic of the business, and business processes contain the action-making logic of the business. Together, business rules and business processes allow the business to become intelligent.

## BPM Requirements on BRM

Business rules answer questions such as: if this client does practice an extreme sport, is it allowed to sell him a life insurance policy worth more than one million dollars? Rules also answer questions such as: what would be the best cell phone contract offer to be proposed to the client knowing what his financial status, and maximum monthly payment tolerance looks like? Rules allow approaching decision making in a declarative way. Rules also make keeping consistency an easier task than it would be without BRMS. In BPM, rules are used to tackle deterministic problems, in the sense that the range of possible decisions is known. In most cases BRs are forced to take one of many previously defined decisions, such as in decision tables. Such methods make constraining and controlling decisions possible, which a big advantage for the business. Other techniques are based on the creation and maintenance of a network of facts and rules, which makes possible to model non-deterministic decisions (such an inference engines based on RETE or LEAPS). Our approach to rule-enabled BPM is based on decision tabled as a mean to capture rules. Not coming up with a decision is not an alternative for running businesses. The order in which rules are defined should also play no role in their definition. As soon as procedural aspects appear in BR modeling, the promised flexibility and the advantages of a true declarative approach are lost. This should be supported by the underlying technology used for implementing business rules.

In Sandifer (2000), criteria for designing BRs are elicited such as (i) Atomicity (each rule concerns one fact), (ii) completeness (each rule tackles one complete fact), (iii) consistency (no contradiction with other BRs) and (iv) clarity (clear and concise statements in proper grammar). Aspects of BRM are tackled and the place of BRM in the software development lifecycle (SDLC) is questioned. However, the concepts explained are only illustrated by ERMs and come from a database-centered observation of BRs. In a BPM-centered observation of BRs other requirements have to be defined. As we have explained in the previous sub-section, BRs should be put into use around a decision-making architecture, which itself relies on BPs as the action-making elements. BRs allow modeling all types of BPM-related decisions. It is important that the expressiveness of the models used to capture these decisions do not limit the scope of decisions that business users can model. Otherwise, the latter would be forced to get a level deeper and modify the implementation of BRs by hand. The goal of a model-driven BRM would be missed. BRs allow governing these decisions: BRM aspects such as BR ownership and rights should me managed in order to efficiently control and track BR use. An adapted organizational roles model, where responsibilities on BR artifacts are defined is necessary for achieving this. Implementation aspects should not be included in BR governance, since this is not the task of the business user. Ideally, BRs would be automatically redeployed and set to production after they have been changed. Decision complexity should also be managed.

Decisions do not all have the same granularity and decisions access or are composed of other decisions. Another requirement on BRs is thus to the possibility to model flows and BR interactions. The next idea that comes to mind when considering this is BR reuse. BRs should be made available in components in order to be used for various purposes and modeling various decisions. Behavioral aspects of the business should also be clearly specified, that guarantees that BPs do what they are expected to do. In BPs, activities contain functionalities, but the desired behavior should be also specified for each state the activity can be executed in. However, modeling this behavior in BPs leads to very complex descriptions. A very simple example can be seen in Figure 3. Due to complex branching conditions, the number of possible paths for further BP execution becomes big. In fact, the number of branching outputs explodes exponentially with the number of conditions (the simplest case being 2^(number of conditions)). Behavioral aspects also contain constraints on BPs and on activities. That is what BRs are about. BRs can declaratively express what behavior is expected from a given activity or BP fragment. Another example of behavioral control is the case where a BP activity executed different actions depending on the state of the BP (e.g. this status is defined by data values of BP attributes). Instead of modeling the many possible actions executed by the activity, letting a BR decide for the activity on which action to execute makes the BP simpler.

In BPs, several types of BRs can be distinguished. The first type of rules are called routing rules. These rules are used to model branching in BP control flows. The second category of rules are functional rules. These rules are enacted on the actions executed by BP activities. They may provoke the execution of different actions depending on the state of the process. In the ARIS framework for example Scheer (1999), data values allow making assertions about the state of the BP. These data values belong to attributes of artifacts part of the definition of BPs (called objects in ARIS) are governed by processes. They can be managed using business rules. Changes to these data values are events in ARIS. Core rules are rules solely destined to be used by other rules. One typical example of core rules are scoring rules in the banking sector, which compute values for destined to be used by other rules based on a set of parameters. Constraining rules are rules that express permissions to execute certain actions. These rules describe what should be allowed to happen in BPs and what should not.

In Table 1, the 4 types of rules that can be modeled for BPs are listed. This classification

*Figure 3. Branching explosion in an EPC – complex branching conditions*

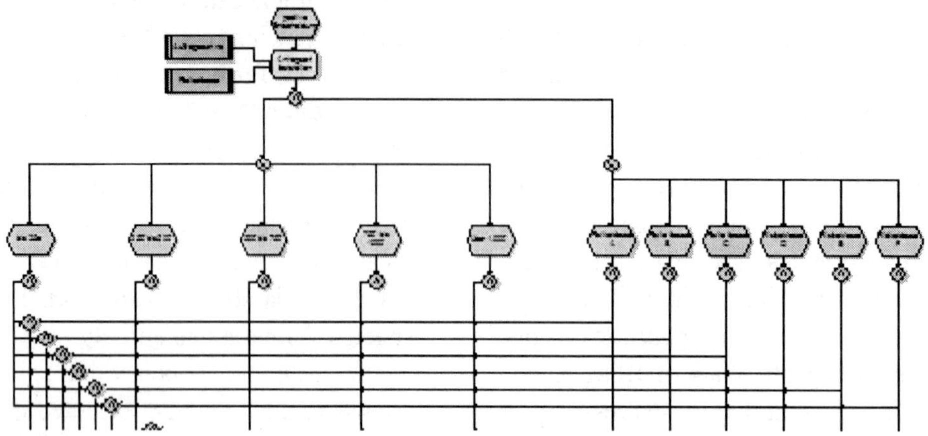

*Table 1. Business process rule types and their exploitation*

| Rules | Modeling | Simulation | Execution | Enforcement |
|---|---|---|---|---|
| Routing | X | X | X | X |
| Functional | X | X | X | X |
| Core | | X | X | |
| Constraining | X | X | X | X |

not to be put in comparison with more technical classifications of business rules (e.g. derivation, integrity, ECA and production rules). It is the result of experience with BPM specific business rules. The different possible exploitations of each rule in the scope of BPs are given. Here again, four types of exploitations have been retained: BP modeling, BP simulation, BP execution and BP run-time BR enforcement. A cross in a cell means that the rule type in the corresponding line is regarded by the exploitation in the corresponding column. Thus, Core rules cannot be enforced and are not part of the BP modeling either. The other rule types are used in all four types of exploitations.

Functional requirements on a rule-based decision framework for BPM can also me stated. BRs need to offer functionalities to model decisions and to test them. Testing decisions means testing formal BRs for correctness as well as testing if BRs really act as the rule modeler expects them to. BR correctness is ensured by testing the rule for:

***Back-Loops***, which is a big problem for BREs: back-looping is when firing a BR leads to the execution of other BRs, and ultimately to the execution of the originally fired BR. A cycle is thus created and the execution of a BR never ends. This is the nightmare of all BR modelers.

***Ambiguity***: checking consistency with other rules composing the decision. Introducing a new BR always contains the risk that the overall consistency of the BRs is not guaranteed, i.e. there is a potential contradiction between the newly introduced BR and the rest of the BRs in the rule base.

***Completeness***: all possible cases that can occur should be modeled in a decision. That ensures that executing a decision is guaranteed to provide an answer: the decision-making problem in this case is decidable.

***Compactness***: the BR set models a decision. The BRMS should assist the user is getting BRs to be as clearly modeled as possible. Compressing the BR set to another smaller (in number of BRs) BR set that just expresses the same decision is a great help for the user.

It must be provable that BRs are effective. One scenario where such functionality is needed is compliance audit. When companies implement controls or policies in order to comply to a regulation, they have to be able to prove this by showing independent auditors that these controls and procedures are put in place and work correctly (El Kharbili (2008)). BRs also need to be deployable in order to be set to production. Here, the more independent the BRs are from implementation languages, the more flexible and reusable the BRs will be across different target implementation systems. In one of the next sections we will shortly introduce a web-service deployment of BRs.

## Modeling Business Rules: A Lifecycle for BPM

The lifecycle presented here is the one used in the ARIS Business Rules Designer (see endnote 11), a product for BRM integrated into the BPM framework ARIS. The lifecycle and the models that will be introduced rely strongly on the Corticon studio and the Corticon Rule Engine, which are two products by Corticon Inc (see endnote 12). the Corticon Rules engine is integrated into the ABRD and is the rules engine used by the ABRD for BRM. The models used ion the ABRD are strongly inspired by those in Corticon Studio. First of all let us have a look at the elements provided by the ABRD. There are 4 ARIS model types that are considered in rule modeling: (i) the vocabulary model, (ii) the Rulesheet model (iii), the Ruleflow model and (iv) the Ruletest model.

Before modelling and managing business rules, business and rule experts have to define and agree on what they seek to model decisions for. Without a clear definition of the aspects of the business rules are enacted on, rules are of no value to any organization. As in Ross (2005), "a fact model establishes the basis for shared operational business knowledge", Ross acknowledges and motivates the need for the definition of these facts and terms models, to form the business vocabulary. A vocabulary is the representation of the business aspects relevant for the modeling of rules commonly agreed on between business analysts. An equivalent to vocabulary usually found in literature about business rules are terms and facts. A vocabulary is formed of entities and relationships, and can be directly created from or mapped to elements of existing data models simply by dragging-dropping UML or ERM models into a Vocabulary model. Vocabulary models present concepts in a tree-like fashion because it is very well adapted to further modeling of BRs. Figure 4 shows how the Vocabulary model looks like.

Once the Vocabulary has been defined, all other ABRD model types can build on it. Rulesheets are the models used in ABRD to model decision units. Each Rulesheet is associated to a Vocabulary and contains a decision table. The decision table is used for modeling sets of BRs. Figure 5 shows an example of a Rulesheet model. A Rulesheet contains rule statements that are linked to elements of the vocabulary the rule is built on. Each column in the Rulesheet is a business rule. It has a set of conditions and corresponding actions. Priorities between rules can be defined using an overriding functionality.

The conditions and actions are all expressed on elements of the vocabulary. Conditions, as well as actions, are conjunctive expressions on Vocabulary element attributes. These attributes are in fact mapped to attributes of UML classes for example. For instance, in Figure 7, the first rule of the Rulesheet reads: ***IF ([Flightplan.Cargo.Volume<100] AND [Flightplan.Cargo.Weight<3000]) THEN (Flightplan.Craft.AircraftType = '747')***. There are many available operators that can be used on the attributes used in de-

*Figure 4. Creating vocabularies from data models. © 2008 IDS Scheer AG. Used with permission.*

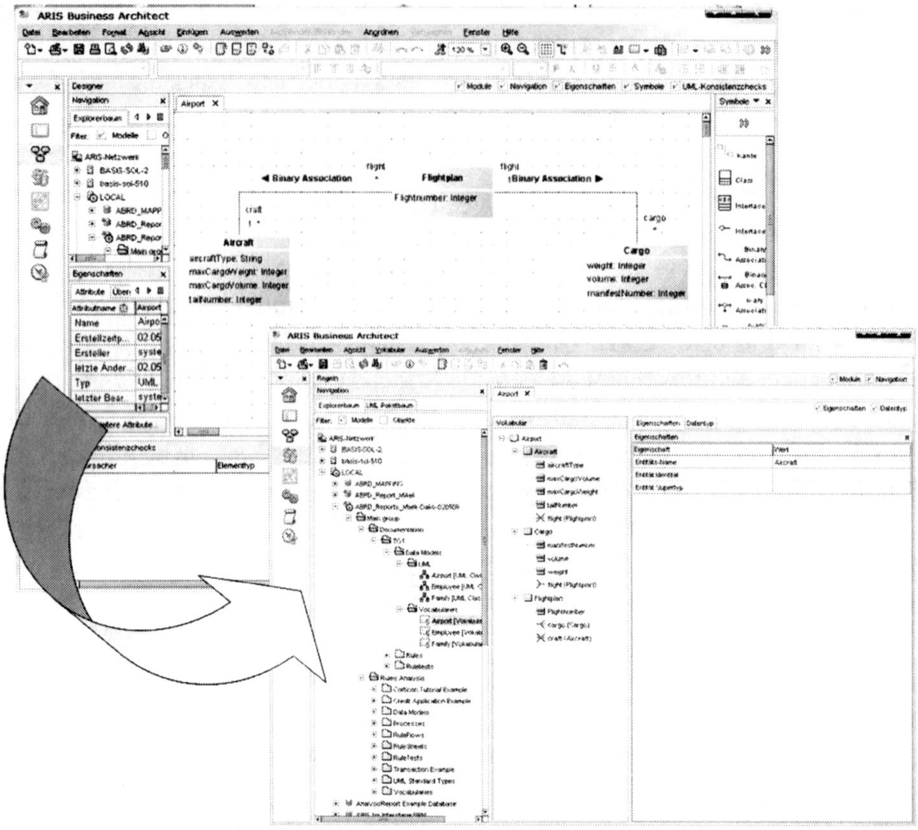

*Figure 5. An example of a Rulesheet model in ARIS. © 2008 IDS Scheer AG. Used with permission.*

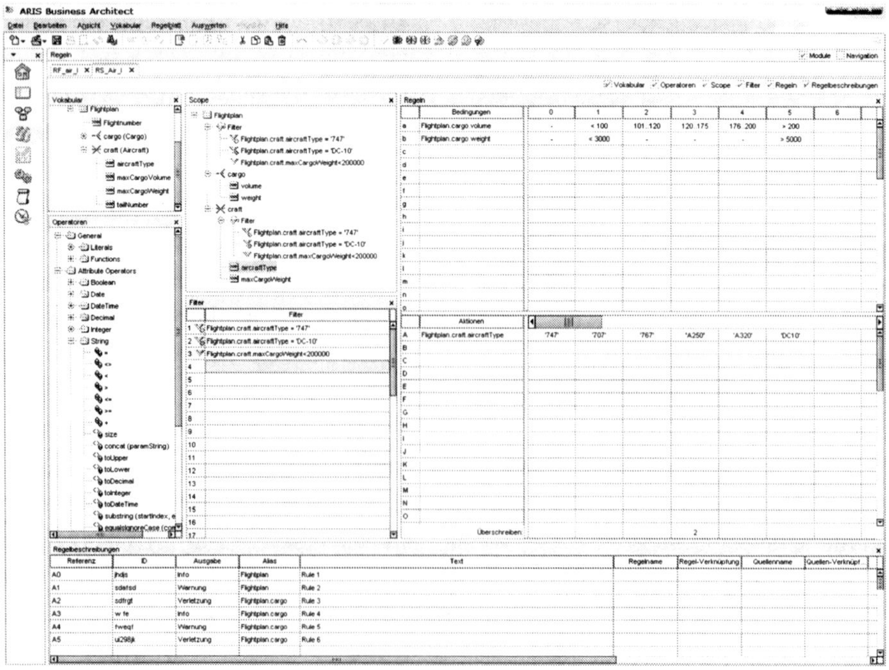

fining rules. For example, it is possible to compare two strings by using the method ".equals(String S)" available under the sub-category "String" of the "attribute operators" category. Besides from attribute operators, general and entity/association operators are available.

Thus, all elements of the Vocabulary can be processed as wanted. The enabler for this is the Corticon Rule Language (CRL), a declarative constraint language based on OMG's OCL (See OCL (2008)). The example of a rule given above is written using in this language. A Rulesheet is thus an aggregation of business rules relating to a certain aspect of the business and operating on a business vocabulary. In a BPMS, business process artifacts are annotated with Rulesheets and hence, the link between business processes and business rules is created. Rulesheets offers necessary functionalities for rule verification. Default rules can also be modeled, which are rules firing without condition, as soon as the Rulesheet is invoked. Rulesheets are used to model decisions, and as we have explained it in previous sections, contain sets of BRs.

In order to allow using the same Vocabulary for several Rulesheets, while keeping the possibility for the business user to only consider subsets of a Vocabulary for a given Rulesheet, a Scope can be defined. A scope can also be use to refine instance data on which the rules engine infers by specifying aliases. Another help for the rules modeler are Filters. A filter makes it possible to constrain the space of possible configurations of scope aliases that are considered by the rules in the Rulesheet as candidates to be processed. For example, the business user can state that he only wants to consider aircrafts in the Rulesheet that have a weight less than 200 tons. The expressions used are logical expressions that are the same that can be used in rules' conditions and actions.

One fundamental difference between the ABRD Rulesheet models (i.e. Corticon too) and other BRMS is that in Rulesheets the rules are checked for inconsistency, back-loops and completeness at design-time. Thus, inference at run-time as it is known from inference-based (e.g. based on the RETE matching algorithm) BRMS, which is the main reason for performance problems, is avoided. That means the rules when run by the Corticon rule engine profit from this and no back-loops or inconsistencies can occur. Additionally, Rulesheets respond to one non-functional criterion identified in the previous section, which is clarity and ease of use. It is of course hard to evaluate such a subjective aspect, but experience has shown that the spreadsheet-like interface for capturing BRs is familiar to and well accepted by business users and helps getting the rules quickly modeled.

Modeling complex decisions can require using more expressive models that Rulesheets and that is why Ruleflows are designed for. Ruleflows model complex decisions by modeling control flows of Rulesheets. An example of a Ruleflow is shown in Figure 6. A Ruleflow can be seen as a business process where atomic activities are all Rulesheets. The requirement of reusing modeled decisions is here fulfilled since Rulesheets are the basic artifacts of Ruleflows. Decisions are verified in Rulesheets, but another requirement from last section on BRMS is BR validation. Ruletests are used for this purpose by allowing the definition of complex business scenarios. Ruletests are used to test decisions modeled in Rulesheets and Ruleflows. In Ruletests, business rule modelers can check that BRs execute conforming to rule specification. They can also check that outputs generated by rules for given input test sets are the same as expected. Figure 7 shows an example of a Ruletest.

The different ABRD models that we have introduced are used to support a Business Rules Management Lifecycle (BRMLC). We will not consider all ABRD models available to manage BRs, since it is not in the scope of this chapter. What is important here is to understand how these models can play together to assist business rule

*Figure 6. ABRD Ruleflow model. © 2008 IDS Scheer AG. Used with permission.*

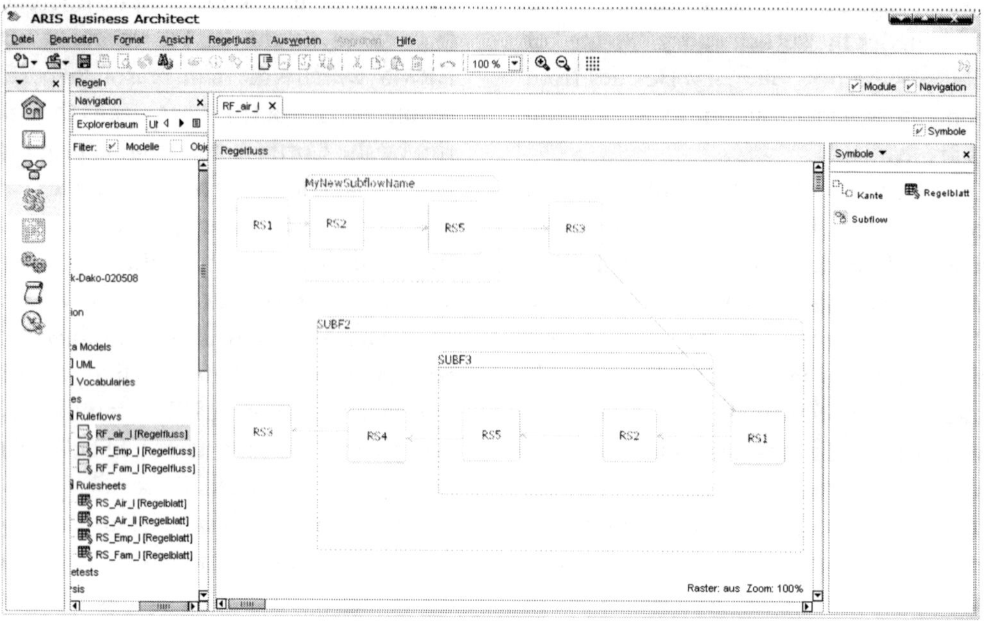


*Figure 7. ABRD Ruletest Model © 2008 IDS Scheer AG. Used with permission..*

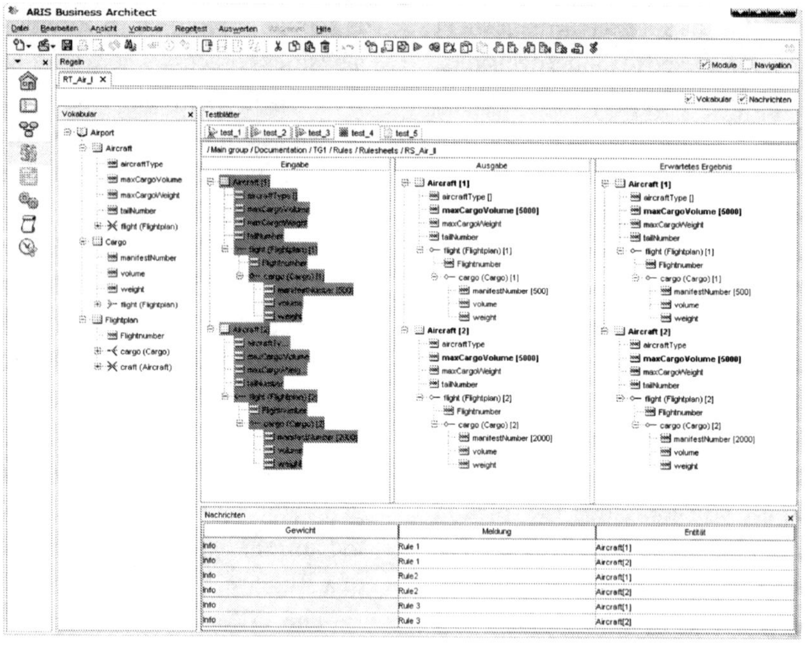


modelers in managing rules. The BRLC is shown in Figure 8 as an EPC process.

The initial step in this lifecycle is auditing BPs for BRs. This step is called decision-point audit. This is done by having business experts parsing the business process that are part of the scope for decision points. Decision points are spots in the BP (branches, functions etc...) where the presence of a BRs in known to both the process expert and communicated to the business rules

*Figure 8. Annotating a business process model artifact with a business rule object*

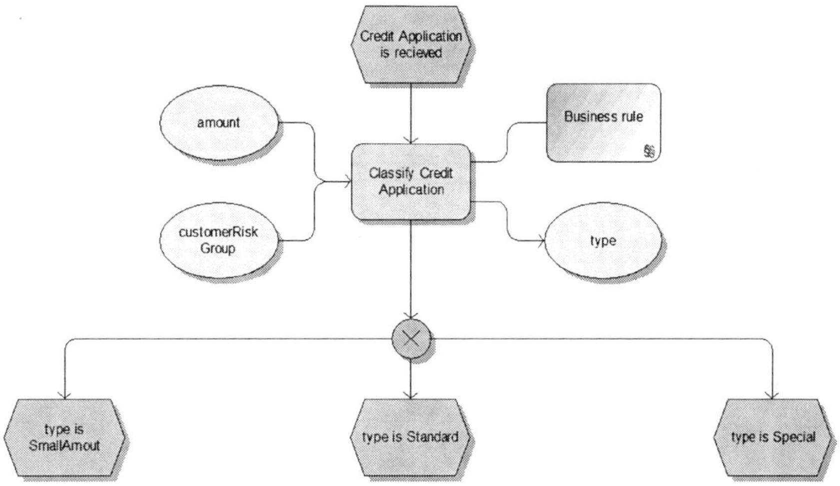


modeler. This step is iterative and is concluded by modifying the audited BP models by including BR artifacts in the selected spots. Decision point audit is further refined in another EPC model illustrated by Figure 9.

Figure 7 shows how this can look like, in the example of a routing (see previous section) decision that is modeled in the BP by annotating the concerned function with the business rule object. The business rule objects can later have either a Ruleflow or a Rulesheet assigned, where the logic of the decision is modeled. This way, the link between BP models and BRs is created.

The next step in the lifecycle is to capture the rules needed to model the decision. Here the business rules modeler is responsible. He has the knowledge about the logic that needs to be modeled. He starts by creating Rulesheet and Ruleflow models, depending on the complexity of the decision to be modeled. In the Rulesheets, he starts by writing down natural language rule statements into the reserved fields. These natural language statements are usually contained in a document called rules specification, where business rules are informally specified. This is the document of work of business analysts.

Then, the rule modeler models the rules contained in the rules specification in the decision table. The different rules modeled in each Rulesheet are then linked to the natural language rule statements. In the example Rulesheet shown in Figure 7, a rule statement for the first rule (1st condition/action) would read: "If the cargo weight to be transported is less than 3000 Kilograms and its volume less than 100 cubic meters then use an aircraft of type 707 for the transport".

The next step is to ensure that the rules are non-ambiguous, do not contain loops and are complete. We call these tests consistency checks. A positive completeness test ensures that the modeled rules cover the whole space of possibly occurring rule invocation states, i.e. all allowed conditions on vocabulary elements have a defined action as an outcome. This rules consistency check task is iterated until all rules in the Rulesheet are consistent. The problem of checking Ruleflows for consistency is equivalent to consistency checks for Rulesheets, since Ruleflow logic is always contained in Rulesheets (Figure 8). Further, rule validation has to be conducted and test scenarios using Ruletests are built. The execution of Rulesheets and Ruleflows is simulated using

Figure 9. A proposed business rules management lifecycle for BPM

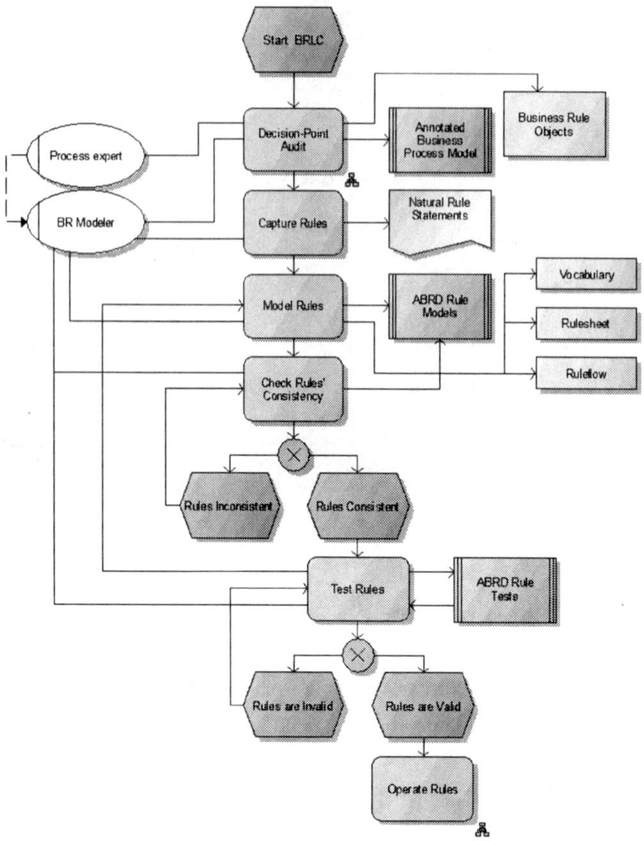


these and the outcome is compared to expected values by the rules modeler. This way, sets of unitary tests can be conducted against Rulesheets and Ruleflows in order to ensure rules have been modeled correctly. In order to build the test scenarios, the rules modeler has to rely on the rules specification.

When this validation is over, operational deployment of rules can be started. Rules deployment can be seen on Figure 10 as a simple EPC. An optional task in this process is reporting. Rule analysis can be done by using reports that are included in the ABRD. These reports document and analyze rule use in BP models, relationships to other BP artifacts, to organizational entities (e.g. process owners or rule experts). Concrete rule deployment is taken in charge by the Corticon Rule Engine. The main idea behind this deployment is making the execution of deployed rules independent from the execution environment. In principle, decisions modeled in the ABRD in Rulesheets/Ruleflows can be exported in any target formalism. Adapted connectors have to be specified. This is allowed by the fact that the design of the ABRD and Corticon decision models sticks to a model driven approach, by avoiding any implementation level details (such as procedural design of decisions).

The vision behind this is that BRs should be executable on any rule engine. Rulesheets/Ruleflows are deployed as web services or as Java services. After BRs have been deployed on the rules server, it becomes possible to invoke BRs as abstract web services. This makes the integration of BRs

*Figure 10. High-level architecture of the ABRD: allying BPM and BRM*

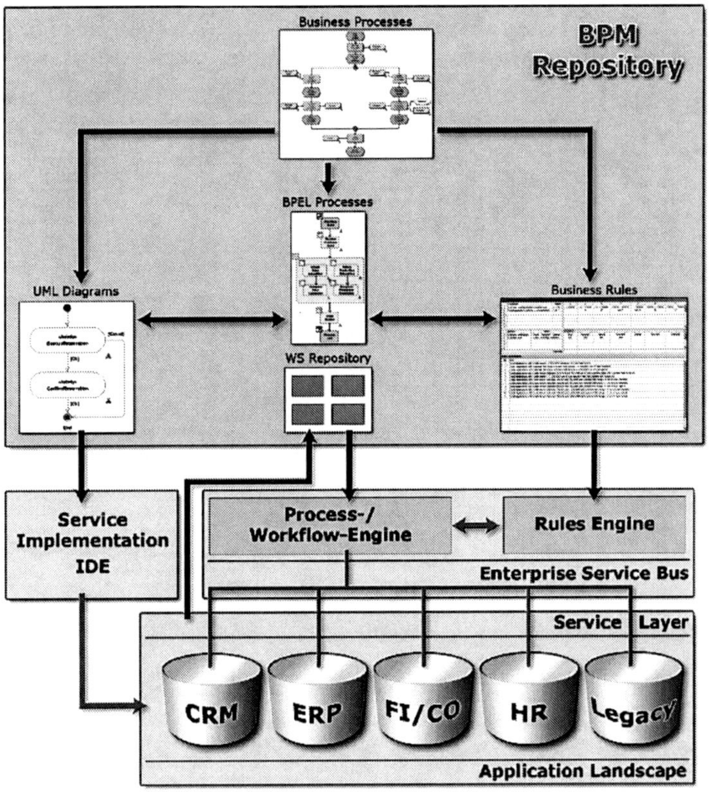


into a web service based SOA possible. The most interesting use in this regard is the transformation of business process models into executable processes, BPEL for example, while propagating information about BRs by incorporating them as web service "invoke" activities in BPEL processes. It should also be possible to monitor the execution of BRs by profiting from SOA governance technologies in order to observe invocations of the BR web services and keeping track of the message exchange with invoking services. Of course, a comprehensive SOA architecture in necessary at this stage and is out of the scope of this chapter. This idea can nevertheless be incepted by having a look at the high-level architecture (see Figure 11) of the solution the ABRD constitutes.

The ARIS framework also makes it possible to generate BPEL processes out of BP models (e.g. EPC) as part of its SOA product functionalities. These executable BPEL process descriptions can then be linked to business rules deployed as web services. These web services are called decision services, as they are generated from Rulesheets or Ruleflows which model decisions. The invocation of a decision service is interpreted by the workflow engine (lower center part of Figure 12) responsible for executing the BPEL descriptions and forwarded to the Corticon rule engine (lower left part of Figure 12). The Corticon rule engine can thus run the corresponding rules that it contains from BR deployment and send an answer message to the workflow execution environment.

*Figure 11. The Rulesheet for the credit application case. © 2008 IDS Scheer AG. Used with permission.*

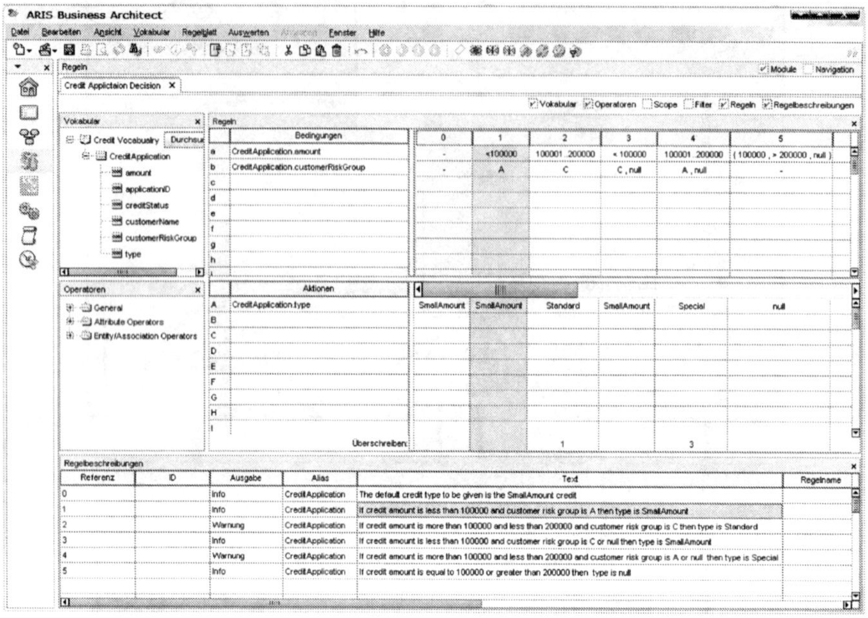

*Figure 12. Validating the rules using a Ruletest model. © 2008 IDS Scheer AG. Used with permission.*

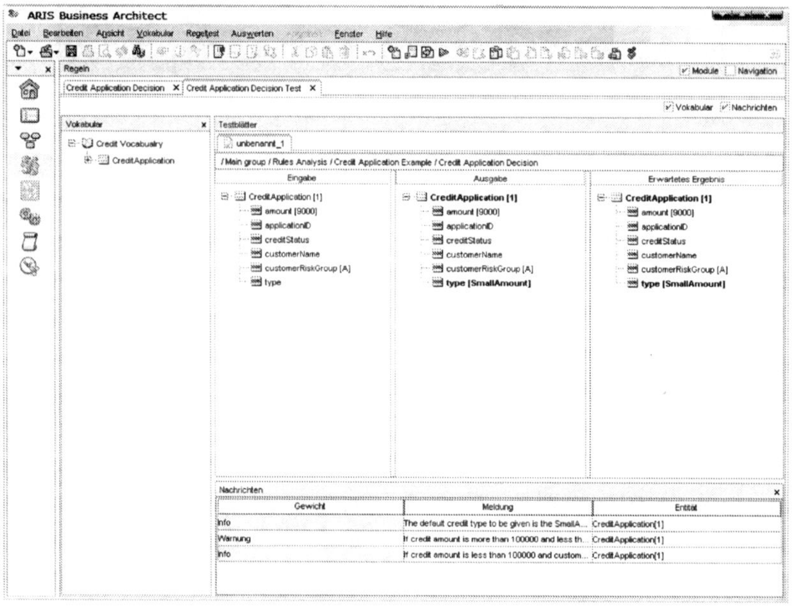

## A Small Example

As we have seen, a number of models, a lifecycle and architecture are proposed as a response to the requirements on a rule based approach to BPM. Using a small business scenario we will now see how we can make the use of the business Rule Lifecycle and the various modeling tools in the context of a given business process: from finding decision points in a process to deploying rules to be executed.

Let's get back to the example in Figure 10. In this process fragment, a branching is reached in the process where a routing decision has to be made. It must be decided which path to take further in the EPC after the XOR branch. To do this, we first annotate the function preceding the branch with a BR. This BR has to decide which of type the credit application is. Relevant input and output data are directly modeled in the EPC fragment in Figure 13. The rule modeler then creates a Rulesheet in order to model the rules of this simple decision. We can start to model the following two rules:

If credit amount is less than 100000 and customer risk group is A then credit type is Small Amount.

If credit amount is more than 100000 and less than 200000 and customer risk group is C then credit type is Standard.

By using the verification functionalities of the Rulesheet we end up with the following set of 6 rules which are shown in Figure 13.

The next step in the BRMLC is to validate the modeled BRs. For this, we use a Ruletest and test the following scenario as an example:

If a credit application's amount is of 9000 euros and the customer's risk group is A then the credit should be of the small amount type.

As shown in Figure 14, the Ruletest's outcome confirms that the Rulesheet was modeled correctly for this test scenario.

## FUTURE TRENDS

Due to the current concentration of research on ontologies, semantic web and semantic web services, it is easy to see that researchers will be trying to move on with results in the field and try to bring them to the next level using semantics. SBVR is an example of this. Also rule exchange is a critical issue. As the sector of BRMS is arriving to maturity with many flagship industrial products which are successful in the market, it will become critical that vendors and the research community both provide a common interchange format for business rules. Such an exchange for-

*Figure 13. The semantic web stack and business rules*

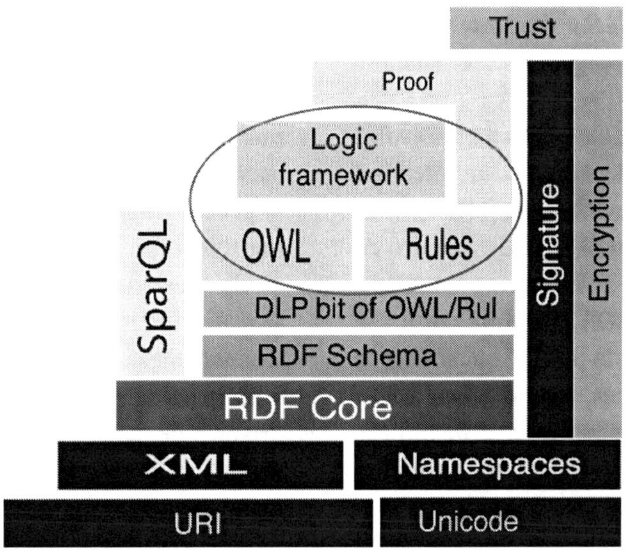

*Figure 14. The future of BRM*

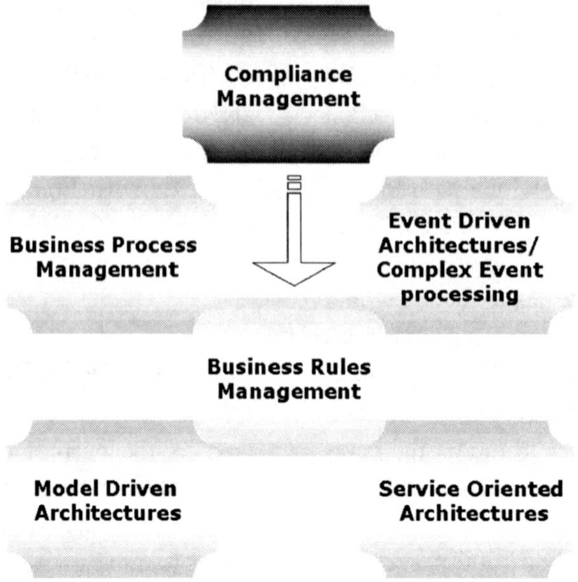


mat is compulsory if BRs are to be deployed in highly collaborative web-based scenarios. The Rule Interchange Format (RIF) (see endnote 6) is one serious candidate for this, because it is being standardized by the W3C. Another possible candidate is the R2ML (Wagner (2006), see endnote 7) languages which has also been designed to be an exchange format and is a comprehensible rule language, containing particularly the RuleML (see RuleML (2004), endnote 8) (portage of the datalog subset of prolog to the semantic web) and the Object Constraint language (OCL, see endnote 9) by the OMG.

The other important dimensions for the evolution of BRM are the service oriented architecture (SOA), the model driven architecture (MDA), business process intelligence (BPI), and event driven architectures (EDA). As BPM moves towards a tighter integration with SOA, the ability of BRM standards to enable rules to profit from advances in SOA and web services, would allow BRM to access to true web scalability and flexibility. This has already been scratched by deploying Rulesheets and Ruleflows (discussed in the previous parts of this chapter) as web services, but SOA is not about web services. When there will be true interoperability between BRM standards and SOA standards, it will be possible to think about distributed frameworks that make business rules accessible on a distributed web of services, and allow rules to communicate together and use inference capabilities to proceed to decisions in an open, distributed environment. A great enabler for this can already be incepted when looking at research on semantic web service composition and negotiation, it is the semantic web.

As for MDA, we have already started going into that direction by understanding that business users do not want to have to handle code in order to get the rules to do what they expect. Generating rule expressions in target formats from rule models designed by the business users and adapted to reasoning is the right way to go. Once we have reached these capabilities and integrated them with existing generic MDA frameworks, model driven rule management will be made available to all. For the moment, tone of the most promising fields for achieving this is certainly BPM as we have shown in this chapter with our model-driven approach to BRM for BPM.

The utilization of BRs in conjunction with event driven architectures is also one promising field of evolution. As BRs not only integrate well with events, in that they can be triggered by or themselves trigger events, events are vital in BR modeling for BPM. Technologies such as complex even processing (CEP) make frameworks allowing event-based application management available. The paradigm of EDA observes the business not as a message-based collaboration (such as in SOA) but as an event-based collaboration. In an EDA, everything is manipulated by events. As we can see in BPM and in the history of research on BRM, events are a capital concept for BRs. Integrating BRM frameworks in an EDA would provide highly-autonomous business automation.

Events also fit very well with process monitoring techniques. BPI includes business process monitoring and process event monitoring. Based on events, which could be fired by alarming values for critical key process indicators (KPI), events can be fired that trigger actions. If the business implements these actions as business rules, even greater agility is achieved in BPI. But BPI techniques do not solely rely on events, in fact for many of them not a all. BPI seeks to provide assistance to the business in taking adequate actions to correct deviations from business goals by monitoring the performance of the BPs. Studying the use of business rules in conjunction with BPI techniques and event is one promising direction of research. Another connex topic to BPI is process simulation. Current known simulation capabilities of BPM frameworks and tools lack simulation that takes "live data" and business rules into account. There are several ways of testing business rules, both for efficiency and effectivity, as we have seen. But these techniques allow to see if the decision-making based on BRs behaves conforming to business expectations. That means that the object of interest here are business dynamics, which is a t a high level. But such testing doesn't permit uncovering hidden side-effects of using BRs, and the influence of these on aspects of BPM such as the performance of BPs, BP interaction with other BP instances, especially in highly collaborative scenarios, which rely on BP choreographies and orchestrations Weske (2007). That makes BP simulation one of the most interesting working directions for BRM in the future.

Last but not least is Compliance Management. Compliance management is the discipline that monitors and enforces compliance of the business to various regulations and standards. Compliance has gotten great attention lately because of the numerous laws voted to ensure it and the high relevance of compliance for the good health of companies and the interest of shareholders. The research community is studying means of formally supporting compliance management through the use of rules. Particularly in BPM, there is a strong community conducting research on this topic.

## CONCLUSION

Organizations have to be quick to react to change, and in order to do this they need to document and manage their policies and rules. By doing this, organizations are able to manage, analyze, communicate and reuse their decision-making artifacts independently from processes. Rules can answer these needs, but we still have to learn how to best model and use rules. We explained why business rules management tailored to the requirements of business process management is needed and the advantages it brings to classical BPM. We then discussed how this can be achieved in the case of business process management and made a corresponding proposal for a business rules management lifecycle. Following this, we detailed what the different phases of this lifecycle consist of and have further shown how these phases work together. Most importantly, we have seen that the several artifacts used throughout this lifecycle fulfill the requirements we defined on such a lifecycle. After having shown how business rules can be put to work productively for business processes, we criticized the perspicacity of our approach and illustrated our ideas in an example from the industry.

Work on allying the power of business rules and the proven adequacy of business process management to provide a business-oriented control of the enterprise (in opposition to an IT-oriented enterprise control) is still at its beginnings. This is partly due to the status of BPM as an emerging research field. But the need for research covering these two complementary fields is emphasized by the high-relevance results from such works could directly have on the way we conduct business. Today, holistic views of the enterprise combining governance, compliance management, business process management and service-oriented architectures seek to enable us to concentrate less on the implementation of business requirements and more on the modeling of these requirements themselves. Business rule management is one core component of this vision.

## REFERENCES


Alves, A. et al. (2007). Web services business process execution language (BPEL) Version 2.0. *OASIS*. Retrieved May 4, 2008 from http://docs.oasis-open.org/wsbpel/2.0/wsbpel-specification-draft.html.

Antoniou, G., Berndtsson, M., Spreeuwenberg, S., Taveter, K., & Wagner, G. (2004). *Report: A first-version visual rule language* (REWERSE Deliverable I1-D1).

Basel Committee (2004). *Basel II revised international capital framework*. Retrieved from http://www.bis.org/publ/bcbsca.htm.

Batory, D. (1994). *The LEAPS Algorithm* (Tech. rep. CS-TR-94-28). Department of Computer Sciences, the University of Texas, Austin, Texas.

Bajec, M., & Krisper M. (2001). *Managing business rules in enterprises.*

Bajec, M., & Krisper, M. (2004). A methodology and tool support for managing business rules in organisations. *Information Systems, 30*(6), 423-443.

Bonatti, P. A., Shahmehri, N., Duma, C., Olmedilla, D., Nejdl, W., Baldoni, M., Baroglio, C., Martelli, A., Patti, V., Coraggio, P., Antoniou, G., Peer, J., & Fuchs, N. E. (2004). *Rule-based policy specification: State of the art and future work* (REWERSE project I2-D1 deliverable).

Corticon Technologies, Inc. (2004) *Under the covers.*

Sarbanes-Oxley Act of 2002, H.R. 3763, 107th Cong., 2nd Sess. (2002).

Cysneiros, L. M., & Sampaio do Prado Leite, J. C. (2000). *Eliciting business rules through iso 9000 documentation: A domain oriented conceptual model.*

Doorenbos, R. D. (1995). *Production matching for large learning systems.* Carnegie Mellon University Pittsburgh, PA, USA.

El Kharbili, M., Stein, S., Markovic, I., & Pulvermueller, E. (2008). Towards a framework for semantic business process compliance management. In *Proceedings of the Workshop on Governance, Risk, and Compliance On Information Systems (GRCIS).*

Endl, R., & Knolmayer, G. et al. (1998). *Modeling processes and workflows by business rules.*

Forgy, C. L. (1982). Rete: A fast algorithm for the many pattern/many object pattern matching problem. *Artificial Intelligence, 19*, 17-37.

Forgy, C. L., & Shepard S. J. (1987). Rete: A fast match algorithm. AI Expert archive, *2*(1), 34-40).

Forgy, C. (n.d.). *OPS5 User's Manual* (Technical Report CMU-CS-81-135). Carnegie Mellon University.

HIPAA Administrative Simplification: Enforcement; Final Rule. 45 C.F.R., 160 & 164 (2006).

Huntington D. (1999). Knowledge-based systems: A look at rule-based systems. In Liebowitz, J. (Ed.), *Knowledge management handbook* (Section IV, pp. 14-1).

*The ISO 17799 IT security standard.* (n.d.) Retrieved May 4, 2008 from http://www.iso.org/iso/iso_catalogue/catalogue_tc/catalogue_detail.htm?csnumber=39612.

*The ISO 27002:2005 IT security standard.* (n.d.). Retrieved May 4, 2008 from http://www.iso.org/iso/iso_catalogue/catalogue_tc/catalogue_detail.htm?csnumber=50297.

*Jess, the rule engine for the Java platform* (n.d.) Retrieved May 4, 2008 from http://herzberg.ca.sandia.gov

Kardasis, P., & Loucopoulos, P. (2004). Expressing and organising business rules. In *Information and software technology, 46*(11), 701-718.

Keller, G., Nüttgens, M., & Scheer, A.-W. (1992). Semantische Prozeßmodellierung auf der Grundlage Ereignisgesteuerter Prozeßketten (EPK). In A.-W. Scheer (Ed.), *Veröffentlichungen des Instituts für Wirtschaftsinformatik* (Vol. 89). Saarbrücken.

Kemsley, S., Straus, D,. & Sinur, J. (2008) Change the rules without changing your business processes: Using business rules management to turbocharge your BPM solution. *Intelligent Enterprise Vendor Perspectives Webcast.* Retrieved February 12, 2008 from http://intelligententerprise.com/webinars.jhtml

Object Management Group (2006). *Business Process Modeling Notation, V1.1.* Official OMG Specification. Retrieved May 4, 2008 from http://www.bpmn.org/Documents/OMG%20Final%20Adopted%20BPMN%201-0%20Spec%2006-02-01.pdf.

*Object Management Group. Object Constraint Language* (OCL). Retrieved February 12, 2008 from http://www.omg.org/technology/documents/modeling_spec_catalog.htm#OCL. Ouyang, C., van der Aalst, W. M. P., Dumas, M., & ter Hofstede, A. H. M. (2006). Translating BPMN to BPEL. *BPM Center.*

Riley, G. (2004). *Clips - A tool for building expert systems.* Retrieved May 4, 2008 from http://clipsrules.sourceforge.net/, retrieved on the 4th May 2008.

Rosca, D., & Wild, C. (2002). Towards a flexible deployment of business rules. *Expert Systems with Applications, 23*(4), 385-394.

Rosca, D., Greenspan, S., Wild, C., Reubenstein, H., Maly, K., & Feblowitz, M. (1995). Application of a decision support mechanism to the business rules lifecycle. In *10th Knowledge-Based Software Engineering Conference* (p. 114).

Rosca, D., Greenspan, S., & Wild, C. (2004). Enterprise modeling and decision-support for automating the business rules lifecycle. *Journal Automated Software Engineering, 9*(4), 361-404.

Rosca, D., Greenspan, S., Feblowitz, M., & Wild, C. (1997). A decision making methodology in support of the business rules lifecycle. In *Proceedings of the Third IEEE International Symposium on Requirements Engineering* (pp. 236-246).

Ross, R. G. (2005). *Business rule concepts: Getting to the point of knowledge* (2nd Ed.).

Ross, R.G. (2003). IT systems perspective - The business rule approach. *Computer, 36*(5), 85-87).

Ross., R. G. (1997). *The business rule book: Classifying, defining and modeling rules* (2[nd] ed.). Boston: Database Research Group, Inc.

RuleML Initiative (2004). *The Rule Markup Initiative Web Site.* Retrieved from, http://ruleml.org

Sandifer, A, & Von Halle, B. (2000). Business rules: Capturing the most elusive information asset. In S. Purba (Ed.), *High-performance databases: Design, development and deployment* (pp. 111-128). CRC Press, Taylor & Francis Group.

SBVR Specification (2008). *Object Management Group.* Retrieved May 4, 2008 from http://www.omg.org/docs/formal/08-01-02.pdf

Scheer, A.-W. (1999). *ARIS - Business process frameworks* (3rd ed.). Springer.

Smith, H., & Fingar, P. (2003). *Business process management: The third wave.*

Stein, S., & Ivanov, K. (2007). EPK nach BPEL Transformation als Voraussetzung für praktische Umsetzung einer SOA Software Engineering 2007. In W.K. Bleek, J. Raasch, & H. Züllighoven, H. (Eds.), *Gesellschaft für Informatik (GI) Vol. 105* (pp.75-80).

Vasilecas, O., & Smaizys, A. (2006). The framework: an approach to support business rule based data analysis. *In Proceedings of the 7th International Baltic Conference on Databases and Information Systems* (pp. 141-147).

Wagner, K., & Klückmann, J. (2006). Prozessdesign als Grundlage von Compliance Management, Enterprise Architecture und Business Rules. In A.-W. Scheer, H. Kruppke, W. Jost & H. Kindermann (Eds.), *Agilität durch ARIS Geschäfprozessmanagement* (pp. 125-136). Springer Berlin/ Heidelberg.

Wagner, G.; Giurca, A. & Lukichev, S. (2006). A usable interchange format for rich syntax rules integrating OCL, RuleML and SWRL. In *Proceedings of the Reasoning on the Web (RoW2006) Workshop at WWW2006.*

Wagner, G., Giurca, A., & Lukichev, S. (2006). A general markup framework for integrity and derivation rules. In F. Bry et al. (Eds.), *Principles and Practices of semantic web reasoning.* Dagstuhl Seminar Proceedings. Internationales Begegnungs- und Forschungszentrum fuer Informatik (IBFI), Schloss Dagstuhl, Germany.

Weske, M. (2007). *Business process management: Concepts, languages, architectures* (1st Ed.). Springer

Zeng, L., Flaxer, D., Chang, H., & Jun-Jang, J. (2002). PLM flow—Dynamic business process composition and execution by rule inference. In *Technologies for e-services* (LNCS 2444, pp. 51-95).

## KEY TERMS AND DEFINITIONS

**Business Process (BP):** in the Merriam-Webster dictionary (see endnote 10), a process is defined as "a series of actions or operations conducing to an end; [...] a continuous operation or treatment especially in manufacture". Business processes have existed as long as businesses have existed. In research, the term comes from two converging directions: research on business economics and research on computer science. Business process refers to the organization of value creating business activities.

**Business Process Management (BPM):** The term is used to appoint the branch of information systems that studies IT techniques and methods that support business processes. It is an evolution of workflow management (WFM) and relies on formal methods and languages for modeling all business related aspects of processes. As shortly stated in Aalst (2003), BPM "includes methods, techniques, and tools to support the design, enactment, management, and analysis of operational business processes". BPM also includes by extension other related and sub-disciplines of BPM such as Business Activity Monitoring (BAM), Business Process Controlling (BPC) and Business Process Intelligence (BPI).

**Business Rule (BR):** is the logic that expresses decision patterns, guidelines or constraints on business aspects. Here, "business aspects" refers to all aspects of companies which are of relevance for the rules. Typically in business process management, "business aspects" are business processes, value chains, organizational charts, resource models and enterprise architecture elements.

**Business Rule Management (BRM):** is the discipline comprehending gathering, auditing, modeling, documentation, organization, communication, implementation, testing, publishing, deployment, enforcement and controlling of business rules throughout the company.

**Business Rule Management System (BRMS):** is a system responsible for supporting a business rule management lifecycle. BRMS allow for the modeling of rules, for keeping the consistency of the rules base, for providing components for deploying rules. BRMS contain necessary graphical user interfaces to allow the capture of business rules. A BRMS usually contains numerous interfaces to external systems in order to allow using the business rules productively. In the case of BPM, a BRMS is the element responsible for managing business rules.

**Rules Engine (RE):** is a framework capable of interpreting business rules. Rule engines usually use forward chaining, backward chaining or both in order to infer on facts, rules and predicated and determine which are applicable. Rule engines are the components of a BRMS that are responsible for the consistency of the business rules base, the simulation of business rules, the inference on business rules. In concrete, rule engines are the core of any BRMS and are the element capable of taking decisions.

**Rulesheet:** is a model allowing the definition of business rules in the form of a decision table. A Rulesheet contains rule statements that are linked to elements of the vocabulary the rule is built on. A Rulesheet is thus an aggregation of business rules relating to a certain aspect of the business and operating on a business vocabulary. In a BPMS, business process artifacts are annotated with Rulesheets and hence, the link between business processes and business rules is created. Rulesheets offers necessary functionalities for rule verification.

**Ruletest:** is a model allowing the simulation of business rules using business scenarios. Whereas Rulesheets allow testing the efficiency of modeled business rules, Ruletests allow testing the effectivity of business rules. This means that Ruletests are used in order to check whether the considered business rules really generate decisions as the business expects it. This is also called rule validation.

**Ruleflow:** is a model used for modeling complex rules, which must be expressed in terms of flows. A Ruleflow can be seen as a business process where the activities are all sets of business rules. Using Ruleflows makes the modeling of complex control flows of business rule execution possible. Of course, modeling Ruleflows relies on modeling Rulesheets, since each Ruleflow activity is in fact specified by a Rulesheet.

**Vocabulary:** is the commonly agreed on representation of the business aspects relevant for the modeling of rules. A vocabulary is formed of entities and relationships, and can be directly created from or mapped to elements of existing data models. An equivalent to vocabulary usually found in literature about business rules are terms and facts.

# Chapter XXIV
# Mapping Policies to Web Rules: A Case of the KAoS Policy Language

**Nima Kaviani**
*University of British Columbia, Canada*

**Dragan Gašević**
*Athabasca University, Canada*

**Marek Hatala**
*Simon Fraser University, Canada*

## ABSTRACT

*Web rule languages have recently emerged to enable different parties with different business rules and policy languages to exchange their rules and policies. Describing the concepts of a domain through using vocabularies is another feature supported by Web rule languages. Combination of these two properties makes web rule languages appropriate mediums to make a hybrid representation of both context and rules of a policy-aware system. On the other hand, policies in the domain of autonomous computing are enablers to dynamically regulate the behaviour of a system without any need to interfere with the internal code of the system. Knowing that policies are also defined through rules and facts, Web rules and policy languages come to a point of agreement, where policies can be defined through using web rules. This chapter focuses on analyzing some of the most known policy languages (especially, KAoS policy language) and describes the mappings from the concepts for KAoS policy language to those of REWERSE Rule Markup Language (R2ML), one of the two proposals to Web rule languages.*

## INTRODUCTION AND MOTIVATION

Rules are among the most frequently used knowledge representation techniques. They can generally be categorized as *reaction rules* (event-condition-actions), *integrity rules* (rules of consistency checking), *derivation rules* (implicitly, derived rules), and *production rules* (Boley, Tabet, & Wagner, 2001). *Facts* can also be regarded as derivation rules with no premises.

Recently, rule markup languages have started to be considered as the vehicle for using rules on the Web and in other distributed systems, forming a new category of rule languages, referred to as

Web rule languages (Wagner, Giurca, & Lukichev, 2006). The main strength of markup languages is their machine readability combined with their inherent potentials to easily be linked to distributed knowledge sources also represented in the form of markup languages (e.g., Data Type Definition (DTD), XML Schema (Fallside & Walmsley, 2004), Resource Description Framework (RDF) (Lassila & Swick, 1999), and Web Ontology Language (OWL) (Smith, Welty, & McGuinness, 2004)). Rule markup languages allow for the reuse, interchange, and publication of rules as well as their communication and execution in a distributed environment. More specifically, rule markup languages allow for specifying business rules as modular, standalone, units in a declarative way, and publishing and interchanging them between different systems and tools (Wagner, Giurca, & Lukichev, 2006). Rule markup languages play an important role in facilitating business-to-customer (B2C) and business-to-business (B2B) interactions over the Internet by enabling information exchange across various stakeholders. For example, they may be used to express derivation rules for enriching Web ontologies by adding definitions of derived concepts, or for defining data access permissions; to describe and publish the reactive behaviour of a system in the form of reaction rules; and to provide a complete XML-based specification of a software agent (Wagner, Giurca, & Lukichev, 2005).

*RuleML* and *REWERSE Rule Markup Language (R2ML)* are two newly and rapidly emerging Web rule languages that help with interchanging various types of rules from one rule domain to another. RuleML represents an initiative for creating a general rule markup language that will support different types of rules and different semantics (Boley, Tabet, & Wagner, 2001). R2ML more or less follows a similar idea to RuleML; however, its design follows Model Driven Architecture (MDA) defined by the Object Management Group (OMG) (Miller & Mukerji, 2003). Moreover, R2ML is a trial to cover a more comprehensive set of atoms and elements required to develop and define rules, thus bringing more flexibility to rule definition. These rule languages are designed to be conformant and compatible with the guidelines and use cases defined by the Rule Interchange Format (RIF) working group (Ginsberg, Hirtle, McCabe, & Patranjan, 2006).

Policies in the domain of autonomous computing are guiding plans that restrict the behaviour of autonomous agents in accessing the resources (Toninelli, Bradshaw, Kagal, & Montanari, 2005). The main advantage in using policies is the possibility to dynamically change the behaviour of the system by adjusting the policies without interfering with the internal code of the system. They can authorize/oblige the users to, or prohibit/dispense them from, accessing the resources or taking particular actions in the system. Policy languages can be considered as instructional sets that enable phrasing and putting systematic guidelines in place for a target agent. KAoS (Uszok, et al., 2003) is one of the most known policy languages that goes beyond the traditional policy systems by giving special care to the context to which the policies are applied. This is done by enabling these policy languages to use domain knowledge (a.k.a, vocabularies) readily available on the Internet and represented in knowledge representation markup languages such as XML-Schemas, RDF, and OWL.

Knowing that policies are also defined through the use of rules and facts that can share online knowledge bases, Web rules and policy languages come to a point of agreement. Web rules and policies get even closer bringing it into the consideration that most of the newly emerging policy languages also have chosen markup syntax for their specifications (e.g., KAoS). Following the hierarchy of rules represented earlier, a policy rule can be considered a reaction rule which generates an action (possibly permission or denial) upon the occurrence of an event and satisfaction of a series of conditions. On the other hand, a policy rule can be considered a derivation rule that leads to generating additional facts once a set of conditions are satisfied (e.g., a policy may result in concluding that a resource is no longer in use, thus adding an extra fact to the knowledge base). Further to this, a policy rule might be defined in the form of an integrity rule to preserve the integrity of information in a system. All this brings Web rules and policies to a joint end where Web rules can be used to describe policies. The inherent flexibility of Web rule languages in using various vocabularies overcomes the intricate problem of reusing policies across various domains. What is more interesting is that the recent policy languages (including KAoS) can be defined over domain knowledge represented in markup languages such as OWL (Smith, Welty, & McGuinness, 2004) and XML Schema (Fallside & Walmsley, 2004), making it a lot easier to integrate them with Web rules. Furthermore, Web rules have been designed to be compatible with different subsets of first order logic (FOL), ranging from description logic to rule-based logic programs. Consequently, the essence and the semantics of the policy rules specified in different policy languages can be captured agnostic to the logical domain in which the language is grounded. This in turn brings more flexibility and compatibility to Web services to be combined with policies defined in the form of Web rules. There are various benefits in providing bidirectional transformations between policies and Web rule languages. First and foremost, it enables sharing of policies between various business processes which in turn helps with policy reuse from a business system to another. Sharing policies across various business processes is another step towards facilitating B2B and B2C interactions, as discussed above. Additionally, the possibility of converting policies from one language to another, while preserving the semantics of the policies, facilitates the deployments of policies in a business system regardless of the underlying technologies used by its business processes.

In order to be able to decide how a markup policy language can best be described through using a Web rule language, there needs to be a deep analysis of the constructs and elements of one domain to be matched with the corresponding elements in the other domain. This chapter focuses on discussing the building blocks of defining policies (from their logics to their concepts) using markup policy languages, and to find the appropriate matches in a corresponding Web rule language. For the policy languages we will mainly focus on KAoS while for the Web rule we will narrow our focus down to R2ML. This will later on lead us to discuss the possibilities for using a unified modeling language to define and interchanging policies between different policy domains, or even more, to rely on a modeling language to facilitate the definition, design, and integration of policies with other parts of a system, including but not limited to business processes.

This chapter is organized as follows. In Section 2 we will provide some background information on different types of underlying logic for defining policies and rules, some of the most known policy languages, the idea behind rule interchange format, and also a background on Web rule languages. Section 3 discusses our policy interchange framework. In Section 4 we present the metamodels for R2ML and KAoS to discuss high level exchange of concepts between the elements of these languages along with the graphical QVT transformations from each language to the other one and back. Section 5 shows a real example of applying the QVT transformations to the concrete representation of a KAoS policy rule which is then followed by a discussion and conclusion to our findings in Section 6.

## BACKGROUND

Before going into the details of how policy languages can be converted to Web rule languages and transformed from one business domain to another, in this section we provide brief background information on the existing policy and Web rule languages. We also discuss how following the principle of modeling can facilitate the process of defining and transforming policy rules from one policy language to another.

## Rule Interchange Format

Most of the proposals on Web rule languages are trying to address the use cases and requirements defined by Rule Interchange Format Working Group (Ginsberg, Hirtle, McCabe, & Patranjan, 2006). *Rule Interchange Format (RIF)* (Ginsberg, Hirtle, McCabe, & Patranjan, 2006) is an initiative to address the problem of interoperability between existing rule-based technologies. RIF is desired to play as an intermediary language between various rule languages and not as a semantic foundation for the purpose of reasoning on the Web. It aims to be a widely adopted W3C consistent standard for exchanging rules among different stakeholders. RIF Working Group has defined 10 use cases which have to be covered by a language compliant to the RIF's properties, three of which are dealing with policies and business rules, namely: *Collaborative Policy Development for Dynamic Spectrum Access, Access to Business Rules of Supply Chain Partners*, and *Managing Inter-Organizational Business Policies and Practices*. *SWRL* (Horrocks, Patel-Schneider, Boley, Tabet, Grosof, & Dean, 2004) and *RuleML* (Boley, Tabet, & Wagner, 2001) are two of the ongoing efforts in this area trying to serve as rule languages to publish and share rule bases on the Web.

## Web Rule Languages

Web rule languages are being developed following the RIF Request for Proposal (RFP) for designing a Web rule language that in the first place provides an intermediary language for exchanging rules between various stakeholders and business processes. The current Web rule languages are not trying to facilitate reasoning on the Web and as a result there is no reasoning engine or infrastructure developed for them. On the other hand, the idea is to have sound transformations from a source rule language to a target rule language, such that the reasoner at the destination (which is of course conformant to the rule language at the destination) can successfully reason over the transformed rules and derive the same conclusions as of the ones at the source. RuleML (Boley, Tabet, & Wagner, 2001) and R2ML (Wagner, Giurca, & Lukichev, 2005) are two of the most known Web rule languages. In this chapter, our focus is more on R2ML, but we briefly skim over RuleML before describing the architecture and of R2ML.

RuleML represents an initiative for creating a general rule markup language that will support different types of rules and different semantics (Boley, Tabet, & Wagner, 2001). It is conceptualized to capture the hierarchy of rule types as we discussed earlier. However, the current version of RuleML covers only some limited forms of rules (Wagner, Damasio, & Antoniou, 2005).

RuleML is built on top of logic programming paradigm of first order logic (i.e. predicate logic). In the tradition of logic programming which is also followed by RuleML, research is focused on computable interpretations of predicate logic, by exploring a great number of semantic extensions and variations. OWL (as well as SWRL (Horrocks, Patel-Schneider, Boley, Tabet, Grosof, & Dean, 2004)) stems from logic-based tradition of artificial intelligence where research is based on classical predicate logic (two-valued) as the one and the only logic. An example of a RuleML rule that uses certain person's attributes to define a rule of "hasMother" and "hasBrother" implies "hasUncle" is shown in Figure 1.

In the next subsection, we skim over the second key web rule language proposal for RIF, known as REWERSE Rule Markup Language (R2ML).

*Figure 1. An example of a RuleML rule showing the "hasUncle" relationship*

```
<Implies>
  <head>
    <Atom>
      <Rel>hasMother</Rel>
      <Var>x1</Var>
      <Var>x2</Var>
    </Atom>
    <Atom>
      <Rel>hasBrother</Rel>
      <Var>x2</Var>
      <Var>x3</Var>
    </Atom>
  </head>
  <body>
    <Atom>
      <Rel>hasUncle</Rel>
      <Var>x1</Var>
      <Var>x2</Var>
    </Atom>
  </body>
</Implies>
```

## REWERSE Rule Markup Language (R2ML)

R2ML is part of the EU-funded REWERSE project, which follows the goal of providing a general rule markup language to make the deployment, execution, publishing, and communication of rules on the Web possible. The approach chosen to develop the R2ML is based on the known Model Driven Architecture (MDA) (Miller & Mukerji, 2003) and Meta Object Facility (MOF) (ODM, 2001) defined by the Object Management Group (OMG). This means that the whole language definition of R2ML can be represented by using UML diagrams as MOF uses UML's graphical notation. The language also has an XML concrete syntax defined by an XML schema.

As shown in Figure 2, R2ML considers three levels of abstraction for rules. At the *computation-independent business domain* level (called CIM in OMG's MDA), rules are statements that express (certain parts of) a business/domain policy (e.g., defining terms of the domain language or defining/constraining domain operations) in a declarative manner, typically using a natural or visual language. At the *platform-independent operational design* level (called PIM in OMG's MDA), rules are formal statements, expressed in some formalism or computational paradigm, which can be directly mapped to executable statements of a software system. At the platform-specific implementation level (called PSM in the OMG's MDA), rules are statements in a language of a specific execution environment such as XSB Prolog (XSB, 2007).

R2ML provides a vocabulary that enables users to define their own world in the form of objects and elements available in the domain of discourse. The vocabulary can be defined as a combination of *Basic Content Vocabulary*, *Relational Content Vocabulary*, and *Functional Content Vocabulary.* Basic Content Vocabulary allows the user to specify the basic elements of the domain such as individual objects and data values, classes and data types, and object and data variables. Relational Content Vocabulary helps to associate different objects from different classes through defining n-ary association and association classes. Finally, Functional Content Vocabulary assists with defining functors that correspond to the standard logic of functions. The functions

*Figure 2. Different abstraction levels considered for rules in R2ML*

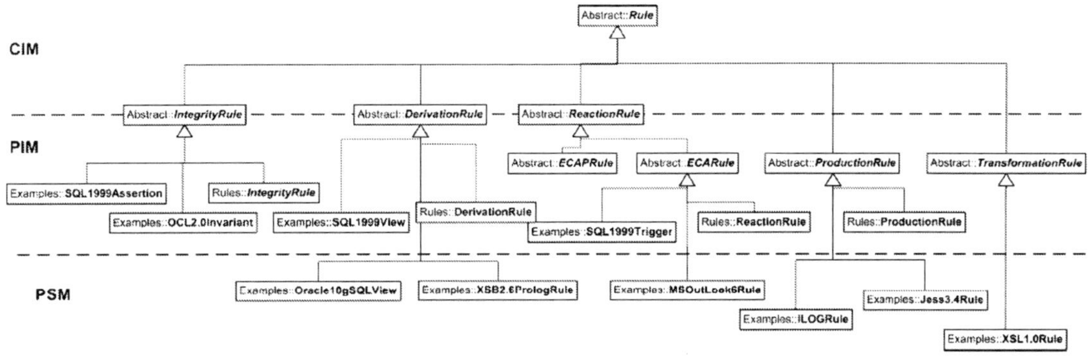


can be data operations to manipulate data values, they can be object operation functions that define object-value operations, or they can be role functions which correspond to functional association (binary association) of the class elements. In (Bradshaw, Dutfield, Benoit, & Woolley, 1997), authors showed how the basic constructs and elements of the OWL language can be transferred and modeled by R2ML atoms and elements. For example, *sameAs* in OWL is equivalent to an *EqualityAtom* in R2ML and *oneOf* in OWL carries the same meaning as *Disjunction* of a set of atoms in R2ML. This means any language with its concepts defined based on OWL (including KAoS and Rei) can be modeled with R2ML constructs elaborately.

Having the objects and concepts of a domain defined, R2ML makes the definition and harmonization of rules over these concepts possible through the use of four different types of rules: *Integrity Rules*, *Derivation Rules*, *Reaction Rules*, and *Production Rules*. Since in this paper we are limited in space, we only review the first two rules and more information about the other rules and constructs of the language can be found in (Wagner, Giurca, & Lukichev, R2ML: A General Approach for Marking-up Rules, 2005).

R2ML integrity rules, also known as (integrity) constraints, consist of a constraint assertion, which is a sentence in a logical language such as first-order predicate logic or OCL (see Figure 3a). R2ML supports two kinds of integrity rules: the *alethic* and the *deontic* ones. The alethic integrity rule can be expressed by a phrase, such as *"it is necessarily the case that"* and the deontic one can be expressed by phrases, such as *"it is obligatory that"* or *"it should be the case that"*. A LogicalStatement is a LogicalFormula that has no free variables, i.e., all the variables from this formula are quantified. In terms of policy languages, integrity rules can be considered as constraints that must hold consistently especially in the level of rule enforcement, e.g. *"it is necessary to give a higher priority to the commands of the administrator than to the commands of the regular users on a system."*

An R2ML derivation rule has conditions and a conclusion (see Figure 3b) with the ordinary meaning that the conclusion can be derived whenever the conditions hold. While the conditions of a derivation rule are instances of the *AndOrNafNegFormula* class, representing quantifier-free logical formulas with conjunction, disjunction and negation; conclusions are restricted to quantifier-free disjunctive normal forms without *NAF* (Negation as Failure, i.e. weak negation). In the context of policies, we consider each deontic policy rule as a single derivation rule with the constraints making the conditions of the derivation rule and the policy decision forming the conclusion of the rule, e.g.

*Figure 3. The metamodel for (a) the integrity rule, and (b) the derivation rule in R2ML*

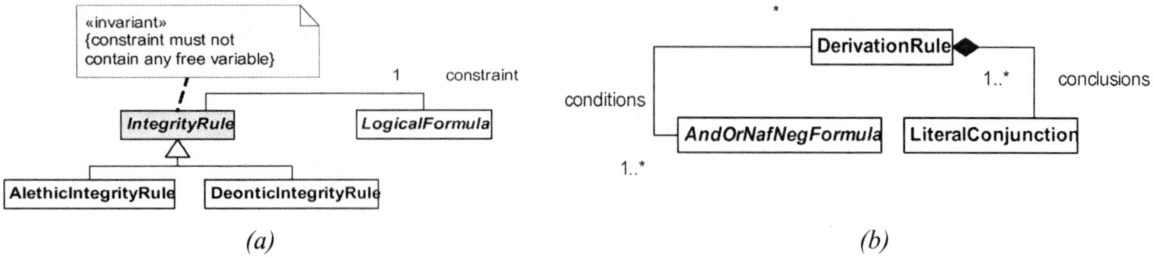


*"If the user is from Simon Fraser University with a valid student ID then give her the permission to enter the area of the university."* It may sound more expressive to define deontic policy rules with deontic integrity rules in R2ML. However, our attempts in doing so showed that deontic rules in the context of policies carry a different meaning from their interpretation in R2ML. In R2ML, a deontic integrity rule represents a constraint that should be satisfied or must hold with a concrete proof for its truthfulness, though a doentic policy demonstrates concerns over performing a duty or obligation as a result of satisfying a series of related conditions.

Atoms are the basic logical constituents of a rule which are compatible with the concepts of OWL, RuleML, and SWRL. Atoms connect objects to values, classes to instances, and objects to objects, put restrictions on the objects and data values, and so on. Here, we briefly represent some of the atoms that are relevant to our purpose of representing policy languages. *ReferencePropertyAtoms* (see Figure 4) associate object terms as subjects with other terms (objects or data values) as objects. A *ReferencePropertyAtom* in R2ML corresponds to an OWL (and similarly a KAoS) object property, or to the OWL concept of value for an individual-valued property. *ObjectDescriptionAtoms* (see Figure 5) are another class of useful atoms for our purpose. They refer to a class as a base type and to zero or more classes as categories, and consist of a number of property/term pairs (i.e., attribute data term pairs and reference property object term pairs). Any instance of such atom refers to one particular object that is referenced by an *objectID*, if it is not anonymous. This atom corresponds to the instantiation of an object from a class in OWL.

## Policy Languages

Policies can be considered as building blocks in enhancing the security, privacy, and usability of a system (Bonatti, et al., 2006). For policies to work effectively, they require to be defined over the domain knowledge that is provided by the domain experts. The process of representing the domain knowledge and defining policy rules over this knowledge, in a machine readable form, is handled by using policy languages. A policy language enables combination of constraints and capabilities, implying rules, over resources that represent policies.

Thus far, there have been various designed policy languages (such as XACML (Godik & Moses, 2003), Ponder (Damianou, Dulay, Lupu, & Sloman, 2001), Protune (Bonatti & Olmedilla, 2005), PeerTrust (Nejdl, Olmedilla, & Winslett, 2004), KAoS (Uszok, et al., 2003), and Rei (Kagal, 2002)) aiming to enable policy engineers to represent their organizational policies in a machine understandable format. In this chapter, we focus on KAoS policy language mainly because *i) it follows a markup format for defining policies*, and *ii) it uses ontologies to represent the domain knowledge*. The first property helps

*Figure 4. R2ML's ReferencePropertyAtom*

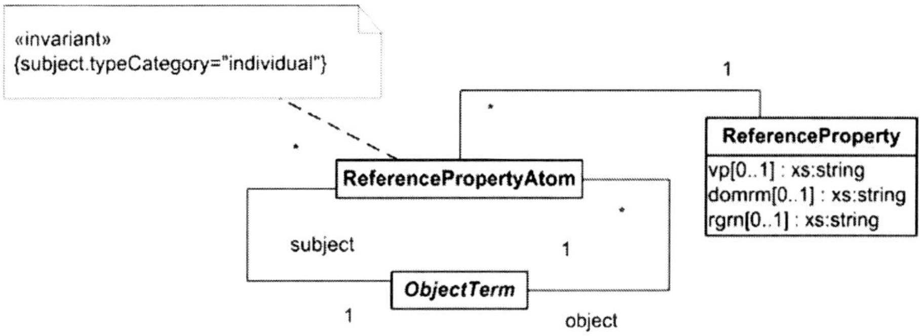


*Figure 5. R2ML's ObjectDescriptionAtom*

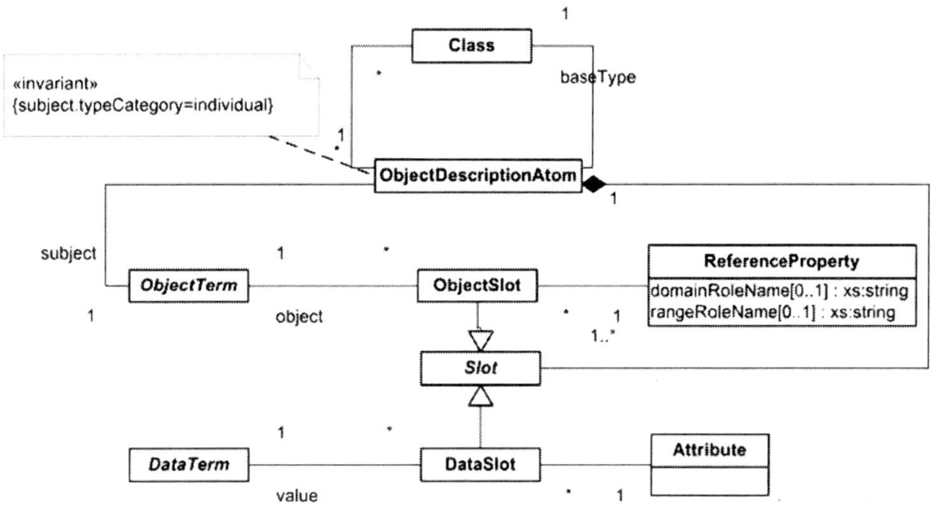


with easier mapping of policies from the KAoS policy languages to a Web rule language, because the source and the target language more or less follow a similar syntactical format. The second property helps with easier transformation of the domain knowledge from the source language to the target language. There have been already efforts on providing accurate mappings between two domain ontologies (Kalfoglou & Schorlemmer, 2003). Such efforts assist with the conversion of the domain knowledge from one conceptual model to a different conceptual model, thus making the domain knowledge easy to be shared across various policy languages and their corresponding reasoning engines.

Rei (Kagal, 2002) and KAoS (Uszok, et al., 2003) are two semantically enriched Web policy languages that use Semantic Web ontologies to define the resources, the behavior, and the users of a domain. Using ontologies enables these two languages to easily adjust themselves to the target system regardless of the number of resources and users in act. KAoS describes the entities and concepts of its world using OWL, while Rei can understand and reason over a domain of concepts defined in either RDF or OWL.

In terms of available policy rules, both KAoS and Rei have four main types. *Permission*, *Prohibition*, *Obligation*, and *Dispensation* in Rei are respectively equivalent to *PosAuthorizationPolicy*,

*NegAuthorizationPolicy*, *PosObligationPolicy*, and *NegObligationPolicy* in KAoS. The defined policy rules in each of the languages are then sent to a reasoner that performs the process of conflict resolution and decision making for the rules that match the current state of the world. This task is done by using Stanford's Java Theorem Prover (JTP) in KAoS and a Prolog engine in Rei version 1.0. Rei version 2.0 has extended its reasoning engine to use F-OWL, an ontology inference engine for OWL, based on Flora and XSB Prolog (XSB, 2007). Although these two policy languages have a lot in common there are dissimilarities between them as well. The main difference between KAoS and Rei is the underlying formalism of the languages. KAoS follows description logic coded in the form of OWL expressions with additional constraints that help with defining policy elements and rules. On the other hand, Rei uses its own language that defines policy rules in terms of Prolog predicates expressed as RDF triples. This way Rei follows semantics close to Prolog's semantics which is itself built on top of the concepts of logic programs.

The process of rule enforcement in KAoS is done by extending its enforcement engine depending on the domain it is going to be used in. In Rei, however, there is no rule enforcement engine. Yet, due to the deterministic properties of declarative logic, reasoning over dynamically determined values in Rei policies is more accurate than KAoS in which chances of dealing with unknown situations are likely to happen. In order for processes and services to communicate remotely, Rei relies on a rich set of Speech Acts. In Rei, Speech Acts are used by a sender to express the request for performing one of the actions: *Delegation*, *Revocation*, *Request*, *Cancel*, *Command*, and *Promise* by the receiver. Conversely, in KAoS the remote communication procedure is done through the message passing of the underlying platform.

Defining KAoS policies as OWL expressions gives the language more flexibility to maneuver over the concepts of the world. Different quantifying expressions, inheritance relationships, cardinality restrictions, etc. can be explicitly expressed in KAoS thanks to the constructs of OWL. It also enables KAoS to perform static conflict resolution and policy disclosure. KAoS has its classes and properties already defined in OWL ontologies, referred to as KAoS Policy Ontologies (KPO) (Uszok, et al., 2003), which are accessible from (Uszok & Bradshaw).

Looking back to the similarities and differences discussed, providing meaningful transformations from/to a web rule language to/from a policy language is not a trivial goal. Aside from all the difference in the syntax of the source and the target language, the transformations should care about the underlying logic that each policy language adheres to. As we discussed above, Rei and KAoS follow different logical formalisms, and thus different mapping considerations require to be done in choosing the appropriate elements in the target language to model the elements of the source language. This helps with capturing the logical semantics of the transformed rules. Furthermore, the concepts that might be missed in the procedure of transformation, due to a logical or conceptual mismatch between the policy language and the web rule language, need to be considered (we discuss it further in the next section). Web rule languages have been design with a broad consideration on the supported logic and required logical elements such that they can capture all various types of rules. However, the mismatch in the logics or the concepts may happen when transferring a rule from a web rule language to a policy language. The concepts that result in a logical or conceptual mismatch must be carefully monitored during the process of transformation. Having the lost information during the process of transformation found, the importance of such elements and the harms and threats that may happen to the resources due to information loss might be taken into consideration.

## Relations between Policies, Rules, and First Order Logic (FOL)

As already mentioned, rules are among the most frequent techniques for representing knowledge. Different knowledge representation methods (including rules in a broader and policies in a narrower sense) typically follow different logical formalisms. A clear use of Web rule languages in place or in conjunction with policies requires a clear understanding of the logical formalisms behind defining rules and policies. There are two major fragments to FOL that are mainly used for representing knowledge, namely, *description logic (DL)* and *computational logic programs (LP)*. Description Logic is a subset of the well-known First Order Logic (FOL) without function symbols (Grosof, Horrocks, Volz, & Decker, 2003), which is similar to a fragment of Horn FOL, *def-Horn*, that also contains no function symbols. LP, however, represents a set of logical constructs that neither includes nor is fully included by FOL but only intersects with it. That is to say, although LP and FOL share some logical constructs, there are concepts in either of the two sets that are missing or do not have any exact equivalent in the other set. For example, FOL can express positive disjunctions (see Grosof, Horrocks, Volz, & Decker, 2003, for details), which are inexpressible in LP. Additionally, various important features and logical elements of LP, which are frequently used in practical rule-based applications, are inexpressible in FOL (e.g. Negation-as-Failure) (Grosof, Horrocks, Volz, & Decker, 2003).

As we mentioned earlier, transformation of rules (or policies) from one language to another requires an accurate syntactic and semantic mapping of the concepts of the source language to the concepts of the target language. Not only does this require a syntactic analysis of the two languages, but also requires capturing the semantics of the rules (or policies), the representation of the domain knowledge, and also the underlying logic of each rule (or policy) language. We have already described that policy languages may come with different logical formalisms, for example KAoS is based on DL while Rei is based on LP. The set of logical constructs for web rule languages is rich enough to capture the concepts of a policy language regardless of whether it follows DL or LP. In such cases, the obtained web rule would admit to the same logical formalisms that the source policy language follows. Nonetheless, when transforming rules from a web rule language to a policy language, it becomes very important to identify the differences in the underlying logic of the source web rule language and the target policy language.

When transforming rules from a DL-based language to an LP-based one, there is no possibility to cover all the concepts. Yet, most of the concepts of these languages are transformable from one to another due to the similarities between DL and def-Horn[1] shown as the intersection of these two sets in Figure 6. Rule transformations, as the major objective followed by RIF and Web rule languages, are not only concerned about a syntactic representation of the elements in a rule language, but they also care about the semantics of the rules. Rule languages can capture the semantics of a domain, first by modeling the domain knowledge using the concepts of the rule language and then by benefiting from the underlying constructs that carry different logical semantics ranging from descriptive elements to declarative ones.

Grosof, Horrocks, Volz, and Decker (2003) provide an elaborate method of mapping the basic elements of description logic to declarative logic programs. OWL as a subset of RDFS corresponds to a fragment of classical FOL. It is shown by Grosof, Horrocks, Volz, and Decker (2003) that OWL elements are convertible to definite Horn FOL elements which in turn are convertible to definite Datalog Logic Programs as a restricted model of Logic Programs (LPs). For example, classes and class expressions are equivalent to FOL formulae with one free variable, and proper-

*Figure 6. Expressive overlap of description logic and logic programs and where Rei and KAoS sit in this classification*

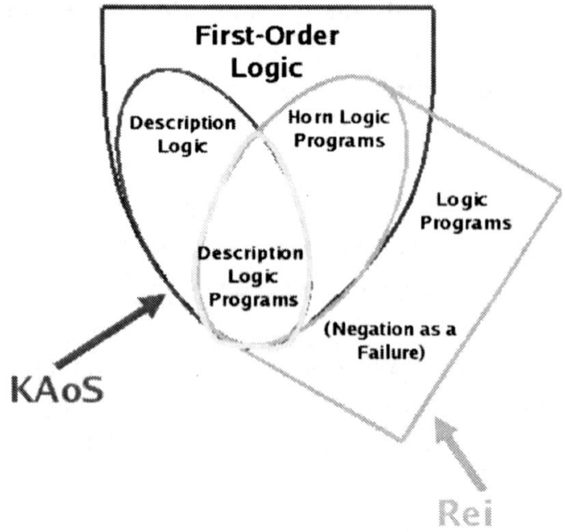


ties (and property expressions when supported by description logic) are equivalent to FOL formulae with two free variables. Classes and property inclusion axioms are also considered as FOL sentences consisting of an implication between two formulae with the free variables universally quantified at the outer level. Table 1 shows a selection of OWL constructs with their corresponding description logic syntax and FOL expressions. Details of the definitions can be found in (Grosof, Horrocks, Volz, & Decker, 2003).

## Model Driven Engineering

The main goal of MDE is to switch the focus from low-level implementation details to problem-specific concepts (Schmidt, 2006). The core activity is then to define languages for particular problem domains (*in our case policies*). Metamodeling is an approach used in MDE for defining languages. A metamodel is a model of a modeling language, that is, a metamodel defines a set of sentences that can be expressed in a modeling language (Kühne, 2006). As such, a metamodel can be regarded as an abstract syntax of a language. Since the goal of MDE is also to provide a common framework for processing different modeling languages, it also defines a technical context in which different modeling languages are defined and used, and on top of which different modeling tools are built.

Metamodeling architectures are used as a solution to this issue. Usually, a metamodeling architecture (e.g., OMG's Model-Driven Architecture) is organized in a layered fashion, where typically there are four layers. The top most layer is called metametamodel (and tagged with M3 or L3), and on this layer a metamodeling languages is defined. Most commonly, this language is defined by itself. In a metamodeling architecture, there is typically one and only one metamodeling language defined. Examples are – OMG's standard Meta-Object Facility (OMG), 2008) and Eclipse's Ecore (Budinsky, Brodsky, & Merks, 2003). While the abstract syntax of two languages is slightly different (i.e., it is a reduced part of UML elements related to class models), the graphical concrete syntax of both languages is borrowed from UML class models. The rationale for having only one M3 language is to have a common space for defining modeling languages on the M2 layer. Thus, various modeling languages can be processed in the same way using the same tool set (e.g., APIs, constraint, query, and transformation languages).

Model transformations are also an intrinsic part of MDE, as the idea is to transform M1 models to different platforms or to allow translation for one type of model (e.g., UML) to another (e.g., ODM). MOF Query/View/Transformation (QVT) is the OMG's official standard for model transformations (OMG, 2005).

## POLICY INTERCHANGE FRAMEWORK

Interchanging the rules in general, and the policies in particular, between different business enterprises is a goal followed by RIF working

*Table 1. Some of the OWL constructors and the equivalent description logic and FOL expressions*

| OWL Constructor | DL Syntax | FOL Expressions |
|---|---|---|
| subClassOf | $C \subseteq D$ | $D \leftarrow C$ |
| transitiveProperty | $P^{+} \subseteq P$ | $\forall x, y, z(P(x, y) \wedge (P(y, z)) \rightarrow P(x, z)$ |
| inverseOf | $P \equiv Q^{-}$ | $\forall x, yP(x, y) \Leftrightarrow Q(y, x)$ |
| intersectionOf | $C_1 \cap ... \cap C_n$ | $C_1(x) \wedge ... \wedge C_n(x)$ |
| unionOf | $C_1 \cup ... \cup C_n$ | $C_1(x) \vee ... \vee C_2(x)$ |
| complementOf | $\neg C$ | $\neg C(x)$ |
| one of | $(a_1,...,a_n)$ | $x = a_1 \vee ... \vee x = a_n$ |
| hasClass | $\exists P.C$ | $\exists y(P(x, y) \wedge C(y))$ |
| toClass | $\forall P.C$ | $\forall y(P(x, y) \rightarrow C(y))$ |

group (Ginsberg, Hirtle, McCabe, & Patranjan, 2006). We already discussed it in the introduction section why interchanging the policies between different business partners is important to achieve. Nonetheless, the variety of policy languages that have been developed so far, the lack of a standard for defining policies, and the limited number of experts in each of the existing policy languages, made us think about designing a policy interchange framework that can easily expand to cope with different policy languages, especially the ones that have not been developed yet. It is worth noting that while interchanging the policies, we deal with their logic, their abstract syntax, their concrete syntax, the domain in which these policies are deployed, and also the semantics that each term or concept in each language or domain carries. Our studies led us to a point where we realized that, in order to have such a framework, we need to start from the logic that the language follows, to the abstract syntax that it represents, having the semantics of its domain concepts in mind.

One way to represent the domain knowledge is to provide an ontological representation of the domain and form the policy rules around these concepts. The ontological representation of domain concepts brings in all the benefits of Semantic Web into the implementation of policy rules and enables agents from different domains to have shared and common understanding of the concepts of each domain. Consequently, the problem of exchanging policies from one domain to another is divided to two sub-problems of i) transferring the concepts of a domain to the concepts of another domain, and ii) capturing and transferring the logic and the semantics of the language from the source language to the target language. The first sub-problem has already undertaken a lot of research in the field of ontology mapping (Kalfoglou & Schorlemmer, 2003). In this section, we basically illustrate how the second sub-problem can be addressed using web rule languages and this is the second sub-problem that the community of RIF is targeting.

The concrete syntax of the languages is probably the least challenging issue to deal with, as there are powerful transformation tools and languages, such as QVT/ATL (OMG, 2005) and XSLT (Clark, 1999), that can mine through the concrete definition of the languages and extract the appropriate concepts. For the abstract syntax of the language, we normally encounter the general concepts that are shared between the languages with similar purposes, and the language-specific concepts that are not shared between all the languages of one single category. The language-specific concepts take more important roles in languages that are in a lower level of abstraction. Policy languages, due to their characteristics in addressing low level domain specific concepts and the differences in the domains that they need to be deployed in, may have various dissimilar concepts that are specific for each language. A powerful policy interchange framework should be able to clearly distinguish between the logic, the language-independent, and the language-specific concepts and try to have general definitions for as many of these concepts as possible.

Figure 7 shows our proposed architecture for a policy interchange framework. As the figure shows, the policy framework, first starts with identifying the similarities and dissimilarities between the underlying logics of the policy languages. The two main logics that are widely used in defining rules and representing the knowledge of a domain are computational logic and descriptive logic and there has been already research on how to map the concepts of these two logics (Grosof, Horrocks, Volz, & Decker, 2003) (Kolovski, Hendler, & Parsia, 2007).

The next level in the architecture is to identify the general concepts that are common across policy languages. Once these concepts are recognized, the rules from one source language to one target language can be mapped to some degree, having the domain knowledge plus the language vocabulary identified and mapped. For the language-specific concepts of a source policy language, either it is possible to define the concepts through using a series of concepts from the target language, or simply the concepts are not convertible. Depending on the importance of the meanings that the unmappable concepts carry, the transformation to the target policy language would be successful or unsuccessful. As Figure 7 shows, by moving to the upper levels of the architecture, the generality of the mappings is significantly decreased, such that reusing it for other policy languages becomes impossible in the top most level. Nonetheless, it is possible for the concepts in the lower levels to be shared between different policy languages. To plug a new policy language to the framework, first and foremost, the underlying logic of the policy language should be identified and mapping rules to cover the corresponding concepts should be developed. Moving to the upper level in the architecture, the common policy concepts are identified and mapped to the concepts of the general policy model used. For the top most level, a detailed review of the source

*Figure 7. The layered policy interchange framework architecture*

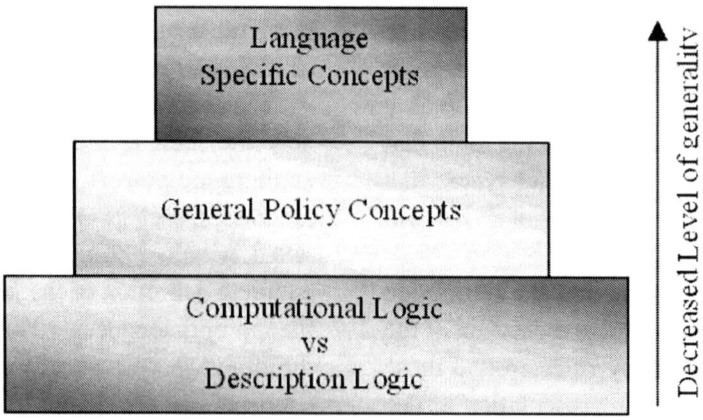

policy language by its experts is required to work around the concepts that are not mappable.

In the following sections of this chapter, we try to extract the abstract syntax of KAoS by providing its metamodel, extract the language-independent and language specific concepts that the KAoS policy language covers, and argue how the mapping between this policy language and R2ML can happen without facing serious information loss.

## METAMODELING FOR POLICIES AND RULE LANGUAGES

Concept modeling is considered a critical step in better understanding and comprehending the constituents of a system in general, and a language in particular. One of our major intentions in providing the metamodel for KAoS is to hide the low-level details of the language in order to focus on its conceptual characteristics. In this section, we present the UML metamodel for KAoS and contrast it against the common concepts in R2ML that can carry the equivalent meanings for those concepts.

As we mentioned earlier, KAoS exploits the use of Web Ontology Language (OWL) to define its concepts. Thus, providing a meta-model for the language can be done through transforming the OWL definition of the languages to a UML metamodel. There have been several efforts in defining standard mappings from ontology languages to UML. In particular, the Ontology Definition Metamodel (ODM) (ODM, 2003) initiative follows the goal of using OMG's Meta Object Facility (MOF) as a metamodeling language to develop a linguistic metamodel to represent the ontology languages. Although ODM, itself, is still undergoing modifications, we chose to use the current state of ODM to represent Rei and KAoS. However, at some points, based on the needs and the real meaning of the concepts in each language, we had to slightly modify the ODM definition for getting a better reflection of the concepts through using UML constructs.

### The Metamodel for KAoS Policy Language

KAoS as a policy language has elements to define and distinguish different types of policies over different resources. Due to space limits, here we only focus on the major concepts in KAoS and show how they can be transformed to their equivalent counterparts in R2ML. Figure 8 represents the types of policies that are covered in KAoS. As Figure 8 shows, a policy in KAoS is either an *AuthorizationPolicy* or an *ObligationPolicy* with each of them further specialized to *PosAuthorizationPolicy*, *NegAuthorizationPolicy*, *PosObligationPolicy*, and *NegObligationPolicy*. It can also be a policy over the policies that conflict, *ConflictedPolicies*, which usually is not considered a policy type, but is just a subclass of the class Policy. An *AuthorizationPolicy* introduces the set of permissions and prohibitions for an actor while dealing with a context. However, KAoS is a context-based policy language that defines permissions and prohibitions not over the roles of the actors but over the context of interactions.

In KAoS, each policy element stands for exactly one policy rule. As shown in Figure 9, a KAoS policy can have a control action, a triggering action, an oblige action, and a series of conditions. It also can define a site to which the policy rule should be applied. As we discussed earlier, KAoS is a context based policy language. *This means that the constraints are defined over the actions that are going to be taken on a resource, rather than the users (or roles) that are going to take these actions.* This can be considered as a major difference between the role-based access control policies and the context-based access control policies.

An action in KAoS, as shown in Figure 10, has a context to which the action is applied, which either is a *dataContext* of any type (i.e. of type

*Figure 8. Different types of policies in KAoS*

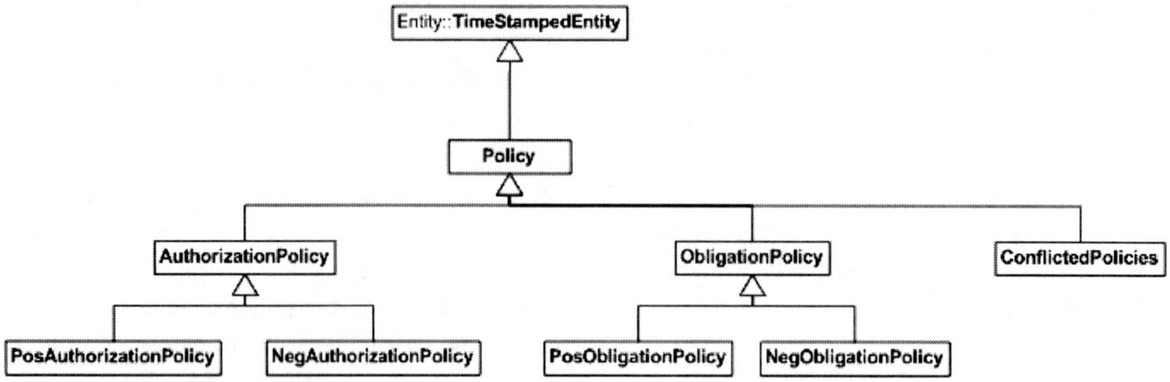


*Figure 9. The policy metamodel for KAoS*

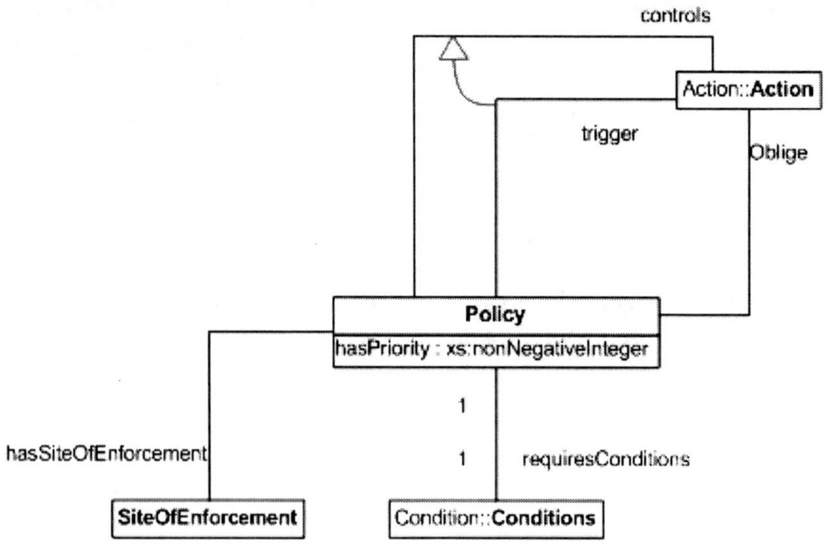


*Figure 10. The metamodel for KAoS Action*

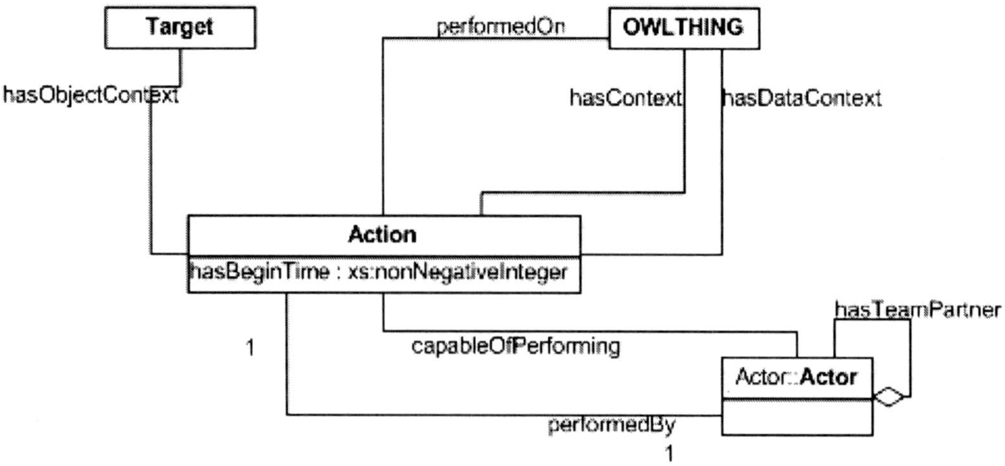

OwlThing), or an *objectContext* of type Target, which is itself an entity. A KAoS action also has one and exactly one Actor, referred to by the *performedBy* property in the definition of the Action. Considering the fact that in context-based policy management Actions are constrained, as opposed to roles in role-based access control, a rich definition for the class Action, its properties, and its attributes helps with having accurate definitions for the constraints and conditions. Having a look at the metamodel in Figure 10, it is clear that KAoS also follows the same objective with defining its actions.

KAoS provides a rich set of actions, inherited from the class Action, that enable better classification and definition of the actions which are going to be performed. This brings more expressivity to the language, enabling a better semantic modeling of each policy. Figure 11 provides a clear classification of all sets of actions that are defined in KAoS. A user can extend any of these classes to make them cope with what she intends to express. *ObligateAction* is one of these important actions in KAoS. An *ObligateAction* makes the active actor to execute the *ObligateAction* once the triggering action happens and the execution constraints for the obligation action are met (see Figure 12).

In KAoS, similar to the hierarchy of actions, a detailed hierarchy of entities and actors have also been defined. Actors in KAoS, represent a complete set of possible agents that might interact with the system (see Figure 13).

Different physical and artificial actors have been defined by the system which helps with precise classification of the actors while working with policies. The only problem with the obtained model, as we have highlighted in Figure 13, is the redundant definition of both Human and Person which seem to be identical. The reason behind distinguishing between these two concepts by the developers of KAoS is not clear to the authors of this article. Furthermore, according to the ontology definition of KAoS for the concept Human, it inherits from both Person and *PhysicalActor*. Since Person itself has been already defined as a subclass of *PhysicalActor*, the inheritance from *PhysicalActor* by Human seems to be redundant.

*Figure 11. The metamodel for various KAoS action types*

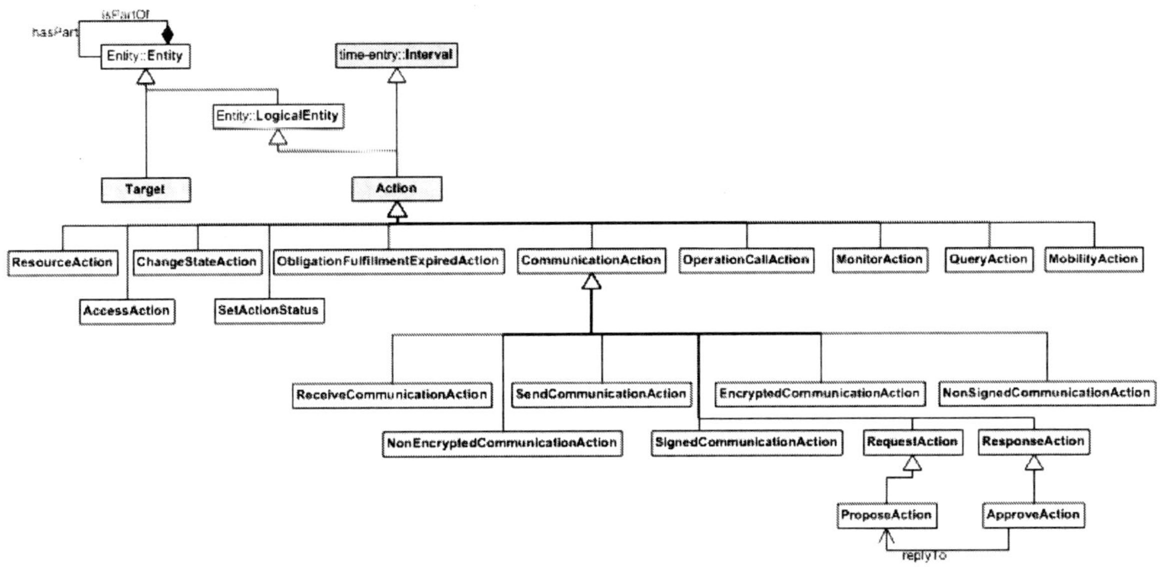

*Figure 12. The metamodel for ObligateAction*

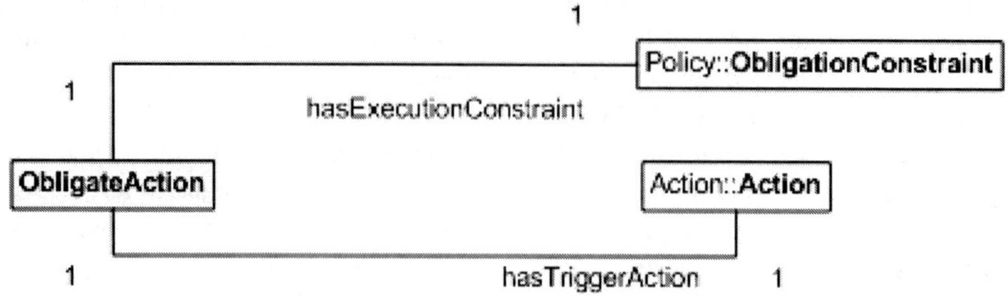


*Figure 13. The metamodel for class Actor in KAoS*

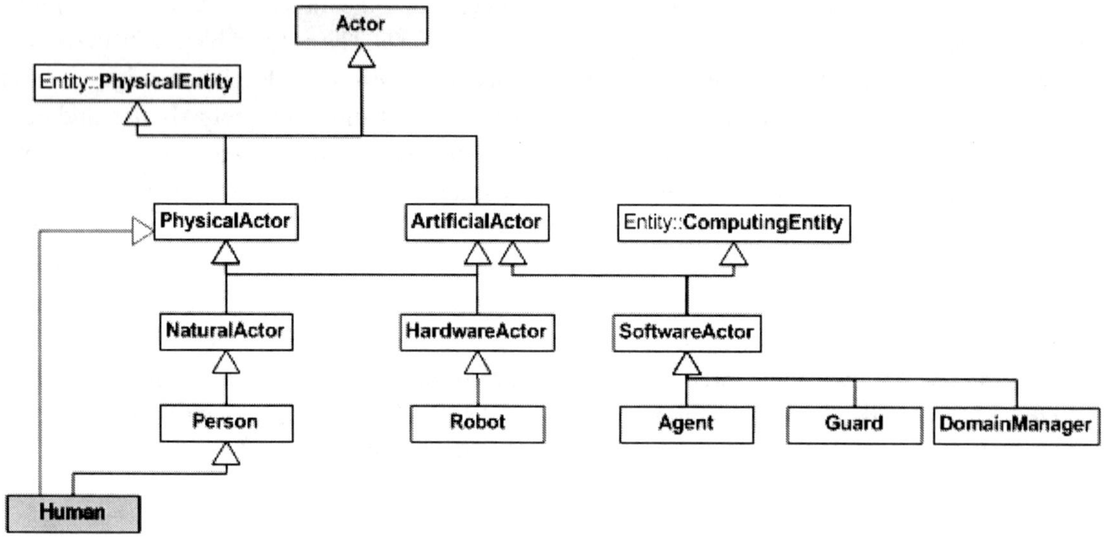


An actor in KAoS can be controlled by a set of policies, and is capable of performing a series of actions. It can also cooperate with other actors as a team (see Figure 14).

For a policy to be able to control the behaviour of a system, conditions are required to be place on either potential roles (in case of role-based access control) or potential actions (in case of context-based access control) in a system. A condition in KAoS represents a situation based on the current State and the history of the events (see Figure 15). As Figure 16 shows, the EventHistory keeps the time-stamped events that are mostly in the form of occurred actions, represented in the *ActionHistory* class. To each *ActionHistory*, relates a *hasRegisteredAction* which represents the current action occurred in each time interval, the actor of the action, and the context to which the action has been applied.

## Conceptual Mapping of KAoS and R2ML

Given a rather comprehensive representation of KAoS and R2ML (both syntactically and semantically), we can take the main step in defining the mappings between these two languages, aiming at providing an exchange method between different concepts of these languages. It should be noted that although we deal with the syntactical mapping

*Figure 14. The metamodel for the class Actor in KAoS*

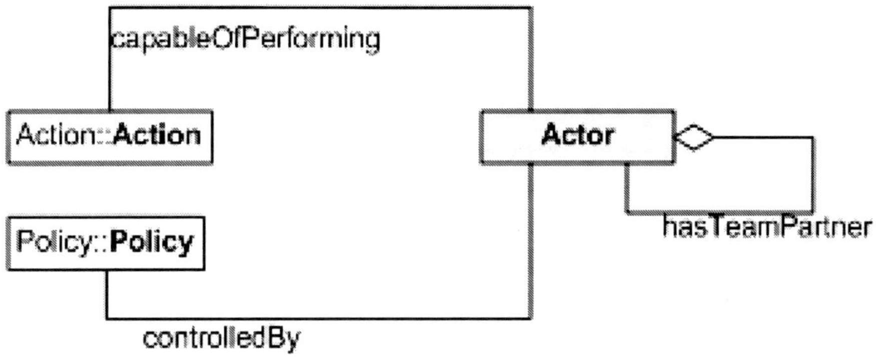


*Figure 15. The metamodel for conditions in KAoS*

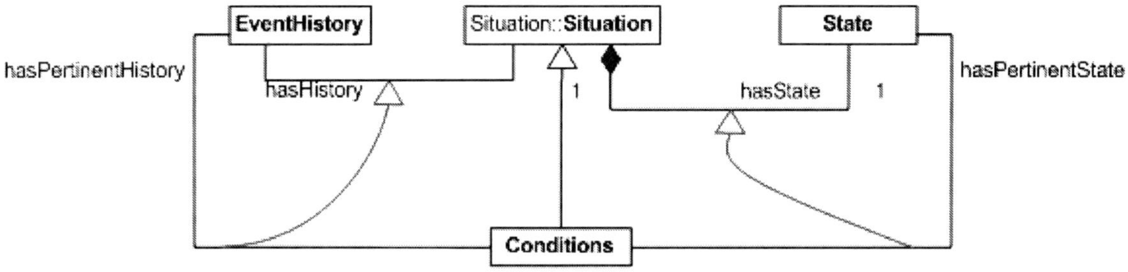


*Figure 16. The metamodel for KAoS Situation*

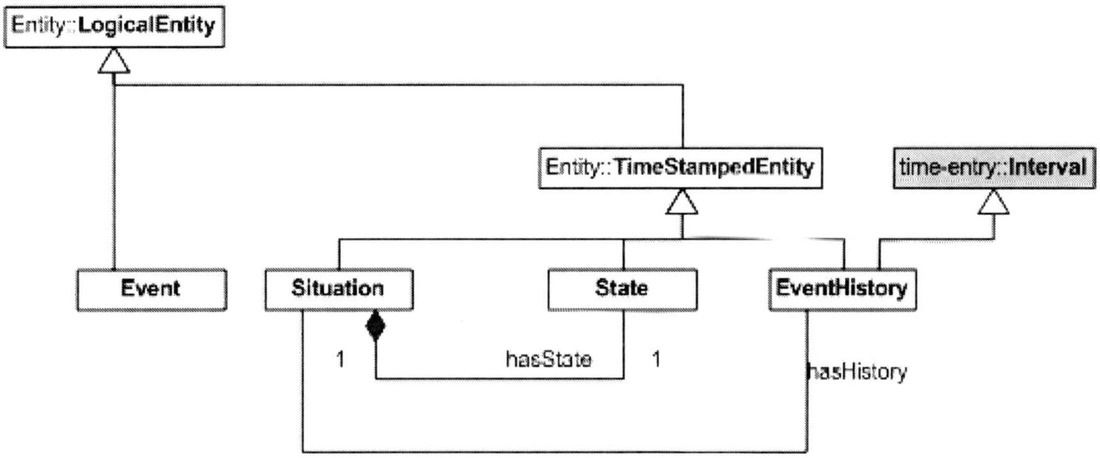


of the elements, the process of mapping considers semantic similarities as well. In this section, we show that, despite having several possibilities to map the elements of one language to the elements of another language, we choose those elements which are most similar semantically. Our mappings in this section follow the *QVT's graphical notations* (QVT, 2005) for transforming between concepts. However, we have included some dashed arrow lines to make the one-to-one mappings easier to understand.

In our introduction to R2ML, we presented the major types of rules that R2ML covers. Among all the rules, *derivation rules* seem to be better options to model the policies, mostly because they entail the meaning of: "obtaining a conclusion upon satisfaction of a set of conditions (or *credentials*)". Using *implication* in *integrity rules* makes them quite close to our intentions too, but as we already mentioned, integrity rules are generally used where the set of conditions of the rule are persistent over time. *Reaction rules* can be used to reflect the pre- and post-conditions of a policy rule, especially while communicating with Web services. Nonetheless, we believe the intention in using policies is closer to the meanings expressed by derivation rules, i.e. deriving the conclusions once the conditions are met. Thereupon, we choose to use *derivation rules* in order to model policy languages. Some other research works covering this same area are available in (Kaviani et al., 2007-1) and (Kaviani et al., 2007-2).

In our representation of different types of policies for KAoS (see Figure 8), we showed that there are generally four main types of policy rules that these languages support, i.e. *PosAuthorizationPolicy*, *NegAuthorizationPolicy*, *PosObligationPolicy*, and *NegObligationPolicy*. To be able to model these concepts in R2ML, we decided to define some general classes for these types of policies using R2ML. Due to the simplicity of the naming for the policies in Rei, we chose the same set of names to define our policy classes in R2ML. We have also mentioned that R2ML has its own mechanisms for defining vocabularies (also called R2ML vocabulary), so it is possible to exploit R2ML vocabulary to define these policy concepts. Figure 17 show the metamodels for the class *Policy* and its subclasses that we have defined in R2ML. As it can be seen in the models, these classes have simple definitions at the moment which currently seem satisfactory for our purpose. However, the metamodels for these concepts are developed in our recent work, aiming at providing a Policy Modeling Language by using R2ML (Wagner, Giurca, & Lukichev, 2005). Again it should be noted that the (Tonti, Bradshaw, Jeffers, Montanari, Suri, & Uszok, 2003) properties of these general R2ML-Policy classes do not have any range and they refer to any available resource. This is mainly to increase the flexibility of our R2ML classes to work with different actions and policies from different policy languages. We may decide to make them more restricted in the future.

*Figure 17. The metamodels for our R2ML Policy classes*

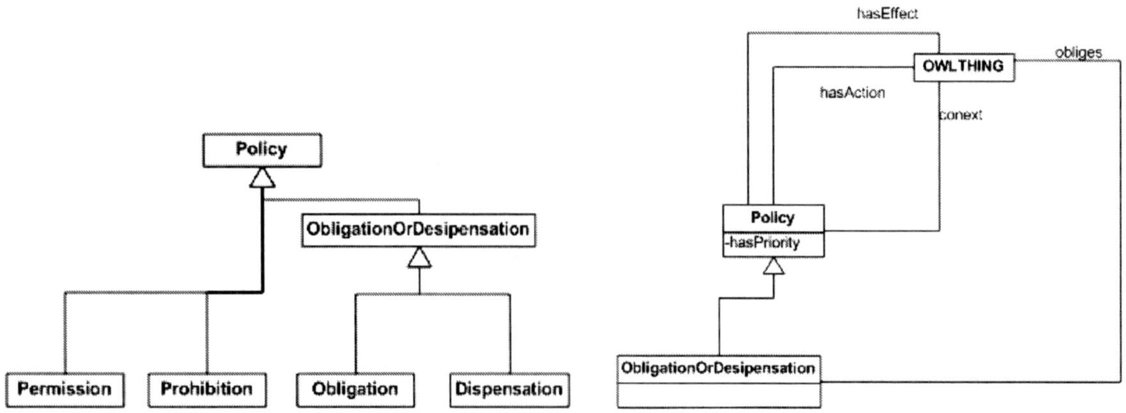


*(a)* *(b)*

While transforming each single policy rule from KAoS to R2ML, we decide about how to map these policies to R2ML policies by choosing the same naming conventions that we introduced at the beginning of this section. Next, we describe how the concepts of R2ML along with the classes that we defined above, are jointly used to provide a meaningful mapping between the concepts of the two policy languages.

We have previously mentioned that a KAoS policy is an object of the Policy class in KAoS Policy Ontology (KPO) (Tonti, Bradshaw, Jeffers, Montanari, Suri, & Uszok, 2003) with its attributes instantiated to a set of users, events, and resources that make the policy fire. Considering the KAoS policy element as a rule, the *control* element is executed upon the occurrence of the events described in the *requiresConditions* element (Figure 9). The *control* element in KAoS refers to an action. It can itself place a series of constraints on the definition of the action that can be executed by the policy. The main advantage in using KAoS is its flexibility of expansion. The *Action* class in KAoS can be easily expanded thanks to the use of OWL, to fully capture the meaning of the final action that is desired to be executed. Placing constraints on different properties of this action can also specialize the meaning that is desired to be transformed.

Once a KAoS policy rule is fired, the decision over whether or not to perform the action of the *control* element is made, an obligation of execution is placed over the *obligation* action, and the *effect* of the action is enforced to the current state of the system. This means, to model a KAoS policy with a derivation rule, we need to place the content of the *controls* element, the *oblige* element, the *hasSiteOfEnforcement* element, and the *effect* element (if there is any) in the conclusion part of the derivation rule. Whatever else that can lead to making such a decision should be placed in the condition part of a derivation rule. This includes the content of the *requiresConditions* element, the *triggerAction*, and also the variables that initialize the values for the elements in the conclusion of a rule (see Figure 18). Figure 18 shows the transformation of a KAoS policy rule to a derivation rule in its most abstract model. The main issue is to decide how and using what R2ML elements this process of transformation can be best achieved. The conclusion of a derivation rule should be only one R2ML *Atom*, now

*Figure 18. Mapping rule of a KAoS Policy to a R2ML derivation rule*

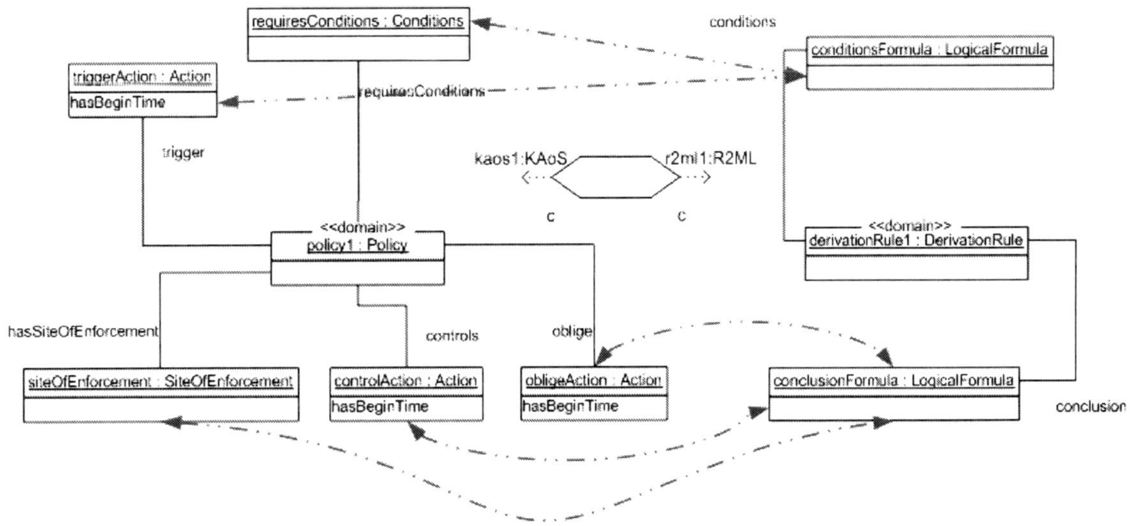

the question is how to manage all the actions in one single *Atom*.

The first step in transforming a KAoS policy rule to a R2ML derivation rule is to transform those elements of the KAoS *Policy* class that are placed in the conclusion part of our R2ML rule, represented with a policy *objectDescriptionAtom*. An instantiation of the derived R2ML policy class will be placed as a conclusion in the derivation rule. Figure 19 shows how this mapping can happen. The *ObjectDescriptionAtom* that captures the meaning of a policy element has four slots. One is the *hasPriority* element which if of type integer and represents a numeric value for the priority of the rule. The other three slots represent object elements that show the action that the policy controls (i.e. *hasAction*), the context to which the policy is applied, and the optional action that the policy may oblige the actor to do.

As earlier argued, an R2ML ObjectDescriptionAtom describes a set of properties and attributes, referred to as *slots*, for an object called the *subject*. As for the policy rules, the conclusion of a policy rule is a policy element (or object) with the values for the appropriate action to be taken, the actor to which giving the permission is allowed or denied, the context to which the policy is applied, and the priority of the policy rule. As a result, an *ObjectDescriptionAtom*, with a policy object as its subject is the best element to model the conclusion of a derivation rule. Figure 20 shows how we instantiate a policy object from our *R2ML Policy* class. It illustrates how the policy object is mapped to the subject of the R2ML *ObjectDescriptionAtom*, the corresponding class for this atom is set to the *Policy* class and different attributes and properties for a policy object are mapped to corresponding *ObjectTerms* or *DataTerms* depending on whether the attribute is a *DataAttribute* or *ObjectAttribute*. It has also been shown in Figure 20 that an *ObjectDescriptionAtom* can have several slots to cover various of data or object attributes for a policy. It makes it the appropriate R2ML element, both syntactically and semantically, to be used for the purpose of transforming between the objects of the policy rules. It should be taken into consideration that, in Figure 20, we are just showing the transformation of the control element to a slot in the policy *ObjectDescriptionAtom* with the property name *hasAction* represented as a *ReferenceProperty* and the corresponding object to this property (*control* in Figure 20) placed in the body of the slot.

However, referring to Figure 8, one should note that the class Policy is an *abstract* class, and so is our R2ML Policy definition of Figure 17. Thus, in Figure 20, the class Policy is usually replaced with one of its concrete sub-classes, i.e. *Permission*, *Prohibition*, *Obligation*, or *Dispensation*

*Figure 19. Mapping a KAoS Policy class to an R2ML policy model*

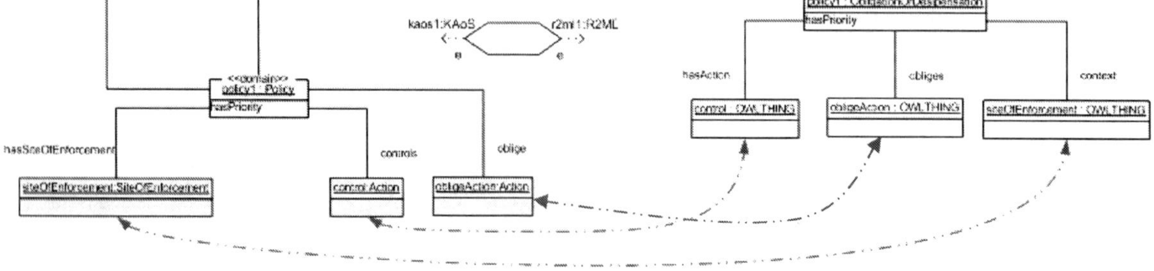

*Figure 20. Mapping a KAoS Policy class to R2ML ObjectDescriptionAtom*

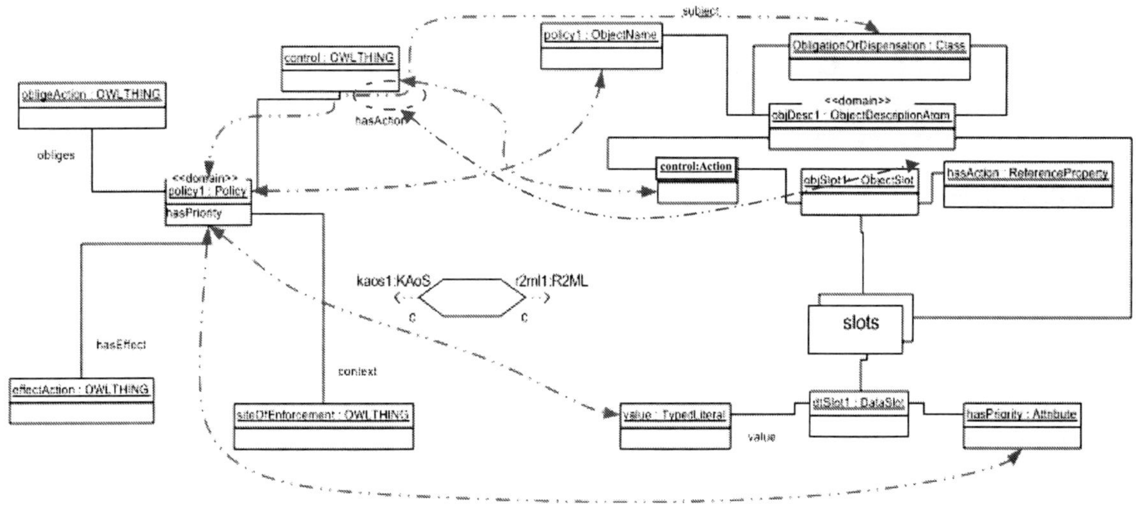


Having the conclusion of our derivation rules constructed, we need to extract its condition part from KAoS policy rules as well. The condition part of a policy derivation rule contains the definition of the variables, actions, actors, conditions, and any other construct that result in deriving the final policy object. Although the conditions of a policy rule could be modeled with *ObjectDescriptionAtoms*, we chose *ReferencePropertyAtoms* mainly to be compliant with other R2ML transformations (e.g., transformations between F-Logic and R2ML also have *ReferencePropertyAtom* in the condition part). It simplifies the later conversions of the policies to other rule languages for which we have R2ML transformations already defined (e.g., Rei, F-Logic, and RuleML). On the other hand, a *ReferencePropertyAtom* triple models a binary predicate. A set of *ReferencePropertyAtoms* with the same subject element can always be combined and converted to any element of higher arity (e.g. *ObjectDescriptionAtom*), and thus using *ReferencePropertyAtom* does not contradict the use of *ObjectDescriptionAtom*. Furthermore, in our case, *ReferencePropertyAtoms* carry even a better semantic meaning for the transformations. Semantically, they are equivalent to an OWL object property, and as KAoS is semantically very close to OWL, they model object properties of KAoS too. Figure 21 shows the conversion of a KAoS property (here, the *performedBy* property) to a *ReferencePropertyAtom* in R2ML. The converted element is then placed in the *conditions* part of a derivation rule.

In some cases, it happens that all the objects are not named-objects, but rather, they refer to *ObjectVariables*. In KAoS, due to the impossibility of representing variables in OWL, a *role-value-map* method is used in which a class representing the set of possible elements for a slot replaces the object variable in order to show all the possible options an element can take. The transformation procedure should recognize these classes and convert them to appropriate R2ML elements. This happens in two steps during our mappings. Once the role-value-mapped class is determined, it is first mapped to an *ObjectClassificationAtom* with a variable name assigned to it, and then the variable is used in place of the required elements for the class, for example in a *ReferencePropertyAtom*. Figure 22 shows the conversion of a role-value-map class to a R2ML *ObjectClassificationAtom*. In this figure, the generated variable *X* shows the

*Figure 21. The abstract transformation of a KAoS property to a R2ML ReferencePropertyAtom*

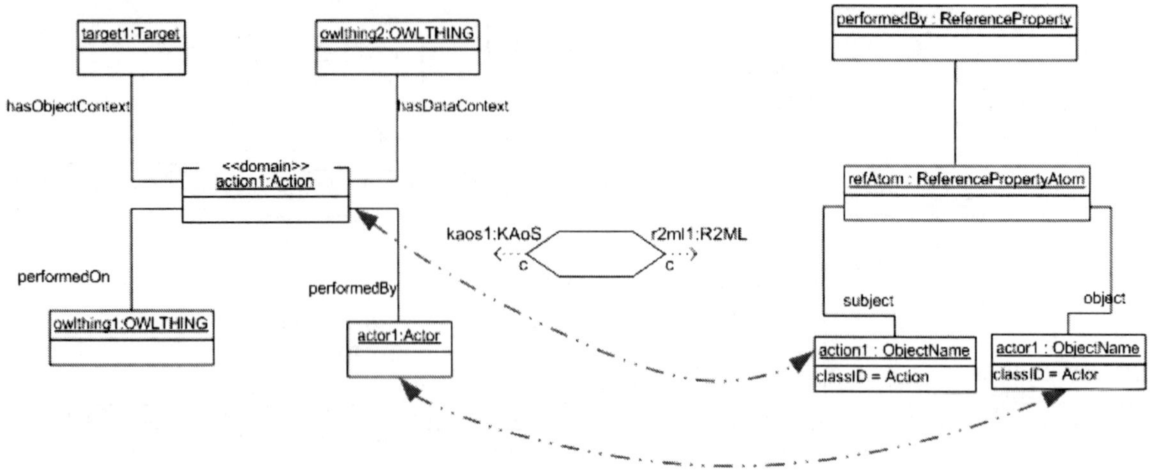


variable that later can be used in other places in the rule.

For example, in case we need to convert it to the *ReferencePropertyAtom* of Figure 21, the only change would be to replace *actor1* with *X*. It should be highly noted that, an *ObjectClassificationAtom*, representing a variable, always needs to be placed in the condition part of the derivation rule, even if the variable needs to be used in the conclusion part. Figure 23 also shows another example of how the combination of *ObjectClassificationAtom* and *ReferencePropertyAtom* can be used to define a class of actors in KAoS

A KAoS policy might also have a *trigger action*. To the best of our knowledge, this element is only used with *NegObligation*- and *PosObligation-Policies* showing a set of events that trigger the occurrence of an action. In our *R2ML Policy* model, we have considered a slot for these actions. Furthermore, the detailed information about these actions is defined as *ReferencePropertyAtoms* in the condition part of a derivation rule. Consequently, the same process shown in Figure 21 is also applied to the trigger element, which itself refers to an action. The *obligeAction* element of a KAoS policy is also considered an *ObjectSlot* in the *ObjectDescriptionAtom* of the R2ML Policy object. The *ObjectSlot* refers to the action that should be performed by the actor as an obligation, once the policy rule is fired.

Having the most important elements of a KAoS policy explained here, we present some of the concrete, XML syntax-based formalization of our transformation rules in Table 2.

Other elements of the KAoS policy language are placed in the body of a derivation rule. As a result, we perform a conversion to *ReferencePropertyAtom* for those elements. The main reason is to be compliant with Rei. Moreover, once we have the relations defined in the form of RDF triples, then conversion to the predicates of higher arity is a lot simpler, as for *n* triples with similar *subject* we can make one ObjectDescriptionAtom of arity ***n*** by combining all these elements.

## AN EXAMPLE OF APPLYING THE TRANSFORMATION RULES TO A KAOS POLICY

To this point, we have conceptualized and formalized our policy transformation framework. Now that the formalization of our transformations is complete, we can apply our definitions to a real

*Figure 22. The abstract transformation of a KAoS role-value property to a R2ML ObjectClassificationAtom*

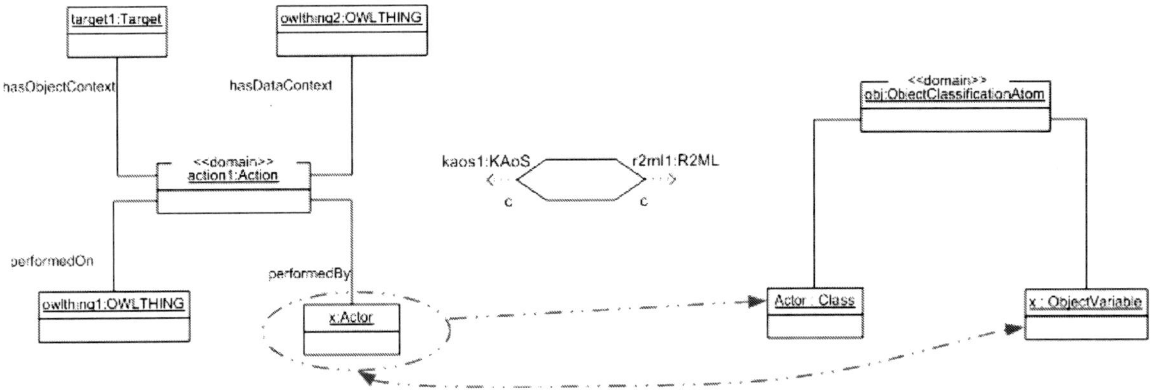


*Figure 23. Using ObjectClassificationAtom and ReferencePropertyAtom to map KAoS variables*

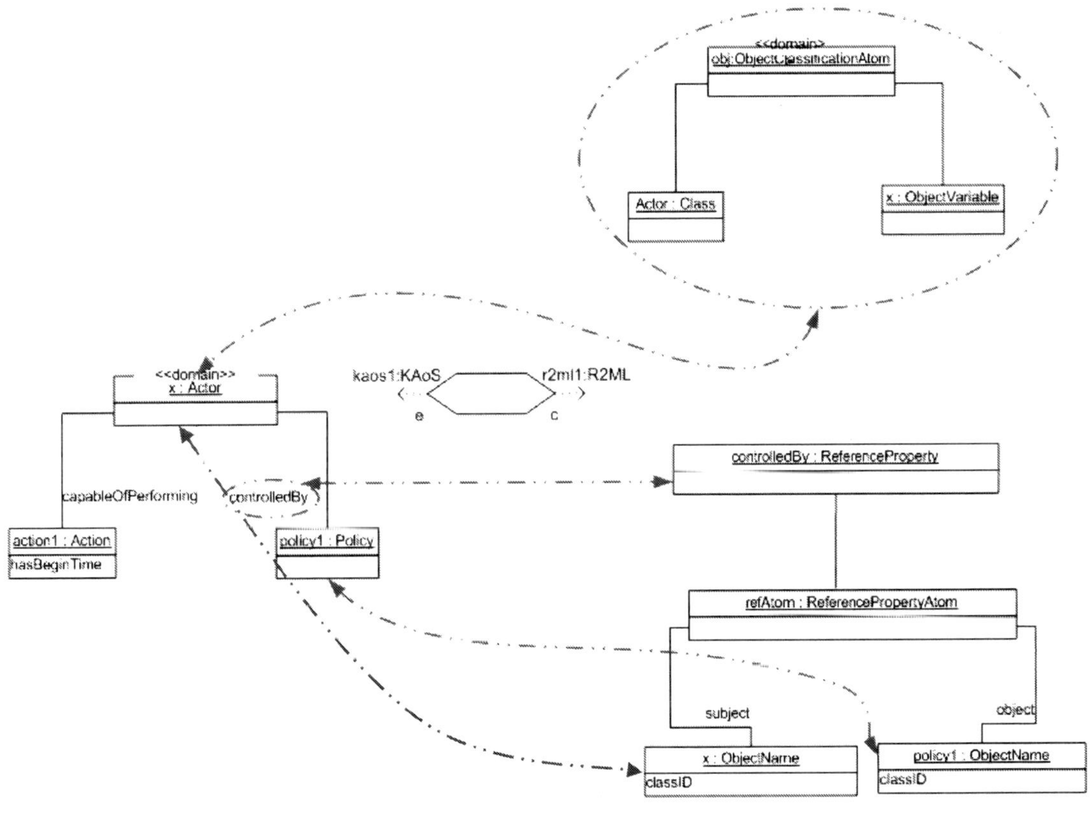


example of the KAoS policy taken from (Toninelli, Bradshaw, Kagal, & Montanari, 2005). Let us consider a travel agency, named Sky Team that plans to promote its services by giving free printing access to those of its customers located in any of the gates of 31 to 57 at the local airport. Table 3a shows the KAoS representation of this policy. The policy is a PosAuthorizationPolicy and instantiates hasSiteOfEnforcement and hasPriority.

*Table 2. The concrete XML-based conversion of some of the KAoS elements to R2ML elements*

| | KAoS Element | R2ML Element |
|---|---|---|
| Policy Element | <policy:**PosAuthorizationPolicy**<br>rdf:ID="Policy_CommunicationCertificate"><br>< policy:requiresConditions rdf:resource="#Policy_ TrustedEntity "/><br><policy:controls<br>rdf:resource="#Policy_CommunicationCertificate_Action"/><br></policy:PosAuthorizationPolicy> | <r2ml:DerivationRule><br><r2ml:conditions><br><!-- The mappings for the condition part goes here --><br></r2ml:conditions><br><r2ml:conclusion><br><r2ml:ObjectDescriptionAtom r2ml:classID="**Permission**"><br><!-- The mappings for the conlcusion part goes here --><br></r2ml:ObjectDescriptionAtom><br></r2ml:conclusion><br></r2ml:DerivationRule> |
| Control Element | <owl:Class rdf:ID="Policy_CommunicationCertificate_Action"><br><owl:intersectionOf><br><!-- the set of constraints on the action in KAoS are defined as intersection of a series of restrictions in OWL --><br></owl:intersectionOf><br></owl:Class> | <r2ml:ObjectDescriptionAtom r2ml:classID="**Permission**"><br><!-- The mappings for the conlcusion part goes here --><br></r2ml:ObjectDescriptionAtom> |
| | <owl:Class><br><owl:Restriction><br><owl:onProperty rdf:resource="http://ontology.ihmc.us/<br>Action.owl#performedBy"/><br><owl:allValuesFrom><br><!-- Defines the set of actors that are responsible for the action--><br></owl:allValuesFrom><br></owl:Restriction><br></owl:Class> | <r2ml:ObjectSlot r2ml:referencePropertyID="<br>http://ontology.ihmc.us/Action.owl#performedBy"><br><!--We can use either r2ml:ObjectName or r2ml:ObjectVariable to define the actors --><br></r2ml:ObjectSlot> |
| | <owl:Restriction><br><owl:onProperty<br>rdf:resource="http://ontology.ihmc.us/Policy.owl#hasPartner"/><br><owl:allValuesFrom<br>rdf:resource="#TrustedServiceProvider"/><br></owl:Restriction> | <r2ml:ObjectSlot r2ml:referencePropertyID="context-N10058"><br><r2ml:object><br><r2ml:ReferencePropertyFunctionTerm<br>r2ml:referencePropertyID="<br>http://ontology.ihmc.us/Policy.owl#hasPartne"><br><r2ml:contextArgument><br><r2ml:ObjectName r2ml:name="#TrustedServiceProvider"/><br></r2ml:contextArgument><br></r2ml:ReferencePropertyFunctionTerm><br></r2ml:object><br></r2ml:ObjectSlot> |

As already stated, we define four main policy classes for our policy rules, including, *Permission*, *Prohibition*, *Obligation*, and *Dispensation*. The first step in providing the mappings is to determine the appropriate R2ML classes and elements that can be used according to our definitions in the previous section. A *PosAuthorizationPolicy* in KAoS is equivalent to our definition of the *Permission* class in R2ML. Having a KAoS policy element necessitates the definition of an *ObjectDescriptionAtom* to describe its properties, which should be placed in the conclusion part of a derivation rule. The *ObjectDescriptionAtom* takes the name of the policy instance as its *subject* and is considered as an instance of the *Permission* class that we defined in the previous chapter. The *hasSiteOfEnforcement* property from a KAoS policy is modeled as an *ObjectSlot* for our *ObjectDescriptionAtom* with its *referencePropertyID* referring to *context* according to our definition of R2ML Policy class. The *hasPriority* attribute in KAoS has an equivalent *hasPriority* element in our R2ML Policy model which is considered as a *DataSlot*.

In our example, we have a PrinterAccessAction as a subclass of AccessAction for which we have the metamodel illustrated in Figure 24. As the figure shows, accessedEntity is the subProp-

*Table 3. (a) A KAoS policy restricting access to a printer only to the community of sky team customers and, (b) its equivalent policy as an R2ML derivation rule*

```
<owl:Class rdf:ID="SkyTeamCustomer">
    <owl:subClassOf
                 rdf:resource="&some-
                 ontology;Customer"/>
    <owl:subClassOf>

       <owl:Restriction>
           <owl:onProperty
                 rdf:resource="&some-
                     ontology;firm"/>
           <owl:allValuesFrom
                rdf:resource="&some-
           ontology;SkyTeamAlliance"/>
       </owl:Restriction>
    </owl:subClassOf>
</owl:Class>

<owl:Class rdf:ID="SkyTeamGate31-
               57PrinterAccessAction">
    <owl:intersectionOf
           rdf:parseType="Collection">

 <owl:Class
           rdf:about=
             "&action;AccessAction"/>
       <owl:Restriction>
            <owl:onProperty
                rdf:resource=
               "&action;performedBy"/>
            <owl:allValuesFrom
                   rdf:resource=
                  "#SkyTeamCustomer"/>
       </owl:Restriction>
       <owl:Restriction>
            <owl:onProperty
            rdf:resource=
            "&action;accessedEntity"/>
            <owl:allValuesFrom
                      rdf:resource=
                      "#Printer31-57"/>
       </owl:Restriction>
    </owl:intersectionOf>
</owl:Class>


<policy:PosAuthorizationPolicy
                rdf:ID="SkyTeamGate31-
                     57PrinterAccess">
    <policy:controls
        rdf:resource="#SkyTeamGate31-
        57PrinterAccessAction"/>
    <policy:hasSiteOfEnforcement
        rdf:resource="&some-
        ontology;TargetSite"/>
    <policy:hasPriority> 10
                  </policy:hasPriority>
</policy:PosAutihorizationPolicy>
```

```
<r2ml:DerivationRule>
  <r2ml:conditions>
     <r2ml:ObjectClassificationAtom
      r2ml:classID="&kaos_action:AccessAction">
        <r2ml:ObjectVariable r2ml:name="X"/>
     </r2ml:ObjectClassificationAtom>

     …

     <r2ml:ReferencePropertyAtom
      r2ml:referencePropertyID=
      "kaos_action:performedBy">

         <r2ml:subject>

      <r2ml:ObjectVariable
       r2ml:name="X"/>

         </r2ml:subject>

         <r2ml:object>

      <r2ml:ObjectVariable
       r2ml:name="Y"/>

         </r2ml:object>

     </r2ml:ReferencePropertyAtom>
       …
     <r2ml:ObjectDescriptionAtom
      r2ml:classID="&some-ontology;Customer">
  <r2ml:subject>
          <r2ml:ObjectVariable
           r2ml:name="Y"
           r2ml:classID="&some-
               ontology;SkyTeamCustomer"/>
         </r2ml:subject>
        <r2ml:ObjectSlot
          r2ml:referencePropertyID="&some-
                           ontology;firm">
          <r2ml:object>
             <r2ml:ObjectVariable
              r2ml:name="W"
         r2ml:classID="&some-
          ontology;SkyTeamAlliance"/>
    </r2ml:Object>
 </r2ml:ObjectSlot>
     </r2ml:ObjectDescriptionAtom>
  </r2ml:conditions>


  <r2ml:conclusion>
     <r2ml:ObjectDescriptionAtom
      r2ml:classID="r2ml_policy:Permission"
        <r2ml:subject>
          <r2ml:ObjectName
           r2ml:objectID="SkyTeamGate31-
                            57PrinterAccess"/>
        </r2ml:subject>
        <r2ml:ObjectSlot
r2ml:referencePropertyID="r2ml_policy:hasAction">
   <r2ml:object>
        <r2ml:ObjectVariable
```

*Figure 24. The KAoS metamodel for AccessAction*

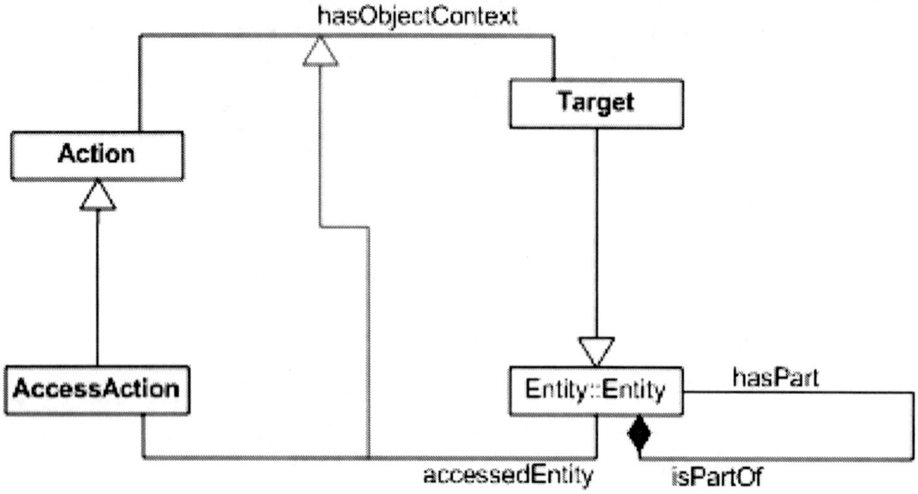


erty of hasObjectContext and thus the class that this property refers to (i.e. Printer31-57 in our example) plays as the context for the PrinterAccessAction.

Having a look at the SkyTeamGate31-57PrinterAccessAction class, as the to-be-controlled class, we realize that SkyTeamCustomer, Printer31-57, and AccessAction are OWL classes. To be able to place these elements in the conclusion part of a derivation rule, we need to associate the classes with variables which show the extensional meaning of the classes. The object for *hasAction* refers to variable *X*. *ObjectClassificationAtoms* seem to be appropriate R2ML constructs to connect a variable to its class name. However, they need to be placed in the condition part of the derivation rule. Getting to this point, we have mapped almost everything in Table 3a to R2ML, but Table 3 provides a description for the class SkyTeamCustomer, considering it as a subclass of the class Customer with its firm property only limited to the class SkyTeamAlliance. As it is a descriptive class, it should be placed in the condition part of a derivation rule, and *ObjectDescriptionAtom* is probably the best element to map this class to. Table 3b shows the full transformation of this KAoS policy to its equivalent R2ML code based on the provided descriptions

## DISCUSSION AND CONCLUSION

In this last section, we analyze the transformations that we have explained so far, we show how these transformations can be used together with other transformations that have been already developed for R2ML based on derivation rules, and we discuss the possibility of using them to convert our policy-driven R2ML rules to the rule languages that are supported by those policy languages

### Compatibility of our Transformations with other R2ML Transformations

As we already stated, R2ML is supposed to be an intermediary language, to/from which other policy languages can be transformed. To achieve this goal several transformations have been defined and developed between R2ML and other rule languages by mapping the rules from these languages to one of four main types of rules that R2ML supports, i.e. *derivation rules*, *integrity rules*, *production rules*, and *reaction rules*. Table 4 shows the current transformations that have been developed between R2ML and other rule languages as well as their corresponding R2ML rules which have been used. As our transformations of policy rules to R2ML are converted to its subset of derivation rules, we can use the currently existing transformation of derivation rules to convert our policies to the rules in any of the target languages (i.e., RuleML, Jess, Jena, and F-Logic).

## Information Loss during Policy Transformation

Information loss happens in most transformations between different policy languages due to the differences in their underlying concepts. However, in our model, regardless of the policy language that is being used, the concepts can be mapped to R2ML thanks to the rich set of elements that R2ML supports for different logical bases. For example, the concept of *Domain* from KAoS can be covered by extending our R2ML Policy model of Figure 17, yet it should be investigated how these concepts are effective and useful when they are shared between different policy languages. Extending R2ML by the concepts from different policy languages under study does not make the whole model more generic, but it reduces the expressivity when coping with the abstract notation of different policy languages. This implies that the more we try to adjust our R2ML policy model to one policy language, the more it will be diverged from adjusting to the concepts of other policy languages. Extensions to a policy model should be done carefully and after deeply reviewing the pros and cons of the newly added concepts. There is a tradeoff between supporting the semantics of one policy language and being able to cope with the semantic of other policy languages, which if not carefully designed, would render the R2ML policy model useless.

To be able to check the feasibility of the transformations in protecting the resources, a deeper analysis should be conducted. The most promising approach would be to deploy the policies on different resources and broker agents and get them to communicate. The level of jeopardizing the resources and the to-be-protected contents should be carefully examined. Policies are the critical rules in a business system which regulates the behavior of the system. Any flaw or miss-interpretation of the policies may result in the non-compostable loss of data and information. This further necessitates a detailed examination of the policies while working on the level of policy exchange. The practical analysis of the transformed policies is part of the upcoming research project that will be conducted in the Laboratory for Ontological Research (LORe) at Simon Fraser University

*Table 4. Transformations between R2ML and other Rule Languages. Right arrows explain transformations from R2ML to these languages while double head arrows explain transformations between R2ML and the language in both directions*

| **R2ML** | **RuleML** | **Jess** | **F-Logic** | **JBoss** | **Jena** | **SWRL** | **OCL** |
|---|---|---|---|---|---|---|---|
| *Derivation* | → | → | → | | → | | |
| Integrity | | | | | | ↔ | ↔ |
| Reaction | | | | | | | |
| Production | → | → | | → | → | | |
| **Transformation Language** | XSLT | | | | | QVT/ATL | |

This book chapter demonstrated the possibility to exchange the policy languages by *1) high level metamodel representation of their concepts, 2) capturing the semantics of the concepts, 3) identifying the similarities and dissimilarities of the modeled concepts*, and *4) applying the transformations to the concrete syntax of the policy language, benefiting from the QVT representation of the transformations*. All the above steps are compliant with the required steps in software design using MDE techniques (Schmidt, 2006). Using MDE techniques to provide the transformations significantly assists with having valid models for the source and the target languages. Furthermore, while using MDE, we are able to detect the inconsistent or missing constructs of different languages. During working with policies we discovered some points of improvement for R2ML which can add to its efficiency. First of all, we found that it is important for rules to have a value as an indicator for their priority. Having a set of rules, as in R2ML derivation or integrity rule sets, we need to provide an indicator of how one rule in this series can be chosen over the other ones. It especially helps with cases where more than one rule might be applicable to a certain situation. Although this issue matters mainly at the level of rule enforcement, where the enforcement engine needs to select among a set of applicable rules, and knowing that the current versions of R2ML are not supposed to be used at any enforcement level, dealing with rule exchange for cases where the source or the target rules need to have priorities makes it necessary to have priority indicators for the rules.

Another point that seems to be open to extend in R2ML is the possibility to use quantifiers to express cardinality. While working with KAoS, we realized that this language supports *min_cardinality* and *max_cardinality*. However, these quantified formulas are not possible to be used in the condition of a derivation rule set as the elements of the condition part are considered quantifier free and should be universally quantified. An extension to R2ML such that cardinalities can be supported in the condition of the derivation rules (and possibly other types of rules) has been also considered as a potential improvement in the next version of R2ML. Further to this, there have been long discussions on whether derivation rules or integrity rules should be used to model the policies. Looking back at the definitions of integrity and derivation rules, we see that integrity rules define that something *must necessarily* hold, or it *should* hold, while derivation rules carry a derivative meaning representing a set of new conclusions based on the presented facts. Both of these types of rules can be used to define the policies depending on the type and the purpose of the policy (i.e. authorization, authentication, and Quality of Service).

## REFERENCES


Boley, H., Tabet, S., & Wagner, G. (2001). Design rationale of RuleML: A markup language for Semantic Web rules. In *International Semantic Web Working Symposium (SWWS)* (pp. 381-402). CA.

Bonatti, P., & Olmedilla, D. (2005). Driving and monitoring provisional trust negotiation with metapolicies. In *IEEE 6th International Workshop on Policies for Distributed Systems and Networks* (pp. 14-23). Stockholm.

Bonatti, P., Duma, C., Fuchs, N., Nejdl, W., Olmedilla, D., Peer, J., et al. (2006). *Semantic Web policies—A discussion of requirements and research issues*. Paper presented at 3rd European Semantic Web Conference (ESWC).

Bradshaw, J., Dutfield, S., Benoit, P., & Woolley, J. (1997). KAoS: Toward an industrial-strength open agent architecture. *Software Agents*, 375-418.

Budinsky, F., Brodsky, S., & Merks, E. (2003). *Eclipse modeling framework*. Pearson Education.

Clark, J. (1999, November 16). XSL transformations (XSLT). *W3C Recommendation.*

Damianou, N., Dulay, N., Lupu, E., & Sloman, M. (2001). *The ponder policy specification language* (LNCS, pp. 18-38.

Fallside, D., & Walmsley, P. (2004, October 28). XML schema part 0: Primer second edition. *W3C Recommendation* .

Ginsberg, A., Hirtle, D., McCabe, F., & Patranjan, P.-L. (2006, July 10). RIF use cases and requirements. *W3C Working Draft* .

Godik, S., & Moses, T. (2003). eXtensible Access Control Markup Language (XACML). *Specification, OASIS.*

Grosof, B., Horrocks, I., Volz, R., & Decker, S. (2003). Description logic programs: Combining logic programming with description logic. In *WWW2003 Conference.* Budapest: ACM.

Horrocks, I., Patel-Schneider, P., Boley, H., Tabet, S., Grosof, B., & Dean, M. (2004, May 21). SWRL: A Semantic Web rule language combining OWL and RuleML. *W3C Member Submission* .

Kagal, L. (2002). *Rei: A policy language for the Me-Centric Project* (Tech. Rep).HP Labs.

Kaviani, N., Gasevic D., & Hatala, M. (2007-1). Web rule languages to carry policies. In *Proceedings of the 2007 IEEE Workshop on Policies for Distributed Systems and Networks (POLICY 2007), Bologna, Italy* (pp. 188-192).

Kaviani, N, Gasevic, D., Hatala, M., Clement, D., & Wagner, G. (2007-2). Integration of rules and policies for Semantic Web services. *International Journal of Advanced Media and Communication (IJAMC), 1*(4), 404-423.

Kalfoglou, Y., & Schorlemmer, M. (2003). Ontology mapping: The state of the art. *The Knowledge Engineering Review, 18*(1), 1-31.

Kolovski, V., Hendler, J., & Parsia, B. (2007). Analyzing Web access control policies. *WWW* (pp. 677-686).

Kühne, T. (2006). Matters of (meta-)modeling. *Software and System Modeling, 5*(4), 369-385.

Lassila, O., & Swick, R. (1999). Resource Description Framework (RDF) model and syntax specification. *W3C Recommendation.*

Miller, J., & Mukerji, J. (2003). MDA guide version 1.0.1. *Object Management Group.*

MOF. (2008, February 20). *Meta Object Facility (MOF) specification 2.0 facility and object lifecycle.*

Nejdl, W., Olmedilla, D., & Winslett, M. (2004). *PeerTrust: Automated trust negotiation for peers on the Semantic Web.* Paper presented at Workshop on Secure Data Management in a Connected World (SDM'04).

ODM. (2001, August 5). *ODM: The Object Management Group ontology definition metamodel.* Retrieved from http://www.omg.org/cgi-bin/apps/doc?ad/05-08-01.pdf

ODM. (2003, March 4). Ontology definition metamodel. *Third Revised Submission to OMG/RFP* (pp. 1-314).

QVT. (2005, June). MOF QVT final adopted specification. *OMG* .

Schmidt, D. (2006). Model-driven engineering. *IEEE Computer, 39*(2), 25-31.

Smith, M., Welty, C., & McGuinness, D. (2004, February 10). OWL Web ontology language guide. *W3C Recommendation* .

Toninelli, A., Bradshaw, J., Kagal, L., & Montanari, R. (2005). *Rule-based and Ontology-based policies: Toward a hybrid approach to control agents in pervasive environments.* Paper preseneted at the Semantic Web and Policy Workshop, Ireland.

Tonti, G., Bradshaw, J., Jeffers, R., Montanari, R., Suri, N., & Uszok, A. (2003). Semantic Web languages for policy representation and reasoning: A comparison of KAoS, Rei, and Ponder. In *The International Semantic Web Conferencel (ISWC)* (pp. 419-437).

Uszok, A., & Bradshaw, J. (n.d.). IHMC ontology and policy management. Retrieved from http://ontology.ihmc.us/

Uszok, A., Bradshaw, J., Jeffers, R., Suri, N., Hayes, P., Breedy, M., et al. (2003). KAoS policy and domain services: Toward a description-logic approach to policy representation, deconfliction, and enforcement. In *IEEE 4th International Workshop on Policies for Distributed Systems and Networks* (pp. 93-96).

Wagner, G., Damasio, C., & Antoniou, G. (2005). Towards a general Web rule language. *International Journal of Web Engineering and Technology, 2*(2), 181-206.

Wagner, G., Giurca, A., & Lukichev, S. (2005). A general markup framework for integrity and derivation rules. In *Principles and practices of Semantic Web reasoning.* Dagstuhl.

Wagner, G., Giurca, A., & Lukichev, S. (2006). A *usable interchange format for rich syntax rules integrating OCL, RuleML and SWRL.* Paper presented at Reasoning on the Web Workshop at WWW2006, Edinburgh, UK.

Wagner, G., Giurca, A., & Lukichev, S. (2005). R2ML: A general approach for marking-up rules. In *principles and practices of Semantic Web reasoning.*

XSB. (2007, October 6). *XSB Prolog.* Retrieved from http://xsb.sourceforge.net/

## KEY TERMS AND DEFINITIONS

**KAoS:** KAOS language is a multi-paradigm specification language with a two-level structure: an outer semantic net layer for declaring concepts, their attributes and links to other concepts and an inner formal assertion layer for formally defining the concept.

**MDA:** an approach to application design and implementation that encourages efficient use of system models in the software development process, and it supports reuse of best practices when creating families of systems.

**MOF 2 :** a metadata management framework, and a set of metadata services that enable the development and interoperability of model- and metadata-driven systems.

**OWL:** Web Ontology Language designed for use by applications that need to process the content of information instead of just presenting information to humans. OWL can be used to explicitly represent the meaning of terms in vocabularies and the relationships between those terms. This representation of terms and their interrelationships is called an ontology.

**QVT:** a standard for model transformation languages

**R2ML:** a MOF-defined general rule language that captures integrity, derivation, production, and reaction rules. It is a well-known proposal for RIF.

**RIF:** Rule Interchange Format is W3C initiative that should define an intermediary language between various rule languages, but it should not provide a formally defined semantic foundation for reasoning on the Web such as OWL for ontologies.

## ENDNOTE

[1] The fragment of Horn FOL with no function symbols (Grossof et al. 2003).

# Chapter XXV
# Languages and Tools for Rule Modeling

**Grzegorz J. Nalepa**
*AGM-University of Science and Technology, Poland*

## ABSTRACT

*This chapter presents selected practical issues of rule modeling. This field combines both classic artificial intelligence methods and software engineering. The chapter gives a concise presentation of selected relevant methods, and approaches, put in an engineering perspective. The modeling language used in the communication between business analysts and experts for analyzing the system requirements should not be too technical. It should allow for visual rule expressions, which can be understood by experts without an extensive technical training. The main goals of this chapter are: to summarize the formal foundations of rules found in the field of AI, including decision tables and trees; discuss main challenges in practical rule design, and modeling; introduce selected recent research in the field of rule design, focusing on visual modeling; and observe some important future trends in rule design and integration. In the chapter it is argued that efficient visual rule modeling methods are crucial for developing complex rule systems.*

## INTRODUCTION AND MOTIVATION

Designing rule-based systems is not a trivial task. Standard software design approaches cannot be used directly, due some fundamental differences between knowledge and software engineering. These include non-procedural declarative knowledge specification, as well as important semantic differences between the rule-based representation and conventional design and programming languages.

The motivation of this chapter is to present selected practical issues of rule modeling. This field combines both classic Artificial Intelligence (AI) (Russell & Norvig, 2003) methods and Software Engineering (SE) (Sommerville, 2004). In AI rules have been studied as a classic knowledge representation method (Brachman & Levesque, 2004, Ligęza, 2006, Harmelen & Lifschitz & Porter, 2007) for expert systems (Liebowitz, 1998, Giarratano & Riley, 2005). Some recent developments in Software Engineering, mainly in business rules systems make it use the AI experiences, putting them in a new context. The chapter gives a concise presentation of selected relevant methods, and approaches, put in an engineering perspective.

Principal objectives of this chapter are:

- to summarize the formal foundations of rules, found in the field of AI, including decision tables and trees,
- discuss main challenges in practical rule design, and modeling,
- introduce selected recent research in the field of rule design, focusing on visual modeling,
- observe some important future trends in rule design and integration.

In this chapter it is argued that efficient visual rule modeling methods are crucial for developing complex rule systems.

## BACKGROUND

### Knowledge Representation

*Rules* are both prime and classic example of a *knowledge representation* method (Brachman & Levesque, 2004, Harmelen & Lifschitz & Porter, 2007). Such methods are developed within *knowledge engineering*. It is a field of Artificial Intelligence applied to building intelligent systems, systems that represent and process knowledge.

Knowledge is often defined as *justified true belief* (Torsun, 1995). So it is then a set of facts or true statements about a world. A representation in a broad sense may be defined as "*the symbolic representation of justified true beliefs or a model of some universe of discourse*"(Torsun, 1995).

It is widely recognized that there is no single formalism suitable to represent knowledge for all purposes. A variety of formalisms and structures is needed. In the field of expert systems (Liebowitz, 1998, Giarratano & Riley, 2005) the *knowledge representation method* is a systematic way of "encoding" what an expert knows about some domain (Jackson, 1999). However "encoding" means here rather "describing" than "encrypting".

Some of the issues arising in a knowledge representation are: syntax, semantics, expressive adequacy, reasoning, completeness, real-world relevance, flexibility. Different representations address these issues in different ways. While there are numerous knowledge representation methods, the logic-based ones are essential to the theory and practice of expert systems and rule systems in general.

In the chapter some fundamental logical rule formats are considered. They are a basis for rule languages. Rules can be practically written and processed in the logic programming paradigm, e.g. in Prolog (Bratko, 2000). Even though the language uses a subset of first order predicate logic (restricted to Horn clauses), it is easy to write meta-interpreters working another languages.

### Decision Rules

*Decision rules* are one of the most important and successful knowledge representation methods. Systems based on rules, the so called rule-based expert systems, or just rule-based systems (RBS) find numerous applications in many fields, from decision support in areas ranging from business to medicine, to intelligent control. This is due to the fact that rules have an intuitive, straightforward semantics, close to the natural language, making them accessible to non-experts. Simultaneously, they also have a clear formal description, which allows for both formalized design and analysis of mission-critical rule-based systems.

A general rule expression corresponds to a *conditional* statement in the natural language. Often this statement is simply presented as:

IF some condition is met THEN some conclusion may be driven

While this may seem simple and straightforward, it is in fact imprecise. The actual semantics depends on the expressive power of the conditional language, as well as the conclusion description.

In RBS, a rule has a general format:

IF condition THEN decision

This format can be used in both forward and backward chaining systems. However, here we focus on the production rule systems, based on the forward chaining paradigm. The power of a rule language stems from the syntax and semantics of the conditional and decision expressions. Number of systems implicitly assume, that this rule format can be extended to the conjunctive normal form (CNF), that is:

IF cond1 AND cond2 AND ... AND condN THEN decision

which in fact corresponds to a Horn clause *F* (Ben-Ari, 2001, Ligęza, 2006), that is:

F = ~p1 v ~p2 v ... ~pk v q

Such a clause can be represented as an implication of the form:

p1 ^ p2 ^ ... ^ pk -> q

which can be regarded as a rule in the above format, where *p*s correspond to conditions and *q* corresponds to the decision. In fact the Prolog language uses a subset of predicate calculus, restricted to Horn clauses (Covington & Nute & Vellino, 1996, Bratko, 2000).

The question is what are the conditional and decision expressions. In a number of systems these correspond to expressions in the propositional calculus, which makes the semantics somehow limited. Some systems try to use some subsets of predicate logic, which gives much more flexibility, but may complicate the RBS design and the inference process. This is the case of Prolog. In XTT (eXtended Tabular Trees) (Nalepa & Ligęza, 2005b) these expressions are in the the attributive logic (Ligęza, 2006). This gives much more power than the propositional logic, but does not introduce problems of the predicate logic-based inference.

Two important factors have to be taken into the consideration: what is the formal calculus used in rules, and how rule are interpreted. Considering the first aspect the propositional and predicate calculi have to be considered (Ben-Ari, 2001). Using quantifiers, it is possible to specify constraint, or integration rules. The second aspect involves forward chaining and backward chaining. Considering these contexts the *production* and *derivation* rules are identified.

Consider a set of production rules (a similar example may be found at openrules.com):

IF age =< 12 THEN greeting = little
IF age > 12 AND sex = male THEN greeting = Mr.
IF age > 12 AND sex = female AND martial = single THEN greeting = Ms.
IF age > 12 AND sex = female AND martial = married THEN greeting = Mrs.

It is a common engineering practice to support the system design with a visual representation. This is also a case with rule-based systems. Giving a proper formulation of rules, it is possible to provide a visual rule representation which is logically equivalent to certain rule sets. Visual forms of rule-based systems incorporate elements of a structure into a rule set. This structure provides an intuitive visualization of the system, often oriented towards a specific application, and it also imposes a specific interpretation rules, especially with respect to the order of rules and preconditions. Two most important representations are discussed, these are: decision tables, and decision trees.

## Decision Tables

A *decision table* represents a set of similar rules grouped with respect to used preconditions and conclusions or actions. The rules covered by a decision table form a kind of a decision unit designed to work in a given context. A classical decision table is a table displaying sequences of conditions which must hold for executing specific actions or drawing specific conclusions. The sequences of conditions are displayed in a readable form, vertically in the classical decision tables, as parts of columns of the decision table, or horizontally, as parts of rows of the table (Brachman & Levesque, 2004, Ligęza, 2006). A decision table corresponding to the set of rules presented in the previous section is shown in Figure 1.

## Decision Trees

*Decision trees* are an intuitive way for specifying decision procedures. They use simple visual representation (an acyclic, directed graph). Decision evaluation is a simple matter of traversing such a tree along a given path. There are various forms of decision trees, such as binary decision trees, attributive decision trees, decision trees for analysis of first-order formulae, etc. (Brachman & Levesque, 2004, Ligęza, 2006). A decision tree corresponding to the set of rules presented in the previous section is shown in Figure 2.

*Binary Decision Trees* (BDT) are perhaps the best known and most popular decision structures used in a number of domains. For intuition, binary decision tree is a structure leading through a series of simple tests to arrive at a certain decision. The tests are limited to evaluation of a propositional statement, which can be true or false, and hence the trees are referred to as binary.

## Business Rules Design

Business rules (BR) approach (Ross, 2003, Halle, 2001) is a new approach to rule design for business applications based on concepts borrowed from knowledge engineering (KE) and rule-based systems. It is becoming an important approach in business application development, especially on the Java platform. A classic description of the main principles of the approach is given in (Ross, 2003). According to it, rules should be: written and made explicit, expressed in plain language, motivated by identifiable and important business factors, single sourced, specified directly by people who have relevant knowledge, managed, and built on facts, and facts should build on concepts. Rules

*Figure 1. A simple decision table*

| AGE | SEX | MARTIAL | GREETING |
|---|---|---|---|
| ≤ 12 | — | — | Little |
| > 12 | female | single | Ms. |
| > 12 | female | married | Mrs. |
| > 12 | male | — | Mr. |

*Figure 2. A simple decision tree*

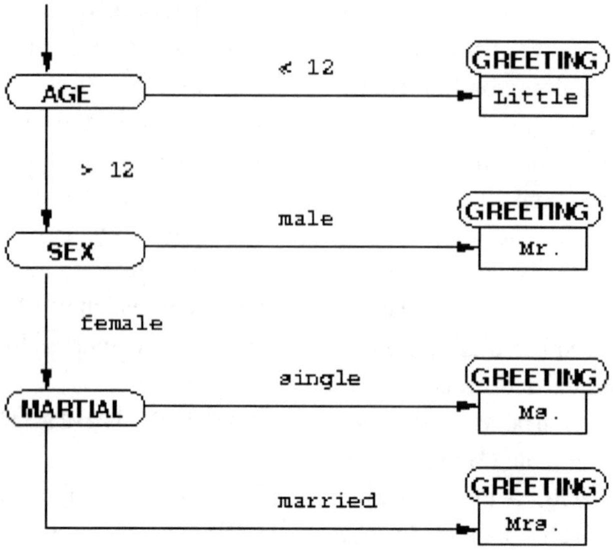


should also exist independent of procedures and workflows.

There are number of *rule types* identified in the BR approach, such as: reactive rules (event-condition-action rules), transformation rules (functional-equational rules), derivation rules (implicational-inference rules), also ones restricted to facts ("premiseless" derivation rules) and queries ("conclusionless" derivation rules), as well as integrity-constraints (consistency-maintenance rules).

From the point of view of formal knowledge engineering, a major issue with current BR methods is related to the *logical foundations* of BR systems. From a point of view of classical KE, a rule-based expert system consists of a knowledge base and an inference engine. The KE process aims at designing and evaluating the knowledge base, and implementing a proper inference engine. The process of building the knowledge base involves the selection of a knowledge representation method, knowledge acquisition, and possibly low-level knowledge encoding. In order to create an inference engine a reasoning technique must be selected, and the engine has to be programmed.

Unfortunately it can be observed, that common approaches to BR tend to mix these formal aspects. While there are attempts to officially define main aspects of the approach, e.g. the *Semantics of Business Vocabulary and Business Rules Specification*, see (OMG, 2006b), these formulation attempts are definitely not satisfactory. The main source of problems seems to be the way in which the rich semantics of business rules is being captured. The problem of defining the business rules vocabulary, seems to fall into the domain of natural language processing. The other problem is the lack of explicit choice of logical calculus. This poses severe problems when it comes to formal description of BR syntax, semantics, and inference.

## Persistent Problems of Rule Modeling

Even though the issue of rule modeling has been present in the AI and engineering research for years, some important practical problems in this area persist. They are especially exposed in the case of business rules modeling. They are partially due to growing complexity of the rule systems, as well as the need for a tighter integration with existing business applications designed with the software engineering methods.

The first problem is related to apparent shortcomings or even lack of a coherent *formal description* of rule systems being modeled. This poses problems with both formal analysis and formally defined translation of rules.

The second problem is related to the *visual representation* used in the design of rule systems. Visual representations used, have scalability problems (it's easy to draw diagrams of several rules, but it becomes very difficult to cope with tens of rules). Lack of well-defined formal foundations of these representations leads to problems with automatic transformation of visual model to the logical one. When introducing a visual representation, it is important to decide what is the main reason for its introduction. Is it just some useful conceptualization tool, helpful in the design; or is it an integral part of the design process.

The third problem concerns the *formal analysis and verification* of rule systems. As the number of rules exceeds even relatively very low quantities, it is hard to keep the rule-base consistent, complete, and correct. These problems are related to knowledge-base verification, and validation The selection of appropriate software tools and programming languages is non-trivial either.

These issues are often neglected in rule modeling. It seems that analysis (where issues such as verification, validation, and evaluation are even not properly separated) is simply considered testing. So the analysis of the knowledge base is implicitly substituted by testing of the implementation. However, in the KE approach, a proper formal analysis of the knowledge base minimizes the need for testing. Such an analysis, along with formal specification of the knowledge representation syntax and semantics, can eliminate not just syntactic errors, but also semantic ones. This is why it should be incorporated into the rule design process as early as possible.

## Rule Modeling

Let us take a closer look at the modeling concept. In this case the knowledge engineering situation is as follows: having a natural language description of certain problem area the designer aims at providing a declarative rule-based description of this area. The rule-based description is somehow formalized (or at least disciplined) compared to the original one. Rules are a knowledge representation method that captures regularities, constraints, and relations. While formalized, this description is a high-level one, close to the original natural language-based one. So the basic sense of rule modeling is to build a rule-based knowledge representation of the problem. It is a classic case of knowledge engineering, where the designer (the knowledge engineer) has to identify, extract, describe and represent knowledge possessed by domain experts, that may be embedded in an information system.

## Modeling Languages

Rule modeling may be considered as one of the classic problems in the field of AI. It is a question of knowledge engineering (KE), and building rule-based expert systems that have strong logical foundations. Within the AI, number of visual knowledge representation methods for rules have been considered. These methods mainly include decision tables, that help combining rules working in the same context, and decision trees, that support visualization of the decision making process. Two important factors for using these methods are:

1. design support - all of these methods help the designer (knowledge engineer) to develop the rule-based model in a more rapid, and scalable way, and

2. logical equivalence - all of them formally correspond to rules on the logical level.

These methods are used to model rules in practical applications.

A common approach to model rule-based systems is to use UML (OMG, 2007a), considered by some an universal modeling language. UML offers a visual or semi-visual method for different aspects of information modeling. Using it, it is possible to model some specific rule types. However, when it comes to practical knowledge engineering it has some major limitations, due to different semantics of rules and object-oriented paradigm. In particular cases, some of these shortcomings can be overcome by the use of the OCL, that allows for constraint specification for UML classes.

OCL (OMG, 2006c) is a declarative language for constraint rules description. OCL rules aim to provide a more specific and coherent UML model. OCL is a textual language providing constraint and object query expressions on any MOF (Meta-Object Facility) model or metamodel, that cannot be expressed otherwise by UML diagrammatic notation. The language is an important component in the OMG recommendations for transforming MOF-based models. More recent developments in this area also include the Alloy Analyzer from MIT (alloy.mit.edu), which aims at overcoming some of the OCL shortcomings.

One area, where UML, or UML-related methods are more useful is the conceptual modeling, that supports practical rule authoring. UML class diagrams are suitable to capture relations between concepts present in rule vocabularies. Example of such a vocabulary is presented in Figure 3. It is a vocabulary for URML-based example (see oxygen.informatik.tu-cottbus.de/rewerse-i1/?q=constraints).

A new specialized method for specifying these kinds of information is SBVR from OMG (OMG, 2006b). It also provides methods that support transforming natural language rule specification, to a formalized rule format, mainly the structured english format.

Since UML is a de facto standard information modeling method in software engineering approaches and tools, it can be treated as a low-level language, on top of which a richer semantics is provided. This is possible thanks to the UML profiles. Based on these a dedicated visual rule modeling language can be built. A prime example of this approach is the URML from the REWERSE I1 group, discussed in the next sections.

Another industry driven effort is OMG Production Rule Representation initiative (OMG, 2007). It aims at providing a unified UML-based representation for rules, aimed at high-level compatibility with existing rule engines, and UML modeling tools. Currently this specification is in

*Figure 3. Rule vocabulary in UML (see the REWERSE WG I1 website)*

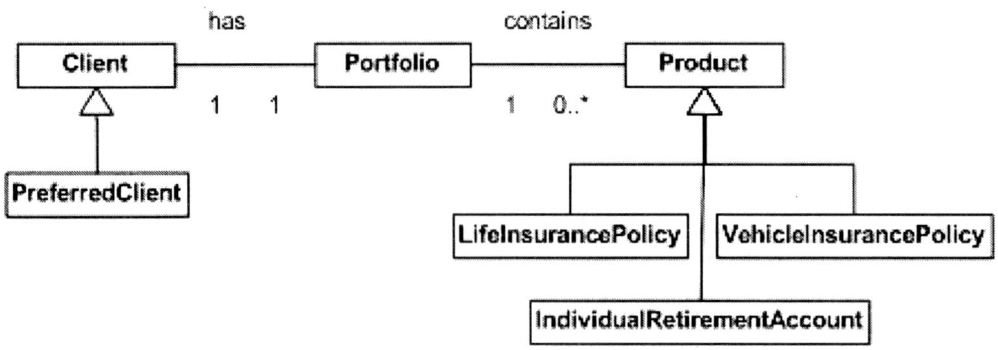

the development. However, it should be considered as an important future standard.

An important community is built around the W3C and the so-called Semantic Web Initiative. The methods built on top of the XML (W3C, 1998) , RDF (Lassila & Swick, 1999) and OWL (McGuinness & Harmelen, 2004) allow also for rule modeling for both web and general purposes. Rule interchange is possible using the XML-based RIF format (www.w3.org/2005/rules). Currently RIF is evolving, being formalized by the RIF Working Group. Another important rule standard from W3C is the SWRL, A Semantic Web Rule Language (Horrocks, et al., 2004), that aims at combining the OWL and RuleML (Boley & Tabet & Wagner, 2001).

The rule-based model can be used as a standalone logical core of a business application. However, in practise this model should be somehow integrated with other models, and components of a heterogeneous application.

## Business Rules and Process Modeling

In business systems rules play an important role in the Business Process Management. It can be defined as: "supporting business processes using methods, techniques and software to design, enact, control and analyze operational processes involving humans, organizations, applications, documents and other sources of information" (Aalst & Hofstede & Weske, 2003). It involves the question of incorporating and representing human knowledge into business systems. Knowledge is an essential factor in the process. Knowledge representation methods need proper syntax, visual representation, and formal foundations. Before applying any of the knowledge representation techniques, knowledge on processes has to be acquired from existing systems and people. While the first task is not very complicated, the second one is not trivial. Most of employees' knowledge is a tacit one (Gronau & Mueller & Uslar, 2004), not easy to be expressed. It may nevertheless not be omitted, as knowledge and business processes are integrated and should be evaluated as a whole. Processes involve collaboration between individuals and/or groups to achieve a goal - in the process there is a need for recognizing the involvement of humans in the execution of business processes to properly model them.

The main features of business process (BP) modeling are: descriptive - describes what happens during a business process, in what way the process has been performed and what improvements have to be made; prescriptive - allows for a definition of a business process and how it should be performed, it lays down rules, guidelines and behavior patterns, and explanatory - links processes with the requirements that are to be fulfilled, explains the rationale of business processes.

These lead to the formulation of the business process model requirements. First of all, a model has to provide a holistic approach dealing with organizational and technical issues (Karagiannis, 1995). Furthermore, models should have a strong formal foundation. It is so because formal models are unambiguous, and increase the potential for analysis (Aalst & Hofstede & Weske, 2003).

There are several techniques for BP model specifications. Some are based on Petri nets, some make use of the UML notation. A common approach is to use some versions of decision trees. It must be nevertheless pointed out that UML, although widely used and adapted, is not expressly designed to map to business execution languages. A new formalized approach to business process modeling is the *Business Process Modeling Notation* (BPMN) (OMG, 2006a, Owen & Raj, 2006).

Recently, a new approach to practical integration of the knowledge representation based on rules, with business process, has been gaining popularity. This is the mentioned previously, business rules approach. In this approach business process descriptions are extended with rules defining business logic executed in a given context. So the process model specifies a higher-level execution flow, while rules implement the lower-level business logic.

### Rule Modeling Process Requirements

Observing different rule modeling languages and approaches, it can be formulated, that the rule representation used in the modeling process should meet certain requirements. It should: be easy to grasp by non-technical people, be possible to process it automatically, and integrate with a certain runtime (rule engine), formalized to some degree, meet certain quality standards (e.g. completeness, lack of redundancy), be suitable for interchange and integration with other systems, be manageable.

The emphasis on these aspects can differ, depending on the goal of providing the rule-based description, e.g. describing system requirements, including constraints, or building a complete system from scratch. Rules can be also thought of as a certain mean of formalized communication.

Finally, the multilayer aspect of the rule language should be considered. A useful and expressive rule language should provide:

- rich, but well-defined semantics,
- formally defined syntax, with clear logical interpretation,
- scalable visual representation, that allows to visualize many rules,
- machine readable encoding, for model interchange and integration.

Rule modeling methods and approaches should also be considered with respect to other modeling methods, such as software engineering methods and methodologies (e.g. UML).

In next sections selected approaches to rule modeling are presented, including industry and community driven proposals from OMG and W3C. Following these, two research projects' proposals are presented. The REWERSE and HeKatE approach have a common starting point: how to visualize rule modeling to simplify the rule design process. However, they use different means and technologies. The REWERSE WG I1 approach is strongly founded in the classic SE methods, such as UML while the HeKatE approach aims at reusing some best concepts from classic AI, such as decision tables and trees. While both of this are experimental research, they may serve as an indication of the development in the area of rule modeling.

## OMG APPROACH

Object Management Group (OMG) (see www.omg.org) is a well-known consortium focused in providing open standards in the areas of object-oriented and model-driven modeling. Some of the best known OMG standards include UML, OCL, MOF, XMI, and MDA. These standard serve as a foundation for the Software Engineering tools and methods.

### Objectives

In the recent years consortium activities included the standardization of the area of Business Strategy, Business Rules and Business Process Management. The standards provided in this area (see www.omg.org/technology/documents/br_pm_spec_catalog.htm) include:

- *Business Motivation Model* (BMM) - which "provides a scheme or structure for developing, communicating, and managing business plans in an organized manner".
- *Business Process Definition Metamodel* (BPDM) - that "provides the capability to represent and model business processes independent of notation or methodology, thus bringing these different approaches together into a cohesive capability".

- *Business Process Maturity Model* (BPMM) - which is "intended for anyone interested or involved in improving an organization's business process related to their products and services - whether the products and services are for internal or external use".
- *Business Process Modeling Notation* (BPMN) - based on the "Business Process Diagram (BPD), which is a Diagram designed for use by the people who design and manage business processes; it also provides a formal mapping to an execution language of BPM Systems".
- *Production Rule Representation* (PRR) - which "provides a standard production rule representation that is compatible with rule engine vendors' definitions of production rules".
- *Semantics of Business Vocabulary and Business Rules* (SBVR) - that "defines the semantics of business vocabulary, business facts, and business rules".
- *Workflow Management Facility* - which is focused on the "interfaces for workflow execution control, monitoring, and interoperability between workflows defined and managed independently of each other".

The OMG's ultimate goal is to provide a single coherent solution based on the above. However, from the rule modeling perspective, and considering maturity and adoption of the above specification, the focus here is on the BPMN, SBVR, and PRR. These three play a growing role in the business rules modeling.

## BPMN

Business Process Modeling Notation (BPMN) (OMG, 2006a, Owen & Raj, 2006) (see www.omg.org/spec/BPMN/1.1) is a new formalized approach to business process modeling. The BPMN is supposed to be a new standard for modeling both business processes and web service processes. BPMN consists of one diagram - the so-called Business Process Diagram (BPD). It consists of several groups of visual elements, namely: Activities, Events, Gateways, Connections, Artifacts, and Swimlanes. BPD is easy to use and understand, allows to model very complex business processes, and can be naturally mapped to business execution languages.

BPMN provides a mapping between the visual notation to the underlying constructs of execution languages including BPEL4WS. On the lower level the subprocess phases may be specified using the rule-based paradigm, by identifying the business rules. These are to be defined with the use of the SBVR and possibly specified using the upcoming PRR standard.

## SBVR

Semantics of Business Vocabulary and Business Rules (SBVR) (OMG, 2006b) (see www.omg.org/spec/SBVR/1.0) aims at documenting both the vocabulary and semantics of business vocabularies, facts, and rules. It provides an XMI schema for the interchange of business vocabularies and rules between software tools. SBVR uses MOF to provide interchange capabilities and enable generating MOF-compliant models. SBVR proposes the so-called Structured English notation as one of possibly many notations that can map to the SBVR Metamodel.

Rules play a very important role in defining business semantics: they can influence or guide behaviors and support policies, responding to environmental situations and events. This means that rules represent the primary means by which an organization can direct its business, defining the operative way to reach its objectives and perform its actions.

Some main business rule types identified in SBVR are

- Structural Business Rules that use two alethic modal operators: it is necessary that and *it is possible that*; and
- Operative Business Rules that use two deontic modal operators: it is obligatory that and *it is permitted that.*

In the SBVR Structured English Notation the following constructs are indicated:

- <u>term</u> - a noun concept (other than an individual concept), one that is part of a vocabulary being used or defined.
- **Name** - an individual concept, a name.
- *verb* - fact type - usually a verb, preposition, or combination thereof.
- keyword - linguistic symbols used to construct statements the words that can be combined with other designations to form statements and definitions.

Some simple rule examples form the (OMG, 2006b) are given below.

- It is obligatory that each rental car *is owned by* exactly one branch.
- A rental must *have* at most three additional drivers.
- It is obligatory that each rental *has* at most three additional drivers.

These examples are a part of the EU-Rent case study provided in the SBVR specification.

## PRR

The Production Rule Representation (PRR) (OMG, 2007) addresses some important requirements related to software systems based on business rules. It provides a standard production rule representation, compatible with rule engine vendors' definitions of production rules. Therefore it can be used for interchange of business rules amongst rule modeling tools. It provides an OMG MDA PIM model and can be included to add production rule capabilities to other OMG metamodels. It is also an example of how the UML with OCL can be used to support production rules in a standardized way. It aims at the integration on the interchange level with W3C's RIF standard.

A simple example of a rule model found in (OMG, 2007) is introduced. Using classes from the model business rules are formulated in the natural language and the described in the PRR OCL, e.g.:

*If the shopping cart contains between 2 and 4 items and either the purchase value is greater than $100 and the customer category is gold or the purchase value is greater than $200 and the customer category is Silver then apply a 15% discount on the shopping cart value.*

```
RULE discount
RULEVARIABLE:
 ?customer: Customer = Customer->any()
 ?shoppingCart: ShoppingCart =
  ShoppingCart->any(c: customer |
c=?customer)
CONDITION:
 (?shoppingCart.containsItemsInRange(2,
4)
 AND
 (((?shoppingCart.items->collect(i:Item|i.
value))->sum()>100
 AND
 ?customer.category == "Gold")
 OR
 ((?shoppingCart.items->collect(value))-
>sum() > 200
 AND
 ?customer.category == "Silver")))
ACTION:
 shoppingCart.discountValue = shopping-
Cart.discountValue+15
```

## Summary

OMG has been a pioneer in the Software Engineering standards for years. With the attempts to standardize the Business Strategy, Rules and Process Management within the MDA framework OMG aims at providing a unified modeling solution for business application, starting from the very general conceptual specification to the implementation. The adoption of the new OMG standards will take time. Currently availability of tools supporting them is limited. However, it should be expected that the situation would change in the future.

# SEMANTIC WEB APPROACH

One of the main goals of the Semantic Web Project is to provide a knowledge layer on top of today's distributed network. In order to achieve this goal a strong structurization solutions, as well as semantic annotation layer are provided on top of the data encoding standards. On these rules can be built, allowing for actual inference on the network. These technologies are being integrated using the Semantic Web stack.

## Objectives

To provide intelligent information processing, the current Web architecture needs to redesigned. A knowledge-based layer needs to be developed on top of data storage technologies. Logic-based inference solution are need to be able to formally verify knowledge stored on the Web.

For allowing rules on the Web the Semantic Web Stack provides the following layers and languages (bootom-up):

1. identifiers and character encoding: URI and Unicode
2. structure and syntax: XML, XML Schema and XML Namespaces
3. data and metadata interchange: RDF
4. taxonomies: RDF Schema
5. ontologies: OWL
6. rules: RIF and SWRL

The upcoming proposals for rule markup and interchange SWRL and RIF use some previous experiences with rule markup languages, including the RuleML effort.

## RuleML

The RuleML language developed by the Rule Markup Initiative (see www.ruleml.org) aimed at unifying existing rule markup language efforts existing around 2000,1. The principal goal was to develop RuleML as the canonical Web language for rules using XML markup, formal semantics, and efficient implementations. The language covers the entire rule spectrum, from derivation rules to transformation rules to reaction rules. RuleML aims for rule interoperation between industry standards (such as JSR 94, SQL'99, OCL, BPMI, WSFL, XLang, XQuery, RQL, OWL, DAML-S, and ISO Prolog) as well as established systems (CLIPS, Jess, etc.). The RuleML specification is a modular one, consisting of several sublanguages, mainly: binary and unary Datalog and Hornlog.

An example rule

*The discount for a customer buying a product is 7.5 percent if the customer is premium and the product is luxury.*

can be represented in the Datalog sublanguage as:

```
<Implies>
 <head>
 <Atom>
 <Rel>discount</Rel>
 <Var>customer</Var>
 <Var>product</Var>
 <Ind>7.5 percent</Ind>
```

```
    </Atom>
    </head>
    <body>
    <And>
    <Atom>
    <Rel>premium</Rel>
    <Var>customer</Var>
    </Atom>
    <Atom>
    <Rel>luxury</Rel>
    <Var>product</Var>
    </Atom>
    </And>
    </body>
</Implies>
```

The RuleML language is an important effort that paved the way for the next generation of rule languages built on top of the existing semantic technologies, such as RDF and OWL. While RuleML gained popularity and some tool support, it now seems it could be gradually replaced by the forthcoming RIF and SWRL standards from W3C.

## RIF

Rules Interchange Format (RIF) (Pashke, et al., 2008) is supposed to be an effective mean of exchanging rules in the industry while being consistent with existing W3C technologies and specifications. RIF is intended to be a specification that builds on the existing W3C specifications. Some of the critical factors in the development of RIF include: alignment with key W3C specifications, wide coverage, encouragement of interoperability, predictability, extensibility, ease of use, and low cost of implementation.

RIF is described by a set of specifications (see www.w3.org/2005/rules/wiki/UCR):

- *RIF-FLD* (Framework of Logic Dialects) document describes a framework of mechanisms for specifying the syntax and semantics of logic-based RIF dialects through a number of generic concepts.
- *RIF-RDF+OWL* (RIF RDF and OWL Compatibility) document specifies the interoperation between RIF and the data and ontology languages RDF, RDFS, and OWL.
- *RIF-DTB* (Data Types and Builtins) document describes RIF data types and built-in functions and predicates
- The *RIF-BLD* (Basic Logic Dialect) document specifies a basic interchange format that allows logic rules (definite Horn rules with equality) to be exchanged.
- The *RIF-PRD* (Production Rules Dialect) document specifies the RIF production rules dialect to enable the interchange of production rules.
- The *RIF-UCR* (Use Cases and Requirements) document describes use cases and requirements for RIF.

RIF is designed as a family of RIF dialects.

- The *RIF Framework for Logic-based Dialects* (RIF-FLD) describes mechanisms for specifying the syntax and semantics of logic-based RIF dialects through a number of generic concepts.
- *RIF-BLD* (Basic Logic Dialect) is a specialization of RIF-FLD capable of representing definite Horn rules with equality enhanced with a number of syntactic extensions to support expressive features such as objects and frames, internationalized resource identifiers (IRIs) as identifiers for concepts, and XML Schema data types.
- *RIF-PRD* (Production Rules Dialect) specifies a production rules dialect to enable the interchange of production rules.

RIF provides a normative syntax based on XML, as well as a non-normative presentation syntax to a more readable and compact presentation.

An example business rule, expressed in the natural language:

*If an item is perishable and it is delivered to John more than 10 days after the scheduled delivery date then the item will be rejected by him.*

Is represented in the ordered representation in RIF BLD Presentation Syntax using Relations:

```
Document( Prefix(pred http://www.w3.org/2007/rif-builtin-predicate#)
 Prefix(func http://www.w3.org/2007/rif-builtin-function#)
 Prefix(ex http://example.org/example#)
 Group
 (
  Forall ?item ?deliverydate ?scheduledate ?diffdays (
   ex:reject("John" ?item) :-
   ex:perishable(?item)
    ex:delivered(?item ?deliverydate "John")
   ex:scheduled(?item ?scheduledate)
    ?diffdays = External(func:days-from-duration(
      External(func:numeric-subtract(?deliverydate ?scheduledate))
   ))
    External(pred:numeric-greater-than(?diffdays 10))
   )
   )
 )
```

RIF, as the general interchange format is accompanied by the SWRL a rule language for the Semantic Web.

## SWRL

The Semantic Web Rule Language (SWRL) (Horrocks, et al., 2004) is based on a combination of the OWL DL and OWL Lite sublanguages of OWL with the Datalog RuleML sublanguages. The SWRL proposal extends the set of OWL axioms to include Horn-like rules. It enables Horn-like rules to be combined with an OWL knowledge base. Three main syntaxes are provided:

- An abstract syntax that extends the OWL abstract syntax.
- An XML syntax for rules based on RuleML and the OWL XML presentation syntax.
- An RDF concrete syntax based on the OWL RDF/XML exchange syntax.

Rules in SWRL are of the form of an implication between an antecedent and consequent. The intended meaning can be read as: whenever the conditions specified in the antecedent hold, then the conditions specified in the consequent must also hold. Both the antecedent (body) and consequent (head) consist of zero or more atoms.

A rule in the SWRL abstract syntax could be:

```
Implies(Antecedent(hasParent(I-variable(x1) I-variable(x2))
  hasBrother(I-variable(x2) I-variable(x3)))
 Consequent(hasUncle(I-variable(x1) I-variable(x3))))
```

One can read from this rule, "*if John has Mary as a parent and Mary has Bill as a brother then John has Bill as an uncle*" (Horrocks, et al., 2004).

## Summary

Developing a flexible and coherent rule layer on top of existing Semantic Web technologies, mainly Description Logics and OWL remains a challenge for the Semantic Web community. Both RIF and SWRL remain work in progress so their final shape is not known now. However, it is clear that a rule standard would eventually evolve. It would then allow for reasoning on the Web.

## REWERSE APPROACH

In this section selected results from the Working Group I1 (oxygen.informatik.tu-cottbus.de/rewerse-i1) of the REWERSE Project (rewerse.net) are presented. The group completed its work in February 2008. The work of the group is focused on rule visual modelling and rule markup, including particular modeling issues such as rule visualization and rule verbalization. The group aimed at using both OMG and W3C technologies, mainly UML for visual rule modeling and markup languages, including RIF and SWRL.

In the area of rule modeling, a UML-based rule modeling language (URML) has been defined and implemented in the graphical rule modeling tool Strelka, which allows to capture rules in the semi-graphical syntax of URML. URML rule expressions can be serialized the in the R2ML format, a markup language for different kinds of rules, addressing the RIF and SWRL proposals.

### Objectives

One of the primary objectives of the REWERSE WG I1 was to provide a general and usable rule language, that would support different rule types, as well as RIF-compatible rule interchange between different rule formats. Another goal was to provide a visual rule modeling support based on this language. The visual representation should be UML-based to extend existing UML modeling tools to support rule languages.

These objectives were met by providing:

- the general rule markup language R2ML and the experimental rule interchange toolset based on it,
- the UML-based rule modeling language URML,
- URML implementation in the Eclipse-based visual rule modeling tool Strelka.

Other results also include an extension of RDF - the reasoning language ERDF, as well as a R2ML-based rule interchange Web service.

### R2ML

R2ML (Wagner & Giurca & Lukichev, 2005) is a general and comprehensive XML-based rule format. It allows for representing multiple rule types, and interchanging rules between different systems and tools. It aims at enriching ontologies by rules, as well as providing connection with visual rule modeling tools. R2ML integrates:

- the Object Constraint Language (OCL) - a standard used in information systems engineering and software engineering (OMG, 2006), with
- the Semantic Web Rule Language (SWRL) - a proposal to extend the Semantic Web ontology language OWL by adding implication axioms (Horrocks, et al., 2004), and
- the Rule Markup Language (RuleML) - a proposal based on Datalog/Prolog (Boley & Tabet & Wagner, 2001).

R2ML covers four main rule categories: derivation rules, production rules, integrity rules and ECA/reaction rules. It allows structure-preserving markup and does not force translation of the rule expressions into a different language paradigm. R2ML provides a rich syntax for expressing rules supporting conceptual distinctions, e.g. between different types of terms and different types of atoms, which are not present in standard predicate logic.

The main goal of R2ML is to provide a representation of rules, targeted at the rule engine platform that is independent of a vendor specific engine. Another requirement is to give an interchange format for rules is that simplifies translator implementation. In order two satisfy this requirement R2ML introduces rich syntax with different type of atoms and terms.

The description of the R2ML condition language semantics uses partial logic (Herre & Jaspars, Wagner, 1995). It is provided with use of the model theory (a similar approach has been used for defining semantics of RDF (Lassila & Swick, 1999) and OWL) (McGuinness & Harmelen, 2004)). The R2ML condition language includes features like different kinds of logical formulas and properties. Since the Web does not operate under the Closed World Assumption, R2ML distinguishes between predicates, which are totally represented (a total assumption corresponds to a predicate-specific Closed-World Assumption) or are partially represented. In the case of a completely represented predicate, negation-as-failure represents falsity, and negation-as-failure and strong negation collapse into classical negation. In the case of a partially represented predicate, negation-as-failure reflects only non-provability, but does not infer the classical negation.

In R2ML the semantics of production rules is defined based on a high-level transition system formalism. The language supports both production rules and reaction rules, so it defines the concept of an action. Following the OMG Production Rule Representation submission (OMG, 2007), an action is either an InvokeActionExpression or an AssignActionExpression or a CreateActionExpression or a DeleteActionExpression. All actions refer to a context which is an R2ML object term. InvokeActionExpression models an object operation invocation. It refers to a UML Operation and contains an ordered, possible empty list of arguments represented as R2ML terms. The execution of this action is done by the corresponding operation-call. A state changing action is either an UpdateActionExpr or a AssertActionExpr or a RetractActionExpr.

R2ML aims to be a generic solution, that can be used to design rules for number of rule engines. With the translation tools provided by the project it is possible to deploy a R2ML-based rule base in a number of open rule engines, such as JBoss rules, or Jena rules. At the same time it can be used for the modeling of Web Services, and other Semantic Web applications.

## URML

The high-level modeling language used in the communication between domain analysts and experts for modeling the system design should not be a technical encoding. Instead, it should allow visual or natural-language-like rule expressions, which can be understood by domain experts without an extensive technical training. UML offers a semi-visual (diagrammatic) language for information modelling allowing a limited number of embedded rule types. Furthermore, the Object Constraint Language (OCL) can be used for expressing integrity rules and derivation rules within UML class diagrams. However, OCL is purely textual and with no visual rule representation. In order to facilitate rule modeling, the REWERSE Working Group I1 has developed a UML-Based Rule Modeling Language (URML), which allows visual rule modeling based on UML class models.

URML (Lukichev & Wagner, 2005) supports the modeling of derivation rules, production rules and reaction rules. In the language a rule is represented graphically as a circle with a rule identifier. Incoming arrows represent rule conditions or triggering events, outgoing arrows represent rule conclusions or produced actions. Following the description at oxygen.informatik.tu-cottbus.de/rewerse-i1/?q=URML the main URML visual elements are:

- Condition arrow refer to a conditioned model element, which is a classifier such as a class or an association. It may come with a filter expression selecting instances from the extension of the condition classifier and with an explicit object variable (or object variable tuple, in the case of an association) ranging over the resulting instance collection.

- Negated condition arrow is crossed at the origin; it denotes a negated condition which has to be conjoined with one or more positive condition arrows such that its variables are covered by them.
- Derivation rules are represented graphically as a circle with an internal label "DR" and a rule identifier attached to it. Incoming arrows represent conditions, outgoing arrows represent conclusions (see Figure 6).
- Production rules are represented graphically as a circle with an internal label "PR" and a rule identifier attached to it. Incoming arrows represent conditions, outgoing arrows with a double arrowhead represent actions, referring either to a class or to an activity (see Figure 7).
- Reaction rules are represented graphically as a circle with an internal label "RR" and a rule identifier attached to it.

In derivation rules conclusion arrows also refer to a classifier model element. Its meaning is to state that the predicate represented by the conclusion classifier applies to any instance that satisfies all rule conditions. In reaction rules there are two kinds of incoming arrows (condition arrows and event arrows) and two kinds of outgoing arrows (action arrows and postcondition arrows, both having a double arrowhead).

Metamodels of R2ML and URML are largely compatible, so URML concepts of rule, condition and conclusion can be directly mapped into R2ML. The so-called OCL filter expressions, used in URML conditions and postconditions to filter instances of a conditioned classifier (e. g. class or association), can also be directly represented in the R2ML. The transformation of URML model into R2ML is implemented in the Strelka tool.

## Strelka Design Tool

In order to allow visual rule modeling, the REWERSE Working Group I1 has developed Strelka, a tool for making graphical URML models. Strelka (oxygen.informatik.tu-cottbus.de/rewerse-i1/?q=node/10) supports serialization of URML models into the XML format R2ML. There are two versions of the tool available. The first one integrates with the well established Eclipse platform (www.eclipse.org). The tool is also available as a plug-in for the open source UML case tool Fujaba. The Fujaba Tool Suite (wwwcs.uni-paderborn.de/cs/fujaba) provides an easy to extend UML, which is based on the Story Driven Modelling and Graph Transformation platform with the ability to add plug-ins. It combines UML class diagrams and UML behavior diagrams to a powerful, easy to use, yet formal system design and specification language.

Strelka allows to capture rules in the semi-graphical syntax of URML and to serialize the URML rule expressions in the R2ML format. There is also an option to generate specific rule code from the R2ML representation for one of the target platforms such as Jena Rules, JBoss Rules.

## Rule Models Example

Let us now present selected rule model examples. A simple derivation rule:

*A bachelor is a male that is not a husband.*

is visualized in URML as shown in Figure 4.

A production rule

*If car is new, then increase premium by $400.*

is modeled in URML as shown in Figure 5. This model is serialized to the R2ML notation as shown below:

*Figure 4. Derivation rule in URML (see the REWERSE WG I1 website)*

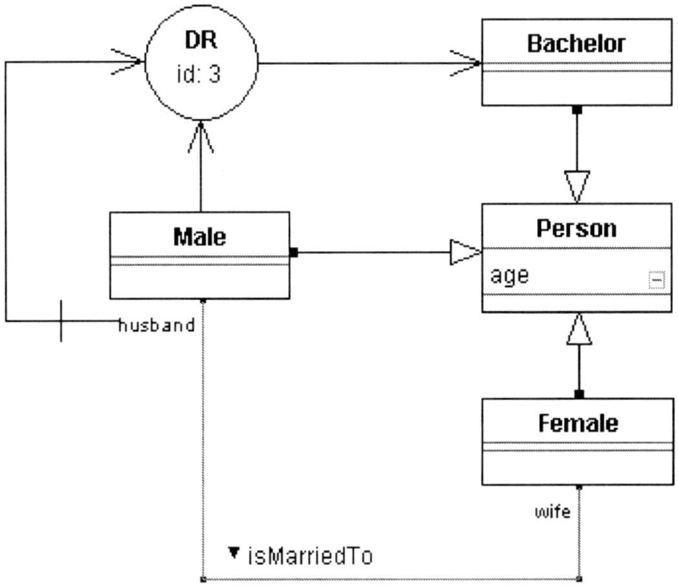


*Figure 5. Production rule in URML (see the REWERSE WG I1 website)*

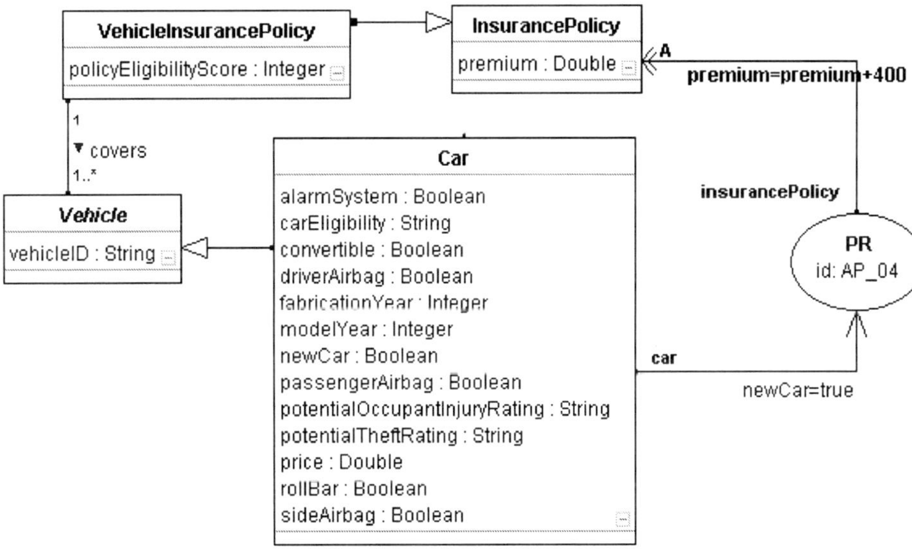


```
<r2ml:ProductionRuleSet r2ml:ruleSetID="
UServProductionRulesSet" xmlns:r2ml="http://
www.rewerse.net/I1/2006/R2ML"
   xmlns:xs="http://www.w3.org/2001/XMLSche-
ma" xmlns:userv="http://www.businessrules-
forum.com/2005/userv#"
   xmlns:op="http://www.xpath2opera-
tions.org/op#" xmlns:swrlb="http://www.
w3.org/2003/11/swrlb">

   <r2ml:ProductionRule
r2ml:ruleID="AP _ 04">
   <r2ml:conditions>
      <r2ml:AttributionAtom
```

```
r2ml:attributeID="userv.Car.newCar">
    <r2ml:subject>
     <r2ml:ObjectVariable  r2ml:name="car"
r2ml:classID="userv:Car"/>
    </r2ml:subject>
    <r2ml:dataValue>
        <r2ml:TypedLiteral
r2ml:datatypeID="xs:boolean"
r2ml:lexicalValue="true"/>
    </r2ml:dataValue>
    </r2ml:AttributionAtom>
    </r2ml:conditions>
    <r2ml:producedAction>
     <r2ml:AssignActionExpression  r2ml:p
ropertyID="userv:VehicleInsurancePolicy.
premium">
    <r2ml:contextArgument>
    <r2ml:ObjectVariable r2ml:name="vehicl
eInsurancePolicy" r2ml:classID="userv:Vehi
cleInsurancePolicy"/>
    </r2ml:contextArgument>
    <r2ml:DatatypeFunctionTerm r2ml:dataty
peFunctionID="op:numeric-add">
    <r2ml:dataArguments>
     <r2ml:AttributeFunctionTerm  r2ml:at
tributeID="userv:VehicleInsurancePolicy.
premium">
    <r2ml:contextArgument>
    <r2ml:ObjectVariable r2ml:name="vehicl
eInsurancePolicy" r2ml:classID="userv:Vehi
cleInsurancePolicy"/>
    </r2ml:contextArgument>
    </r2ml:AttributeFunctionTerm>
        <r2ml:TypedLiteral
r2ml:lexicalValue="400"
r2ml:datatypeID="xs:integer"/>
    </r2ml:dataArguments>
    </r2ml:DatatypeFunctionTerm>
    </r2ml:AssignActionExpression>
    </r2ml:producedAction>
    </r2ml:ProductionRule>

   <r2ml:ProductionRuleSet>
```

The serialization can be done automatically by the Strelka tool. Using the provided translators the JBossRules representation is as follows:

```
    no-loop true
    agenda-group "auto _ premiums"

    when
      $car:Car(newCar    ==    true,
$carPremium:premium)
    then
    $car.setPremium($carPremium.doubleVal-
ue() + 400);
    modify($car);

   end
```

The whole UServ (BRForumBRForum2005) case study can be found at oxygen.informatik.tu-cottbus.de/userv.

## Summary

The approach presented in this section addresses some important rule modeling process requirements identified previously. It offers a visual representation of rules that facilitates the use of rule-based technologies. It shows that even with a relatively small and simple extension of UML metamodel, rule modeling can be implemented. It allows for connecting widely used OMG MOF methodologies with emerging Semantic Web technologies. UML case tools with support for rules may be used for modeling of Semantic Web applications, which include ontologies and rules. The proposed method for visual rules modeling and verbalization introduces the possibility for rule-based software development, which is a powerful paradigm for special classes of software applications.

This approach focuses on modeling different kinds of rules, with a visual representation transparently presenting single rules. This representation is defined in terms of MOF and remains coherent with UML. The rule execution model close to classic rule engines, as well as the translation facilities allow for direct modeling of rules for number of systems.

However, some limitations can also be pointed out. The visual representation provided by the URML while transparent, does not scale up to larger rule sets. Currently, there is no clear support for an on-line verification of the R2ML/URML-based rules.

## HEKATE APPROACH

In this part a short introduction to rule modeling provided by the HeKatE project is given (hekate.ia.agh.edu.pl). A principal idea in this approach is to model, represent, and store the logic behind the software (sometimes referred to as business logic) using advanced knowledge representation methods taken from knowledge engineering. The logic is then encoded with use of a declarative representation. The logic core would be then embedded into a business application, or an embedded control system. The remaining parts of the business or control applications, such as interfaces, or presentation aspects, would be developed with a classic object-oriented or procedural programming languages such as Java or C. The HeKatE project should eventually provide a coherent runtime environment for running the combined Prolog and Java code. The project officially ends in November 2009.

### Objectives

Some of the main goals of HeKatE are to provide an integrated design and implementation process, thus closing the semantic gap in the system design and implementation (Mellor & Balcer, 2002), and automating the implementation, providing an executable solution, that includes an on-line formal analysis of the model, during the design.

HeKatE uses methods and tools in the areas of:

- knowledge representation, for the visual design,
- knowledge translation, for the automated implementation,
- knowledge analysis, for the formal verification.

The approach is based on the classic knowledge engineering concepts.

### HeKatE Rule Language

In the HeKatE project an extended rule language is proposed. It is based on the original XTT (eXtended Tabular Trees) language described in (Nalepa & Ligęza, 2005b). The version used in the project is called XTT^2 (referred here as XTT). The XTT rule language is based on the classic concepts of rule languages for rule-based systems (Liebowitz, 1998, Ligęza, 2006), with certain important extensions and features, such as: strong formal foundation based on attributive logic, explicit rulebase structurization, extended rule semantics.

In XTT there is a strong assumption, that the rule base is explicitly structured. The rules with same sets of attributes are grouped within decision tables. On the rule level explicit inference control is allowed. In this way, a set of tables is interconnected using links, corresponding to inference control. This makes up a decision-tree like structure, with tables in the tree nodes. In a general case, the XTT is a graph, with cycles optionally allowed. Let us now move to attributive logic that provides a formal foundation for the rule language.

Attributive logics (AL) constitute a simple but widely-used tool for knowledge specification and inference. In fact in a large variety of applications in various areas of Artificial Intelligence (AI) (Russell & Norvig, 2003) and Knowledge

Engineering (KE) attributive languages constitute the core knowledge representation formalism. The most typical areas of applications include rule-based systems (Liebowitz, 1998, Ligęza, 2006), expert systems (ones based on rule formalism) (Jackson, 1999, Torsun, 1995) and advanced database and data warehouse systems with knowledge discovery applications (Klösgen & Zytkow, 2002) and contemporary business rules and business intelligence components (e.g. Jess, Drools).

The very basic idea in Set Attributive Logic (SAL) is that attributes can take atomic or set values. After (Ligęza, 2006) it is assumed that the atomic formulae of SAL can have three forms, where the attribute takes single value, all the values of given set (the so-called internal conjunction), and some of the values of a given set (the so-called internal disjunction). ALSV(FD) (Attributive logic with Set Values over Finite Domains) has been introduced with practical applications for rule languages in mind. In fact, the primary aim of the presented language is to extend the notational possibilities and expressive power of the XTT-based tabular rule-based systems. XTT rules use ALSV(FD) for the formulation of the rule condition.

Knowledge representation with XTT incorporates extended attributive table format. Further, similar rules are grouped within separated tables, and the whole system is split into such tables linked by arrows representing the control strategy. Consider a set of *m* rules incorporating the same attributes *A1, A2,...,An*. In such a case the preconditions can be grouped together and form a regular matrix. Together with the conclusion part this can be expressed as a decision table with rules having attributes *A1...An* as table rows.

## Rule Prototyping

The first stage in HeKatE rule modelling process is rule prototyping. The ARD method (Ligęza, 2006, Nalepa & Wojnicki, 2008) aims at capturing relations between attributes in terms of Attributive Logic (Ligęza, 2006). Attributes denote certain system properties. A property is described by one or more attributes. ARD captures functional dependencies among these properties. A simple property is a property described by a single attribute, while a complex property is described by multiple attributes. It is indicated that particular system property depends functionally on other properties. Such dependencies form a directed graph with nodes being properties.

There are two kinds of attributes adapted by ARD: conceptual attributes and physical attributes. A conceptual attribute is an attribute describing some general, abstract aspect of the system to be specified and refined. Conceptual attribute names are capitalized, e.g.: WaterLevel. Conceptual attributes are being finalized during the design process, into possibly multiple physical attributes. A physical attribute is an attribute describing a well-defined, atomic aspect of the system. Names of physical attributes are not capitalized, e.g. theWaterLevelInTank1. Physical attributes cannot be finalized, they are present in the final rules capturing knowledge about the system.

There are two transformations allowed during the ARD design. These are: finalization and split. Finalization transforms a simple property described by a conceptual attribute into a property described by one or more conceptual or physical attributes. It introduces a more specific knowledge about the given property. Split transforms a complex property into a number of properties and defines functional dependencies among them.

During the design process, upon splitting and finalization, the ARD model grows. This growth is expressed by consecutive diagram levels, making the design more and more specific. This constitutes the hierarchical model. Consecutive levels make a hierarchy of more and more detailed diagrams describing the designed system. The implementation of such hierarchical model is provided through storing the lowest available, most detailed diagram level at any time, and additional information needed to recreate all of the higher levels, the so-called Transformation Process History (TPH). TPH forms a tree-like structure then, denoting what particular property is split into or what attributes a particular property attribute is finalized into.

After the conceptual system design a rule prototyping algorithm is used. The input of the algorithm is the most detailed ARD diagram, that has all of the physical attributes identified (in fact, the algorithm can also be applied to higher level diagrams, generating rules for some parts of the system being designed). The output is a set of rule prototypes in a very general format:

rule: condition attributes | decision attributes

The ARD design method, is supported by a software tool called VARDA (ai.ia.agh.edu.pl/wiki/hekate:varda). It is a purely declarative prototype written in Prolog, that supports automated visualization at any diagram level using GraphViz.

## Rule Modeling

The actual rule modeling is carried out with use of the XTT method (Nalepa & Ligęza, 2005b). The main goal of the methodology is to move the design procedure to a logical level, where knowledge specification is based on the use of abstract rule representation. The design can be automatically translated into a low-level Prolog code. On the other hand, selected formal system properties can be automatically analyzed on-line during the design, so that its characteristics are preserved. The generated Prolog code constitutes a prototype implementation of the system. Since it is equivalent to the visual design specification it can be considered an executable specification.

The main idea behind the XTT representation and design method aims at combining selected existing approaches, namely decision trees and decision tables, by building a special hierarchy of Object-Attribute-Tables (Ligęza & Wojnicki & Nalepa, 2001, Ligęza, 2006). It allows for a hierarchical visual representation of the OAT tables linked into tree-like structure, according to the control specification provided. XTT as a design and knowledge representation method offers a transparent, high density knowledge representation as well as a formally defined logical, Prolog-based interpretation, while preserving flexibility with respect to knowledge manipulation.

A prototype CASE tool for the XTT called HQEd has been developed. It supports XTT-based visual design methodology, with an integrated, incremental design and implementation process, providing the possibility of the on-line, incremental, verification of formal properties (see ai.ia.agh.edu.pl/wiki/hekate:hqed).

## Rule Model Example

The HeKatE design process can be easily presented using the following example. It is a reworked Thermostat case, found in (Ligęza, 2006, Nalepa & Ligęza, 2005a). The main problem described there is to create a temperature control system for an office.

The design process is shown in Figure 6. First, it is stated that there is a system to be designed which is described by a single conceptual attribute Thermostat. It is so called level 0 of the design. Refining knowledge about what is the purpose of the system makes a transition to the diagram at level 1; it is a finalization. It is stated that the thermostat controls temperature and this control has something to do with time. That is why Thermostat is finalized into Time and Temperature. Furthermore, at level 2, it is stated that Temperature depends on Time. There are two properties identified in the system and a functional dependency between them. Then at level 3 a general concept of time is refined. It is described by four attributes: Date, Hour, season, operation, through the finalization transformation. Attributes Date and Hour remain conceptual ones, since it is not precisely known at this point what

*Figure 6. Thermostat example design*

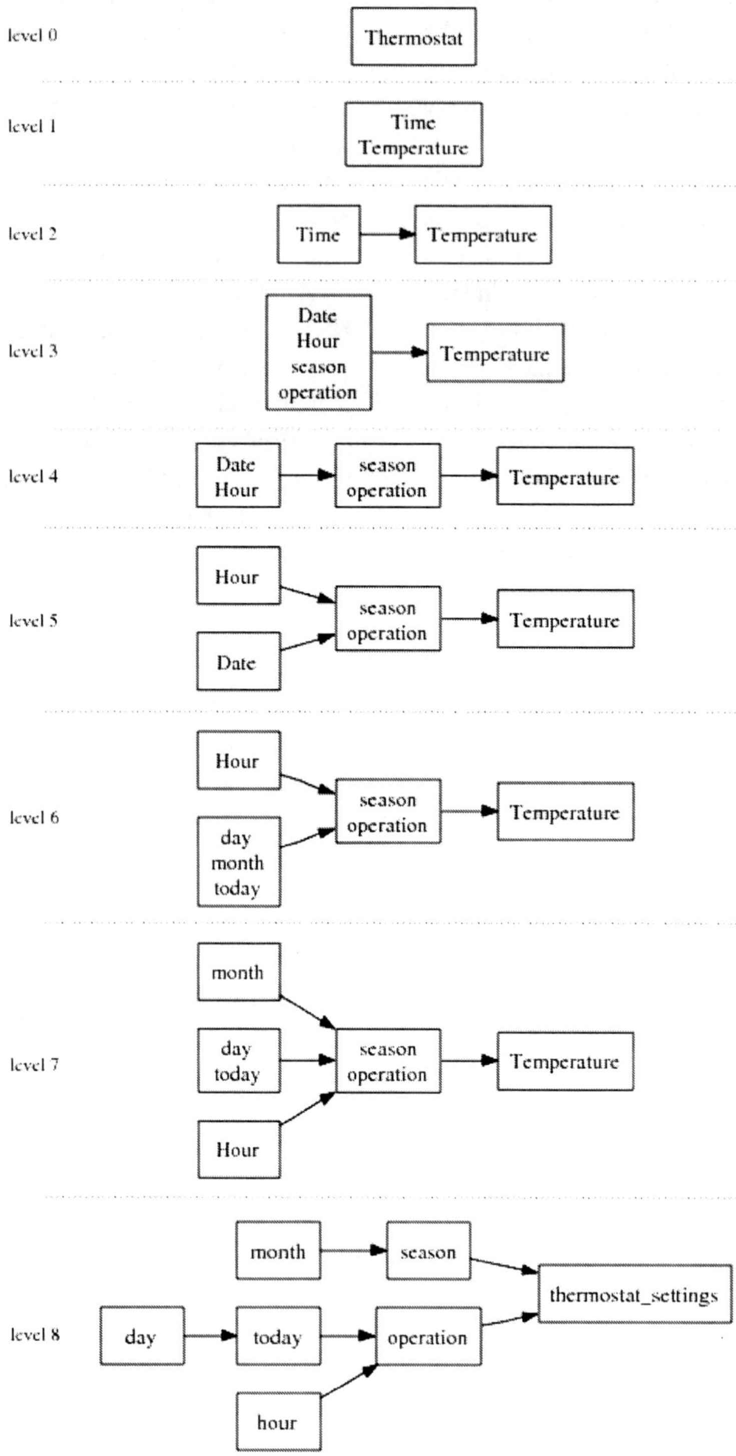

exactly they mean, in terms of their representation and perhaps further processing, whereas season and operation are physical attributes. Properties described by such physical attributes cannot be finalized any more. These attributes denote current season and whether the system operates within business hours or not. The design process ends if all properties of the system are described by single physical attributes. At level 5 a split can be observed; it means that Hour is independent from Date.

A complete Transformation Process History diagram is given in Figure 7. It takes into consideration every transformation. Transformations are expressed by directed edges. Vertices represent property state before and after the transformation according to the indicated direction. Using the ARD model rule prototypes for XTT are generated, as observed in Figure 8.

The actual rule modeling is carried out with the use of the XTT method within the HQed tool, see Figure 9. Basing on the rule prototypes XTT table schemes are provided. The tables are filled with rows corresponding to rules. Tables can be linked to control the inference process. Rule analysis features (e.g. completeness check) are provided on-line, during the design (Ligęza & Nalepa, 2005). Using the visual form, a Prolog prototype can be automatically generated, and run using a custom shell.

## Summary

The main focus in the HeKatE rule design process is to provide a hierarchical rule modeling method based on classic AI tools, such as decision tables and decision trees. The approach offers a compact rule representation with an expressive rule syntax. It also includes possibility of an automatic translation of the design specification into a low-level code, a well-defined system semantics and the possibility of on-line system verification.

Compared to the REWERSE WGI1 approach it uses AI methods instead of the SE ones. It puts emphasis on the rule base design and structure visualization. Thanks to strong formal foundations it allows for a formal rule analysis. However, at the current stage the integration of this approach with UML-based systems poses more problems than in the case of URML.

## MODELING PROCESS FOR RULES

Considering the languages described in this chapter some practical guidelines for an effective modeling process for rules can be given. The process should provide means for:

- building rule vocabularies, identifying concepts, and their semantics,
- determining high-level structure ruleflow, rulebase contexts, allowing for partial rule prototyping,
- building rules capturing the knowledge,
- integrating the ruleset,
- analyzing the quality of the model.

The process should incorporate analysis and verification, as well as means to support practical integration of the rule model. The role of the rule modeling language is crucial in this process.

## FUTURE TRENDS

Expected future developments in the area of rule modeling include:

- extended visual rule design,
- tighter integration of the rule model with the application interface,
- integrated early analysis, including validation, and verification,
- semi-automated rule exchange through XML-based formats, including RIF compliance,

*Figure 7. Thermostat TPH diagram*

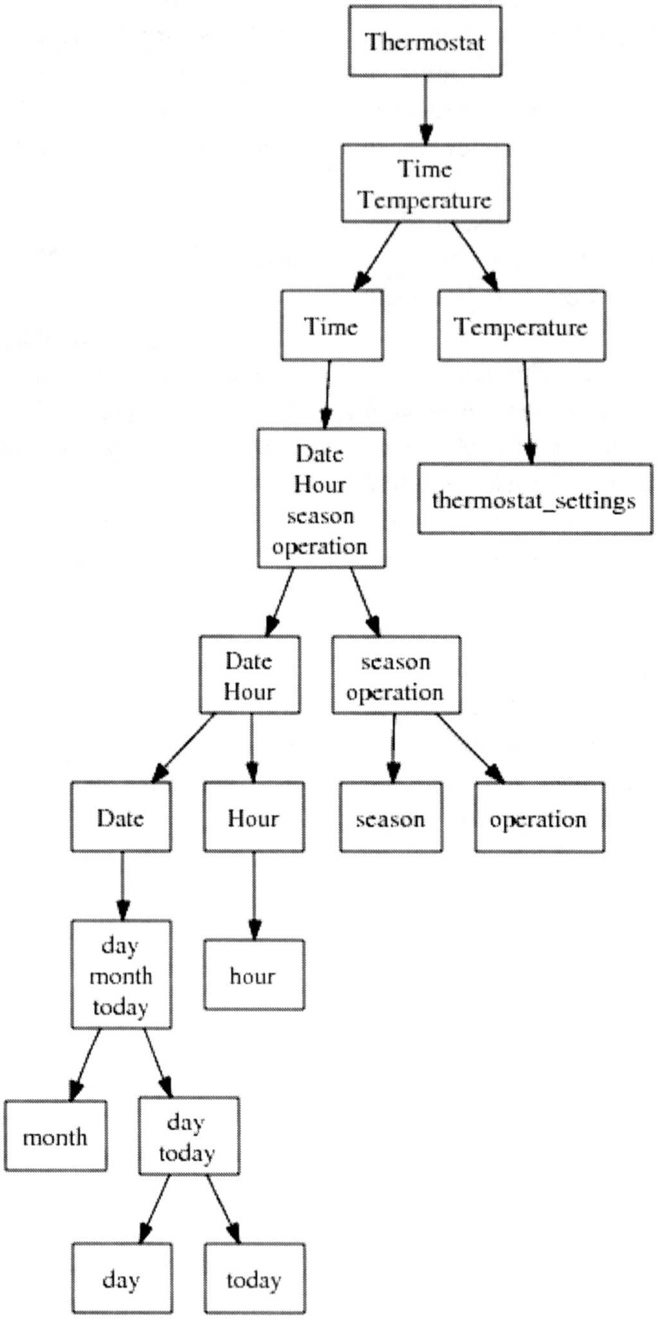


- MOF-based solutions for embedding rules into applications designed with UML,
- web services orchestration with rules.

## CONCLUSION

In this chapter some classic rule modeling methods were discussed. The focus of the chapter was on

*Figure 8. Rule prototypes for XTT*

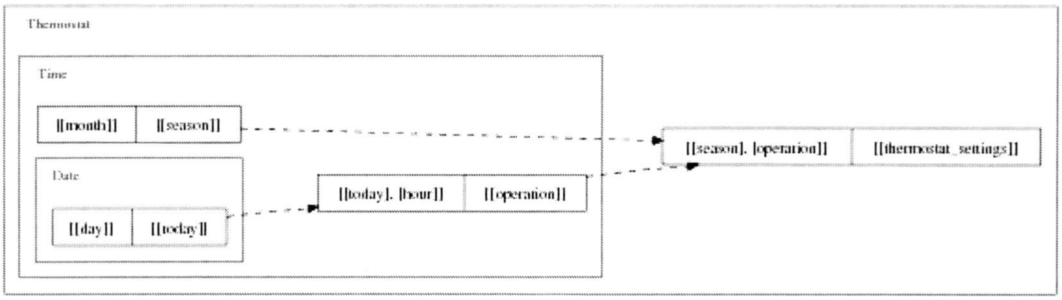


*Figure 9. Structured rulebase in XTT*

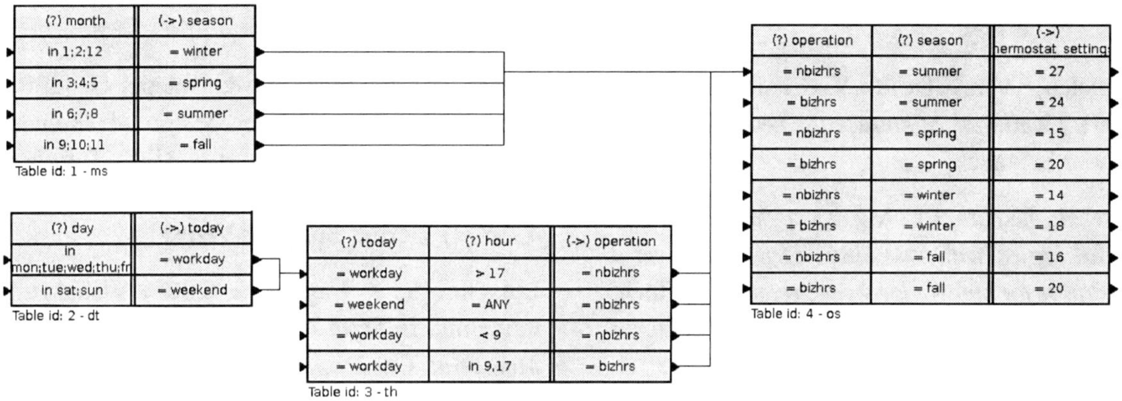


the visual rule modeling methods, especially in the field of business rules design.

Currently three strong and influential communities provide different solutions in the field of rule modeling. These are: the classic AI community, advocating extensions of classic AI methods and tools, such as knowledge representation methods and logic programming, the SE community with the OMG as the representing body, with UML language and the MDA framework, the Semantic Web community with W3C as the representing body, providing tools to build and interchange ontologies, and rules build on top of them. At the moment it is unclear what are the final solutions that emerge. It would be beneficial for the rule community if these three communities agree on common rule standards.

## REFERENCES

Aalst, W. M. P. van der, Hofstede, A. H. M. ter, & Weske, M. (2003). Business process management: A survey. In *Proceedings of the International Conference on Business Process Management (BPM 2003), Eindhoven, The Netherlands* (p. 1-12). Springer.

Ben-Ari, M. (2001). *Mathematical logic for computer science.* London: Springer-Verlag.

Boley, H., Tabet, S., & Wagner, G. (2001). Design rationale of ruleml: A markup language for Semantic Web rules. In *Swws'01.* Stanford.

Brachman, R., & Levesque, H. (2004). *Knowledge representation and reasoning* (1st ed.). Morgan Kaufmann.

Bratko, I. (2000). *Prolog programming for artificial intelligence* (3rd ed.). Addison Wesley.

BRForum. (2005). *Userv product derby case study* (Tech. Rep.). Business Rules Forum.

Covington, M. A., Nute, D., & Vellino, A. (1996). *Prolog programming in depth*. Prentice-Hall.

Giarratano, J. C., & Riley, G. D. (2005). *Expert systems*. Thomson.

Halle, B. von. (2001). *Business rules applied: Building better systems using the business rules approach*. Wiley.

Harmelen, F. van, Lifschitz, V., & Porter, B. (Eds.). (2007). *Handbook of knowledge representation*. Elsevier Science.

Herre, H., Jaspars, J. O. M., & Wagner, G. (1995). *Partial logics with two kinds of negation as a foundation for knowledge-based reasoning*. (Tech. Rep. CS-R9574). Centrum voor Wiskunde en Informatica (CWI).

Horrocks, I., Patel-Schneider, P. F., Boley, H., Tabet, S., Grosof, B., & Dean, M. (2004, May 21). Swrl: A Semantic Web rule language combining owl and ruleml. *W3C member submission*.

Jackson, P. (1999). *Introduction to expert systems* (3rd ed.). Addison Wesley.

Karagiannis, D. (1995, August). Bpms: Business process management systems. *SIGOIS Bulletin, 16*(1), 10-13.

Klösgen, W., & ytkow, J. M. (Eds.). (2002). *Handbook of data mining and knowledge discovery*. New York: Oxford University Press.

Lassila, O., & Swick, R. R. (1999). *Resource description framework (rdf ) model and syntax specication* (Tech. Rep.). World Wide Web Consortium. Retrieved from http://www.w3.org/TR/REC-rdf-syntax

Liebowitz, J. (Ed.). (1998). *The handbook of applied expert systems*. Boca Raton: CRC Press.

Ligęza, A. (2006). *Logical foundations for rule-based systems*. Berlin, Heidelberg: Springer-Verlag.

Ligęza, A., & Nalepa, G. J. (2005). *Visual design and on-line verication of tabular rule-based systems with xtt*. In K. P. Jantke, K.-P. Fähnrich, & W. S. Wittig (Eds.), *Marktplatz internet: Von e-learning bis e-payment: 13. leipziger informatik-tage, lit 2005* (pp. 303312). Bonn: Gesellschaft fur Informatik.

Ligęza, A., Wojnicki, I., & Nalepa, G. (2001). Tab-trees: a case tool for design of extended tabular systems. In H. M. et al. (Ed.), *Database and expert systems applications* (Vol. 2113, pp. 422-431). Berlin: Springer-Verlag.

Lukichev, S., & Wagner, G. (2005). Visual rules modeling. In *Sixth International Andrei Ershov Memorial Conference Perspectives of System Informatics, Novosibirsk, Russia*. Springer.

McGuinness, D. L., & Harmelen, F. van. (2004, February 10). Owl Web ontology language overview. *W3C recommendation*.

Mellor, S. J., & Balcer, M. (2002). *Executable UML: A foundation for model-driven architectures*. Boston: Addison-Wesley Longman Publishing Co., Inc.

Gronau, N, Mueller, C., & Uslar, M. (2004). The kmdl knowledge management approach: Integrating knowledge conversions and business process modeling. In *Proc. 5th international conference on practical aspects of knowledge management, Vienna, Austria* (LNCS 3336, pp. 1-10).

Nalepa, G. J., & Ligęza, A. (2005a). Conceptual modelling and automated implementation of rule-based systems. In T. S. Krzysztof Zieliński (Ed.), *Software engineering: Evolution and emerging technologies* (Vol. 130, pp. 330340). Amsterdam: IOS Press.

Nalepa, G. J., & Ligęza, A. (2005b). A graphical tabular model for rule-based logic programming and verification. *Systems Science, 31*(2), 8995.

Nalepa, G. J., & Wojnicki, I. (2008). *Towards formalization of ARD+ conceptual design and refinement method.* In D. C. Wilson & H. C. Lane (Eds.), *Flairs-21: Proceedings of the twenty-rst international Florida artificial intelligence research society conference: 15-17 May 2008, Coconut Grove, Florida, USA* (pp. 353-358). Menlo Park, California: AAAI Press.

Object Management Group. (2006, May). *Object constraint language version 2.0* (Tech. Rep.). OMG.

OMG. (2006a). *Business process modeling notation (bpmn) specification* (Tech. Rep. No. dtc/06-02-01). Object Management Group.

OMG. (2006b). *Semantics of business vocabulary and business rules (sbvr)* (Tech. Rep. No. dtc/06-03-02). Object Management Group.

OMG. (2006c). *Object Constraint Language, version 2.0.* Object Management Group.

OMG. (2007a). *OMG Unified Modeling Language, Infrastructure, V2.1.2.* Object Management Group.

OMG. (2007b). *Production rule representation beta1* (Tech. Rep.). Object Management Group.

Owen, M., & Raj, J. (2006). *Bpmn and business process management. Introduction to the new business process modeling standard.* (Tech. Rep.). OMG.

Paschke, A. et. al. (2008). *Rif use cases and requirements.* W3C.

Ross, R. G. (2003). *Principles of the business rule approach* (1st ed.). Addison-Wesley Professional.

Russell, S., & Norvig, P. (2003). *Artificial intelligence: A modern approach* (2nd ed.). Prentice-Hall.

Sommerville, I. (2004). *Software engineering* (7th ed.). Pearson Education Limited.

Torsun, I. S. (1995). *Foundations of intelligent knowledge-based systems.* London, San Diego, New York, Boston, Sydney, Tokyo, Toronto: Academic Press.

W3C. (1998). *Extensible markup language (xml) 1.0* (Tech. Rep.). W3C recommendation.

Wagner, G., A.Giurca, & Lukichev, S. (2005). R2ml: A general approach for marking up rules. In F. Bry, F. Fages, M. Marchiori, & H. Ohlbach (Eds.), *Principles and practices of Semantic Web reasoning, Dagstuhl seminar proceedings 05371.*

## KEY TERMS AND DEFINITIONS

**Business Rules:** an approach for applying rule methods in business applications.

**Decision Rule**: classic example of a knowledge representation method, rule corresponds to a conditional statement.

**Decision Table:** represents a set of similar rules grouped with respect to used preconditions and conclusions or actions.

**Decision Tree:** is intuitive way for specifying decision procedures, a simple visual representation (an acyclic, directed graph).

**Knowledge Representation:** a systematic way of encoding what an expert knows about some domain.

**PRR:** Production Rule Representation standard from OMG.

**RIF:** Rule Interchange Format markup from the W3C

**Rule Modeling:** designing rules, possibly in a visual way.

**SBVR:** stands for the Semantics of Business Vocabulary and Business Rules specification from OMG.

**SWRL:** The Semantic Web Rule Language is a proposal for a Semantic Web rules language combining sublanguages of OWL and RuleML.

**URML:** UML-Based Rule Modeling Language from the REWERSE WG I1, it extends the UML metamodel with a notion of a rule and defines a visual rule notation.

**XTT:** a hybrid knowledge representation for rules modeling, from the HeKatE project based on the use of decision tables and trees.

# Chapter XXVI
# Automated Interpretation of Key Performance Indicators by Using Rules

**Bojan Tomic**
*University of Belgrade, Serbia*

## ABSTRACT

*Business reporting is an essential task for every enterprise. In order to make appropriate decisions, decision makers need quality reports. Some recent articles suggest that reports generated by BI (Business Intelligence) systems contain mostly data (key performance indicator values) and little or no information. Data has no meaning and must be interpreted in order to become information. Information is, naturally, much more useful because it directly contributes to recipients' knowledge and can be acted upon. The consequence is that it is left to the decision maker to manually analyze large quantities of data presented in individual reports in order to derive information. A potential solution for automated business data interpretation is presented in this chapter. It proposes using rules to capture and formalize business knowledge and then utilizing these rules to infer information from data automatically.*

## INTRODUCTION

Decision makers need accurate and relevant data and information in order to react to various business situations appropriately. Business goals, strategies and policies reflect current state of business, and the only way to get informed is through reports. Therefore, it can be said that one of the key elements of any business information system is the reporting system.

Reporting systems have evolved through time from paper-based, spreadsheet manual reporting to modern, computerized systems where reports concerning any aspect of business can be generated automatically at the touch of a button. Nowadays, users can specify themselves what kind of report they want and how it should look and then get it in a few seconds. This is a big improvement because the whole process is more efficient and users are not constrained to a small set of predefined reports.

But, what about effectiveness? Are these reports useful? Some recent magazine articles and interviews with professionals in this area reveal that there might be a problem of "too much data and too little information" (Gnatovich, 2005; Whiting, 2002; Wise, 2008). Furthermore, they suggest that in order for a decision maker to acquire information and gain an insight into business performance, he/she must manually sift through large amounts of data, and that customized reports are just not enough. Although these claims have not yet been scientifically proven, they could be seen as an indicator of what is wrong. In order to better explain a view of this problem, it is important to distinguish data from information.

Most scientists agree that *data* becomes *information* when *meaning* is added to it (Floridi, 2005, pp. 353). Also, it must be pointed out that information contributes to the knowledge of the recipient.

Business reports contain key performance indicator values presented in a suitable form. *Key performance indicator* (KPI onward) is "a significant measure used on its own, or in combination with other key performance indicators, to monitor how well a business is achieving its quantifiable objectives" (Georgetown University, 2008). Examples are: profit (by product, by market etc.), expenses, market share etc. KPI values can be presented using a spreadsheet, graph etc. But, no matter how it is presented, KPI consists mostly of data and has little or no meaning. Therefore, KPIs themselves contain very little or no information. This can be demonstrated by using a very simple example.

Let's say that a product manager wishes to see how well some product is doing on the market. The first KPI that needs to be analyzed is profit. The manager must look at the value of overall profit for the current year, the past year, for each individual market and segments within, and then deduct the real status of the product. Each of these profit aspects can be presented by a graph or table, so he/she needs to analyze them individually in order to gain an insight and see whether profit is on the rise, stagnant or declining, whether product sales concerning some market segment are declining etc. This applies to all other KPIs as well: product expenses, product market share, etc. So, it becomes obvious that KPIs, no matter how they are presented, contain mostly data and little or no information. **The user is the one who manually analyzes data, adding meaning to it and transforming it into information.** So, is there another approach to creating information from KPIs that can substitute the current one?

It seems that an ideal solution would be to automate this "data-to-information" transformation process. **It would then be up to the reporting system to create reports that contain information as well as data.** This implies that information would be derived without human involvement, making it accessible anytime, anywhere and at a touch of a button. On the other hand, decision makers would be greatly relieved of the data analysis burden. The key to understanding how this can be achieved is by realizing what is necessary to create information from data.

This chapter summarizes the results of our ongoing research activities in this area. The main objective is to introduce readers with a **potential solution for improving the quality of reports generated by reporting systems by raising the quantity of information they contain.** Individual objectives include:

- Providing a short "state-of-the-art" in the area of reporting systems
- Presenting a new approach for automated transformation of business data (KPI values) into information.
- Presenting appropriate system architecture for integrating the proposed solution into a reporting system.
- Presenting some implementation notes concerning development of a prototype.
- Providing discussion about the proposed solution, it's viability and potential future research

## BACKGROUND

In order to better explain "state-of-the-art" in the area of reporting systems it is important to define what *reporting* is. An informal definition could be that it is a process which involves querying data sources with different logical models to produce a human readable report. Since the goal of every report is to aid decision makers, it is not surprising that reporting systems are usually an integral part of *decision support systems* (DSS onward). According to Alter (1980) a DSS is a computerized system for helping make decisions – it helps people working alone or in a group to gather intelligence, generate alternatives and make choices.

Reporting systems support only the intelligence process. They help decision makers to get informed about the current state of business. But that's not all. Reporting systems do not address every aspect of intelligence e.g. gathering and storing of data. There is, however, a popular term which describes systems that do – *business intelligence* (BI onward). According to Rossetti (2006), BI is "a broad category of applications and technologies for gathering, storing, analyzing, and providing access to data to help enterprise users make better business decisions". Finally, it can be concluded that BI systems are a part of DSS and that reporting systems are an integral part of BI.

So, how do modern reporting systems work and what do they consist of? Since BI and reporting systems are very intertwined and boundaries between the two are not clear, one must analyze BI architecture as a whole in order to get a clear perspective (Figure 1).

Before any reports can be generated or analysis performed, BI needs to acquire business data. Usually, this is not easy as business information systems tend to be very heterogeneous, and there can be many different sources of data. Data sources include databases, files or other applications. Methods for accessing the source are usually SQL queries, file access procedures, or direct calls to applications. The whole process is called *extract-transform-load* (ETL onward) and can be summarized as "extracting data from their operational data sources or external data sources, transforming the data which includes cleansing, aggregation, summarization, integration, as well as basic transformation, and loading the data into some form of data warehouse" (California State University, 2007).

*Data warehouse* (DW onward) is "a copy of transaction data specifically structured for querying and reporting" (Greenfield, 2008). DW can be implemented in many ways: as a relational database, hierarchical database, object database etc. The motivation behind copying *transactional data* (data generated by operational processes) is that performance of transactional databases can be degraded if they are simultaneously used for reporting. This is supported by the fact that data models in such databases are normalized which can dramatically slow down complex queries and analytical calculations. DW data is different from transactional data in that it is: *subject-oriented, time-variant, non-volatile* and *integrated* (Inmon, 1995).

Although integrated, DW data is still relatively atomic. Reports and analysis require complex, derived and summarized data as well as atomic data. In addition, there is a need for customized querying which is not supported by DW. Performing complex calculations every time a report is needed would have a very negative effect on system performance. The solution is to summarize data in advance, divide it into multiple dimensions and introduce indexing. A generic term for techniques, methods and applications that enable this is *on-line analytical processing* (OLAP onward). It is defined as "ability to consolidate, view, and analyze data according to multiple dimensions, in ways that make sense to one or more specific enterprise analysts at any given point in time" (Codd, Codd & Salley 1993).

OLAP acquires DW data, summarizes it and divides it into dimensions. For example, there can

Figure 1. BI architecture

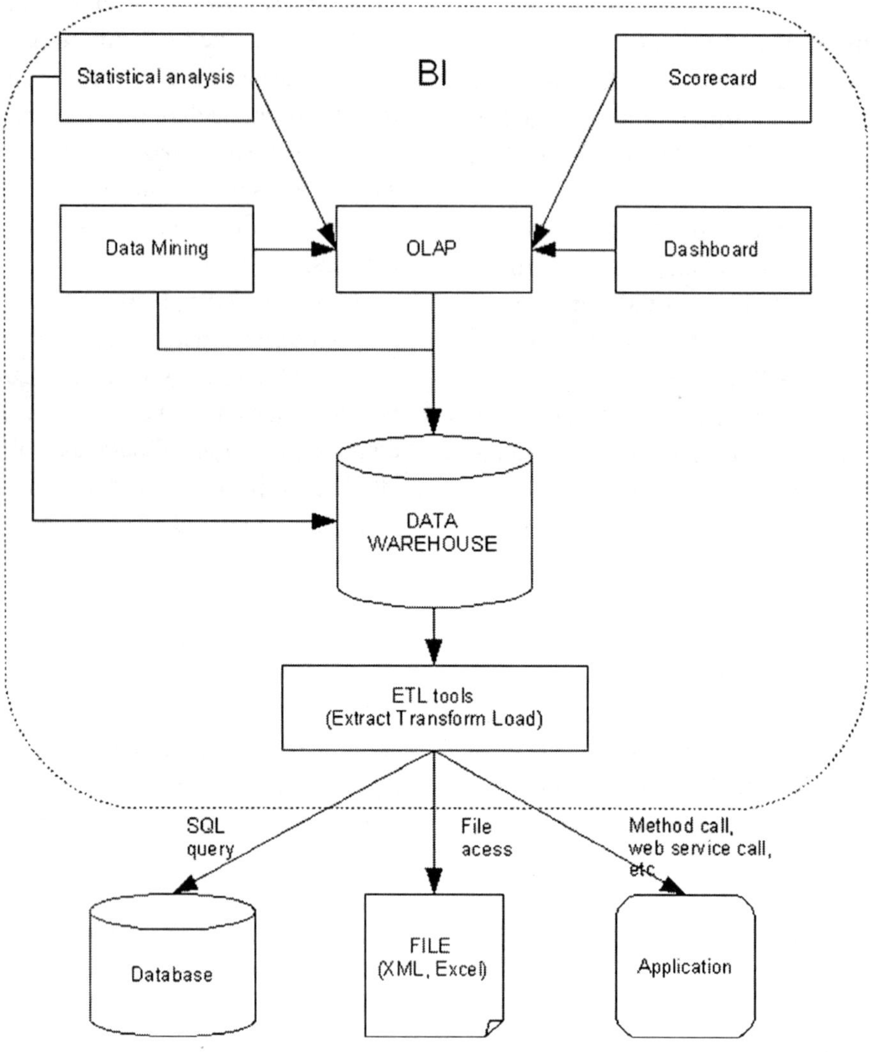


be a time dimension, product dimension, market dimension etc. This implies that a special data model has to be used, the so-called *star schema*. When data is organized in this way, it is very easy to get a customized report containing, for example, "sales data concerning product X for the third quarter of the current year. The result of populating a star schema with actual data is an *OLAP cube* (Data cube). If it has more than three dimensions it can be called a *hyper cube*.

In order to get a report, the user must perform OLAP queries. *Multidimensional expressions* (MDX onward) is a query language used for that purpose (Microsoft , 2007) and is a part of the *XMLA standard* (XML for Analytics) for data acquisition, query and manipulation in analytical systems (XMLA council, 2007). It is similar to SQL and utilizes a SQL like syntax, but queries can return multidimensional results (e.g. data cubes) and not just relational tables.

Complex data analysis can be performed using statistical methods: correlation analysis, methods for proving normal (Gaussian) distribution, basic statistical indicators etc. Modern BI systems can contain implementation of these methods in the form of statistical modules. Input data can be acquired from DW or OLAP. The only prerequisite is that the user (in this case business analyst) must have some presumption concerning relations between business data which has not yet been proven.

Statistical methods, however, cannot be applied for all complex analysis. This refers to situations where analysts cannot make presumptions about business data relations because: data is too broad or complex, data concerns some new aspect of business etc. The problem is that there can be abundance of implicit relations that will never be discovered simply because they are unexpected. In order to remedy this problem, BI often has *data mining* (DM onward) capabilities. DM can be defined as "nontrivial extraction of implicit, previously unknown, and potentially useful information from data" (Frawley, Piatetsky-Shapiro & Matheus 1992, pp. 214).

The most important thing to acknowledge is that DM works automatically, so every implicit relation, no matter how unexpected or well hidden, can be discovered. Output generated by DM can be presented in the form of rules, decision trees, spreadsheets, graphs etc. For example, a DM generated rule can state that "in 90% of situations, when buying summer shirts customers buy sunscreen lotion as well".

OLAP queries can result in generation of various reports. The problem with this approach is that users often have to view and interpret a lot of less relevant graphs and tables in order to get informed. Additionally, these reports are usually displayed on different screens, so they cannot be viewed at once. *Dashboards* provide a solution. According to Few (2004), a dashboard is "a visual display of the most important information needed to achieve one or more objectives; consolidated and arranged on a single screen so the information can be monitored at a glance".

Besides arranging all important graphs and tables on a single screen, dashboards usually allow boundary values to be set for each KPI. Gauges and dials are used to visualize data, and they are often divided into green, yellow and red sections (zones). Section limits are defined by boundary values, so users can immediately spot which KPIs are out of bounds (because indicators on their displays move into the red section).

Reports can also provide strategic data. By monitoring this data, decision makers can see whether business strategies and goals are being achieved or not. The solution is to define strategies and goals in the form of quantifiable business objectives. These can then be transformed into a set of goal KPI values, which can be monitored automatically. This solution is called the *balanced scorecard*. The Balanced Scorecard Institute (2007) defines balanced scorecards as "strategic planning and management systems that are used extensively in business and industry, government, and nonprofit organizations worldwide to align business activities to the vision and strategy of the organization, improve internal and external communications, and monitor organization performance against strategic goals". In BI terminology, these are just called *scorecards*.

BI users can be divided into two groups: *decision makers* and *business analysts*. Each group has its own goals, needs and expectations regarding BI and, therefore, utilizes different parts of the system (Figure 2).

Business analysts are oriented toward finding new, previously unknown relations hidden in business data. They might want to find, for example, what are the characteristics of expensive toothpaste buyers. OLAP queries result in reports presented in spreadsheet or graphical form. Business analyst can use these to get the "feel" of data and possible relations. If they notice something valuable, they can prove or disprove it by using statistical methods. Reports generated by these methods are also in spreadsheet or graphical form. Finally, business analysts want to find hidden relations as well, so they employ DM to do the job automatically for them.

*Figure 2. BI users*

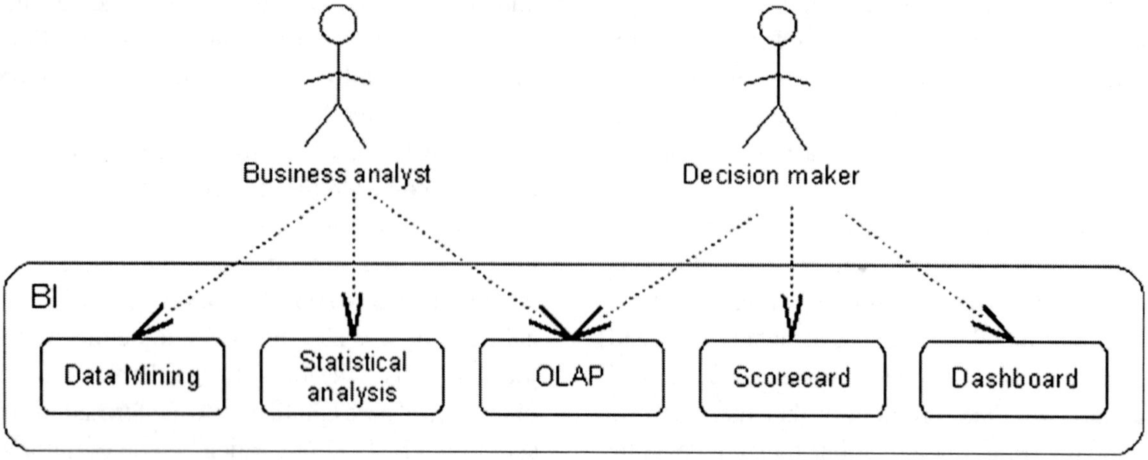


Decision makers, on the other hand, just want to confirm expected relations in business data: whether profit is rising, stagnant or declining, if the costs are within predefined boundaries etc. They monitor KPI values in order to get informed about the current state of business, and use this knowledge as feedback concerning previously made decisions. OLAP queries decision makers perform and reports they get contain all necessary data needed to find current trends and their causes. Dashboards take reporting a step further because data is presented via gauges and dials, so boundary values can be set. When a decision maker looks at a dashboard, he/she can immediately notice which KPIs have abnormal values because, for example, appropriate gauge indicators are in the red zone. Scorecards provide an efficient way of merging strategic objectives with KPI values. Therefore, scorecards can help decision makers determine whether business is reaching its objectives without having to look at numerous individual OLAP reports.

Currently, BI systems are moving toward integration with transactional systems. The idea is to promote reporting down to the operational level, thus introducing a third group of BI users – *line of business users* (LOB). LOBs include low level executives and employees in charge of specific, ongoing operational processes. This means that reporting could be available to everyone and not just decision makers. BI systems that support this feature are called *business activity monitoring systems* (BAM onward). BAM is "a system that provides real-time access to critical business performance indicators in order to improve the speed and effectiveness of business operations" (McCoy, 2002, pp. 2).

When describing future trends in BI systems, two popular terms emerge: *BI 2.0* (Nicholls, 2006) and *real-time BI* (White, 2004). Both of them refer to BI systems where traditional BI is merged with BAM in order to shorten the *action distance* - "the distance (latency, space, or gap) between the set of information generated by the BI system and the set of actions appropriate to a specific business situation" (Hackathorn, 2004). According to Nicholls (2006) BI 2.0 systems should be:

- *Event driven* – this means analyzing data, event by event, either in parallel with the business process or as an implicit process step.
- *Real time* – in order to react swiftly, BI 2.0 must work in real time.

- *With automated analysis* – to be able to automatically interpret data, dynamically, in real time.
- *Forward looking* – understanding the impact of any given event on an organization needs.
- *Process oriented* – to be embedded within a process in order to make the process inherently smarter.
- *Scalable* – this is critical because event streams can be unpredictable and occur in very high volumes.

On the other hand, some recent magazine articles and interviews with professionals in this area reveal that there might be a problem of "too much data and too little information". Whiting (2002) made this point by acknowledging that a high percentage of data is there and that the challenge to overcome is extracting it and presenting it in a way that's useful for managers to make decisions. Gnatovich (2005) stated that "one would think this avalanche of data would be delivering huge benefits, but executives report that they are no closer to making good use of this mountain of information". Similarly, Wise (2008) concluded that "BI is so data heavy that it's often very difficult in terms of being able to get the information that people want". As stated, these claims are not yet scientifically proven, but can be used as an indicator of what is wrong.

## PROBLEM IDENTIFICATION AND ANALYSIS

Business analysts want to find new, previously unknown relations hidden in business data. Besides business knowledge, this task requires a great deal of experience and intuition. Additionally, a lot of technical knowledge is needed because methods business analysts utilize tend to be complex to use. However, all these demands can be partially justified by the fact that finding new relations is like a "journey into the unknown".

Decision makers, on the other hand, use BI just to get informed about the current state of business. Simply put, their goal is to determine whether business is going good or bad and why. Unfortunately, it seems that even this relatively structured and repetitive task cannot be accomplished easily. Abundance of data and lack of information contained in BI reports might be the problem. This can be demonstrated with a simple example.

Let's say that a sales manager wants to get informed about profit concerning four products under his supervision. Initially, he gets a predefined report containing one table and four graphs (Figure 3). The table contains profit values for the current month, and the same month of the previous year. Monthly profit values for the current year are presented in graphical form.

The manager reviews this report and, after some thought, notices that profit values concerning product A are much lower than they were last year (80$ versus 150$). By looking at the appropriate graph, he realizes that product A profits are constantly declining and that this trend started in March and continued in April. He, therefore, concludes that product A is not doing so good. In order to confirm his suspicions, the sales manager makes an OLAP query. As result, he gets a graph containing monthly product A profit values for the current and last year (Figure 4).

After reviewing this time series, the manager can confirm that product A is not doing so well since profit is constantly declining month after month. The next step is to determine what the root cause of this negative trend is. After making another OLAP query, the manager gets a graph containing profit values for product A, but grouped by stores they were sold in (Figure 5).

This final report points to the cause of the problem. The manager realizes that most of the product A profit decline is caused by low profits made in store 1, especially when compared to

*Figure 3. BI report example*

| | Profit | |
|---|---|---|
| Product | April 2008.($) | April 2007.($) |
| Product A | 80 | 150 |
| Product B | 24 | 10 |
| Product C | 330 | 500 |
| Product D | 25 | 26 |

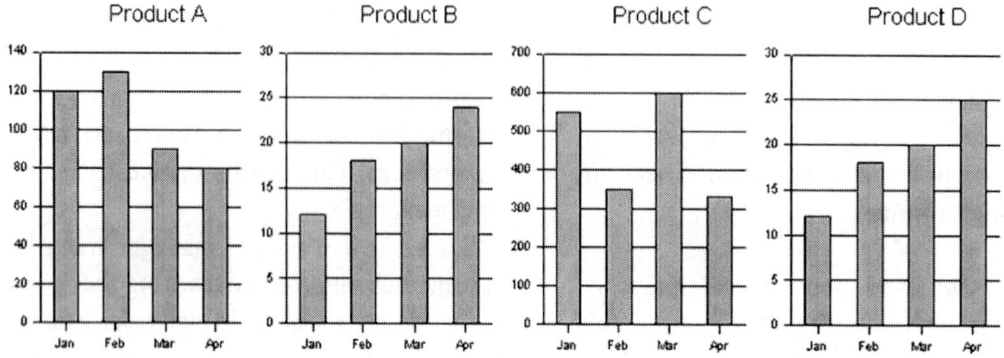


*Figure 4. BI report example 2*

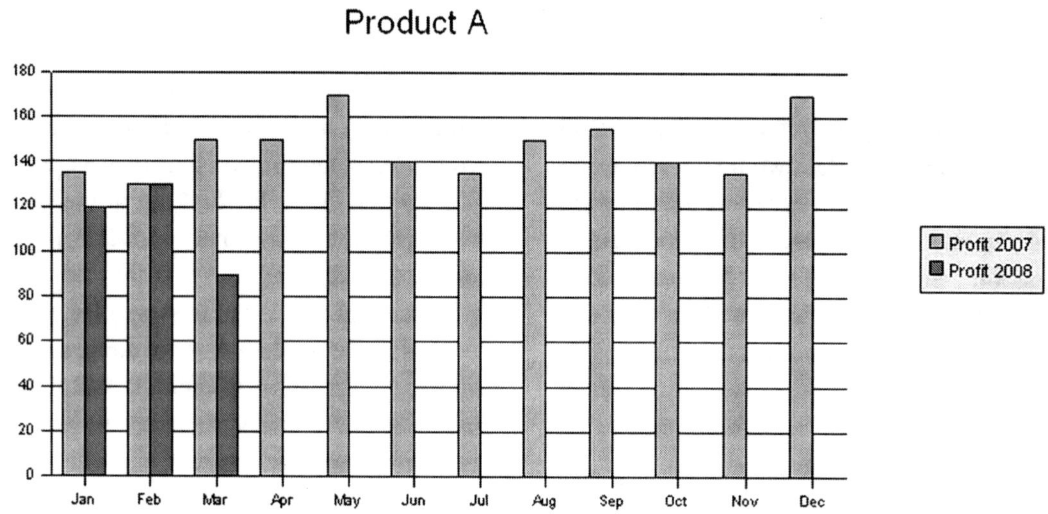


March and April of the previous year. He can also see that stores 2 and 3 are doing relatively well.

In order to see how other products are doing, the sales manager must repeat this procedure three more times. This is a simple example, but it demonstrates that it is not easy to acquire information from customized reports although all necessary data is present.

Dashboards and scorecards somewhat improve information acquiring because certain unexpected, abnormal or out of bound values can be automatically detected. If profits were represented by one or more scorecards, the manager would be alarmed every time profit values fail to reach goal values. On the other hand, if dashboards were used, low profit would be indicated by a gauge with its index in the red zone and therefore

*Figure 5. BI report example 3*

be easily detected. Confirming the problem and finding root cause, however, would still rely on manual OLAP querying and report reviewing. Therefore, it can be concluded the problem is that **decision makers need to manually analyze a lot of individual reports in order to get informed about the current state of business**.

The consequences are numerous. First of all, decision makers that don't have appropriate technical skills will never utilize the full potential of BI. Dashboards and scorecards are relatively straightforward to use, but making OLAP queries and report formatting can be difficult. Unfortunately, even technically qualified users can have problems. If, for example, the number of products or organizational units that need to be monitored via KPIs is great, getting informed would mean sifting through a very large number of reports. But, perhaps the biggest issue is that lack of concentration, business knowledge or experience on behalf of the decision maker can lead to inadequate report analysis and thus wrong decisions. For example, if the user were tired while reviewing reports, he might fail to notice some important trend.

So, what seems to be the cause? Why do decision makers need to do manual report analysis? One possible explanation is that reports contain mostly data and little or no information. The key to understanding this issue is differentiating between data and information. *Data* can informally be defined as one or more values represented in numeric, symbolic or any other form. Some popular scientific definitions of information are:

- "Information is data that has been processed into a form that is meaningful to the recipient." (Davis & Olson, 1985, pp. 200)
- "Data is the raw material that is processed and refined to generate information." (Silver & Silver, 1989, pp. 6)
- "Information equals data plus meaning." (Checkland & Scholes, 1990, pp. 303)

Although there is no consensus on a single, unified definition of information, it is generally accepted that "information = data + meaning" (Floridi, 2005, pp. 353).

Examples of data are: blue, 32$, 17C etc. Data has little or no meaning by itself. In order to gain meaning, multiple pieces of data must be contextually related and analyzed. For example, if one was to get informed about the weather, knowing that the temperature is 17C does not mean that much. But, when this data is contextually related with the fact that it is December and that the location is Moscow, there is sufficient ground for the user to infer that the weather is uncommonly warm for this location and time of year. This finding constitutes information, as it contributes to the knowledge of the recipient. Furthermore, this information is "actionable" without the need for any additional analysis i.e. when the weather is so unpredictable it is wise to wear multiple layers of clothes. Information is, therefore, much more useful than data.

KPI values are data. Since BI reports contain these values presented in the form of graphs or tables, it becomes obvious that they too contain only data. The difference is that, in reports, data gets grouped and contextually related. Graphs and tables are not, by themselves, in any way actionable because key KPI values and trends are not indicated and have yet to be discovered through reviewing and analysis. Dashboards and scorecards can contain some information concerning reaching KPI goal or boundary values, but this information is very limited. In order to make it actionable, the user must find root causes by analyzing further reports.

Finally, it can be deducted that the cause of the problem is that **the decision maker is the one who (manually) analyzes data in BI reports, adding meaning to it and transforming it into information.**

## PROPOSED SOLUTION

There is a course of action that can be taken in order to remedy this problem. The ideal solution would be to automate this "data-to-information" transformation process. It would **then be up to the BI system to create reports that contain information as well as data**. The key to understanding how this can be achieved is by realizing what is necessary to create information from data.

The process of creating information from data can be divided into three steps:

- Data acquisition
- Making contextual data relations
- Creating information

Data acquisition starts when the decision maker wishes to get informed. He/she starts the appropriate application and receives some initial, predefined reports. These usually contain only basic, individual pieces of data.

At this point, the decision maker starts to analyze data. Basic reports are usually not enough, so he/she makes additional queries and, by looking at various graphs, tables etc. puts different pieces of individual data into context with one another, for example: this year profit is 100$ and last year it was 80$. Although contextually related, this remains just data, because it has very little meaning.

The final step is to create information. In order to do this, the decision maker must interpret data adding meaning to it. This is done by using appropriate interpretation (business) knowledge: "when last year's profit is greater than this year's profit, business is going good". Finally, the information he/she creates looks like this: "Since last year's profit was 80$ and this year's profit is 100$, our business is going good". This is information since it is actionable and contributes to the users' knowledge – the decision manager is informed and can take appropriate actions. The whole process can be represented as an equation:

DATA + INTERPRETATION KNOWLEDGE = INFORMATION

Since it is agreed that:

DATA + MEANING = INFORMATION

It can be concluded that:

INTERPRETATION KNOWLEDGE = MEANING

So, in order to automate the "data-to-information" transformation process, interpretation knowledge must be formalized and implemented in such a way that it can be used by a machine.

The first question that needs to be answered is: what knowledge representation techniques can be used for this purpose? It seems that rules might be adequate. For example, the following interpretation knowledge can be formalized like this:

"When last year's profit is greater than this year's profit, business is going good."

IF
Current year company profit > Last year company profit
THEN
Business is going good

The number of these rules can be great, so rule facts cannot be presented just by using variables and relations. There is a need for a more structured approach and objects provide a good solution. Previously stated interpretation knowledge can be presented by using rules and objects in combination:

IF
Company.getProfit (CURRENT_YEAR) > Company.getProfit (LAST_YEAR)
THEN
Company.setProfitStatus (GOOD_STATUS);

When relaying business information, one can use fuzzy terms in order to describe his/her opinion. The difference between "good" and "excellent" profit is one example of such ambiguity. Therefore, it appears necessary to include fuzzy sets and operations in addition to objects and rules when formalizing interpretation knowledge.

There is, however, one point to be made: these KPI interpretation rules are not business rules. Business *rule* is "a statement that defines or constrains some aspect of business; it is intended to assert business structure or to control or influence the behavior of the business" (Business Rules Group, 2000, pp. 4).

Here are two examples of business rules: "regular customers get 5% discount on all purchases" and "any customer that has had more than 20 purchases in the last year can be considered a regular customer". The first rule influences sale processes, whereas the other defines one aspect of business – regular customers. Business rules can, of course, be formalized as if-then statements. KPI interpretation rules, on the other hand, neither define, control nor influence behavior of business. They represent business knowledge, not business constraints.

KPI interpretation rules represent expert knowledge in the business domain, so it can be said that the system containing these rules is an *expert system* (Durkin, 1994, pp. 7) and not just a *rule-based system*.

So, how should the conclusions (information) be derived exactly? Since it is agreed that knowledge should be presented by using rules, only two techniques qualify for the job: forward chaining and backward chaining. Prolog is a well known programming language that uses the backward chaining inferencing technique to prove or disprove goal facts (Fisher, 2008). Forward chaining is, on the other hand, commonly used in modern expert system development environments like Jess to derive as much information as possible from available data (Sandia National Laboratories, 2008). It seems that forward chaining is the technique of choice for automated KPI interpretation because decision makers already have abundance of data at their disposal and need all the information that can be derived from that data.

The final question is: what techniques can be used to represent automatically derived information? If a decision maker was to get, for example, an information like "Company.status = GOOD_STATUS" he/she would not know what it actually means.

The solution appears in the form of *explanation facilities*. They are considered to be a very important part of any expert system because by "using this facility, an expert system can provide explanation to the user about *why* is it asking a question and *how* it reached some conclusion" (Durkin, 1994, pp. 30). Automated KPI interpretation doesn't require both types of explanations as communication with the user is not in the form of questions and answers. But, explaining how some conclusion (information) was derived, would be of outmost importance.

Explanation facilities often utilize *"canned text"* - predefined, natural-language-like sentences. A set of these sentences is linked to each important conclusion and added to the final explanation every time that conclusion is made. Explanation is, therefore, formed gradually and varies according to derived information, but most importantly: it is in a natural-language-like form and can simply be read without the need for any further interpretation. Information presented in this way can immediately increase users' knowledge.

To summarize, the **proposed solution for automated KPI interpretation consists of the following:**

- Using rules, objects and fuzzy sets and operations to formalize KPI interpretation knowledge.
- Using forward chaining to inference information from business data by correlating it with formalized KPI interpretation knowledge.
- Using an explanation facility to present this derived information in a natural-language-like format so it can be embedded in a report and easily understood.

The goal is to **improve the quality of reports generated by BI systems by raising the quantity of information they contain.**

So, how would reporting look like if this solution was implemented? Let's say that the sales manager wishes to analyze profit concerning some product. If profit interpretation knowledge were formalized, it would look something like this:

Rule group - "Overall yearly profit interpretation"

Rule 1

IF Product.getProfit(2008) > Product.getProfit(2007)

THEN Product.setProfitStatus (GOOD);

Product.addProfitExplanation(GOOD);

Rule 2

IF Product.getProfit(2008) < Product.getProfit(2007)

THEN Product.setProfitStatus (BAD);

Product.addProfitExplanation(BAD);

Rule 3

IF Product.getProfit(2008) == Product.getProfit(2007)

THEN Product.setProfitStatus (ACCEPTABLE);

Product.addProfitExplanation(ACCEPTABLE);

Rule group - "Monthly profit interpretation"

Rule 4

IF Product.getProfitStatus(2008)==GOOD AND

Product.allMonthlyProfitsLargerThanLastYear(2008)

THEN Product.setMonthlyProfitStatus (EXCELLENT); Product.addMonthlyProfitExplanation(EXCELLENT);

Rule 5

IF Product.getProfitStatus(2008)==GOOD AND

Product.MostMonthlyProfitsLargerThanLastYear(2008)

THEN Product.setMonthlyProfitStatus (GOOD);

```
Product.addMonthlyProfitExplanation(GOOD);

Rule 6
IF Product.getProfitStatus(2008)==GOOD
AND
NOT (Product.MostMonthlyProfitsLargerThanLastYear(2008))

THEN Product.setMonthlyProfitStatus
(RISES_IN_LEAPS);
     Product.addMonthlyProfitExplanation(
RISES_IN_LEAPS);
```

Before going on any further, some explanation is needed. The previous example displays profit interpretation knowledge formalized by using rules and objects. But where do objects like "Product" come from and how do they relate to business data? It is already stated that DW and OLAP contain "raw" or semi-summarized business data. Since they are usually implemented by using relational databases, this data is in the relational form (relational tables and rows). This is fine for storage and query purposes, but data manipulation requires data to be presented in some other form, usually by using objects. In order for relational data to be extracted and presented as objects, one can employ various relational-object mapping frameworks (see the following section). Therefore, objects listed in the previous example refer to business data extracted from DW or OLAP and presented by using objects.

When the sales manager wishes to get informed, he/she just presses a few buttons and waits for a report. In the meantime, the system performs the following steps:

- Acquires product profit values.
- Deducts information using profit values and profit interpretation knowledge.
- Forms an explanation regarding derived information and creates a report.

Before any analysis can be performed the system needs to acquire relevant data. In this case, it is product A profit data (Table 1).

Then, the system matches this data with appropriate interpretation rules. First it checks whether total profit is greater, equal or lesser than last year's. Since it is greater, rule 1 gets executed (fired) thus creating information about the overall profit status – profit is good. An explanation describing this finding is simultaneously added to the final report. Monthly profits are interpreted in the next inference cycle. The system finds that most of this year's monthly profits are larger than last year's – this is the case for 7 months out of 12. Previous and current findings induce rule 5 to be fired thus leading to another conclusion and appropriate explanation: monthly profit is good. Finally, the sales manager gets a report like this (Figure 6).

The system deducted that overall profit is on the rise (which is good) and that monthly profit data can confirm that it is a steady rise (which is also good). This report contains information as well as data. Information is presented using natural-language like sentences, so it is easy to read. Furthermore, no additional queries nor profit analysis are needed as the user is fully informed, and can take appropriate actions.

This research was inspired by an article written by Liebowitz (2006) where he stated that BI systems are not intelligent and that they should incorporate AI techniques in order to make them more efficient. The actual solution was derived as analogy to *knowledge-driven data mining* (Graco, Semenova & Dubossarsky, 2007). The difference is that knowledge-driven data mining approach proposes using knowledge to enhance the data mining process rather than the reporting process. The analogy is that both use knowledge to automate creating information from data.

*Table 1. Profit data example*

| Product A profit (mil. $) | Jan | Feb | Mar | Apr | May | Jun | Jul | Aug | Sep | Oct | Nov | Dec | Tot. |
|---|---|---|---|---|---|---|---|---|---|---|---|---|---|
| 2008 | 10 | 12 | 15 | 13 | 8 | 2 | 1 | 2 | 4 | 10 | 15 | 20 | 112 |
| 2007 | 9 | 10 | 14 | 15 | 7 | 2 | 1 | 2 | 3 | 11 | 14 | 17 | 105 |

*Figure 6. Automatically generated report containing information as well as data*

| | PROFIT | |
|---|---|---|
| Product A | 2008 | 2007 |
| Total | 112 | 105 |
| Average | 9,33 | 8,75 |

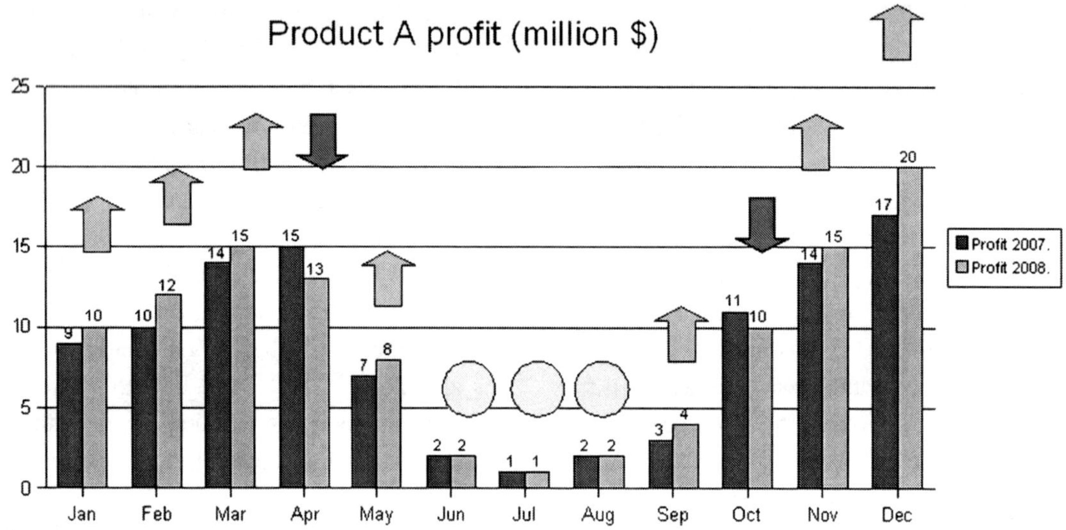


Conclusion:
Product A exibits a stable rise in profit.

Explanation:
Product A overall profit is on the rise when compared to last year. Last year it was 105 million and this year it is 112 million $. This is a 6.7% increase.

Product A monthly profit shows a stable rising trend. Most of this year's monthly profits are larger than last year's. This is the case for 7 months out of 12 (Jan,Feb,Mar,May,Sep,Nov,Dec).

## PROPOSED ARCHITECTURE AND IMPLEMENTATION NOTES

Besides reporting, BI systems enable gathering, transformation and storing of data. They also provide tools and methods for deep data analysis used by business analysts. The proposed solution aims at improving the reporting process alone. It is, therefore, clear that it should be integrated into existing BI systems, and not used as a separate application. Modified BI architecture is shown in the following figure (Figure 7).

The majority of BI architecture elements remain unchanged. ETL tools and DW perform the same function because data still needs to be extracted, transformed, loaded and stored before it can be used. Statistical methods, data mining and OLAP cannot be excluded because business

*Figure 7. Modified BI architecture*

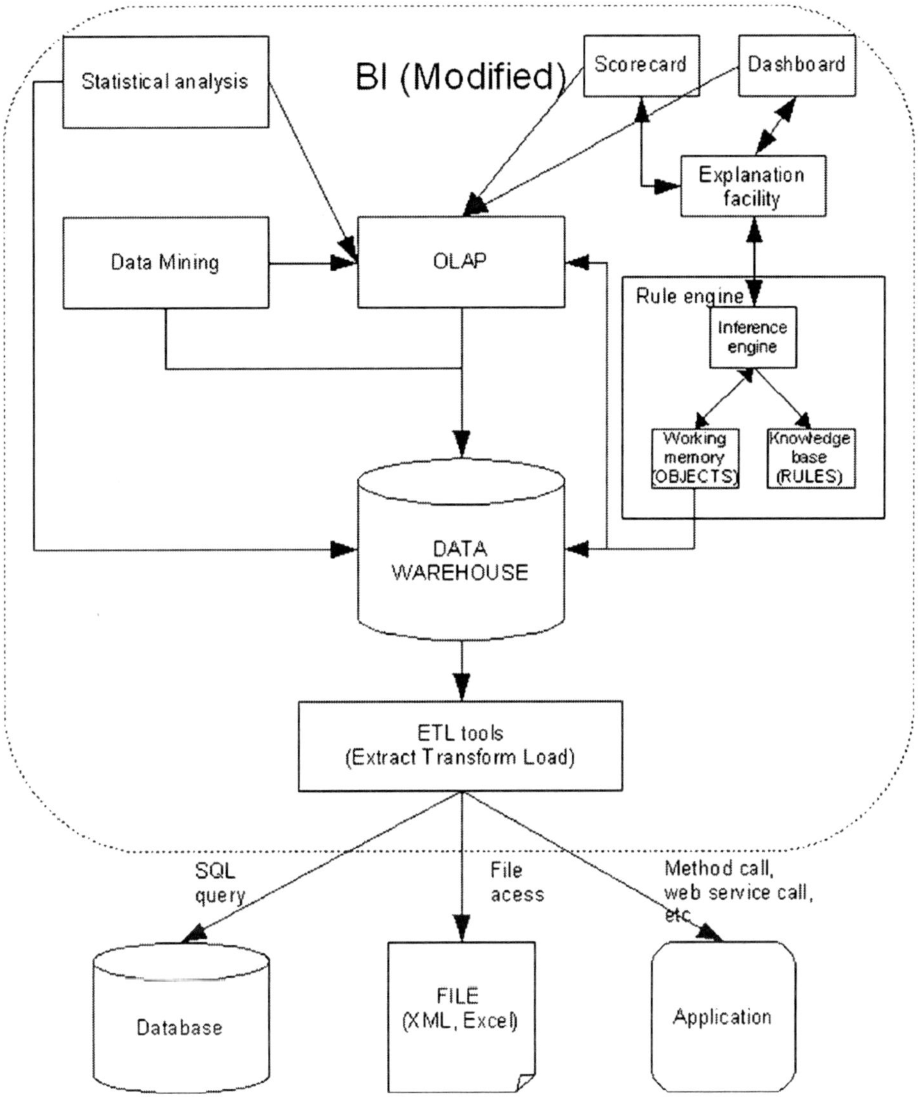


analysts perform deep data analysis using these modules.

Modifications begin by embedding a rule engine into BI. This is necessary as there is no other way of executing KPI interpretation rules on business data. Every rule engine consists of three core elements: knowledge base, working memory and inference engine. The knowledge base would, in this case, contain KPI interpretation knowledge presented as rules.

Modern rule engines, like Jess (Sandia National Laboratories, 2008), store initial and derived facts as objects in the working memory. So, in order to enable making information from business data, business data would need to be presented as objects. DW and OLAP usually get implemented by using relational databases. The challenge would then be to extract and transform this business data from relational to object form. Fortunately, there is abundance of relational-object mapping frameworks i.e. Hibernate (JBoss, 2008) that do just that.

To summarize, the solution would be to extract initial facts (business data) automatically from OLAP or DW using relational-object mapping frameworks, and insert them as objects into the working memory. The inference engine would then use forward chaining to extract information from this data.

An addition to the rule engine would be an explanation facility. It would provide natural-language-like explanations concerning conclusions (information) inferenced by the rule engine. This facility is portrayed as a relatively independent module, but in reality, it would probably be very closely linked to the rule engine.

Finally, BI would incorporate scorecard and dashboard modules, but they would be used in a different manner. Instead of just acquiring data from OLAP, these modules would acquire properly presented information from the explanation facility as well. This way, users would get reports that contain data, but also the interpretation of this data. Graphs, tables, scorecards and dashboards would still occupy a significant part of every report providing a way of ensuring that the rule engine deducted correct information.

The implementation can be performed in several steps:

1. Acquire or develop a rule engine.
2. Develop an explanation facility.
3. Develop an integration framework.
4. Acquire and formalize KPI interpretation knowledge.
5. Form explanations.

First of all, a rule engine has to be acquired or developed. It must conform to previously made requirements: knowledge presented as rules (this is trivial as it is a rule engine), facts presented as objects and forward chaining as inference technique. For example, two modern rule engines that conform to these requirements are Jess (Sandia National Laboratories, 2008) and Drools (Drools, 2008).

The second step is to develop an explanation facility and connect it to the rule engine. This facility must be able to present explanations in the form of "canned text". It should be closely linked to the rule engine so that the engine could issue commands to it while executing rules.

The third step is to produce a framework which integrates the rule engine and the explanation facility with a BI system. This framework is supposed to contain two connection points. It should enable the rule engine to acquire data from OLAP or DW in the form of objects. Since DW and OLAP are usually implemented by using relational databases, this can be achieved by employing relational-object mapping frameworks such as Hibernate (JBoss, 2008) or Java Persistence API (Sun Microsystems, 2008). Additionally, it should enable the explanation facility to insert explanations into reports. This means connecting it to the scorecard and dashboard modules.

Next, KPI interpretation knowledge must be acquired. This can be done by interviewing decision makers, but also by reviewing literature (books, journals, case studies etc.). This can prove to be somewhat a problem because decision makers may not be available or ready to share their knowledge, but the literature should provide a solid starting point. When acquired, KPI interpretation knowledge must be formalized (by using rules and objects) in order to make it accessible to the rule engine.

The final step is to define explanations in the form of "canned text" statements. Each explanation should consist of one or more sentences and should be linked to some important conclusion. That way, every time an important conclusion is made, explanation concerning that finding can be added to the report.

Thorough testing should follow each step of implementation. Rule engine, explanation facility and integration framework can be tested using regular software testing techniques. KPI interpretation knowledge testing is somewhat different. The set of rules representing this knowledge should conform to functional requirements, but it should also be verified for consistency. Among other things, consistency checking involves identifying and eliminating redundant, conflicting and unsatisfiable rules. Finally, explanations should be tested in order to verify that they appear in the desired form and order.

At this point in our research, we have developed a simple (but working) prototype. It includes a DW developed by using MySQL 5.0 (MySQL, 2008). The Drools rule engine performs the inference process, and Java Persistence API is used to extract data directly from the DW and transform it into objects (which are, in turn, inserted into the working memory). A small explanation facility framework was implemented by using regular Java classes and it is being enhanced in order to enable internationalization. Reports that this system generates appear in the form of regular text documents at this point, and are not integrated into BI reports. The system analyzes student exam data by monitoring the following KPIs: average student mark, percentage of failed students, percentage of passed students and number of students. In order to derive information, each KPI value is compared to corresponding values from previous periods, and values from other related KPIs.

## DISCUSSION ABOUT THE PROPOSED SOLUTION

This research is at its beginning stage, so results presented in this chapter just include theory and explanation of the new approach, proposed architecture, and some implementation notes concerning the simple prototype that was developed. The evaluation can, therefore, only be limited to a discussion about perceived positive and negative aspects of the prototype in relation to current BI solutions. At some point, when a full-featured prototype is developed, a proper evaluation with a group of users will be possible.

Positive aspects of the proposed solution could be summarized as follows:

- Decision makers would be relieved of the data analysis burden
- Less technical knowledge would be required
- Large quantities of data could be interpreted automatically
- Information would be readily available
- Information would be objective
- Information quality would be constant

One of the most obvious benefits would be that decision makers wouldn't have to analyze reports manually which would lead to significant time savings. And since analysis would be automatic and information presented in natural-language-like sentences, there would be no need for additional manual OLAP querying. As a result, even users with very little technical knowledge could fully realize the potentials of the solution.

Monitoring would be performed automatically, so large quantities of data could be efficiently analyzed thus solving the "mountains of data" problem. Motivation to gather as much data as possible would peak as it would effect in more information being deducted rather more data being piled.

But most importantly, information would be available anytime, anywhere, at the touch of a button. By excluding the human factor in data-to-information transformation, information would become objective and of constant quality. It wouldn't depend on decision makers' concentration, knowledge, experience or bias.

The solution could produce some negative aspects as well:

- Decision makers might not be comfortable with loss of control regarding data analysis

- Quality of interpretation knowledge would dictate quality of derived information
- Quantity of interpretation knowledge would dictate quantity of derived information

If the proposed solution was implemented, decision makers wouldn't need to do manual analysis anymore. One negative aspect is that they could look at the new system with skepticism, expressing doubts about the quality of automatically derived information. This could be overcome by providing actual data together with information thus enabling users to verify all findings.

KPI interpretation knowledge must be of highest quality in order for the solution to function properly. If it was incomplete, inconsistent or irrelevant, the system would not derive quality information. Fortunately, there is abundance of objective sources from which this knowledge can be harvested – books, journals, case studies etc.

Another issue is the quantity of interpretation knowledge. Even if it were of highest quality, low quantities of knowledge could lead to very few conclusions being made. These conclusions, albeit important, would not inform the user fully about the state of business.

The core difference between current BI systems and the proposed solution is in their orientation. Current BI systems are "data oriented". Their primary focus is on extracting, transforming, loading, summarizing and appropriately presenting data. BI 2.0 and Real-Time BI systems exhibit these tendencies even more – data is supposed to be processed so fast that reports should be available almost immediately. The proposed solution is, on the other hand, "information oriented". Although reports would still contain data, the focus would be on extracting information from this data and presenting it to the user. Furthermore, data would not be treated as output, but rather as input to the reporting process.

## FUTURE TRENDS

Solution viability is also a subject for discussion. From an engineering perspective, successful implementation seems feasible. Modern rule engines are object-oriented and can, therefore, be easily integrated into existing applications. ETL, DW and OLAP already provide necessary data extraction, transformation, loading and storage capabilities, so development can be focused only on those modules needed for automated data-to-information transformation. There are also lots of reporting tools which enable custom report creation from code and can be suitable for integration with the explanation facility.

Things are not so clear when viewed from end-user perspective. Solution benefits are very obvious – no more manual report analysis, abundance of automatically generated high quality information, objective reporting etc. Unfortunately, so are the potential dangers. If the implemented solution was incomplete, inconsistent or flawed in any way, it would result in severe consequences regarding decision quality. Fear of losing control over data analysis is another important factor. Decision makers may not be ready to accept automatically generated information, and may want to manually verify every finding.

Finally, if the solution was properly implemented and gradually introduced into BI systems, it could have the potential to change reporting altogether. Once collected and formalized, KPI interpretation knowledge could be reused, broadened and multiplied as needed therefore reproducing the same quality of reports in every BI system. Information would be generated without human involvement, leaving to decision makers just to verify its findings. This means that BI systems implementing this solution could, in certain circumstances, infer more information than decision makers themselves. The decision making process would be refocused on design and choice stages instead of intelligence.

In the future, there is a potential of improving the proposed solution by embedding other types of knowledge as well. Besides KPI interpretation, the system could provide suggestions regarding appropriate actions to be taken in a particular business situation. If, for example, product profit was low, and profit margins were high, the system could suggest lowering profit margins in order to provide a more competitive price.

This is an ongoing research, so future trends involve our current and future efforts regarding this research. Research results presented in this chapter are primarily theoretical and need to be empirically verified. There is no better way to do this than by creating a prototype and then using it to test all hypotheses. Since a simple prototype has already been developed, our current efforts are directed toward developing a full-featured software prototype in accordance with the proposed architecture and implementation notes.

Our future research is aimed at conducting a thorough evaluation of the full-featured prototype. It would include verifying whether embedding this solution in a BI system really raises the quality of reporting. At this point, it is clear that an experiment should be conducted where both participant groups (control and experimental) would get the same amounts of training regarding BI. In the first round of the experiment, both groups would get the same series of BI generated reports and a test requiring them to notice some trends in data (information) and to answer a few questions regarding what they deducted. In the second round, the control group would get a series of BI generated reports, and the experimental group would get the same reports but broadened with information automatically generated by the prototype. As in the previous round, both groups would get a test requiring them to notice some trends and to answer a few questions. Response time would be measured throughout the experiment. Then, a comparation of response times and test results should provide enough facts to prove or disprove that reports generated by the prototype inform users better than BI generated reports.

## CONCLUSION

A solution for automated KPI interpretation presented in this chapter is aimed at solving the problem of "too much data too little information" in business reports generated by BI systems. The core of this problem is that BI reports, no matter how well presented, contain mostly data (KPI values) and little or no information. Since data consists of one or more values and has no meaning, end users (decision makers) are forced to manually analyze large numbers of reports in order to get informed. As a consequence, they can be overwhelmed by large quantities of data and therefore miss to notice some important information.

The idea is to automate this data-to-information transformation process by capturing knowledge needed to interpret data and then using it to create information. Since the process is supposed to be automated, captured knowledge must be formalized. Rules, objects and fuzzy sets and operations are used as knowledge representation techniques. Inference includes matching formalized knowledge with data in order to create information. Being data oriented, forward chaining is used as inference technique. Finally, an explanation facility is utilized in order to present the derived information appropriately – in the form of natural-language-like sentences.

Automated KPI interpretation is aimed at improving the reporting process alone, so it should be integrated into existing BI systems. Integration architecture includes embedding a rule engine and explanation facility into a BI system and connecting them so data can be acquired from DW or OLAP, entered into the rule engine, transformed to information, appropriately presented by the explanation facility and then inserted into a report.

Positive aspects of the solution are numerous. Decision makers would be relieved of the data analysis burden and would need less technical knowledge in order to utilize the system. Additionally, large quantities of data could be interpreted automatically. But most importantly, information would be readily available, objective and of constant quality.

One negative aspect of the solution is that decision makers may not be comfortable with loss of control regarding data analysis. Because of this, reports should include data as well as information so users can manually verify all findings. Another issue is that quality and quantity of interpretation knowledge dictates the quality and quantity of derived information. Therefore, knowledge needs to be of highest quality, and sufficient in volume.

Finally, it can be concluded that the solution seems viable from an engineering standpoint. From end user perspective, things are not so clear. This chapter summarizes the results of an ongoing research, so both of these claims have yet to be empirically tested. Current and future research efforts are oriented toward creating a full-featured prototype and then conducting a series of studies to prove or disprove the viability of the proposed approach.

## REFERENCES

Alter, S. L. (1980). *Decision support systems: Current practice and continuing challenges.* Reading, Mass: Addison-Wesley Pub.

Balanced Scorecard Institute. (2007). *What is the balanced scorecard?* Retrieved January 9, 2008 from http://www.balancedscorecard.org/BSCResources/AbouttheBalancedScorecard/tabid/55/Default.aspx

Business Rules Group. (2000). *Defining business rules – What are they really?* Retrieved January 9, 2008 from http://www.businessrulesgroup.org/first_paper/BRG-whatisBR_3ed.pdf

California State University. (2007). *Data warehouse glossary.* Retrieved February 15, 2008, from http://it.csumb.edu/site/x7101.xml

Checkland, P. B., & Scholes, J. (1990). *Soft systems methodology in action.* New York, NY: John Wiley & Sons.

Codd, E. F., Codd, S.B. & Salley, C.T. (1993). *Providing OLAP to user analysts: An IT Mandate.* Hyperion Solutions Corporation. Retrieved January 15, 2008, from http://dev.hyperion.com/resource_library/white_papers/providing_olap_to_user_analysts.pdf

Davis, G. B., & Olson, M. H. (1985). *Management information systems: Conceptual foundations, structure, and development.* New York: McGraw-Hill.

Drools. (2008). *Drools business rules management system.* Retrieved November 14, 2008, from http://www.jboss.org/drools/

Durkin, J. (1994). *Expert systems – design and development.* New York: Macmillan Publishing Company.

Few, S. (2004). Dashboard confusion: A clear understanding of dashboards requires delving beneath the marketing hype. *Intelligent enterprise.* Retrieved January 1, 2008, from http://www.intelligententerprise.com/showArticle.jhtml?articleID=18300136

Fisher, J. R. (2008). *Prolog tutorial.* California State Polytechnic University. Retrieved November 13, 2008, from http://www.csupomona.edu/~jrfisher/www/prolog_tutorial/contents.html

Floridi, L. (2005). Is semantic information meaningful data? *Philosophy and Phenomenological Research, 70*(2), 351-370.

Frawley, W., Piatetsky-Shapiro, G. & Matheus, C. (1992). Knowledge discovery in databases: An overview. *AI Magazine*, 213-228.

Georgetown University. (2008). *Data warehouse glossary*. Retrieved February 1, 2008, from http://uis.georgetown.edu/departments/eets/dw/GLOSSARY0816.html

Gnatovich, R. (2005). Closing the analysis gap. *Imedia Connection*. Retrieved January 10, 2008, from http://www.imediaconnection.com/content/6060.asp

Graco, W., Semenova, T. & Dubossarsky, E. (2007). Toward knowledge-driven data mining. In *Proceedings of the 2007 international workshop on Domain driven data mining: Conference on Knowledge Discovery in Data* (pp. 49-54). New York: ACM.

Greenfield, L. (2008). The data warehousing information center. *LGI International*. Retrieved February 10, 2008, from http://www.dwinfocenter.org/defined.html

Hackathorn, R. (2004). Minimizing action distance. *The Data Administration Newsletter*. Retrieved February 10, 2008, from http://www.tdan.com/i025fe04.htm

Inmon, W. (1995). *Tech topic: What is a data warehouse*? New York: Prism Solutions.

JBoss Corporation. (2008). *Hibernate – Relational persistence for Java and .NET*. Retrieved November 13, 2008, from http://www.hibernate.org/

Liebowitz , J. (2006). Where is the "intelligence" behind "business intelligence"? *Business Intelligence*. Retrieved January 7, 2007, from http://www.businessintelligence.com/ex/asp/code.128/pagenum.1/xe/article.htm

McCoy, D. (2002). Business activity monitoring: Calm before the storm. *Gartner Group*. Retrieved May 15, 2007, from http://www.gartner.com/resources/105500/105562/105562.pdf

Microsoft. (2007). *SQL Server 2005 books online: Multidimensional expressions (MDX) reference*. Microsoft SQL development center. Retrieved May 10, 2007, from http://msdn2.microsoft.com/en-us/library/ms145506.aspx

MySQL. (2008). *MySQL – The world's most popular open-source database*. Retrieved November 13, 2008, from http://www.mysql.com/

Nicholls, C. (2006). BI 2.0: The next generation. *DM Review Magazine*. Retrieved June 7, 2007, from http://www.dmreview.com/issues/20061101/1066763-1.html

Rossetti, L. (2006). *What is business intelligence*? Retrieved January 10, 2008 from http://searchdatamanagement.techtarget.com/sDefinition/0,,sid91_gci213571,00.html

Sandia National Laboratories. (2008). *Jess, the Rule Engine for the Java Platform*. Retrieved November 13, 2008 from http://herzberg.ca.sandia.gov/

Silver, G. A. & Silver, M. L. (1989). *Systems analysis and design*. MA: Addison Wesley.

Sun Microsystems. (2008). *Java Persistence API*. Retrieved November 13, 2008, from http://java.sun.com/javaee/overview/faq/persistence.jsp

White, C. (2004). Building the smart business: In-Line real-time BI. *DM Review Magazine*. Retrieved June 10, 2007, from http://www.dmreview.com/issues/20040901/1009281-1.html

Whiting, R. (2002). Analysis gap. *Information Week*. Retrieved January 10, 2008, from http://www.informationweek.com/story/IWK20020418S0007

Wise, L. (2008). With broader BI Usage, who owns the data? *IT Business Edge*. Retrieved January 09, 2008, from http://www.itbusinessedge.com/item/?ci=37391

XMLA council. (2007). *XML for analysis*. Retrieved May 11, 2007 from http://www.xmlforanalysis.com/index.htm

## KEY TERMS AND DEFINITIONS

**Business intelligence (BI):** A broad category of applications and technologies for gathering, storing, analyzing, and providing access to data to help enterprise users make better business decisions.

**Data:** One or more values represented in numeric, symbolic or any other form.

**Explanation facility:** A module which enables the expert system to provide explanation to the user about why is it asking a question and how it reached some conclusion.

**Expert system:** A computer program designed to model the problem solving ability of a human expert.

**Information:** Information equals data plus meaning.

**Key performance indicator (KPI):** A significant measure used on its own, or in combination with other key performance indicators, to monitor how well a business is achieving its quantifiable objectives.

**Reporting:** A process which involves querying data sources with different logical models to produce a human readable report.

# Chapter XXVII
# Mining Association Rules

**Mihai Gabroveanu**
*University of Craiova, Romania*

## ABSTRACT

*During the last years the amount of data stored in databases has grown very fast. Data mining, also known as knowledge discovery in databases, represents the discovery process of potentially useful hidden knowledge or relations among data from large databases. An important task in the data mining process is the discovery of the association rules. An association rule describes an interesting relationship between different attributes. There are different kinds of association rules: Boolean (crisp) association rules, quantitative association rules, fuzzy association rules, etc. In this chapter, we present the basic concepts of Boolean and the fuzzy association rules, and describe the methods used to discover the association rules by presenting the most important algorithms.*

## INTRODUCTION

The progress of Information Technology has determined companies to record a large quantity of information in huge databases. Due to the fact that a lot of useful knowledge is hidden in these databases, the companies need to extract this knowledge in order to help the decision-making process.

Data mining, also known as knowledge discovery in databases (Chen, Han, & Yu, 1996), provides efficient automated techniques for discovering potentially useful, hidden knowledge or relations among data from large databases. Data mining functions include classification, clustering, prediction, regression, and link analysis (associations), etc.

Mining association rules represent an unsupervised data mining method that allows identifying interesting associations, correlations between items, and frequent patterns from large transactional databases. This problem was first introduced by Agrawal et al. in (Agrawal, Imielinski, & Swami, 1993).

The original motivation for searching association rules came from the need to analyze the *market-basket data* that stores items purchased on a per-transaction basis, in order to identify customer behaviors by finding different items often purchased together. A typical example of an association rule has the following statement:

*popcorn* ⇒ *beer* [*support* = 7%, *confidence* = 74%] (1)

This rule expresses a relation between *popcorn* and *beer* namely *"customers who purchase popcorn also purchase beer in the same transaction"*. The *support* and *confidence* are two measures used to show the rule interestingness. The support of 7% for rule presented above states that *popcorn* and *beer* appeared together in 7% of all recorded transactions. The confidence measure describes the chance that there is beer in a transaction provided that there is also *popcorn*. In this case, 74% of all transactions involving *popcorn* also involved *beer*.

Such rules provide a source of information for retailers to develop marketing strategies, catalog design and store layout. Thus, the company can decide that items frequently purchased together to be placed in proximity in order to encourage the sale of such items together. Other strategy is placing the items frequently purchased together at opposite ends of the store in order to attract customers that purchase such items to pick up other items along the way.

The association rules can be categorized in different ways as follows (Han & Kamber, 2006):

1. Based on the types of values handled in the rule:
   a. *Boolean (crisp) association rules*: the association is between the presence or absence of items. Rule (1) is an example of such a Boolean association rule.
   b. *Quantitative association rules*: the rules describe associations between quantitative items or attributes. Here the attribute values are partitioned into intervals. An example of such rule is the following:

$$age(X,"20...30") \wedge income(X,"2000...3000") \Rightarrow cars(X,"0..1") \quad (2)$$

   c. *Categorical association rules*: the attributes referred by rules are categorical, such as *occupation* or *sex*. An example of such a rule is the following:

$$occupation(X,"student") \wedge sex(X,"male") \Rightarrow plays(X,"basket") \quad (3)$$

2. Based on the number of data dimension involved in the rule:
   a. *Single-dimensional rules:* the items or attributes from rules refer only one dimension. An example of such rule is the following:

$$buys(X,"beer") \wedge buys(X,"cola") \Rightarrow buys(X,"popcorn") \quad (4)$$

   The dimension referred is buys.
   b. *Multidimensional rules:* the rules refer to two or more attributes. The rule (2) is an example of multidimensional rule. The dimensions referred to are age, income and cars.

3. Based on the abstraction levels of data involved in the rule:
   a. *Single-level rules:* the rules refer attributes or items at the same level of abstraction, i.e., at the lowest level in

*Figure 1. Drinks hierarchy*

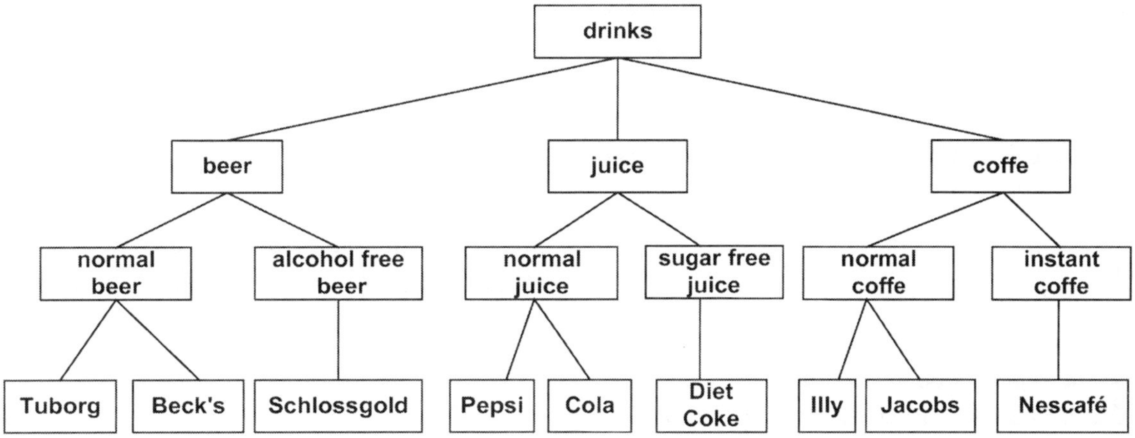


a hierarchy. An example of such rule is the following:

$$buys(X,"Schlossgold") \wedge buys(X,"Pepsi") \Rightarrow buys(X,"Nescafé") \quad (5)$$

b. *Multi-level rules:* the rules refer attributes or items organized at different levels of abstraction in a hierarchy. For instance, if we consider the concepts hierarchy from Figure 1 for drink items from a grocery, then an example of multi-level association rule is the following:

$$buys(X,"alcohol\ free\ \ beer") \Rightarrow buys(X,"sugar\ free\ juice") \quad (6)$$

A problem of classical association rules is that not every kind of data can be handled. Several extensions to the classical relational model have been proposed to support quantitative data (Rakesh&Agrawal, 1996). This algorithm discretizes the domains of quantitative attributes into intervals and the problem of mining quantitative association rules is reduced to the mining Boolean association rules. These intervals may not be concise and meaningful for human beings. Linguistic terms expressed by fuzzy sets can be used instead of using intervals. Fuzzy association rules can handle both quantitative and categorical data and are expressed in linguistic terms, which are more natural and understandable for human beings. An example of fuzzy association rules is the following:

$$age(X, young) \wedge income(X, high) \Rightarrow cars(X, many) \quad (7)$$

Here, the notions of *young*, *high* and *many* can be modeled as fuzzy sets.

In the following sections we focus on the Boolean and fuzzy association rules. We provide the basic concepts as presented in the specialized literature and we present the methods used to discover this kind of association rules by presenting the most important algorithms.

## BOOLEAN ASSOCIATION RULES

### Basic Concepts

The formal model for Boolean association rules was first proposed by Agrawal (Agrawal, Imielinski, & Swami, 1993).

Let $I = \{i_1,...,i_m\}$ be a set of *m* distinct Boolean attributes, also called *items*. A set of items $X = \{i_1,\ldots,i_k\} \subseteq I$ is called *itemset* or *k-itemset* if it contains *k* items. A *transaction T* over *I* is a set of items, $T \subseteq I$, associated with a unique identifier called TID. Let $D = \{t_1,...,t_n\}$ be a database containing a set of transactions over *I*. A transaction *T* is said to contain itemset *X* if and only if $X \subseteq T$.

**Definition 1 (Support count and support).** The ***support count*** (also known as frequency, absolute support or simple count) of an itemset *X* is the number of transactions from *D* that contain itemset *X*. Formally, support count for itemset *X* is defined as:

$$\text{support\_count}(X) = |\{T \in D| \, X \subseteq T\}|$$

The ***support*** or ***relative support*** of an itemset *X* is defined as proportion/percentage of transactions from *D* that contain itemset *X* (the occurring probability of the itemset *X* in database *D*):

$$\text{support}(X) = P(X) = \frac{|\{T \in D| \, X \subseteq T\}|}{|D|}$$

**Definition 2 (Association Rule).** Let *X* and *Y* two itemsets. An ***association rule*** is an implication of the form $X \Rightarrow Y$, where $X \subset I, Y \subset I$, and $X \cap Y = \varnothing$. *X* and *Y* are called *the antecedent*, respectively *the consequent* of the rule.

Formally speaking, this means that transactions including *X* will include *Y* as well, with a high probability.

The quality of an association rule is expressed by several measures. Two of them, namely the *support* and the *confidence* are essential (Agrawal, Imielinski, & Swami, 1993)

**Definition 3 (Rule support).** The *support* of association rule $X \Rightarrow Y$ is defined as the *proportion/percentage of transactions in D that contains both X and Y* (the probability of both *X* and *Y* co-occurring in *D*):

$$\text{support}(X \Rightarrow Y) = P(X \cup Y) = \frac{|\{T \in D \,|\, X \cup Y \subseteq T\}}{|D|} = \frac{\text{support\_count}(X \cup Y)}{|D|} \quad (8)$$

**Definition 4 (Rule confidence).** The ***confidence*** of association rule $X \Rightarrow Y$ is defined as the *proportion (percentage) of transactions in D containing X which also contain Y* (conditional probability of *Y* given *X*):

$$\text{confidence}(X \Rightarrow Y) = P(Y / X) = \frac{|\{T \in D|X \cup Y \subseteq T\}|}{|\{T \in D|X \subset T\}|} = \frac{\text{support\_count}(X \cup Y)}{\text{support\_count}(X)} \quad (9)$$

Usually, the support and confidence values are expressed as percentages between 0% and 100%, rather than 0 to 1.0.

**Example 1:** *Let us consider transactional database D showed in Table 1 containing few customer transactions.*

*Here, the set of items is* $I = \{bread, butter, eggs, milk\}$, *and database transactions are* $D = \{T_1, T_2, T_3, T_4, T_5, T_6\}$.

*An example of an association rule is the following:*

*Table 1. Example of market basket data*

| TID | Items List |
|---|---|
| 1 | *butter, eggs, milk* |
| 2 | *bread, butter* |
| 3 | *bread, eggs, milk* |
| 4 | *bread, butter, eggs, milk* |
| 5 | *eggs, milk* |
| 6 | *bread, eggs* |

$\{eggs\} \Rightarrow \{milk\}$

*This means that some of the customers that bought milk, also bought and eggs. The support of this association rule is calculated as:*

$$\text{support}(\{eggs\} \Rightarrow \{milk\}) = \frac{|\{T_1, T_3, T_4, T_5\}|}{|D|}$$

$$= \frac{4}{6} = 0.66$$

*which means that: "66% of customers bought eggs and milk together"*

*The confidence of this association rule is calculated as:*

$$\text{confidence}(\{eggs\} \Rightarrow \{milk\}) = \frac{|\{T_1, T_3, T_4, T_5\}|}{|\{T_1, T_3, T_4, T_5, T_6\}|}$$

$$= \frac{4}{5} = 0.80$$

*and it means that: "from all customers that bought eggs, 80% of them also bought milk"*

**Definition 5 (Strong Association Rule).** Rules having support and confidence greater than or equal to a user-specified minimum support threshold (*minsup*) and a minimum confidence threshold (*minconf*) respectively are named ***strong association rules***.

**Definition 6 (Frequent itemset).** An itemset $X$ is called ***frequent itemset*** or ***large itemset*** if its support is greater than or equal to a minimum support threshold (*minsup*). In what follows, we call *minimum support count* the number of transactions required for an itemset to be frequent.

**Problem (Mining Association Rules Problem).** *Given a set of items I, a transactional database D over I, a minimum support threshold (minsup) and a minimum confidence threshold (minconf), extract all strong association rules.*

## Mining Boolean Association Rules

Since, the support of an association rule $X \Rightarrow Y$ is equal with the support of itemset $X \cup Y$, and we are interested only in those that satisfy minimum support and minimum confidence, the problem of mining association rules can be decomposed in two steps:

1. *find all frequent itemsets*
2. *generate all strong association rules from the frequent itemsets founded at step 1*

**Example 2**: *Let us consider transactions from Table 1, minimum support of 30% and minimum confidence of 60%. Table 2 shows all frequent itemsets with respect to the minimum support threshold.*

*Table 3 shows the list of all strong association rules with respect to a minimum support and confidence threshold from D.*

*Table 2. Frequent itemsets from D*

| Itemset | Support Count | Support % |
|---|---|---|
| *{bread}* | 4 | 66.66% |
| *{butter}* | 3 | 50.00% |
| *{eggs}* | 5 | 83.33% |
| *{milk}* | 4 | 66.66% |
| *{bread, butter}* | 2 | 33.33% |
| *{bread, eggs}* | 3 | 50.00% |
| *{bread, milk}* | 2 | 33.33% |
| *{butter, eggs}* | 2 | 33.33% |
| *{butter, milk}* | 2 | 33.33% |
| *{eggs, milk}* | 4 | 66.66% |
| *{bread, eggs ,milk}* | 2 | 33.33% |
| *{butter, eggs, milk}* | 2 | 33.33% |

*Table 3. Strong association rules from D*

| Rule | Confidence (%) | Support (%) |
|---|---|---|
| $\{eggs\} \Rightarrow \{milk\}$ | 100.0% | 66.66% |
| $\{milk\} \Rightarrow \{eggs\}$ | 80.0% | 66.66% |
| $\{eggs\} \Rightarrow \{bread\}$ | 75.0% | 50.00% |
| $\{bread\} \Rightarrow \{eggs\}$ | 60.0% | 50.00% |
| $\{eggs\} \Rightarrow \{bread, milk\}$ | 100.0% | 33.33% |

The computational performance of mining association rules algorithms is determined by the finding of frequent itemsets step. In the following we focus on this step.

A naïve approach for finding frequent itemsets is to generate all $2^m$ possible itemsets from *I*, and scan the database *D* in order to determine the itemsets support. The itemsets with support greater than or equal to the minimum support threshold represent frequent itemsets. The complexity of this algorithm is $O(2^m n)$. The main problem arises from the number of possible itemsets $2^m$, which grows exponentially with the number of items. For example, if $m$=100, then the number of possible itemsets is about $1.27 \times 10^{30}$. This number of itemsets is too huge to store or compute support count.

Since association rules were introduced, many algorithms for mining association rules have been proposed.

The AIS (Agrawall, Imielinski, Swami) algorithm was the first algorithm proposed for mining association rules in (Agrawal, Imielinski, & Swami, 1993). This algorithm generates only the one-item consequent association rules, which means that rules have the following form $\{i_{k_1},\ldots,i_{k_p}\} \Rightarrow \{i_q\}$. The main drawback of the AIS algorithm is related to the too many candidate itemsets, and the fact that it requires too many pass over the entire database.

The Apriori (Agrawal & Srikant, 1994) algorithm is one of most popular and efficient algorithm used for finding frequent itemsets. Based on the Apriori algorithm, many algorithms were designed: Apriori-TID and Apriori-Hybrid (Agrawal & Srikant, 1994), DHP (Jong Soo Park, 1995), PARTITION (Savasere, Omiecinski, & Navathe, 1995), DIC (Brin, Motwani, Ullman, & Tsur, 1997). These algorithms try to improve the efficiency by making some modifications in order to reduce the number of database scans, pruning the candidates by different techniques.

The FP-growth (Frequent Pattern growth) (Han, Pei, & Yin, 2000) is an algorithm for mining frequent itemsets without candidate generation. It compresses the database representing the frequent items into FP-Tree. The frequent patterns are generated from this FP-Tree.

In the following we present the Apriori and FP-growth algorithms in details.

## Apriori Algorithm

The Apriori algorithm was proposed by Agrawal and Srikant in 1994 in (Agrawal & Srikant, 1994). The algorithm uses *prior knowledge* of frequent itemsets properties, in order to reduce the number of candidate itemsets presented below.

**Lemma 1 (Apriori Property):** *All nonempty subsets of a frequent itemset must also be frequent.*

*Proof.* Let *X* be a frequent itemset. If $Y \subseteq X$ is not frequent, then this means that the number of transactions containing *Y* is less of the minimum support count. But, *X* is obtained from *Y* by adding a few items, this means that the number of transactions that contain *X* is less than or equals to the number of transactions containing *Y*. Therefore, the support count of itemset *X* is less of the minimum support count, i.e. *X* is a not frequent itemset. This result, contradicts the initial assumption, therefore *Y* is frequent itemset. □

The Apriori property is an *antimonotone* property in the sense that if a set cannot pass a test, then all its supersets will also fail the same test.

**Corollary 1:** *If itemset X contains a sub-itemset that is not frequent, then itemset X is not frequent.*

**Corollary 2:** *If a k-itemset contains an (k-1)-itemset that is not frequent, then it is not frequent.*

In what follows we will denote the set of all frequent *k*-itemsets with $L_k$.

The Apriori algorithm consists in an iterative search, where frequent *k*-itemsets are used to explore frequent ($k$+1)-itemsets. First, the database $D$ is scanned in order to compute the support of each 1-itemset. The items that satisfy the minimum support are collected and form the set of frequent 1-itemsets, $L_1$. Next, $L_1$ is used to find $L_2$, the set of frequent 2-itemsets, which is used to find $L_3$, etc. The process stops when the set $L_k$ became empty. At each step $k$ of the algorithm the database is scanned once in order to determine the frequent $k$-itemsets.

The finding process of $L_k$ from $L_{k-1}$, for $k \geq 2$, consists of two steps:

**Step 1: (Join Step)** At this step, a set of candidate $k$-itemsets, denoted $C_k$, is generated by joining $L_{k-1}$ with $L_{k-1}$. We assume that items within a transaction or itemset are sorted in lexicographic order. So, the $C_k$ is obtained as in Box 2.

Condition $x_{k-1} < y_{k-1}$ ensures that no duplicates are generated.

**Step 2: (Prune Step)** For each itemset from $C_k$ must be determine the support in order to determine the set of frequent $k$-itemsets, $L_k$. A full scan of database $D$ is required to do this work. The number of candidate itemsets from $C_k$ can be reduced using the Apriori property, before starting the database scan. So, based on Corollary 2, we remove all $k$-itemsets containing some ($k$-1)-subsets that is not present in $L_{k-1}$. This subset testing can be quickly performed maintaining a hash tree of all frequent itemsets. The generated set $C_k$ is a superset of $L_k$.

**Example 3:** *Let us consider a small database D with five items I={I1,I2,I3,I4,I5} and eight transactions as shown in Table 4. Let us suppose that the minimum support count required for an itemset to be frequent is 2 (minsup=2/8=0.25). Figure 2 shows how the Apriori algorithm works for finding frequent itemsets from D.*

- Initially, each item from $I$ is considered as a frequent 1-itemset candidate. Therefore, $C_1$ has five 1-itemsets candidates {*I1*}, {*I2*}, {*I3*}, {*I4*} and {*I5*}. The algorithm scans the database $D$ and count for each 1-itemset the number of occurrences (see Figure 2a). The set of frequent 1-itemsets, $L_1$, consists of those 1-itemsets from $C_1$ that have support count greater than or equal with minimum support count 2. In this case, all candidates in $C_1$ satisfy minimum support count (see Figure 2b).
- The candidate set of 2-itemsets, $C_2$, used to generate the set of frequent 2-itemsets is obtained by joining $L_1$ with itself, and removing itemsets which have subsets not in $L_1$. In this case, no candidates are removed from $C_2$

*Table 4. Database sample*

| TID | List of items ID |
|---|---|
| T1 | I1, I2, I3, I5 |
| T2 | I2, I3 |
| T3 | I1, I2, I4 |
| T4 | I3, I4, I5 |
| T5 | I1, I2, I3 |
| T6 | I1, I3 |
| T7 | I1, I3, I4 |
| T8 | I3, I4 |

*Box 1.*

$$C_k = L_{k-1} \triangleright\triangleleft L_{k-1} = \{\{x_1,\ldots,x_{k-1},y_{k-1}\} \mid X \in L_{k-1}, Y \in L_{k-1}, \\ X = \{x_1,\ldots,x_{k-1}\}, Y = \{y_1,\ldots,y_{k-1}\}, \\ x_1 = y_1 \wedge \ldots \wedge x_{k-2} = y_{k-2} \wedge x_{k-1} < y_{k-1}\}$$

*Figure 2. Discovering frequent itemsets using the Apriori Algorithm*

| Itemset | Support Count |
|---|---|
| {I1} | 5 |
| {I2} | 4 |
| {I3} | 7 |
| {I4} | 4 |
| {I5} | 2 |

*a) C1*

| Itemset | Support Count |
|---|---|
| {I1} | 5 |
| {I2} | 4 |
| {I3} | 7 |
| {I4} | 4 |
| {I5} | 2 |

*b) L1*

| Itemset | Support Count |
|---|---|
| {I1, I2} | 3 |
| {I1, I3} | 4 |
| {I1, I4} | 2 |
| {I1, I5} | ~~1~~ |
| {I2, I3} | 3 |
| {I2, I4} | ~~1~~ |
| {I2, I5} | ~~1~~ |
| {I3, I4} | 3 |
| {I3, I5} | 2 |
| {I4, I5} | ~~1~~ |

*c) C2*

| Itemset | Support Count |
|---|---|
| {I1, I2} | 3 |
| {I1, I3} | 4 |
| {I1, I4} | 2 |
| {I2, I3} | 3 |
| {I3, I4} | 3 |
| {I3, I5} | 2 |

*d) L2*

| Itemset | Support Count |
|---|---|
| {I1, I2, I3} | 2 |
| {I1, I3, I4} | ~~1~~ |

*e) C3*

| Itemset | Support Count |
|---|---|
| {I1, I2, I3} | 2 |

*f) L3*

because all 1-subitemsets of each candidate 2-itemset are in $L_1$ (see Figure 2c).

- The database *D* is scanned in order to determine the support count of candidates 2-itemses from $C_2$. The set of frequent 2-itemsets, $L_2$ (see Figure 2d), is obtained by removing from $C_2$ all of the 2-itemsets that do not satisfy the minimum support count threshold. In this case, the following itemsets are removed: {*I1*, *I5*}, {*I2*, *I4*}, {*I2*, *I5*}, {*I4*, *I5*}.
- Next, the candidate set of 3-itemsets, $C_3$, is generated:
- **Join step:** Initially, C3 is obtained by joining L2 with itself

$$C_3 = L_2 \triangleright\triangleleft L_2 = \{\{I1, I2, I3\}, \{I1, I2, I4\}, \{I1, I3, I4\}, \{I3, I4, I5\}\}$$

- **Prune step:** Based on the Apriori property we remove from C3 all 3-itemsets that contains 2-subitemsets which are not in L2. Thus, we have:
  - the itemset {*I1*, *I2*, *I4*} is removed because it contains the 2-itemset {*I2*, *I4*} that is not member of $L_2$
  - the itemset {I3, I4, I5} is removed

because it contains the 2-itemset {*I*4, *I*5} that is not member of $L_2$. Therefore, $C_3 = \{\{I1, I2, I3\}, \{I1, I3, I4\}\}$

- The database *D* is scanned in order to determine $L_3$. This is obtained by removing from $C_3$ all of the 3-itemsets that do not satisfy the minimum support count threshold. Therefore, $L_3 = \{\{I1, I2, I3\}\}$ (see Figure 2c). Next, $C_4$ is obtained by joining $L_3$ with itself, in this case we obtain $C_4 = \emptyset$ and algorithm stops because $L_4 = \emptyset$.

Pseudo-code for the Apriori algorithm is given in Algorithm 1. Initially, the candidate set to frequent 1-itemsets, $C_1$, is formed from all items from *I* (Step 2). While the set of candidates, $C_k$ is not empty the database *D* is scanned in order to determine the support count of each candidate from $C_k$ (steps 6 to 12). The set of frequent *k*-itemsets, $L_k$, consists of those *k*-itemsets from $C_k$ with support count greater than or equal to minimum support count (Step 13). Next, $C_k$ is generated calling apriori_gen procedure based on $L_{k-1}$, for $k \geq 2$. Finally, the set of all frequent itemsets *L* is returned as result.

The apriori_gen procedure presented in Algorithm 2 has two steps. During the first step (*join step*), $L_{k-1}$ is joined with $L_{k-1}$ to obtain the set $C_k$ of potential frequent candidates (steps 2 to 6). In the second step (*prune step*) the candidates containing infrequent (*k-1*)-subitemsets are not added to $C_k$ (steps 8-16).

In this algorithm the Apriori property plays an important role by substantially reducing the number of candidate itemsets. By example, in the database presented on Table 4 the number of the candidate 3-itemsets that must be checked, using the Apriori algorithm, is $|C_3|=2$ (when the minimum support count is 2 while using the naïve approach, the number of the candidate 3-itemsets that must be checked is

*Algorithm 1. Apriori*

**Input:** *D* : the database with transactions
*minsup*: the minimum support threshold
**Output:** *L*: the set of frequent itemsets in *D*
**Method:**

1: $L = \emptyset$
2: $C_1 = \{\{i\} \mid i \in I\}$
3: $k = 1$
4: **while** $C_k \neq \emptyset$ **do**
5: *//compute support count for each candidate itemset*
6: **for each** transaction $T \in D$ **do**
7: **for each** candidate $X \in C_k$ **do**
8: **if** $X \subseteq T$ **then**
9: $X.count++$
10: **end if**
11: **end for**
12: **end for**
13: $L_k = \{X \in C_k \mid X.count \geq minsup \times |D|\}$
14: $L = L \cup L_k$
15: $k++$
16: $C_k$ = apriori_gen( $L_{k-1}$ )
17: **end while**
18: **return** *L*

*Algorithm 2. apriori_gen*

```
Input:   L_{k-1} : the set of frequent (k-1)-itemsets
Output:  C_k : the set of candidate k-itemsets

function apriori_gen(L_{k-1})
C_k = ∅
for each (k-1)-itemset X = {x_1,...,x_{k-1}} ∈ L_{k-1} do
    for each (k-1)-itemset Y = {y_1,...,y_{k-1}} ∈ L_{k-1} do
        if x_1 = y_1 ∧ ... ∧ x_{k-2} = y_{k-2} ∧ x_{k-1} < y_{k-1} then
            //join step
            Z = {x_1,...,x_{k-1}, y_{k-1}}
            //prune step
            has_infrequent_subset = false
            for each (k-1)-itemset Z' ⊂ Z do
                if Z' ∉ L_{k-1} then
                    has_infrequent_subset = true
                    break
                end if
            end for
            if not has_infrequent_subset then
                C_k = C_k ∪ Z
        end if
    end for
end for
return C_k
```

$\binom{5}{3}=10.$

We summarize a comparison between the number of itemsets candidates that must be checked for both cases (i.e. naïve and Apriori) in Table 5.

One of the most important factor that affects the Apriori algorithm complexity is the minimum support threshold. The number of candidate itemsets tends to increase if the minimum support threshold is reduced. Other factors that affect the computational complexity of Apriori algorithm are: the number of items, the number of transactions and the average transaction width.

The Apriori algorithm has two bottlenecks:

1. multiple scans of the database required to check the frequencies of candidate itemsets. If $n$ is the length of longest frequent itemset then $n$+1 scans is required. For very large databases these scans involve many I/O operations that are time consuming.
2. the generation process of candidates can produce a huge number of candidate itemsets. For example, $10^4$ frequent 1-itemsets will generate $10^7$ candidate 2-itemsets.

A depth analysis of Apriori algorithm complexity is presented in (Tan, Steinbach, & Kumar, 2005).

## FP-Growth Algorithm

Han et al. propose the FP-growth, Frequent Pattern growth, algorithm (Han, Pei, & Yin, 2000) for mining frequent itemsets that break the bottlenecks

*Table 5. Comparative number of candidates*

| Candidate Itemsets Length | Number of candidates | |
|---|---|---|
| | Naïve | Apriori (minimum support count = 2 |
| 1 | 5 | 5 |
| 2 | 10 | 10 |
| 3 | 10 | 2 |
| 4 | 5 | 0 |
| 5 | 1 | 0 |

of Apriori algorithm. The frequent itemsets are generated without candidate generation and only two passes over the database are required. The algorithm includes two steps:

1. encoding the database into a compact data structure called FP-Tree and
2. generating the frequent itemsets directly from FP-Tree data structure using a divide-and-conquer strategy

The FP-Tree is a compressed representation of the database.

**Definition 7** (FP-Tree) (Han, Pei, Yin, & Mao, 2004). A ***FP-Tree*** (Frequent-Pattern Tree) is a tree structure defined as follows:

1. It consists of one root labeled as "*null*", a set of *item-prefix subtrees* as the children of the root, and a *frequent-item-header* table.
2. Each node in the item-prefix subtree consists of three fields: *item-name*, *count*, and *node-link*, where *item-name* registers which item this node represents, *count* registers the number of transactions represented by the portion of the path reaching this node, and *node-link* links to the next node in the FP-tree carrying the same item-name, or null if there is none.
3. Each entry in the frequent-item-header table consists of two fields: item-name and head of node-link (a pointer pointing to the first node in the FP-tree carrying the item-name).

*Algorithm 3. FP-Tree*

**Input:** *D*: the database with transactions
*minsup*: the minimum support threshold
**Output**: *T*: the frequent pattern tree, FP-tree

**Method**:

1. The database *D* is scanned once in order to generate the set of frequent 1-itemsets. Let be *F* the above mentioned set. The frequent 1-itemsets are sorted in a descending order of their support counts. Let be *L* the ordered list of frequent 1-itemsets.
2. Create the root node of an FP-Tree, *T*, and label it as *null*. For each transaction *Trans* from the database *D* do the followings:
   a) select and sort the frequent items in *Trans* according to the order of *L*. Let the sorted frequent items in *Trans* be [*p*|*P*], where *p* is the first frequent item and *P* is the remaining frequent items list.
   b) call function *insert_tree*([*p*|*P*], *T*) which is performed as follows: if *T* has a child *N* such that *N.item-name=p.item-name*, then increment the count of *N* by 1, else new node *N* is created and *N.item-name=p.item-name*, *N.count*=1, its parent link is linked to *T* and *node-link* is linked to the node with the same *item-name* via a node-link structure. If *P* is nonempty call *insert_tree*(*P*, *N*) recursively.

The Algorithm 3 describes the construction process of FP-tree.

**Example 4:** *Let us consider the database shown in Table 4 and let suppose that the minimum support count required for an itemset to be frequent is 2. In the following we present how it works the construction process of the FP-tree associated with the above mentioned database:*

1) First, the database is scanned once in order to determine the support count of each item. The infrequent items are ignored and the frequent items are sorted in a decreasing support count. In this case all items are frequent. We denote with *L* this ordered list. Thus, we obtain *L*={{*I3*:7}, {*I1*:5}, {*I2*:4}, {*I4*:4}, {*I5*:2}}.
2) The root of the FP-tree, *T*, is created and is labeled with *null*. Next, the algorithm scans the transactions database for the second time in order to construct the FP-tree. The items from each transaction are processed in *L* order (see Table 6). The Figure 3 shows the FP-trees obtained after insertion of the first four transactions. Let us consider that

*Figure 3. The FP-tree construction process*

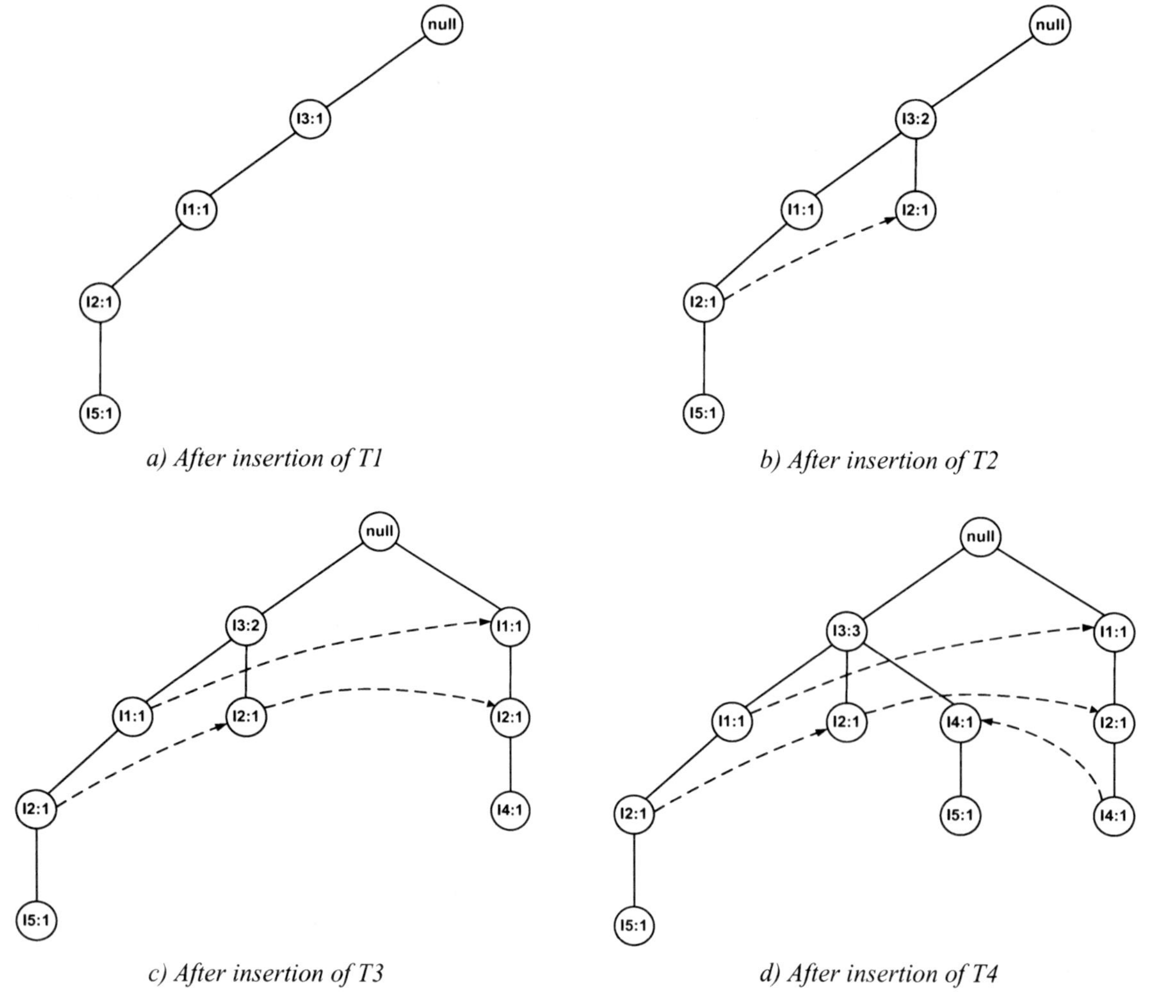


*a) After insertion of T1*

*b) After insertion of T2*

*c) After insertion of T3*

*d) After insertion of T4*

the first transaction, *T*1, is scanned. The items from the transaction are processed in the following order *I3, I1, I2, I5*. In this case *p* is *I3*, and *P* is [*I1, I2, I5*]. The *insert_tree*([*p*|*P*], *T*) function is called. The function tries to determine if *T* has a child, such that *N.item-name=I*3. Because *T* does not have any child, according to the rules, a new node labeled *I*3 is created, its support count is 1 and node is linked to *null*. Next, the *insert_tree*([*I1*|[*I2,I5*]], *I3*) function is called and node *I1:1* is added as child of *I3:1*. The obtained FP-tree after processing of all the items from T1 is presented in Figure 3a. Briefly, as a conclusion the transaction *T1* is encoded into the following path *null* → *I3* → *I4* → *I*2 → *I*5. Every node along the path has the frequency count equal with 1. Further on, the *insert_tree* function is called for the second transaction, *T2* = [*I3, I2*]. Now, *p* is *I3*, and *P* is [*I2*]. Because there is a child of *T* with item-name equal with *I3*, the support count of this child is incremented with 1. On next step, the *insert_tree*([*I*2|[]], *I*3) function is called and node *I*2:1 is added as child of *I*3:2. After processing all the items from *T*2, we obtain the FP-Tree that is presented in the Figure 3b. In this case, the transaction *T2* is mapped into the following path *null* → *I3* → *I2* that shares a common prefix item, *I3*, with the first transaction. As the labeled node *I3* is shared by both transactions i.e. *T1* and *T2*, the frequency count of this node is increased to 2. The process continues until every transaction from D is processed. The Figure 4 shows the FP-tree associated with database *D*.

The frequent patterns are mined from the FP-tree by the **FP-growth** procedure (Han, Pei, & Yin, 2000) presented in Algorithm 4.

The algorithm starts from each frequent length-

*Table 6. The ordered transactions list*

| TID | Ordered List of items ID |
|---|---|
| T1 | I3, I1, I2, I5 |
| T2 | I3, I2 |
| T3 | I1, I2, I4 |
| T4 | I3, I4, I5 |
| T5 | I3, I1, I2 |
| T6 | I3, I1 |
| T7 | I3, I1, I4 |
| T8 | I3, I4 |

*Figure 4. FP-tree database representation*

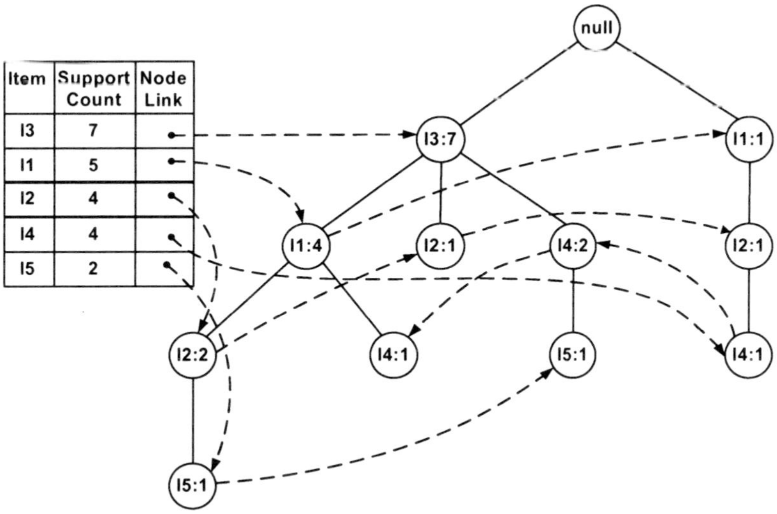

*Algorithm 4. FP-growth*

```
Input:      Tree: the FP-tree to be mined
Output:     the set of frequent patterns
Method: call FP-growth(Tree, null)

procedure FP-growth(Tree, α)
if Tree contains a single path P then
    for each combination, denoted as β, of the nodes in the path P do
        generate frequent pattern β ∪ α with support = minimum support of nodes in β
    end for
else
    for each a_i in the header of Tree do
        generate frequent pattern β = a_i ∪ α with support = a_i.support
        construct β's conditional pattern base and then β's conditional FP-tree Tree_β
        if Tree_β ≠ ∅ then
            call FP-growth(Tree_β, β)
        end if
    end for
end if
return
```

*Figure 5. Construction process of conditional FP-tree for I5*

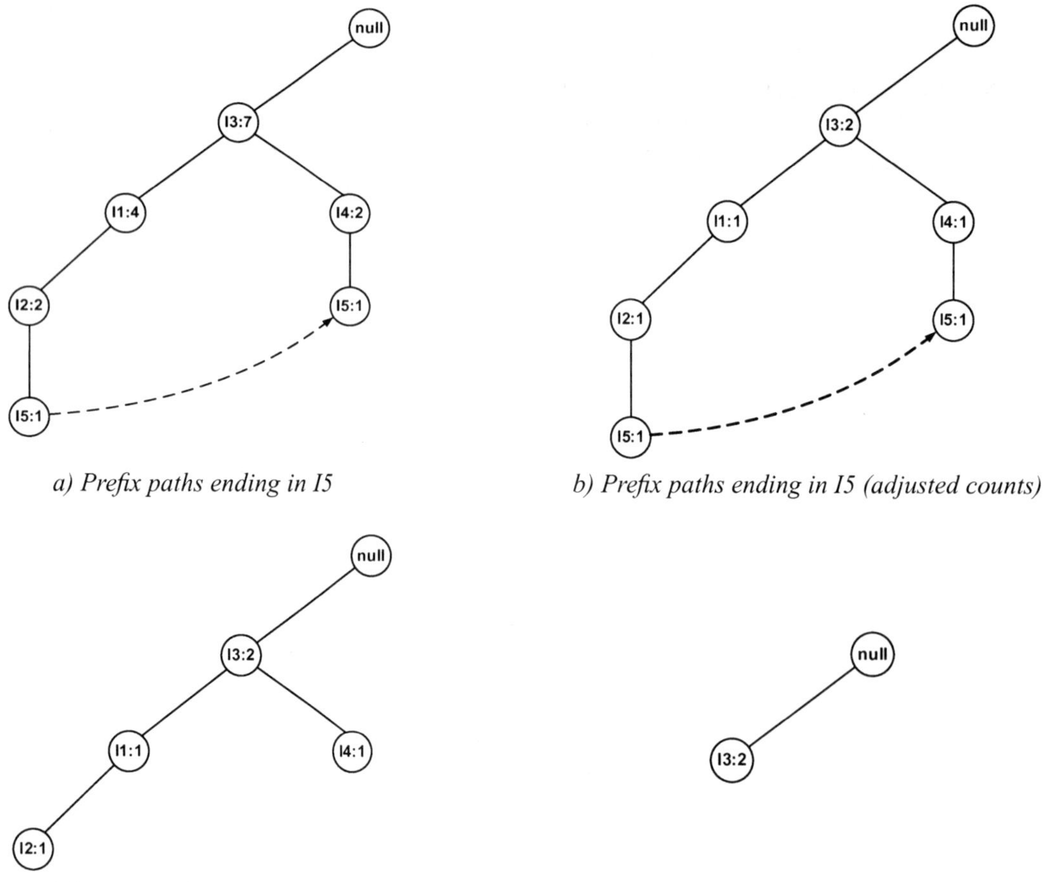


*a) Prefix paths ending in I5*

*b) Prefix paths ending in I5 (adjusted counts)*

*c) Conditional pattern base I5*

*d) Conditional FP-Tree for I5*

1pattern, as an *initial suffix pattern*, constructs its *conditional pattern base* (a sub-database consisting of the set of prefix paths in FP-tree co-occurring with it), then constructs its conditional FP-tree, and performs mining recursively on such tree. In this way, the initial problem is split into smaller subproblems.

**Example 5**: *Let us consider the FP-tree showed in Figure 4. The FP-growth algorithm tries to find frequent itemsets ending with I5, i.e. the last element from frequent-item-header table, followed by I4, I2, I1, and finally with I3. Let us start from I5. The task of finding the frequent itemsets ending with I3 is presented below:*

- First, all paths containing node *I5* are extracted. These paths are quickly accessed using node-link pointers associated with *I5*. These paths are called prefix paths and are showed in Figure 5a.
- Summing the support count of all nodes labeled with *I5* we obtain the support count for itemset {*I5*}. In this case the support count of itemset {*I5*} is 2 that is greater than or equal the minimum support count. So, the itemset {*I5*} is a frequent itemset.
- Next, the algorithm tries to solve the finding frequent itemsets ending in {*I4,I5*}, {*I2,I5*}, {*I1,I5*} and {*I3,I5*} subproblems. In order to solve these subproblems the prefix paths must be converted to a *conditional FP-tree*, which is a particular FP-tree used to find frequent itemsets ending with a particular suffix. This is constructed as follows:

a. The support counts along the prefix paths must be update because some of the counts include transactions that do not contain item *{I5}*. For example, the path *null*$\rightarrow$*I3:7*$\rightarrow$*I4:2*$\rightarrow$*I5:1* includes transactions *T8 ={I3,I4}* that does not contain *I5*. In Figure 5b is showed the tree obtained.
b. The prefix paths are truncated by removing nodes labeled with *I5* (see Figure 5c).
c. After updating the support counts along the prefix paths, some of the items may be infrequent and can be removed. In this case, the nodes *I1*, *I2* and *I3* are removed because their support count

*Algorithm 5. GenerateStrongRules*

**Input:** $D$ : the database with transactions
*minconf*: the minimum confidence threshold
$L$: the set of frequent itemsets In $D$
**Output:** $\Re$ : the set of strong association rules in $D$
**Method:**

1: $\Re = \emptyset$
2: $k = 1$
3: **for each** frequent itemset $Z \in L$ **do**
4:     **for each** $X \subset Z, X \neq \emptyset$ **do**
5:         $\text{confidence}(X \Rightarrow Z \setminus X) = \frac{\text{support\_count}(Z)}{\text{support\_count}(X)}$
6:         **if** $\text{confidence}(X \Rightarrow Z \setminus X) \geq minconf$
7:             $\Re = \Re \cup \{X \Rightarrow Z \setminus X\}$
8:         **end if**
9:     **end for**
10:**end for**
11:**return** $\Re$

is 1 less than the minimum support count 2.

The Figure 5d shows the conditional FP-tree for *I5*. From this conditional FP-tree that contains only a single path is generated the frequent itemset *{I3, I5}* with support count 2.

The performance studies (Han, Pei, Yin, & Mao, 2004) show that FP-growth algorithm is about an order of magnitude faster than the Apriori algorithm.

## Generating Association Rules

Having the set of frequent itemsets in database *D*, we can easy generate strong association rules from them. Recall that strong association rules are those that satisfy a user defined minimum support and minimum confidence threshold. If we use only frequent itemsets for generating association rules the minimum support threshold is already guaranteed. Then, we only need to generate from these the rules and kept only rules that satisfy the minimum confidence.

Pseudo-code for generating strong association rules is presented in Algorithm 5. Each frequent itemset *Z* is partitioned in two non-empty subsets, *X* and *Z\X,* and from these subsets the rule $X \Rightarrow Z \setminus X$ is generated. If this rule has confidence greater than or equal to the minimum confidence threshold (*minconf*), the rule is kept as strong association rule. Note that, the confidence computing of generated rule from frequent itemset *Z* does not require supplementary scans of the database *D* because both *X* and *Z\X* are frequent itemsets and the support count of these itemsets are already computed during frequent itemset generation.

*Table 7. Association rules generated from {I1, I2, I3}*

| Rule | Confidence | Confidence (%) |
|---|---|---|
| {I1, I2} ⇒ {I3} | 2/3 | 66.66% |
| {I1, I3} ⇒ {I2} | 2/4 | 50.00% |
| {I2, I3} ⇒ {I1} | 2/3 | 66.66% |
| {I1} ⇒ {I2, I3} | 2/5 | 40.00% |
| {I2} ⇒ {I1, I3} | 2/4 | 50.00% |
| {I3} ⇒ {I1, I2} | 2/7 | 28.57% |

**Example 6:** *Let us consider the frequent itemsets generate for database Table 4. Table 7 shows the rules that can be generated from frequent itemset Z={I1,I2,I3}.*

*If we assume that the minimum confidence threshold is 50%, then the following strong association rules are generated from this itemset:*

{*I*1, *I2*} ⇒{*I3*}, {*I*1, I3}⇒{*I2*}, {*I2*, *I3*} ⇒ {*I*1}, {*I2*} ⇒ {*I*1, *I*3}.

## Measures of Interestingness

Often, if the number of association rules generated is too big, then the support and confidence measures are not enough in order to select interesting rules. Numerous alternative measures of interestingness were proposed (Brin, Motwani, Ullman, & Tsur, 1997; Piatetsky-Shapiro, 1991; Klemettinen, Mannila, Ronkainen, Toivonen, & Verkamo, 1994). These measures can be divided into two categories: objective and subjective measures. Objective measures can be applied independently of a particular application while the subjective measures relate the interestingness of a rule to the specific information needs of a user in a specific context. Support and confidence are objective measures. Other popular objective measure of interestingness is ***lift*** defined as follows:

$$\mathrm{lift}(X \Rightarrow Y) = \frac{\mathrm{confidence}(X \Rightarrow Y)}{\mathrm{support}(Y)} = \frac{P(Y/X)}{P(Y)} = \frac{P(X \cup Y)}{P(X)P(Y)} \quad (10)$$

This is a correlation measure. Lift values greater than 1 indicate that $X$ and $Y$ are positively correlated, meaning that the occurrence of one implies the occurrence of the other. Values less than 1 indicate that $X$ is negatively correlated with occurrence of $Y$, meaning that the occurrence of $X$ inhibits the occurrence of Y. If lift value is equal to 1, then X and Y are independent and there is no correlation between them.

Lift measures how many times more often $X$ and $Y$ occur together than expected if they where statistically independent.

**Example 7**: *Let us consider transactional database D showed in Table 1 that contains few transactions from a grocery store. Let us consider the following discovered association rule:*

$$\{bread\} \Rightarrow \{eggs\} \text{ [support } = 50\% \text{ and confidence } = 60\%] \quad (11)$$

*If we consider minimum support of 30% and minimum confidence of 60%, then this rule is a strong association rule. But, this rule is misleading because the probability of purchasing eggs is 83.33% which is even larger than 60% the confidence of {bread} ⇒ {eggs}. This means that eggs and milk are negatively associated, i.e. buying one of these items actually decreases the probability of purchasing the other. The lift measure assesses the degree to which the occurrence of one decrease/increase the occurrence of the other.* In this case

$$\text{lift}(\{bread\} \Rightarrow \{eggs\}) = \frac{\text{confidence}(\{bread\} \Rightarrow \{eggs\})}{\text{support}(\{eggs\})} = \frac{0.60}{0.83} = 0.72$$

*Because this value is less than 1, there is a negative correlation between the occurrence of {bread} and {eggs}.*

Thus, the lift measure aids to filter out misleading strong association rules.

Shapiro in (Piatetsky-Shapiro, 1991) propose the ***leverage*** measure defined as follows:

$$\text{leverage}(X \Rightarrow Y) = \text{support}(X \Rightarrow Y) - \text{support}(X) \cdot \text{support}(Y)$$

Leverage measures the difference of $X$ and $Y$ appearing together in the data set and what would be expected if $X$ and $Y$ where statistically dependent.

## FUZZY ASSOCIATION RULES

Usually, into a database the attributes are not restricted to binary values, they also contain numerical or categorical values. The problem of mining quantitative association rules is introduced and an algorithm has been provided in (Rakesh&Agrawal, 1996). This algorithm first determines a number of partitions of each quantitative attribute; next, the original attributes are replaced by attribute-interval pairs, and then maps the values of each attribute value into the corresponding interval. Thus, the problem of mining quantitative association rules is reduced to mining Boolean association rules.

**Example 8**: *Let us consider a database sample with quantitative and categorical attributes presented in Table 8. As it can be seen, the attributes Age, Income and Weight are quantitative attributes and the attribute Married is a categorical attribute.*

*If we decide to partition Age into three partitions "0...29", "30...45", and "46...99", the attribute Income in 0...3000, 3001...10000, and attribute Weight in 0...80, 81...200, then the mapping values attributes to binary values is presented in Table 9.*

*An example of a quantitative association rule is:*

$$\langle Age, \text{"30...45"} \rangle \Rightarrow \langle Income, \text{"3001...1000"} \rangle$$

*Table 8. Database with quantitative and categorical attributes*

| TID | Age | Income | Weight | Married |
|---|---|---|---|---|
| 1 | 25 | 1200 | 55 | no |
| 2 | 31 | 4000 | 80 | yes |
| 3 | 40 | 4850 | 96 | yes |
| 4 | 48 | 6000 | 90 | yes |
| 5 | 60 | 3000 | 80 | no |
| 6 | 65 | 1900 | 95 | yes |

*Table 9. Mapping quantitative attributes*

| TID | Age | | | Income | | Weight | | Married | |
|---|---|---|---|---|---|---|---|---|---|
| | 0...29 | 30...45 | 46...90 | 0...3000 | 3001...10000 | 0...80 | 81...200 | no | yes |
| 1 | 1 | 0 | 0 | 1 | 0 | 1 | 0 | 1 | 0 |
| 2 | 0 | 1 | 0 | 0 | 1 | 1 | 0 | 0 | 1 |
| 3 | 0 | 1 | 0 | 0 | 1 | 0 | 1 | 0 | 1 |
| 4 | 0 | 0 | 1 | 0 | 1 | 0 | 1 | 0 | 1 |
| 5 | 0 | 0 | 1 | 1 | 0 | 1 | 0 | 1 | 0 |
| 6 | 0 | 0 | 1 | 1 | 0 | 0 | 1 | 0 | 1 |

Kouk (Chan Man Kuok & Wong, 1998) have point out the sharp boundary problem, and argue that the sharp partitioning of many quantitative attributes does not suite our data intuitive understanding. For example, if we consider a set representing people of middle age, ranging from 30 to 45 years old, then a person aged 29 years would be a 0% representative and a 31 year old would be 100%. Obviously, the difference between those ages is not that great. Implementing fuzziness can overcome this problem. Thus, the age attribute may be given by three fuzzy sets: *young*, *middle* and *old*.

A fuzzy set (Zadeh, 1965) is defined by a *membership function* that assigns to the each attribute value a value between 0 and 1 to indicate how much this value belongs to the fuzzy sets. For example, the membership function of *middle* age may assign the value of 1.0 to age 31, 0.9 to the age 29 and 0 to the age 70.

Mining association rules based on fuzzy sets can handle both quantitative and categorical data, providing the necessary support to use uncertain data types with existing algorithms; it can also create smoother transition boundaries between partitions for numerical values, therefore constituting a perfect solution for both well-defined and imprecise data.

In what follows we present the formal definition of fuzzy association rules and one of the most representative algorithms used to mine this kind of rules.

## Basic Concepts

The basic problem of finding fuzzy association rules was introduced in (Chan Man Kuok & Wong, 1998).

Let $D = \{t_1, \ldots, t_n\}$ be a database characterized by a set $I = \{i_1, \ldots, i_m\}$ of categorical or quantitative attributes (items). For each attribute $i_k$, ($k$ = 1,...,$m$), we will consider $n(k)$ associated fuzzy sets. Let $F_{i_k} = \{F_{i_k}^1, \ldots, F_{i_k}^{n(k)}\}$ be the set of all these fuzzy sets.

For an attribute $i_k$ and a fuzzy set $F_{i_k}^j$, the membership function, denoted $\mu_{F_{i_k}^j}$, is defined as:

$$\mu_{F_{i_k}^j} : \text{dom}(i_k) \to [0,1],\ k = 1,\ldots,m,\ j = 1,\ldots,n(k)$$

Typically, a domain expert provides the set of fuzzy sets associated with attributes, along with their corresponding membership functions. A method for finding fuzzy sets based on clustering techniques is presented in (Fu et al., 1998).

Using standard notation from fuzzy sets theory, the fuzzy sets $F_{i_k}^j$ are defined as:

a. if $i_k$ is a quantitative attribute , then

$$F_{i_k}^j = \begin{cases} \sum_{\text{dom}(i_k)} \mu_{F_{i_k}^j}(v_k)/v_k, & \text{if } i_k \text{ is discrete} \\ \int_{\text{dom}(i_k)} \mu_{F_{i_k}^j}(v_k)/v_k, & \text{if } i_k \text{ is continous} \end{cases}$$

b. if $i_k$ is a categorical attribute, $\text{dom}(i_k) = \{v_{1k}, \ldots, v_{n(k)k}\}$, where $n(k)$ is the number of categories, then

$$F_{i_k}^j = \frac{1}{v_{jk}}$$

For a specified transaction $t_j \in D$ and an attribute $i_k$, we denote with $t_j[i_k]$ the value of attribute $i_k$ in transaction $t_j$.

**Example 9**: *Let us consider D = $\{t_1, t_2, t_3, t_4, t_5, t_6\}$ the database sample with quantitative and categorical attributes presented in Table 8. As it can be seen, the set of attributes is I={Age, Income, Weight, Married}. For the quantitative attribute Weight we can take into consideration the following three fuzzy sets: "thin", "middle" and "fat" (see Figure 6). Thus, we have $F_{Weight}$ = {thin, middle, fat}. If we consider that a person is "fat" if his weight is greater than 90 kilograms, then the membership function of fuzzy set "fat" can be defined as:*

$$\mu_{fat} : [0,200] \to [0,1].$$

*Figure 6. Membership functions for fuzzy sets associated with Age*

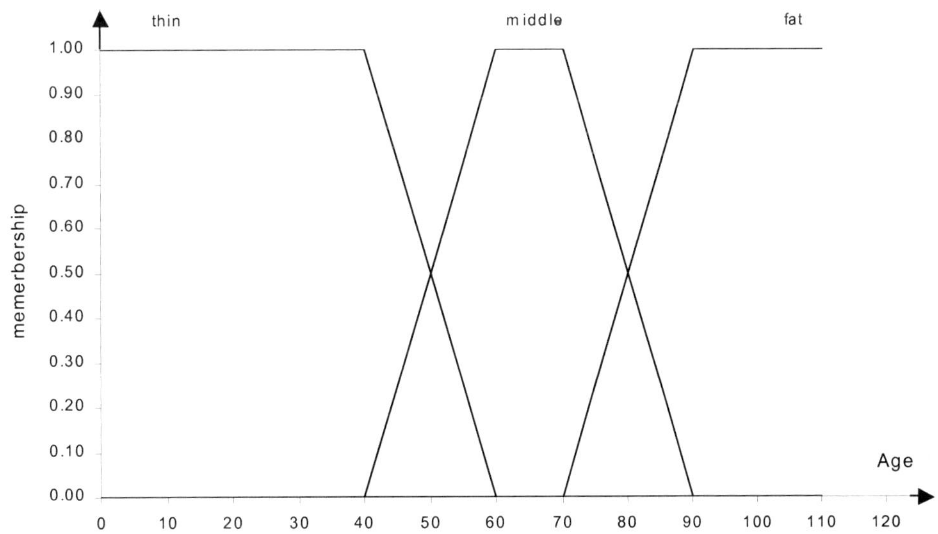

$$\mu_{fat}(x) = \begin{cases} 0, & \text{if } x < 70 \\ \dfrac{x-70}{90-70}, & \text{if } 70 \le x < 90 \\ 1, & \text{if } x \ge 90 \end{cases}$$

*The attribute Married is a categorical attribute with the domain of {no,yes}. For this attribute we consider two fuzzy sets {no,yes} defined as follows:*

$\mu_{no} : \{no, yes\} \rightarrow [0,1]$,

$$\mu_{no}(x) = \begin{cases} 1, & \text{if } x = no \\ 0, & \text{otherwise} \end{cases}$$

$\mu_{yes} : \{no, yes\} \rightarrow [0,1]$

$$\mu_{yes}(x) = \begin{cases} 1, & \text{if } x = yes \\ 0, & \text{otherwhise} \end{cases}$$

*Then,*

$\mu_{fat}(t_2) = \mu_{fat}(t_2[Weight]) = \mu_{fat}(80) = 0.5$ and =
$\mu_{yes}(t_2) = \mu_{yes}(t_2[Married]) = \mu_{yes}(yes) = 1$

*Similarly, let $F_{Age}$ = {young, middle, old}, $F_{Income}$ = {low, medium, high} be the sets of fuzzy sets associated with attributes Age and Income, respectively.*

**Definition 8 (Fuzzy itemset).** We call **fuzzy itemset** the tuple $\langle X, F_X \rangle$, where $X \subseteq I$ is a set of attributes and $F_X$ is a set of fuzzy sets associated with attributes from *X*. A fuzzy itemset $\langle X, F_X \rangle$ is called **k-fuzzy itemset** if the number of attributes from *X* is *k*.

For example, the tuple⟨{*Age*, *Income*}, {*young*, *high*}⟩ represents a fuzzy itemset for the database given in Example 9.

**Definition 9 (Fuzzy association rule):** A ***fuzzy association rule*** is an implication of the following form

$$X \in F_X \Rightarrow Y \in F_Y,$$

where $X, Y \subseteq I, X \cap Y = \emptyset, X = \{x_1, \ldots x_p\}, Y = \{y_1, \ldots y_q\}$, and $F_X = \{a_1, \ldots a_p\}$, are fuzzy sets related to attributes from *X* and *Y* respectively. More exactly, $a_i \in F_{x_i}$, ($i$=1,…,$p$) and $b_i \in F_{y_i}$, ($i$=1,…,$q$).

The left side $X \in F_X$ of the rule is called *antecedent,* while the right side of the rule $Y \in F_Y$ is called *consequent.* We denote this rule with:

$$\langle X, F_X \rangle \Rightarrow \langle Y, F_Y \rangle$$

The intuitive signification of this fuzzy association rule, $\langle X, F_X \rangle \Rightarrow \langle Y, F_Y \rangle$, is: "*if a transaction (tuple) satisfies the property $X \in F_X$, then it will also satisfy the property $Y \in F_Y$ with a high probability*".

**Example 10:** A*n example of a fuzzy association rule for the database given in Example 9 is "If Age is middle and Income is high then Weight is fat"*

*Here, X={Age, Income}, Y={Weight}, $F_X$ = {middle, high}, $F_Y$ = {fat} and the rule can be represented as:*

⟨{*Age, Income*}, {*middle, high*} ⟩ ⇒ ⟨{*Weight*} {*fat*} ⟩

Several extensions of the classic quality measures (support and confidence) have been proposed in (Chan Man Kuok & Wong, 1998) in order to express the quality of a fuzzy association rule.

**Definition 10 (Itemset fuzzy support value).** The ***fuzzy support value*** of itemset fuzzy itemset $\langle X, F_X \rangle$ in *D* is:

$$FS_{\langle X, F_X \rangle} = \frac{\sum_{t_i \in D} \prod_{x_j \in X} \alpha_{a_j}(t_i[x_j])}{|D|}$$

where

$$\alpha_{a_j}(t_i[x_j]) = \begin{cases} \mu_{a_j}(t_i[x_j]), & \text{if } \mu_{a_j}(t_i[x_j]) \ge \omega \\ 0, & \text{otherwise} \end{cases}$$

and ω is a user specified minimum threshold for the membership function. Thus, the values of membership functions less than this minimum threshold, ω, are ignored.

**Definition 11 (Frequent fuzzy itemset).** A fuzzy itemset $\langle X, F_X\rangle$ is called a ***frequent (large) fuzzy itemset*** if its fuzzy support value is greater than or equal to the minimum support threshold (*minsup*), namely $FS_{\langle X,F_X\rangle} \geq minsup$.

**Definition 12 (Rule fuzzy support value).** *Let* $\langle X, A\rangle \Rightarrow \langle Y, B\rangle$ *be a fuzzy association rule. The* ***fuzzy support value of the rule*** *is defined as fuzzy support value of the itemset* $\langle X, Y\rangle$, $\langle A, B\rangle$:

$$FS_{\langle X,A\rangle \Rightarrow \langle Y,B\rangle} = FS_{\langle\{X,Y\},\{A,B\}\rangle}$$

**Definition 13 (Frequent rule).** *A fuzzy association rule is called a* ***frequent rule*** *if its fuzzy support value is greater than or equal to the minimum support threshold (minsup), namely* $FS_{\langle X,A\rangle \Rightarrow \langle Y,B\rangle} \geq \min \sup$.

**Definition 14 (Rule Fuzzy Confidence).** *Let* $\langle X, A\rangle \Rightarrow \langle Y, B\rangle$ *a fuzzy association rule. The* ***fuzzy confidence value*** *of the rule is defined as:*

$$FC_{\langle X,A\rangle \Rightarrow \langle Y,B\rangle} = \frac{FS_{\langle Z,C\rangle}}{FS_{\langle X,A\rangle}}$$

*where* $Z = \{X, Y\}$ *and* $C = \{A, B\}$.

The confidence of the rule is defined as the fraction between the value of fuzzy support of the fuzzy itemset $\langle Z, C\rangle$ and the value of fuzzy support of the fuzzy itemset $\langle X, A\rangle$.

*Table 10. The membership functions values*

| TID | Age | | | Income | | | Weight | | | Married | |
|---|---|---|---|---|---|---|---|---|---|---|---|
| | young | middle | old | low | medium | high | thin | middle | fat | no | yes |
| 1 | 1.00 | 0.00 | 0.00 | 0.60 | 0.20 | 0.00 | 0.25 | 0.75 | 0.00 | 1 | 0 |
| 2 | 0.40 | 0.60 | 0.00 | 0.00 | 1.00 | 0.00 | 0.00 | 0.50 | 0.50 | 0 | 1 |
| 3 | 0.00 | 1.00 | 0.00 | 0.00 | 0.15 | 0.85 | 0.00 | 0.00 | 1.00 | 0 | 1 |
| 4 | 0.00 | 0.70 | 0.60 | 0.00 | 0.00 | 1.00 | 0.00 | 0.00 | 1.00 | 0 | 1 |
| 5 | 0.00 | 0.00 | 1.00 | 0.00 | 1.00 | 0.00 | 0.00 | 0.50 | 0.50 | 1 | 0 |
| 6 | 0.00 | 0.00 | 1.00 | 0.00 | 0.90 | 0.00 | 0.00 | 0.00 | 1.00 | 0 | 1 |

*Box 2.*

$$FS_{\langle\{Age,Income\},\{old,medium\}\rangle \Rightarrow \langle\{Weight\},\{fat\}\rangle} = \frac{0.00\times 0.20\times 0.00 + 0.00\times 1.00\times 0.50}{6} + \frac{0.00\times 0.15\times 1.00 + 0.60\times 0.00\times 1.00 + 1.00\times 1.00\times 0.50 + 1.00\times 0.90\times 1.00}{6} = 0.23$$

*Box 3.*

$$FC_{\langle\{Age,Income\},\{old,medium\}\rangle \Rightarrow \langle\{Weight\},\{fat\}\rangle} = \frac{FS_{\langle\{Age,Income,Weight\},\{old,medium,fat\}\rangle}}{FS_{\langle\{Age,Income\},\{old,medium\}\rangle}} = \frac{0.23}{0.32} = 0.73$$

**Example 11:** *Let us consider the database D presented in Table 8 and the corresponding membership functions values associated with database attributes presented in Table 10.*

*Having the fuzzy association rule ⟨{Age, Income}, {old, medium}⟩ ⇒ ⟨{Weight} {fat}⟩, and considering* $\varpi = 0.1$, *then the fuzzy support value is seen in Box 2.*

*The fuzzy confidence value for this rule is seen in Box 3.*

A fuzzy association rule is considered as *interesting* if it has enough support and high confidence value. This association rule can be encountered under the name of ***strong fuzzy association rule***.

**Problem (Mining Fuzzy Association Rules).** *Given the database D, characterized by a set of attributes I, the fuzzy sets associated with attributes from I, the minimum support threshold for membership function ω, the minimum support threshold (minsup) and the minimum confidence threshold (minconf), extract all interesting fuzzy association rules.*

Just as in the Boolean association rule case, the problem of the fuzzy association rules mining can be divided in two subproblems:

1. *find all frequent fuzzy itemsets.*
2. *generate the fuzzy association rules from the frequent fuzzy itemsets founded.*

The majority of algorithms for mining fuzzy association rules (see (Gyenesei, 2000b) and (Hong, Kuo, Chi, & Wang, 2000) are based on the Apriori algorithm.

## MINING FREQUENT FUZZY ITEMSETS

In this section we present an extension of the Apriori algorithm (Gyenesei, 2000a) for the extraction of the frequent fuzzy itemsets. The Apriori property can be extended to fuzzy itemsets such as:

**Lemma 2 (Fuzzy Apriori Property).** *If a fuzzy itemset* $\langle X, F_X\rangle$ *is a frequent fuzzy itemset in D and* $Y \subseteq X$, $F_Y \subseteq F_X$, *then the fuzzy itemset* $\langle Y, F_Y\rangle$ *is also frequent in D.*

*Proof:* Let $Y \subseteq X$ and $t_i \in D$. Because $\alpha_{a_j}(t_i[x_j]) \in [0,1], \forall x_j \in X$, then we have the results that can be seen in Box 4.

But, $\langle X, F_X\rangle$ is a frequent fuzzy itemset $\Leftrightarrow FS_{\langle X, F_X\rangle} \geq minsup \Rightarrow FS_{\langle Y, F_Y\rangle} \geq minsup$, i.e. $\langle Y, F_Y\rangle$ is a frequent fuzzy itemset. □

From the above lemma we can draw the conclusion that any fuzzy subitemset of a frequent fuzzy itemset is also frequent.

*Box 4.*

$$\prod_{x_j \in X} \alpha_{a_j}(t_i[x_j]) \leq \prod_{y_j \in Y \subseteq X} \alpha_{a_j}(t_i[y_j]) \Rightarrow$$

$$\sum_{t_i \in D} \prod_{x_j \in X} \alpha_{a_j}(t_i[x_j]) \leq \sum_{t_i \in D} \prod_{y_j \in Y \subseteq X} \alpha_{a_j}(t_i[y_j]) \Rightarrow$$

$$\frac{\sum_{t_i \in D} \prod_{x_j \in X} \alpha_{a_j}(t_i[x_j])}{|D|} \leq \frac{\sum_{t_i \in D} \prod_{y_j \in Y \subseteq X} \alpha_{a_j}(t_i[y_j])}{|D|} \Rightarrow$$

$$FS_{\langle X, F_X\rangle} \leq FS_{\langle Y, F_Y\rangle}$$

*Algorithm 6. FuzzyApriori*

**Input:** *D*: the database
*I*: the set of attributes
$F_{i_1}, \ldots, F_{i_m}$: the set of fuzzy sets associated with attributes from $I$
*minsup*: the minimum support threshold
$\omega$: minimum threshold for the membership function
**Output:** *L*: the set of frequent fuzzy itemsets in $D$
**Method**:

$L = \varnothing$
$C_1 = \left\{ \langle \{i_k\}, \{F_{i_k}^j\} \rangle \middle| i_k \in I, F_{i_k}^j \in F_{i_k} \right\}$
$k = 1$
**while** $C_k \neq \varnothing$ **do**
    *//compute fuzzy support count for each candidate itemset*
    **for each** transaction $t \in D$ **do**
        **for each** canditate $\langle X, A \rangle \in C_k, X = \{x_1, \ldots, x_k\}, A = \{a_1, \ldots, a_k\}$ **do**
            $FS_{\langle X, A \rangle} += \prod_{x_j \in X} \alpha_{a_j}(t_i[x_j])$
        **end for**
    **end for**
    $L_k = \{\langle X, A \rangle \in C_k \mid FS_{\langle X, A \rangle} \geq minsup \times |D|\}$
    $L = L \cup L_k$
    $k++$
    $C_k$ = fuzzy_apriori_gen($L_{k-1}$)
**end while**
**return** $L$

**Corollary 3:** *If a k-fuzzy itemset contains a (k-1)-fuzzy itemset that is not frequent, then it is not frequent.*

The algorithm, similarly with Apriori algorithm, starts the computation of the set $L_1$ of frequent 1-fuzzy itemsets, then $L_1$ is used to generate the set $L_2$ of frequent 2-fuzzy itemsets, which is used to find $L_3$, and so on.

The pseudo-code for this algorithm is presented in Algorithm 6.

The algorithm used to generate the set $L_{k-1}$ $L_k$ of frequent *k*-fuzzy itemsets from $L_{k-1}$, $k \geq 2$, contains two steps:

**Step 1: (Join Step)** Generate the set $C_k$ of candidate *k*-fuzzy itemsets by joining $L_{k-1}$ with $L_{k-1}$. We assume that all items in the itemsets are sorted in lexicographic order. The $C_k$ is obtained as in Box 5.

**Step 2: (Prune Step)** Based on Corollary 3, we remove from $C_k$ all *k*-fuzzy itemsets containing some (*k-1*)-fuzzy sub-itemsets that are not present in $L_{k-1}$. In this way, the number of candidates is reduced. The generated set $C_k$ is a superset of $L_k$.

**Example 12**: *Let us consider that $L_2$ consists of the following itemsets: ⟨{Age, Income}, {old, medium}⟩, ⟨{Age, Income}, {old, high}⟩, ⟨{Age, Weight}, {old, fat}⟩, ⟨{Income, Weight}, {high, medium}⟩.*

*Then the set $C_3$ of candidate 3-fuzzy itemsets obtained by joining $L_2$ with $L_2$ consists of the following fuzzy itemsets:*

*Box 5.*

$$
\begin{aligned}
C_k = L_{k-1} \bowtie L_{k-1} = \{ \langle Z, C \rangle \mid & \langle X, A \rangle \in L_{k-1}, \langle Y, B \rangle \in L_{k-1}, \\
& \langle Z\ C \rangle = \langle \{x_1, \ldots, x_{k-2}, x_{k-1}, y_{k-1}\}, \{a_1, \ldots, a_{k-2}, a_{k-1}, b_{k-1}\} \rangle, \\
& \langle X\ A \rangle = \langle \{x_1, \ldots, x_{k-2}, x_{k-1}\}, \{a_1, \ldots, a_{k-2}, a_{k-1}\} \rangle, \\
& \langle Y\ B \rangle = \langle \{x_1, \ldots, x_{k-2}, y_{k-1}\}, \{a_1, \ldots, a_{k-2}, b_{k-1}\} \rangle, \\
& x_{k-1} < y_{k-1} \}
\end{aligned}
$$

*Algorithm 7. fuzzy_apriori_gen*

**Input:** $L_{k-1}$ : the set of frequent (*k-1*)-fuzzy itemsets
**Output:** $C_k$ : the set of candidate *k*-fuzzy itemsets
**function** fuzzy_apriori_gen( $L_{k-1}$ )
$C_k = \varnothing$
**for each** (*k-1*)-itemset $\langle X, A \rangle \in L_{k-1}, X = \{x_1, \ldots, x_{k-1}\}, A = \{a_1, \ldots, a_{k-1}\}$ **do**
  **for each** (*k-1*)-itemset $\langle Y, B \rangle \in L_{k-1}, X = \{y_1, \ldots, y_{k-1}\}, A = \{b_1, \ldots, b_{k-1}\}$ **do**
    **if** $((x_1 = y_1) \wedge (a_1 = b_1)) \wedge \ldots \wedge ((x_{k-2} = y_{k-2}) \wedge (a_{k-2} = b_{k-2})) \wedge x_{k-1} < y_{k-1}$ **then**
      *//join step*
      $\langle Z, C \rangle = \langle \{x_1, \ldots, x_{k-1}, y_{k-1}\}, \{a_1, \ldots, a_{k-1}, b_{k-1}\} \rangle$
      *//prune step*
      $has\_infrequent\_subset = false$
      **for each** (*k*-1)-fuzzy itemset $\langle Z', C' \rangle$ of $\langle Z, C \rangle$ **do**
        **if** $\langle Z', C' \rangle \notin L_{k-1}$ **then**
          $has\_infrequent\_subset = true$
          **break;**
        **end if**
      **end for**
      **if not** $has\_infrequent\_subset$ **then**
        $C_k = C_k \cup \{\langle Z, C \rangle\}$
    **end if**
  **end for**
**end for**
**return** $C_k$

⟨{*Age, Income, Weight*}, {*old, medium, fat*}⟩,

⟨{*Age, Income, Weight*}, {*old, high, fat*}⟩.

In Algorithm 7 it is presented the pseudo-code for fuzzy_apriori_gen procedure used to generate candidate *k*-fuzzy itemsets, $C_k$. As input we provide the set $L_{k-1}$ of frequent (*k*-1)-fuzzy itemsets.

## Generating Fuzzy Association Rules

After finding the frequent fuzzy itemsets, we will derive from them the strong fuzzy association rules. From a frequent fuzzy itemset $\langle Z, F_Z \rangle$, we consider all subsets of $Z$, $X \subset Z$, and derive rules of the following form $\langle X, F_X \rangle \Rightarrow \langle Z \backslash X, F_{Z \backslash X} \rangle$. Because the fuzzy support value of

*Algorithm 8. GenerateStrongFuzzyRules*

**Input:** $D$ : the database with transactions
*minconf*: the minimum confidence threshold
$L$: the set of frequent fuzzy itemsets in $D$
**Output:** $\Re$ : the set of strong fuzzy association rules in $D$
**Method:**

$\Re = \varnothing$
$k = 1$
**for each** frequent fuzzy itemset $\langle Z, F_Z \rangle \in L$ **do**
    **for each** subset $X \subset Z, X \neq \varnothing$ **do**
        **if** $FC_{\langle X,F_X \rangle \Rightarrow \langle Z\setminus X, F_{Z\setminus X}\rangle} \geq minconf$ **then**
            $\Re = \Re \cup \{\langle X, F_X \rangle \Rightarrow \langle Z \setminus X, F_{Z\setminus X} \rangle\}$
        **end if**
    **end for**
**end for**
**return** $\Re$

generated fuzzy association rule, $FS_{\langle X,F_X\rangle} \Rightarrow \langle Z\setminus X, F_{Z\setminus X}\rangle$, is equal with fuzzy support value $FS\langle Z, F_Z\rangle$ of the fuzzy itemset $\langle Z, F_Z\rangle$, then the rule satisfies the minimum support threshold. It remains only to compute the rule confidence and to check if this value is larger than the user-specified minimum confidence threshold. In Algorithm 8 we present the pseudo-code used to describe the procedure that generates strong fuzzy association rules.

## CONCLUSION

Mining association rules is one of most important task in data mining used to discover associations among items from large databases. We presented in this chapter the theoretical concepts related to this task. The Apriori and FP-growth algorithms used to mining association rules have been highlighted. The fuzzy association rules have also been described. These rules can handle quantitative and categorical data and are expressed in linguistic terms, which are more natural and understandable for human beings.

To the best of our knowledge, the related literature comprises many papers that focus their topics on various extensions of association rules (i.e., quantitative association rules, negative association rules, weighted association rules, generalized association rules, temporal association rules, etc). Also, new algorithms or improvements of existing algorithms for mining association rules have been developed, introduced and tested into practice.

The associations rules can be used in different domains: market and risk management, medicine (Ordonez, Santana, & Braal, 2000), Web mining (Pei, Han, Mortazavi-Asl, & Zhu, 2000), document analysis (Holt & Chung, 1999), network intrusion detection (Zhao & Lu, 2004), (Lee, Stolfo, & Mok, 2000), etc.

## REFERENCES

Agrawal, R., Imielinski, T., & Swami, A. N. (1993). Mining association rules between sets of items in large databases. In P. Buneman & S. Jajodia (Eds.), *Proceedings of the 1993 ACM SIGMOD International Conference on Management of*

*Data, Washington, D.C., May 26-28, 1993* (pp. 207-216). ACM Press.

Agrawal, R., & Srikant, R. (1994). Fast algorithms for mining association rules. In J. B. Bocca, M. Jarke, & C. Zaniolo (Eds.), *Proc. 20th Int. Conf. Very Large Data Bases, (VLDBb)* (pp. 487-499). Morgan Kaufmann.

Brin, S., Motwani, R., Ullman, J. D., & Tsur, S. (1997). Dynamic itemset counting and implication rules for market basket data. In *Proceedings ACM SIGMOD International Conference on Management of Data* (pp. 255-264). ACM Press.

Chan Man Kuok, A. F., & Wong, M. H. (1998). Mining fuzzy association rules in databases. *SIGMOD Rec., 27*(1), 41-46.

Chen, M.-S., Han, J., & Yu, P. S. (1996). Data mining: An overview from a database perspective. *IEEE Transactions on Knowledge and Data Engineering, 8*(6), 866-883.

Dubois, D., Hullermeier, E., & Prade, H. (2006). A systematic approach to the assessment of fuzzy association rules. *Data Mining and Knowledge Discovery, 13*(2), 167=192.

Fu, A., Wong, M., Sze, S., Wong, W., Wong, W., & Yu, W. (1998). Finding fuzzy sets for the mining of fuzzy association rules for numerical attributes. In *Proceedings of the First International Symposium on Intelligent Data Engineering and Learning (IDEAL'98)* (pp. 263-268).

Gyenesei, A. (2000a). *A fuzzy approach for mining quantitative association rules* (Tech. Rep. No. TUCSTR-336).

Gyenesei, A. (2000b). Mining weighted association rules for fuzzy quantitative items. In *Principles of data mining and knowledge discovery* (p. 416-423).

Han, J., & Kamber, M. (2006). *Data mining: Concepts and techniques* (2nd ed.). Morgan Kaufmann.

Han, J., Pei, J., & Yin, Y. (2000). Mining frequent patterns without candidate generation. In W. Chen, J. Naughton & P.A. Bernstein (Eds.), *2000 ACM SIGMOD Intl. Conference on Management of Data* (pp.1-12). ACM Press.

Han, J., Pei, J., Yin, Y., & Mao, R. (2004). Mining frequent patterns without candidate generation: A frequent-pattern tree approach. *Data Mining and Knowledge Discovery, 8*(1), 53-87.

Holt, J. D., & Chung, S. M. (1999). Efficient mining of association rules in text databases. In *CIKM'99: Proceedings of the eighth international conference on information and knowledge management* (pp.234–242). New York: ACM.

Hong, T.-P., Kuo, C.-S., Chi, S.-C., & Wang, S.-L. (2000). Mining fuzzy rules from quantitative data based on the AprioriTid Algorithm. In *Sac '00: Proceedings of the 2000 ACM symposium on applied computing* (pp. 534-536). New York: ACM Press.

Jong Soo Park, P. S. Y., Ming-Syan Chen. (1995). An effective hash-based algorithm for mining association rules. In *Proceedings of the 1995 ACM SIGMOD International Conference on Management of Data* (pp.175-186). San Jose, Canada.

Klemettinen, M., Mannila, H., Ronkainen, P., Toivonen, H., & Verkamo, A. I. (1994). Finding interesting rules from large sets of discovered association rules. In N. R. Adam, B. K. Bhargava, & Y. Yesha (Eds.), *Third International Conference on Information and Knowledge Management (CIKM'94)* (pp. 401-407). ACM Press.

Lee,W., Stolfo, S. J., & Mok, K.W. (2000). Adaptive intrusion detection: A data mining approach. *Artificial Intelligence Review, 14*(6), 533-567.

Ordonez, C., Santana, C. A., & Braal, L. de. (2000). Discovering interesting association rules in medical data. In *ACM SIGMOD workshop on research issues in data mining and knowledge discovery* (p. 78-85).

Pei, J., Han, J., Mortazavi-Asl, B., & Zhu, H. (2000). Mining access patterns efficiently from web logs. In *PADKK '00: Proceedings of the 4th Pacific-Asia Conference on Knowledge Discovery and Data Mining, Current Issues and New Applications* (pp. 396-407). London, UK: Springer-Verlag.

Piatetsky-Shapiro, G. (1991). Discovery, analysis, and presentation of strong rules. In *Knowledge discovery in databases* (pp. 229-248). AAAI/ MIT Press.

Rakesh Agrawal, R. S. (1996). Mining quantitative association rules in large relational tables. In H. V. Jagadish & I. S. Mumick (Eds.), *Proceedings of the 1996 ACM SIGMOD international conference on management of data* (pp. 1-12). Montreal, Quebec, Canada.

Savasere, A., Omiecinski, E., & Navathe, S. B. (1995). An efficient algorithm for mining association rules in large databases. In U. Dayal, P. M. D. Gray, & S. Nishio (Eds.), *Proceedings of 21th International Conference on Very Large Data Bases (VLDB'95)* (p. 432-444). Morgan Kaufmann.

Tan, P.-N., Steinbach, M., & Kumar, V. (2005). *Introduction to data mining.* Addison Wesley.

Zadeh, L. (1965). Fuzzy sets. *Information and Control*, *8*(3), 338-353.

Zhao, D., & Lu, Y.-s. (2004). Association rule mining in intrusion detection systems. In S. J. B. Yoo, K.-w. Cheung, Y.-C. Chung, & G. Li (Eds.), *Network Architectures, Management,* (Vol. 5282, p. 577-581).

## KEY TERMS AND DEFINITIONS

**Apriori Alorithm:** The seminal algorithm for mining frequent itemsets for Boolean association rules based on Apriori property

**Association Rule:** An implication of the form $X \Rightarrow Y$, means that transactions including $X$ will include $Y$ as well, with a high probability.

**Association Rules Mining:** The finding strong association rules.

**Confidence:** A quality measure that indicates the strength or the reliability of the associations detected. The ***confidence*** of association rule $X \Rightarrow Y$ is defined as the *proportion (percentage) of transactions containing X that also contain Y* (conditional probability of occurring $Y$ given $X$).

**Data Mining:** The process of nontrivial extraction of implicit, previously unknown and potentially useful information (such as knowledge rules, constraints, regularities) from data in databases.

**Frequent Itemset:** An itemset with support is greater than or equal to minimum support threshold

**Itemset:** A collection of items. For example, all items bought by one customer during one visit to a department store.

**Market Basket Analysis:** Analysis of customer habits by searching for the set of items that are frequently purchased together.

**Strong Association Rule:** An association rule having support and confidence greater than or equal to a user-specified minimum support threshold and respectively a minimum confidence threshold.

**Support:** A quality measure used to reflect the usefulness of discovered rules. The ***support*** of association rule $X \Rightarrow Y$ is defined as the *proportion/ percentage of transactions that contains both X and Y* (the probability of both $X$ and $Y$ cooccurring in transactions)

# Compilation of References


Aalst, W. M. P. van der, Hofstede, A. H. M. ter, & Weske, M. (2003). Business process management: A survey. In *Proceedings of the International Conference on Business Process Management (BPM 2003), Eindhoven, The Netherlands* (p. 1-12). Springer.

Abdelmoty, A. I., & Smart, P. D., & Jones, C. B., Fu, G., & Finch, D. A. (2005). Critical evaluation of ontology languages for geographic information retrieval on the Internet. *In Journal of Visual Languages & Computing* (pp., 331--358).

Abiteboul, S. & Kanellakis, P. C. (1989). Object identity as a query language primitive. In *Proceedings of the ACM SIGMOD International Conference on Management of Data* (pp. 159–173). Portland, Oregon, USA.

Abiteboul, S., & dos Santos, C. (1995). IQL(2): A model with ubiquitous objects. In *Proceedings of the International Workshop on Database Programming Languages*, Gubbio, Italy. (LNCS 760).

Abiteboul, S., & Vianu, V. (1991). Datalog extensions for database queries and updates. *Journal of Computer and System Science, 43*, 62-124.

Abiteboul, S., Lausen, G., Uphoff, H., & Waller, E. (1993). Methods and rules. In *Proceedings of the ACM SIGMOD International Conference on Management of Data* (pp. 32-41). Washington, D.C: ACM Press.

Adi, A., & Etzion, O. (2002). The situation manager rule language. In *Proceedings RuleML 2002.*

Adi, A., & Etzion, O. (2004). Amit - The situation manager. *The VLDB Journal, 13*(2), 177-203.

agent-based discrete event simulation. In P. Giorgini, B. Henderson Sellers & M. Winikoff (Eds.), *Proceedings of Agent-Oriented Information Systems, 5th International Bi-Conference Workshop, AOIS 2003, Melbourne, Australia and Chicago, IL, USA* (LNCS 3030, pp. 174-188).

Agrawal, R., & Srikant, R. (1994). Fast algorithms for mining association rules. In J. B. Bocca, M. Jarke, & C. Zaniolo (Eds.), *Proc. 20th Int. Conf. Very Large Data Bases, (VLDBb)* (pp. 487-499). Morgan Kaufmann.

Agrawal, R., Cochrane, R., & Lindsay, B (1991). On maintaining priorities in a production rule system. In *Proceedings International Conference on Very Large Data Bases* (pp. 479-487).

Agrawal, R., Imielinski, T., & Swami, A. N. (1993). Mining association rules between sets of items in large databases. In P. Buneman & S. Jajodia (Eds.), *Proceedings of the 1993 ACM SIGMOD International Conference on Management of Data, Washington, D.C., May 26-28, 1993* (pp. 207-216). ACM Press.

Aiken, A., Widom, J., & Hellerstein, J.M. (2004). Behaviour of database production rules: Termination, confluence and observable determinism. In *Proceedings of Int. Conf. on Management of Data*. ACM.

Albert, L. (1989). Average case complexity analysis of rete pattern-match algorithm and average size of join in databases. *Rapport de recherche no. 1010*, INRIAROC-QUENCOURT.

Alferes, J., Damasio, C., & Pereira, L. (2003). Semantic Web logic programming tools. (LNCS 2901, pp. 16-32).

Alferes, J.J., et al. (2002). Evolving logic programs. In *JELIA'02*.

Allen, P., Angele, J. Barker, K., Baxter, D., Chaudhri, V., Chaw, S., et al. (2004). Project Halo: Towards a digital Aristotle. *AI Magazine 25(4),* 29-48.

Alter, S. L. (1980). *Decision support systems: Current practice and continuing challenges*. Reading, Mass: Addison-Wesley Pub.

Alves, A. et al. (2007). Web services business process execution language (BPEL) Version 2.0. *OASIS*. Retrieved May 4, 2008 from http://docs.oasis-open.org/wsbpel/2.0/wsbpel-specification-draft.html.

Andert, E. P. (1992). Integrated knowledge-based system design and validation for solving problems in uncertain environments. *Int. J. of Man-Machine Studies, 36*, 357-373.

Angele, J., & Boley, H., & Bruijn, J., & Fensel, D. (2005). *Web Rule rule Language language (WRL).* Retrieved 2008, from http://www.w3.org/Submission/WRL

Antoniou, G, & van Harmelen, F. (2004). *A Semantic Web primer.* MIT Press.

Antoniou, G., Berndtsson, M., Spreeuwenberg, S., Taveter, K., & Wagner, G. (2004). *Report: A first-version visual rule language* (REWERSE Deliverable I1-D1).

Antoniou, G., Billington, D., Governatori G., & Maher, M., J. (2001). Representation results for defeasible logic. *ACM Transactions on Computational Logic, 2*(2), 255-287.

Antoniou, G., Damasio, C., V., Grosof, B., Horrocks, I., Kifer, M., Maluszynski, J., et al. (2005). *Combining rules and ontologies: A survey.* Deliverables I3-D3, REWERSE. Retrieved from, http://rewerse.net/deliverables/m12/i3-d3.pdf.

Antoniou, G., Damasio, C.V., Grosof, B., Horrocks, I, Kifer, M, Maluszynski, J., et al. (2005). *Combining rules and ontologies — A survey.* Deliverables I3-D3, REWERSE. Retrieved from, http://rewerse.net/deliverables/m12/i3-d3.pdf

Arroyo, S., Toma, I., Roman, D., Drumm, C., Dimitrov, M., Spork, M., Nagypal, G., Domingue, J., & Henke, J. (2004). *Report on state of the art and requirements analysis (Deliverable 3.1).* DIP EU FP6 Integrated Project.

Astefanoaei, L., Mol, C., Sindlar, M., & Tinnemeier, N. (2008).Going for gold with 2APL. In *Proceedings of the Fifth International Workshop on Programming Multi-Agent Systems.* Springer.

Atkinson, M. P., & Buneman, O. P. (1987). Types and persistence in database programming languages. *ACM Computing Surveys, 19*(2), 105-190.

ATLAS group, Nantes. (ver. 0.7) (n.d.). *Atlas transformation language - User manual.* Retrieved from http://www.eclipse.org/gmt/atl/doc/ATL_User_Manual[v0.7].pdf

ATLAS Transformation Language (ATL). (2008). Retrieved from http://www.sciences.univ-nantes.fr/lina/atl.

Baader, F. (2003). *The description logic handbook : Theory, implementation and applications.* Cambridge University Press.

Baader, F., & Sattler, U. (2001). An overview of tableau algorithms for description logics. *Studia Logica, 69*(1), 5-40.

Baader, F., Calvanese, D., McGuiness, D., Nardi, D., & Patel-Schneider, P. (2003). *The description logic handbook: Theory, implementation, and applications*. Cambridge: Cambridge University Press.

Baader, F., Calvanese, D., McGuinness, D., Nardi, D., & Patel-Schneider, P.F. (Eds.) (2002). *The description logic handbook*. Cambridge: University Press.

Baba, P. (2006). Using ORM-based models as a foundation for a data quality firewall in an advanced generation data warehouse. In *Proceedings of ORM'06* (LNCS).

Baclawski, K., Kokar, M., Kogut, P., Hart, J., Smith, J., Holmes, W., Letkowski, J., & Aronson, M. (2001). Extending UML to support ontology engineering for the Semantic Web. In *Proc. of 4th International Conference on UML* (pp. 342-360).

Baclawski, K., Kokar, M.K., Kogut, P., Hart, L., Smith, J.E., Letkowski, J., Emery, P. (2002). Extending the Unified Modeling Language for ontology development. *International Journal*

Baeten, J. C. (2005). A brief history of process algebra. *Theor. Comput. Sci. 335*(2-3), 131-146.

Baeten, J.C.M., & Weijland, W.P. (1990). Process Algebra. In *Cambridge tracts in theoretical computer science*. Cambridge University Press.

Bailey, J., Bry, F., Furche, T., & Schaffert, S. (2005). Web and Semantic Web query languages: A survey. In N. Eisinger & J. Małuszynski (Eds.), *Reasoning Web, First International Summer School 2005* (LNCS 3564). Berlin: Springer.

Bajec, M., & Krisper M. (2001). *Managing business rules in enterprises*.

Bajec, M., & Krisper, M. (2004). A methodology and tool support for managing business rules in organisations. *Information Systems, 30*(6), 423-443.

Bal, R., & Balsters, H. (1993). A deductive and typed object-oriented language. In S. Ceri, K. Tanaka, & S. Tsur (Eds.), *Proceedings of the International Conference on Deductive and Object-Oriented Databases* (pp. 340–359). (LNCS 760).

Balanced Scorecard Institute. (2007). *What is the balanced scorecard?* Retrieved January 9, 2008 from http://www.balancedscorecard.org/BSCResources/AbouttheBalancedScorecard/tabid/55/Default.aspx

Baldoni, R., Contenti, M. & Virgillito, A. (2003). The evolution of publish/subscribe communication systems. In *Future directions in distributed computing*.

Baley, J., et al. (1997). *Abstract interpretation of active rules and its use in termination analysis*. Paper presented at Int. Conf. on Database Theory.

Ball, M., Boley, Hirtle, H.D., Mei, J., & Spencer, B. (2005). The OO jDREW reference implementation of RuleML. In *Proceedings of Rules and Rule Markup Languages for the Semantic Web* (RuleML-2005) (pp. 218-223). (LNCS 3791).

Ball, P. (1996). *Introduction to discrete event simulation*. University of Strathclyde.

Baral, C. (2003). *Knowledge representation, reasoning, and declarative problem solving*. Cambridge. Cambridge University Press.

Baralis, E., & Widom, J. (1994). An algebraic approach to rule analysis by means of triggering and activation graphs. In *VLDB 94*.

Barbuceanu, M., Gray, T., & Mankovksi, S. (1999). Roles of obligations in multiagent coordination. *Applied Artificial Intelligence, 13*(1), 11-38.

Barja, M. L., Fernandes, A. A. A., Paton, N. W., Williams, M. H., Dinn, A., & Abdelmoty, A. I. (1995). Design and implementation of ROCK & ROLL: A deductive object-oriented database system. *Information Systems, 20*(3), 185-211.

Barr, V. (1999). Applications of rule-base coverage measures to expert system evaluation. *Knowledge-Based Systems, 12*, 27-35.

Basel Committee (2004). *Basel II revised international capital framework*. Retrieved from http://www.bis.org/publ/bcbsca.htm.

Bass, L., Clements, P., & Kazman, R. (2005). *Software architecture in practice*. Addison-Wesley.

Bassiliades, N., Antoniou, G., Vlahavas, I. (2006). A defeasible logic reasoner for the Semantic Web, *International Journal on Semantic Web and Information Systems 2(1), 1-4.*

Bassiliades, N., Vlahavas, I., & Elmagarmid, A. K. (2000). E-DEVICE: An extensible active knowledge base system with multiple rule type support. *IEEE Transactions on Knowledge and Data Engineering, 12*(5), 824-844.

Batory, D. (1994). *The LEAPS Algorithm* (Tech. rep. CS-TR-94-28). Department of Computer Sciences, the University of Texas, Austin, Texas.

Bauer, B., Muller, J., Odell, J., (2001). Agent UML: A formalism for specifying multiagent interaction. In P. Ciancarini & M. Wooldridge (Eds.), *Agent-oriented software engineering* (pp. 91-103). Springer-Verlag.

Baumeister, J. (2004). *Agile development of diagnostic knowledge systems*. Amsterdam, The Netherlands: IOS Press.

Baumeister, J., & Puppe, F. (2008). Web-based knowledge engineering using knowledge wikis. In *Proc. of the 2008 AAAI Spring Symposium on "Symbiotic Relationships between Semantic Web and Knowledge Engineering*" (pp. 1-13). Stanford University, USA.

Baumeister, J., Bregenzer, J., & Puppe, F. (2006). Grey-box robustness testing of rule systems. In *Proc. of the 29th Annual German Conference on Artificial Intelligence (KI-2006)* (LNAI 4314, pp. 346-360).

Baumeister, J., Kleemann, T., & Seipel, D. (2007). Towards the verification of ontologies with rules. In *Proc. of the 20th International Florida Artificial Intelligence Research Society Conference* (pp. 524-529). Menlo Park, USA: AAAI Press.

Baumeister, J., Knauf, R., & Puppe, F. (2005). Semi-automatic generation of test cases by case morphing. In *Proc. 18th Intl. FLAIRS Conference* (pp. 814-815). Menlo Park, USA: AAAI Press.

Baumeister, J., Menge, M., & Puppe, F. (2008). Visualization techniques for the evaluation of knowledge systems. In *Proc. of the 21th International Florida Artificial Intelligence Research Society Conference (FLAIRS-2008)*. Menlo Park, USA: AAAI Press.

Baumeister, J., Puppe, F., & Seipel, D. (2004). Refactoring methods for knowledge bases. *Engineering Knowledge in the Age of the Semantic Web: 14th International Conference, EKAW 2004* (LNAI 3257, pp. 157-171).

Bechhofer, S., Moller, R., & Crowther, P. (2003). The DIG description logic interface. In D. Calvanese, G. de Giacomo, & F. Franconi (Eds.), *Proc. of the 2003 Int. Workshop on Description Logics (DL 2003)* (Vol. 81 of CEUR Workshop Proceedings), Rome, Italy.

Bechhofer, S., van Harmelen, F., Hendler, J., Horrocks, I., McGuinness, L., D., Patel-Schneider, F., P., & Stein, A., L. (2004). OWL Web Ontology Language Reference. *W3C Recommendation*. Retrieved from http://www.w3.org/TR/owl-ref/.

Beck, K (1999). *Extreme programming explained: Embrace change*. Menlo Park, USA: Addison-Wesley.

Becker, M.Y., & Nanz, S. (2008). The role of abduction in declarative authorization policies. In *Proceedings of the 10th International Symposium on Practical Aspects of Declarative Languages (PADL)* (pp. 84-99).

Beckett, D. (Ed). (2004). RDF/XML Syntax Specification (Revised). *W3C Recommendation*, http://www.w3.org/TR/rdf-syntax-grammar/.

Beckett, D., & McBride, B. (2004). *RDF/XML syntax specification* (Revised). Retrieved April 14, 2008, from http://www.w3.org/TR/rdf-syntax-grammar/.

Behrends, E., Fritzen, O., May, W., Schenk, F. (2008). Event algebras and process algebras in ECA rules. *Fundamenta Informaticae, 82*, 237-263.

Behrends, H. (1994). Simulation-based debugging of active databases. In *Proceedings of IEEE Workshop "Research Issues in Database Engineering, RIDE 94"*. Houston, TX.

Behrends, H. (1995). *A description of event based activities in database related information systems.* Report 3/1995. Univ. of Oldenburg.

Bellifemine, F., Bergenti, F., Caire, G., & Poggi, A. (2005). JADE - A Java agent development framework. In R. Bordini, M. Dastani, J. Dix, A. El Fallah Seghrouchni (Eds.), *Multi-agent programming: Languages, platforms and applications* (pp. 125-147). Springer.

Bellifemine, F., Caire, G., & Greenwood, D. (2005). *Developing multiagent systems with JADE.* Chichester, UK: John Wiley and Sons.

Ben-Ari, M. (2001). *Mathematical logic for computer science.* London: Springer-Verlag.

Ben-Ari, M. (2001). *Mathematical logic for computer science.* London: Springer-Verlag.

Bennett, B., & Galton, A.P. (2004). A unifying semantics for time and events. *Artificial Intelligence, 153*(1-2), 13-48.

Bergenti, F., Gleizes, M.-P., & Zambonelli, F. (Eds.). (2004). *Methodologies and software engineering for agent systems.* In *Multiagent systems, artificial societies, and simulated organizations* (Vol. 11). Kluwer Academic Publisher.

Berners-Lee, T., Hendler, J., & Lassila, O. (2001). The Semantic Web. *Scientific American Magazine, 284*(5), 34-43.

Berners-Lee, T., Lassila, O., & Hendler, J. (2001, May). The Semantic Web. *Scientfic American.*

Bertino, E. & Montesi, D. (1992). Towards a logical object-oriented programming language for databases. In *Proceedings of the International Conference on Extending Database Technology* (pp. 168-183) (LNCS 580).

Bertino, E., Catania, B., & Zarri, G.P. (2001). *Intelligent database systems.* London: Addison-Wesley and ACM Press.

Berztiss, A., & Bubenko, J. (1995). A software process model for business reengineering. In *Proceedings of ISDO'95.* Chapman and Hall.

Bézivin, J. (2005). On the unification power of models. *Software and System Modeling, 4*(2), 171-188.

Bézivin, J. (2005). On the unification power of models. *Software and System Modeling,4*(2), 171-188.

Bezivin, J., Hammoudi, S., Lopes, D., & Jouault, F. (2004). Applying MDA approach for Web Service platform. In *Proc. of the 8th IEEE Int. Conf. on Enterprise Distributed Object Computing Conf.* (pp. 58–70).

Bierman, G.M., & Wren, A. (2005). First-class relationships in an object-oriented language. In A.P. Black (Ed.), *Proceedings of ECOOP'2005* (LNCS 3586, pp. 262-286).

Bird, L., Goodchild, A., & Halpin, T. (1999). Object role modelling and XML-schema. In *Proc. of the 19th Int. Conf. on Conceptual Modeling ER'02.* Springer LNCS.

Björn, V., & Parrow, J. (1998). Concurrent constraints in the fusion calculus. In *Proceedings of ICALP.*

Bjorner D. & Jones, C., (1978). The Vienna development method: The meta-language (LNCS 61).

Bloesch, A., & Halpin, T. (1996). ConQuer, a conceptual query language. In *Proc. of ER'96* (LNCS 1157).

Boley, H. (2003). *Object-oriented RuleML: User-level roles, URI-grounded clauses and order-sorted terms* (LNCS 2876, pp. 1-16).

Boley, H. (2006). The RuleML family of Web rule languages. In *Proceedings of the Fourth Workshop on Principles and Practice of Semantic Web Reasoning,* Budva, Montenegro (LNCS 4187, pp. 1-15).

Boley, H. (2007). *Are your rules online? Four Web rule essentials.* RuleML (pp. 7-24).

Boley, H., Kifer, M., Patranjan, P., L., & Polleres, A. (2007). Rule interchange on the Web. In *Reasoning Web 2007* (No. 4636, pp. 269-309). Springer.

Boley, H., Tabet, S., & Wagner, G (2001). Design rationale for RuleML: A markup language for Semantic Web rules. In *Proceedings of the International Semantic Web Working Symposium (SWWS'2001)* (pp. 381-401).

Boley, H., Tabet, S., & Wagner, G. (2001). Design rationale of ruleml: A markup language for Semantic Web rules. In *Swws'01*. Stanford.

Boley, H., Tabet, S., & Wagner, G. (2001, July/August). Design rationale of RuleML: A markup language for Semantic Web rules. In I.F. Cruz, S. Decker, J. Euzenat, & D.L. McGuinness (Eds.), *Proc. SWWS'01(The first Semantic Web Working Symposium), Stanford.*

Bonatti, P. A., Shahmehri, N., Duma, C., Olmedilla, D., Nejdl, W., Baldoni, M., Baroglio, C., Martelli, A., Patti, V., Coraggio, P., Antoniou, G., Peer, J., & Fuchs, N. E. (2004). *Rule-based policy specification: State of the art and future work* (REWERSE project I2-D1 deliverable).

Bonatti, P., & Olmedilla, D. (2005). Driving and monitoring provisional trust negotiation with metapolicies. In *IEEE 6th International Workshop on Policies for Distributed Systems and Networks* (pp. 14-23). Stockholm.

Bonatti, P., Duma, C., Fuchs, N., Nejdl, W., Olmedilla, D., Peer, J., et al. (2006). *Semantic Web policies—A discussion of requirements and research issues.* Paper presented at 3rd European Semantic Web Conference (ESWC).

Bonatti, P.A., Olmedilla, D., & Peer, J. (2006). Advanced policy explanations on the Web. In *Proceedings of the 17th European Conference on Artificial Intelligence (ECAI 2006)* (pp. 200-204).

Bonner, A. J., & Kifer, M. (1993). Transaction logic programming. In *Proceedings of the International Conference on Logic Programming* (pp. 257-279). Budapest, Hungary: MIT Press.

Bonner, A.J., & Kifer, M. (1995). *Transaction logic programming (or a logic of declarative and procedural knowledge).* University of Toronto.

Bordini, B., Dastani, M., Dix, J., & El Fallah Seghrouchni, A. (Eds.) (2005). *Multi-agent programming: Languages, Platforms and Applications.* Springer.

Bordini, R. H., Hübner, J. F., & Vieira, R. (2005). Jason and the Golden Fleece of agent-oriented programming. In R. H. Bordini, M. Dastani, J. Dix, & A. El Fallah Seghrouchni (Eds), *Multi-agent programming: Languages, platforms and applications* (pp. 3-37). Springer-Verlag.

Bordini, R. H.,Wooldridge,M., & Hübner, J. F. (2007). *Programming multi-agent systems in agentspeak using Jason (Wiley Series in Agent Technology).* John Wiley & Sons.

Bordini, R., Hübner, J. F., & Vieira, R. (2005). Jason and the golden fleece of agent-oriented programming. In R. Bordini, M. Dastani, J. Dix, A. El Fallah Seghrouchni (Eds.), *Multi-agent programming: Languages, platforms and applications* (pp. 3-37). Springer.

Borgida, A. (1988). Modeling class hierarchies with contradictions. In *Proceedings of the ACM SIGMOD International Conference on Management of Data* (pp. 433-443). Chicago, IL: ACM Press.

Boswell R., & Craw C. (1999). Organizing knowledge refinement operators. In *Validation and verification of knowledge based systems* (pp. 149-161). Amsterdam, The Netherlands: Kluwer.

Botia, J. (2005). *Debugging huge multi-agent systems: group and social perspectives.* Paper presented at Programming Multi-Agent Systems (PROMAS), Agentlink Technical Forum 3, Budapest.

Brachman, R., & Levesque, H. (2004). *Knowledge representation and reasoning* (1st ed.). Morgan Kaufmann.

Bradshaw, J. M. (1997). Introduction to software agents. In J.M. Bradshaw (Ed.). *Software agents*, (pp. 3-46). AAAI Press.

Bradshaw, J., Dutfield, S., Benoit, P., & Woolley, J. (1997). KAoS: Toward an industrial-strength open agent architecture. *Software Agents*, 375-418.

Brahe, S., & Osterbye, K. (2006). Business process modeling: Defining domain specific modeling languages by use of UML profiles. In *ECMDA-FA 2006* (LNCS 4066, pp. 241-255). 2006.

Brambilla, M., Ceri, S., Comai, S., & Tziviskou, C. (2005). Exception handling in workflow-driven Web applications. In *Proc. of the 14th International WWW Conference* (pp. 170–179).

Bratko, I. (2000). *Prolog programming for artificial intelligence* (3rd ed.). Addison Wesley.

Bratman, M. (1987). *Intention, plans, and practical reason*. Harvard University Press.

Bratman, M., Israel, D., & Pollack, M. (1988). Plans and resource-bounded practical reasoning. *Computational Intelligence, 4*(4), 349-355.

Braubach, L., Pokahr, A., & Lamersdorf, W. (2005). *Extending the capability concept for flexible BDI agent modularization*. In *Proceedings of the 3th international workshop on programming multiagent systems (ProMAS'05)*.

Braubach, L., Pokahr, A., & Lamersdorf, W. (2006). Tools and standards. In S. Kirn, O. Herzog, P. Lockemann, O. Spaniol (Eds.), *Multiagent engineering - Theory and Applications in enterprises* (pp. 503-530). Springer.

Braubach, L., Pokahr, A., Lamersdorf, W., & Moldt, D. (2005). Goal representation for BDI agent systems. In R. Bordini, M. Dastani, J. Dix, A. El Fallah Seghrouchni (Eds.), *Proceedings of the 2nd International Workshop on Programming Multiagent Systems, Languages and Tools (PROMAS 2004)*, 3rd International Joint Conference on Autonomous Agents & Multi-Agent Systems (AAMAS'04) (pp. 46-67). New York: Springer.

BR Forum. (2005). *Userv product derby case study* (Tech. Rep.). Business Rules Forum.

BR Forum. (2005). *Userv product derby case study* (Tech. Rep.). Business Rules Forum.

Brickley, D., & Guha, R.V. (2004). *RDF vocabulary description language 1.0: RDF schema*. Recommendation 10 February 2004, W3C. Retrieved April 22, 2008, from http://www.w3.org/TR/rdf-schema/.

Brin, S., Motwani, R., Ullman, J. D., & Tsur, S. (1997). Dynamic itemset counting and implication rules for market basket data. In *Proceedings ACM SIGMOD International Conference on Management of Data* (pp. 255-264). ACM Press.

Brna, P., Brayshaw, M., Esom-Cook, M., Fung, P., Bundy, A., & Dodd, T. (1991). An overview of prolog debugging tools. *Instructional Science, 20*(2), 193–214.

Brockmans, S., & Haase, P. (2006). *A Metamodel and UML Profile for Rule-extended OWL DL Ontologies - A Complete Reference* (Tech. Rep.). Universität Karlsruhe (TH).

Brockmans, S., Volz, R., Eberhart, A., & Loeffler, P. (2004). Visual modeling of OWL DL ontologies using UML. In *Proceedings of the Third International Semantic Web Conference*, (pp. 198-213).

Broekstra, J., Kampman, A., & van Harmelen, F. (2002). Sesame: A generic architecture for storing and querying rdf and rdf schema. *International Semantic Web Conference* (pp. 64-68).

Brooks, R. A. (1989). How to build complete creatures rather than isolated cognitive simulators. In K. VanLehn (Ed.), *Architectures for intelligence* (pp. 225-239).

Brownston, L., Farrell, R., Kant, E., & Martin, N. (1985). *Programming expert systems in OPS5*. Addison-Wesley.

Bruijn, J., & Lara, R., & Polleres, A., & Fensel, D. (2005). OWL DL vs. OWL flight: Conceptual modeling and reasoning for the semantic Semantic Web. In *Proceedings of the 14th international conference on World Wide Web* (pp. 623-632). New York: ACM Press. New York, NY, USA.

Bry, F., & Marchiori, M. (2005). *Ten theses on logic languages for the Semantic Web*. PPSWR (pp. 42-49).

Buchanan, B. G., Barstow, D., Bechtal, R., Bennett, J., Clancey, W., Kulikowski, C., Mitchell, T., & Waterman, D. A. (1983). Constructing an expert system. In F. Hayes-Roth, D. A. Waterman & D. B. Lenat (Eds.), *Building expert systems* (pp. 127-167). Menlo Park, USA: Addison-Wesley.

Buchanan, B.G., & Shortliffe, E.H. (1984). *Rule-based expert systems: The MYCIN experiments of the Stanford heuristic programming project*. Menlo Park, USA: Addison-Wesley.

Budde, R., Kautz, K., Kuhlenkamp, K, & Züllinghoven, H. (1992). *Prototyping: An approach to evolutionary system development*. Berlin, Germany: Springer.

Budinsky, F., Brodsky, S., & Merks, E. (2003). *Eclipse modeling framework.* Pearson Education.

Budinsky, F., Brodsky, S.A., Merks, E., Ellersick, R., & Grose, T.J. (2003). *Eclipse modeling framework.* Addison Wesley Professional.

Bui Bach, D., Meersman, R., Spyns, P. & Trog, D. (2007). Mapping OWL/DL into ORM/RIDL. In *Proc of ORM Workshop'07, On the Move to Meaningful Internet Systems 2007: OTM 2007 Workshops, Vilamoura, Portugal* (LNCS 4805).

Burdescu, D.D., & Brezovan, M. (2001). Algorithms for high level petri nets simulation and rule-based systems. *Acta Universitaris Cibiniensis, XLIII*, 33-39.

Burmeister, B., Steiert, H.-P., Bauer; T., & Baumgärtel H. (2006). Agile processes through goal- and context-oriented business process modeling. In *Proceedings of Business Process Management Workshops*, Vol. 4103, (pp. 217-228). Springer.

Buscher, G., Baumeister, J., Puppe, F., & Seipel, D. (2005). User-centered consultation by a society of agents. In *Proc. of the 3rd International Conference on Knowledge Capture (K-CAP 2005), Banff, Alberta, Canada.* New York, USA: ACM Press

Busetta, P., Howden, N., Rönnquist, R., & Hodgson, A. (2000). Structuring BDI agents in functional clusters. In *ATAL '99: 6th International Workshop on Intelligent Agents VI, Agent Theories, Architectures, and Languages (ATAL)* (pp. 277-289). London: Springer-Verlag.

Business Rule Forum. (2005). *UServ product derby case study.* Retrieved June 1, 2007, from http://www.business-rulesforum.com/2005_Product_Derby.pdf.

Business Rules Group. (2000). *Defining business rules – What are they really?* Retrieved January 9, 2008 from http://www.businessrulesgroup.org/first_paper/BRG-whatisBR_3ed.pdf

Butterworth, P., Otis, A., & Stein, J. (1991). The gemstone object database management system. *Communications of the ACM, 34*(10), 64-77.

Byrne, C., & Edwards, P. (1996). Refinement in agent groups. In G. Weiß (Ed.), *Proceedings of the IJCAI'95 Workshop on Adaption and Learning in Multi-Agent Systems* (pp. 22-39). Springer.

Cabeza, D., Hermenegildo, M., & Varma, S. (1996). The PiLLoW/CIAO library for Internet/WWW programming using computational logic systems. In *Proceedings of the 1st Workshop on Logic Programming Tools for INTERNET Applications* (pp. 72-90), September 2-6. Bonn, Germany.

Cabitza, F., Sarini, M., & Seno, B. D. (2005). Djess - A context-sharing middleware to deploy distributed inference systems in pervasive computing environments. In *Proceedings of the International Conference on Pervasive Services (ICPS'05)* (pp. 229-238). IEEE Computer Society.

Cacace, F., Ceri, S., Crepi-Reghizzi, S., Tanca, L., & Zicari, R. (1990). Integrating object-oriented data modelling with a rule-based programming paradigm. In *Proceedings of the ACM SIGMOD International Conference on Management of Data* (pp. 225-236). Atlantic City, NJ: ACM Press.

California State University. (2007). *Data warehouse glossary.* Retrieved February 15, 2008, from http://it.csumb.edu/site/x7101.xml

Campbell, L., Halpin, T., & Proper, H. (1996). Conceptual schemas with abstractions – Making flat conceptual schemas more comprehensible. *Data & Knowledge Engineering, 20*(1), 39-85.

Cardelli, L., & Gordon, A.D. (1998). Mobile ambients. In M. Nivat (Ed.), *Proceedings of the First international Conference on Foundations of Software Science and Computation Structure (March 28 - April 4, 1998)* (LNCS 1378, 140-155).

Cardoso, H. L. (2007). *Integrating JADE and Jess.* Retrieved from http://jade.tilab.com/doc/tutorials/jade-jess/jade_jess.html

Cardoso, J. (2007). The Semantic Web vision: Where are we? *Intelligent Systems, 22*(5), 84-88.

Carey, M., DeWitt, D., & Vanderberg, S. (1988). A data model and query language for EXODUS. In *Proceedings of the ACM SIGMOD International Conference on Management of Data* (pp. 413-423). Chicago, IL: ACM Press.

Carroll J. J., Dickinson, I., Dollins, C., & Reynolds, D. (2003). *Jena: Implementing the Semantic Web recommendations.* (Internet Report). Retrieved 2007 from http://dsonline.computer.org/0211/f/wp6jena.htm.

CaseTalk. (2006, October). Retrieved from, http://www.casetalk.com/php/

Cattell, R. G. G. & Barry, D. (Eds.) (1997). *The object database standard: ODMG 2.0.* Los Altos, CA: Morgan Kaufmann.

Cattell, R. G. G. (Ed). (1996). *The object database standard: ODMG-93, Release 1.2.* Los Altos, CA: Morgan Kaufmann.

CEP (2008) Event processing glossary, event processing technology society. Retrieved from http://complexevents.com/?p=361.

Ceri, S., Gottlob, G., & Tanca, T. (1990). *Logic programming and databases.* Springer-Verlag.

Cervesato, I., Chittaro, L., & Montanari, A (1996). A general modal framework for the event calculus and its skeptical and credulous variants. *ECAI'96.*

Chakravarthy, S., et al. (1994). Composite events for active databases: semantics contexts and detection. *VLDB 94.*

Chalupsky, H., & Russ, T. (2002). Why not: Debugging failed queries in large knowledge bases. In *Proceedings of the Fourteenth Innovative Applications of Artificial Intelligence Conference (IAAI-02)* (pp. 870–877).

Chan Man Kuok, A. F., & Wong, M. H. (1998). Mining fuzzy association rules in databases. *SIGMOD Rec.*, *27*(1), 41-46.

Chandy, K.M. (2006). *Event-driven applications: costs, benefits and design approaches.* Paper presented at Gartner Application Integration and Web Services Summit 2006.

Chang, C. L., Combs, J. B., & Stachowitz, R. A. (1990). A report on the expert systems validation associate (EVA). *Expert Systems with Applications, 1*(3), 217-230.

Chang, C.-L., & Lee, R. C.-T. (1973). *Symbolic logic and mechanical theorem proving.* New York & London: Academic Press.

Charfi, A., & Mezini, M. (2004). Hybrid Web service composition: Business processes meet business rules. In *Proc. of the 2nd International Conference on Service Oriented Computing* (pp. 30-38).

Checkland, P. B., & Scholes, J. (1990). *Soft systems methodology in action.* New York, NY: John Wiley & Sons.

Chen, H., Fellah, S., & Bishr., Y. A. (2005). Rules for geospatial Semantic Web applications. In *W3C Workshop on Rule Languages for Interoperability.*

Chen, M.-S., Han, J., & Yu, P. S. (1996). Data mining: An overview from a database perspective. *IEEE Transactions on Knowledge and Data Engineering, 8*(6), 866-883.

Chen, P. P. (1976). The entity-relationship model – Toward a unified view of data. *ACM Transactions on Database Systems.*

Cheng, A. M., Browne, J. C., Mok, A. K., & Wang, R. (1993). Analysis of real-time rule-based systems with behavioral constraint assertions specified in Estella. *IEEE Trans. Softw. Eng., 19*(9).

Chinnici, R., Moreau, J., Ryman, A., & Weerawarana, S. (2007). Web Services Description Language (WSDL) Ver. 2.0 Part 1: Core Language. *W3C Recommendation.* Retrieved from http://www.w3.org/TR/2007/REC-wsdl20-20070626.

Cirstea, H., Kirchner, C., Moossen, M., & Moreau P.-E. (2004). Production systems and rete algorithm formalisation. *Rapport de recherche n° inria-00280938 – version 1.*

Clancey, W.J. (1983). The epistemology of a rule based expert system - a framework for explanation. *Artificial Intelligence, 20*(3), 215-252.

Clancey, W.J. (1984). Use of mycin's rules for tutoring. In *Rule-based expert systems*. Addison-Wesley.

Clark, J. (1999, November 16). XSL transformations (XSLT). *W3C Recommendation.*

Clark, K.L., & Tärnlund, S.-A. (Eds). (1982). *Logic programming*. London: Academic Press.

Clement, L., Hately, A., Riegen, C., & Rogers, T. (2004). *UDDI Ver. 3.0.2*. Retrieved from http://www.oasis-open.org/committees/uddi-spec/doc/spec/v3/uddi-v3.0.2-20041019.htm.

*CLIPS: A tool for building expert systems*. (2008). Retrieved April 22, 2008, from http://clipsrules.sourceforge.net/.

Clocksin, W.F., and Mellish, C.S. (1981) *Programming in Prolog*. Berlin, Germany: Springer-Verlag.

Clynch, N., & Collier, R. (2007). SADAAM: Software agent development – An agile methodology. In *Proceedings of the Workshop of Languages, Methodologies, and Development tools for multi-agent systems (LADS'007), Durham, U.K.*

Cockcroft, S.A. (1997). Taxonomy of spatial data integrity constraints. In *Geoinformatica*, *4*(1), 327-343.

Codd, E. (1971). Alpha: A data base sublanguage founded on the relational calculus of the database relational model. In *Proceedings of ACM SIG-FIDET Workshop on Data Description, Access and Control,* San Diego, CA.

Codd, E. F. (1970). A relational model of data for large shared data banks. *Communications of the ACM, 13*(6), 377-397.

Codd, E. F., Codd, S.B. & Salley, C.T. (1993*). Providing OLAP to user analysts: An IT Mandate.* Hyperion Solutions Corporation. Retrieved January 15, 2008, from http://dev.hyperion.com/resource_library/white_papers/providing_olap_to_user_analysts.pdf

Coenen, F. (2000). Validation and verification of knowledge-based systems: Report on EuroVav'99. *Knowledge Engineering Review 15*(2), 187-196.

Coenen, F., & Bench-Capon, T. (1993). *Maintenance of knowledge-based systems*. Academic Press.

Cohen, J. (1988). A view of the origins and development of Prolog. *Communications of the ACM, 31(1)*, 26-36.

Cohen, P., & Levesque, H. (1990). Intention is choice with commitment. *Artificial Intelligence 42,* 213-261.

Cohn, A. G., Bennett, B., Gooday, J., & Gotts, N. M. (1997). Qualitative spatial representation and reasoning with the region connection calculus. *GeoInformatica, 1*, 275-316.

Collet, C., & Coupaye, T. (1996). Composite events in NAOS. *Dexa. 1996*. Zürich, Switzerland.

Collier, R., Rooney, C., & O'Hare, G.M.P. (2004). A UML-based software engineering methodology for agent factory. In *Proceedings of the 16th International Conference on Software Engineering and Knowledge Engineering (SEKE-2004), Banff, Alberta, Canada.*

Collier, R., Ross, R., & O'Hare, G.M.P. (2005). A role-based approach to reuse in agent-oriented programming. In *AAAI Fall Symposium on Roles, an interdisciplinary perspective (Roles 2005), November 3-6, Hyatt Crystal City, Arlington, Virginia, USA.*

Collier, R.W. (1996). *The realisation of agent factory: An environment for the rapid prototyping of intelligent agents.* Master of Philosophy Thesis, University of Manchester Institute of Science and Technology (UMIST), England.

Collier, R.W. (2001). *Agent factory: A framework for the engineering of agent-oriented applications.* Doctoral Thesis, University College Dublin, Ireland.

Collier, R.W., O'Hare G.M.P., Lowen, T., Rooney, C.F.B. (2003). Beyond prototyping in the valley of the agents. In *Multi-Agent Systems and Applications III: Proceedings of the 3rd Central and Eastern European Conference on Multi-Agent Systems (CEEMAS'03), Prague, Czech Republic* (LNCS 2691*)*.

Colmerauer, A. (1982). Prolog and infinite trees. In K. L. Clark & S.-A. Tarnlund (Eds.), *Logic Programming, Volume 16* (pp. 231-251). London: Academic Press.

Colmerauer, A. (1985). Prolog in 10 figures. *Communications of the ACM, 28*(12), 1296-1310.

Colmerauer, A., & Roussel, P. (1996). The birth of Prolog. In T. J. Bergin, Jr. & R. G. Gibson, Jr. (Eds.), *History of programming languages—II* (pp. 331-367). New York: ACM.

Common Situation Data Format (2003). *The common base event.* Submitted to OASIS, 2003.

Corticon Technologies, Inc. (2004) *Under the covers.*

Covington, M. A., Nute, D., & Vellino, A. (1996). *Prolog programming in depth.* Prentice-Hall.

Coyle, F. (2002). *XML, Web services, and the data revolution.* Addison Wesley.

Cragun, B. J., & Steudel, H. J. (1987). A decision-table-based processor for checking completeness and consistency in rule-based expert systems. *Int. J. Man-Machine Studies, 26*, 633-648.

Cranefield, S. (2001). UML and the Semantic Web. In *Proceedings of the Semantic Web Working Symposium,* (pp. 113-130).

Cranefield, S., Haustein, S., & Purvis, M. (2001). UML-based ontology modeling for software agents. In *Proc. of the Workshop on Ontologies in Agent Systems, 5th International Conference on Autonomous Agents, Montreal* (pp. 21-28).

Craw, S., & Boswell, R. (2002). Debugging knowledge-based applications with a generic toolkit. *In Proceedings of the 12th IEEE International Conference on Tools with Artificial Intelligence (ICTAI 2002)* (pp. 182–185).

Craw, S., & Sleeman, D. (1996). *Knowledge based refinement of knowledge based systems* (Tech. rep.). The Robert Gordon University.

Cugola, G., Nitto, E.D., & Fuggeta, A. (1998). *Exploiting an event-based infrastructure to develop complex distributed systems.* Paper presented at Int. Conf. on Software Engineering.

Cullot, N. Ghawi, R., & Yetongnon, Y. (2007). DB2OWL: A tool for automatic database-to-ontology mapping. In M. Ceci, D. Malerba, & L. Tanca (Eds.), *Proceedings of the Fifteenth Italian Symposium on Advanced Database Systems (SEBD 2007)* (pp. 491-494).

Cysneiros, L. M., & Sampaio do Prado Leite, J. C. (2000). *Eliciting business rules through iso 9000 documentation: A domain oriented conceptual model.*

Czarnecki, K., Helsen, S. (2006). Feature-based survey of model transformation approaches. *IBM Syst. J. 45*(3), 621-645.

Damianou, N., Dulay, N., Lupu, E., & Sloman, M. (2001). *The ponder policy specification language* (LNCS, pp. 18-38.

DaRaedt, L. (1992). *Interactive theory revision.* London: Academic Press.

Dastani, M. (2008). 2APL: a practical agent programming language. *Autonomous Agents and Multi-Agent Systems, 16*(3), 214-248.

Dastani, M., & Meyer, J.-J. (2007). A practical agent programming language. In *Proceedings of the fifth international workshop on programming multi-agent systems (ProMAS'07).*

Dastani, M., Dignum, V., & Dignum, F. (2003). Role-assignment in open agent societies. In *Proceedings of the Second International Conference on Autonomous Agents and Multiagent Systems (AAMAS'03), Melbourne.* ACM Press.

Dastani, M., Grossi, D., Meyer, J.-J., & Tinnemeier, N. (2008). Normative multi-agent programs and their logics. In *Proceedings of the Workshop on Knowledge Representation for Agents and Multi-Agent Systems (KRAMAS'08).*

Dastani, M., Hobo, D., Meyer, J.-J. (2007). Practical extensions in agent programming languages. In *Proceedings of the Sixth International Joint Conference on Autonomous Agents and Multiagent Systems (AAMAS'07)* (pp. 923-925). ACM Press.

Dastani, M., Mol, C. & Steunebrink, B. (2008). Modularity in agent programming languages: An illustration in extended 2APL. In *Proceedings of the 11th Pacific*

*Rim International Conference on Multi-Agents* (PRIMA 2008).

Datamaster (2008). *DataMaster*. Retrieved November, 10th, 2008 from http://protegewiki.stanford.edu/index.php/DataMaster

Davies, W., & Edwards, P. (1994). Agent-K: An integration of AOP and KQML. In *Proceedings of the Third International Conference on Information and Knowledge Management (CIKM'94)*. ACM Press.

Davis, G. B., & Olson, M. H. (1985). *Management information systems: Conceptual foundations, structure, and development*. New York: McGraw-Hill.

Davis, R. (1979). Interactive transfer of expertise. *Artificial Intelligence, 12*, 121–157.

Davis, R. (1984). Interactive transfer of expertise. In *Rule-based expert systems*. Addison-Wesley.

Dayal, U., Buchmann, A., & Chakravarty, S. (1996). The HiPAC Project. In. J. Widom & S. Ceri (Eds.), *Active database systems*. Morgan Kaufmann.

de Bruijn, J., Eiter, T., & Tompits, H. (2008). *Embedding approaches to combining rules and ontologies into autoepistemic logic*. Paper presented at the 11th International Conference on Principles of Knowledge Representation and Reasoning (KR2008), Sydney, Australia, September 16-19.

de Bruijn, J., Lara, R., Polleres, A., & Fensel, D. (2005). OWL DL vs. OWL flight: Conceptual modeling and reasoning for the Semantic Web. In *Proceedings of the International Conference on World Wide Web* (pp. 623-632). ACM Press.

de Bruijn, J., Lausen, H., Krummenacher, R., Polleres, A., Predoiu, L., & Kifer, M., et al. (2005). The Web service modeling language WSML (Tech. Rep. WSML. WSML Final Draft D16.1v0.21). Retrieved from http://www.wsmo.org/TR/d16/d16.1/v0.21/.

de Bruijn, J., Lausen, H., Polleres, A., & Fensel, D. (2006, June). The Web service modeling language: An overview. In *Proc. 3rd European Semantic Web Conf.*

de Bruijn, J., Polleres, A., Lara, R., & Fensel, D. (2005). OWL DL vs. OWL flight: Conceptual modeling and reasoning on the Semantic Web. In *Proc. WWW2005*, (pp. 623-632).

de Laborda, C.P., & Conrad, S. (2005). RelationalOWL - A data and schema representation format based on OWL. *Conceptual Modeling, 43*, 89-96.

De Troyer, O., Casteleyn, S., & Plessers, P. (2005). Using ORM to model Web systems. In *Proceeding of ORM'05*.

Dean, M., Schreiber, G., Bechhofer, S., van Harmelen, F., Hendler, J., Horrocks, I., et al. (2004). *OWL Web ontology language reference*. W3C Recommendation 10 February 2004. Retrieved April 22, 2008, from http://www.w3.org/TR/owl-ref/.

Declambre, L.M.L., & Etheredge, J.N. (1988). *A self-controlling interpreter for the relational production language*. Paper presented at the ACM Sigmod Int. Conf. on the Management of Data.

Degen, W., Heller, B., Herre, H., & Smith, B. (2001). GOL: Towards an axiomatized upper level ontology. In B. Smith & N. Guarino (Eds.), In *Proceedings of FOIS'01*. Ogunquit, Maine, USA, October 2001. ACM Press.

Demey, J., Jarrar, M., & Meersman, R. (2002). A conceptual markup language that supports interoperability between business rule modeling systems. In *Proc. of the 10th Int. Conf. on Cooperative Information Systems (CoopIS 02)* (LNCS 2519).

Demey, J., Jarrar, M., & Meersman, R. (2002). Markup language for ORM business rules. In *Proc. of International Workshop on Rule Markup Languages for Business Rules on the Semantic Web (RuleML-ISWC02 workshop)*.

DeMillo, R.A., Lipton, R. J., & Sayward, F. G. (1978). Hints on test data selection: Help for the practicing programmer. *IEEE Computer, 11*(4), 34–41.

Denecker, M., Missiaen, L., & Bruynooghe, M. (1992). Temporal reasoning with abductive event calculus. *ECAI'92*. Vienna, Austria.

Dennett, D. (1971). Intentional systems. *Journal of Philosophy, 68*, 87-106.

Dershowitz, N., & Lee, Y.-J. (1987). Deductive debugging. In *Proceedings of the Symposium on Logic Programming* (pp. 298–306).

Deux, O. et al. (1991). The O2 system. *Communications of the ACM, 34*(10), 35-48.

Díaz, O., & Jaime, A. (1997). EXACT: An extensible approach to active object-oriented databases. *The VLDB Journal 6*(4), 282-295.

Dietrich, J. (2004). A rule-based system for ecommerce applications. In M.G. Negoita, R.J. Howlett, & L.C. Jain (Eds.), *Proceedings KES'2004* (LNCS 3213, pp. 455-463).

Dietrich, J., et al. (2003). Rule-based agents for the Semantic Web. *Journal on Electronic Commerce Research Applications.*

Dietrich, J., Hiller, J., & Schenke, B. (2007). Take - A rule compiler for derivation rules. In A. Paschke, & Y. Biletskiy (Eds.), *Proceedings of RuleML'2007* (LNSC 4824, pp. 134-148).

Dietrich, J., Kozlenkov, A., Schroeder, M., & Wagner, G. (2003). Rule-based agents for the Semantic Web. *Electronic Commerce Research Applications, 2*(4), 323-338.

Dietrich, J., Schenke, B., & Wagner, G. (2008). On code generation for derived associations. In *Proceedings ASWEC'2008*. IEEE Computer Society.

Dinn, A., Paton, N.W., Williams, M.H., & Fernandes, A.A.A. (1996). An active rule language for ROCK & ROLL. In *Proceedings of the 14th British National Conference on Databases*. Springer-Verlag.

Dobbie, G. & Topor, R. (1995). On the declarative and procedural semantics of deductive object-oriented systems. *Journal of Intelligent Information System, 4*(2), 193-219.

Doherty, P. et al. (1998). TAL: Temporal action logics language specification and tutorial. *Linköping Electronic Articles in Computer and Information Science, 3*(015).

Donini, F. M., Lenzerini, M., Nardi, D., & Schaerf, A. (1998). AL-log: Integrating datalog and description logics. *In Journal Intelligent Information Systems, 10*(3), 227-252.

Doorenbos, R. (1995). *Production matching for large learning systems*. PhD Thesis, Carnegie Mellon University, UMI Order No. GAX95-22942.

Doorenbos, R. D. (1995). *Production matching for large learning systems*. Carnegie Mellon University Pittsburgh, PA, USA.

Drabent, W., Henriksson, J., & Maluszynski, J. (2007). HD-rules: A hybrid system interfacing prolog with DL-reasoners. In *Proceedings of Applications of Logic Programming to the Web, Semantic Web and Semantic Web Services* (Vol. 287) (pp. 76-90). CEUR-WS.

Dragone, M., Holz, T., & O'Hare, G.M.P. (2007). Using mixed reality agents as social interfaces for robots. In *Proceedings of the 16th IEEE International Symposium on Robot and Human Interactive Communication (RO-MAN 2007), Jeju Island, Korea.*

*Drools* (2008). Retrieved April 22, 2008, from http://www.jboss.org/drools/.

Drools. (2008). *Drools business rules management system*. Retrieved November 14, 2008, from http://www.jboss.org/drools/

Dubois, D., Hullermeier, E., & Prade, H. (2006). A systematic approach to the assessment of fuzzy association rules. *Data Mining and Knowledge Discovery, 13*(2), 167–192.

Ducasse, M., & Noy, J. (1994). Logic programming environments: Dynamic program analysis and debugging. *Journal of Logic Programming, 19*(20), 351–384.

Ducasse, S., Lienhard, A., & Renggli, L. (2007). Seaside: A flexible environment for building dynamic Web applications. *IEEE Software, 24*(5), 56-63.

Duffy B.R., O'Hare G.M.P., Campbell A.G., Stafford J.W., & O'Grady M.J. (2005). NEXUS: fusing the real & the virtual. In *Proceedings of EUROCON 2005 - The International Conference on "Computer as a Tool", Belgrade, Serbia & Montenegro.*

Dumas, M., Aldred, L., Heravizadeh, M., & terHofstede, A. (2002). Ontology markup for Web forms generation. *WWW'02 Workshop on Real-World Applications of RDF and the Semantic Web.*

Dung, P.M., & Mancaralle, P. (2002). Production systems with negation as failure. *IEEE Transactions onf Knowledge and Data Engineering, 14*(2).

Dung, P.M., & Mancarella, P. (1996). Production systems need negation as failure. In *Proceedings of 13th National Conference on AI.* AAAI Press.

Durkin, J. (1994). *Expert systems – design and development.* New York: Macmillan Publishing Company.

Dustdar, S., & Schreiner, W. (2005). A survey on Web Services composition. *International Journal of Web and Grid Services, 1*(1), 1-30.

EclipseDoc (2002). *How abstract components are created when reading a schema – an animation in 9 slides (complexType).* Retrieved from http://help.eclipse.org/help33/index.jsp?topic=/ org.eclipse.xsd.doc/references/diagrams/images/xsd-diagram-ct02/img0.html

Egenhofer, M. J. (2002). Toward the semantic geospatial web. In *Proceedings of the tenth ACM international symposium on Advances in geographic information systems* (pp. 1-4). ACM Press.

Egenhofer, M.J., Mark, D.M. & Herring, J. (1999). *The 9-Intersection: Formalism and its use for natural-language spatial predicates.* Report, National Center for Geographic Information and Analysis.

Eisenstadt, M., & Brayshaw, M. A. (1990). Knowledge engineering toolkit: The first of a two-part series presenting a knowledge-engineering toolkit for building expert systems. *Byte Magazine, 15*(10) pp. 268-282.

Eisenstadt, M., Brayshaw, M., & Paine, J. (1991). *The transparent prolog machine: Visualizing logic programs.* Springer.

Eiter, T. et al. (2001). A framework for declarative update specification in logic programs. *IJCAI.*

Eiter, T., & Gottlob, G., & Mannila, H. (1997). Disjunctive Datalog. *ACM Transactions on Database Systems, 22* (pp. , 364-418) Vol 22.

Eiter, T., & Lukasiewicz, T., & Schindlauer, R., & Tompits, H. (2004). Combining answer set programming with description logics for the semantic Semantic Web. In *KR2004: Principles of Knowledge Representation and Reasoning* (pp. 141–151). Menlo Park, CA: AAAI Press, Menlo Park, California.

Eiter, T., Faber, W., Leone, N., Pfeifer, G., & Polleres, A. (2004). A logic programming approach to knowledge-state planning: Semantics and complexity. *ACM Trans. Comput. Logic 5*(2), 206-263.

Eiter, T., Gottlob, G., & Mannila, H. (1997). Disjunctive datalog. *ACM Transactions on Database Systems, 22*(3), 364-418.

Eiter, T., Ianni, G., B., Polleres, A., Schindlauer, R., & Tompits, H. (2006). Reasoning with rules and ontologies. In P. Barahona, F. Bry, E. Franconi, U. Sattler, & N. Henze (Eds.), *Reasoning Web, Second International Summer School 2005, Tutorial Lectures* (LNCS, pp. 93-127).

Eiter, T., Ianni, G., B., Schindlauer, R., & Tompits, H. (2005). A uniform integration of higher-order reasoning and external evaluations in answer set programming. In *Proc. IJCAI 2005.* Morgan Kaufmann.

Eiter, T., Ianni, G., B., Schindlauer, R., & Tompits, H. (2006). dlvhex: A system for integrating multiple semantics in an answer-set programming framework. In M. Fink, H. Tompits, & S. Woltran (Eds.), *Proceedings 20th Workshop on Logic Programming and Constraint Systems (WLP 06)*, (pp. 206-210).

Eiter, T., Lukasiewicz, T., Schindlauer, R., & Tompits, H. (2004). Combining answer set programming with description logics for the Semantic Web. In *Proceedings of the International Conference of Knowledge Representation and Reasoning* (pp. 141-151). Morgan Kaufmann.

El Kharbili, M., Stein, S., Markovic, I., & Pulvermueller, E. (2008). Towards a framework for semantic business process compliance management. In *Proceedings of the*

*Workshop on Governance, Risk, and Compliance On Information Systems (GRCIS).*

El-Geresy, B.A., & Abdelmoty, A.I. (2004). SPARQS: A qualitative spatial reasoning engine. *In Journal of Knowledge-Based Systems, 17,* (pp. 89-102): Vol 17.

Endl, R., & Knolmayer, G. et al. (1998). *Modeling processes and workflows by business rules.*

Ensing, M. (2007). *The rule manager: A graphical business rules environment.* Demo at RuleML-2007. Retrieved from, http://2007.ruleml.org/docs/Acumen%20 Business%20-%20RuleML.pdf

Erikson, J. (1993). *CEDE: Composite event detector in an active database.* University of Skövde.

Erl, T. (2005). *Service-oriented architecture: Concepts, technology, and design.* Prentice Hall.

Esposito, F., Semeraro, G., Fanizzi, N., & Ferilli, S. (2002). Multistrategy theory revision: Induction and abduction in inthelex. *Machine Learning, 38*(1-2), 133-156

Fages, F., & Lissajoux, R. (1992). S'emantique op'erationnelle et compilation des syst emes de production. *Revue d'intelligence articielle, 6*(4), 431-456.

Fagin, R, (1977). Multivalued dependencies and a new normal form for relational databases. *ACM Trans. on Database Systems, 2*(3).

Falkovych, K., Sabou, M., & Stuckenschmidt, H. (2003). UML for the Semantic Web: Transformation-based approaches. In B. Omelayenko & M. Klein (Eds.), *Knowledge transformation for the Semantic Web* (pp. 92-106). Amsterdam: IOS Press.

Fallside, D., & Walmsley, P. (2004, October 28). XML schema part 0: Primer second edition. *W3C Recommendation .*

Farrell, J., & Lausen, H. (2007). *Semantic annotations for WSDL and XML schema.* W3C Recommendation, W3C, August 2007. Retrieved April 22, 2008, from http://www.w3.org/TR/sawsdl/.

Few, S. (2004). Dashboard confusion: A clear understanding of dashboards requires delving beneath the marketing hype. *Intelligent enterprise.* Retrieved January 1, 2008, from http://www.intelligententerprise.com/showArticle.jhtml?articleID=18300136

*FIC Basic Logic Dialect* (2008). Retrieved November 10, 2008 from http://www.w3.org/TR/rif-bld/

Fiege, L., Mühl, G., Pietzuch, P. (2006). *Distributed event-based systems.* Springer.

Fielding, R.T. (2000). *Architectural styles and the design of network-based software architectures.* PhD thesis. University of California, Irvine.

Fikes, R., Hayes, P., & Horrocks, I. (2004). OWL-QL – A language for deductive query answering on the Semantic Web. *Journal of Web Semantics: Science, Services and Agents on the World Wide Web, 2,* 19-29.

Fikes, R.E., & Nilsson, N.J., (1971). STRIPS: A new approach to the application of theorem proving to problem solving. *Artificial Intelligence, 2,* 189-208.

Finin T., Labrou, Y., & Mayfield, J. (1997). KQML as an agent communication language. In J. Bradshaw (Ed.), *Software agents.* Cambridge, MA: MIT Press.

Finin, T., Weber, J., Wiederhold, G., Genesereth, M., McKay, D., & Fritzson, R. et al. (1993). *Specification of the KQML agent-communication language – plus example agent policies and architectures.* (EIT TR 92-04).

FIPA (2000). *FIPA agent communication language.* Retrieved December 2001, from http://www.fipa.org/

FIPA (2000). *The foundation for intelligent physical agents standards.* Retrieved May 1, 2008, from http://www.fipa.org

Fisher, J. R. (2008). *Prolog tutorial.* California State Polytechnic University. Retrieved November 13, 2008, from http://www.csupomona.edu/~jrfisher/www/prolog_tutorial/contents.html

Fisher, M. (1994). A survey of concurrent METATEM—The language and its applications. In D. M. Gabbay & H. J. Ohlbach (Eds.), *Temporal Logic—Proceedings of the First International Conference (LNAI Vol. 827)* (pp. 480–505). Heidelberg, Germany: Springer-Verlag.

Fishman, D. H., Beech, B., Cate, H. P., Chow, E. C., Connors, T., Davis, J. W., et al. (1987). Iris: An object-oriented database management system. *ACM Trans. on Office Information Systems, 5*(1), 48-69.

Flesca, S., & Greco, S. (1998). Declarative semantics for active rules. In G. Quirchmayr, E. Schweighofer, & T. J. Bench-Capon (Eds.), *Proceedings of the 9th international Conference on Database and Expert Systems Applications* (LNCS 1460, pp. 871-880).

Floridi, L. (2005). Is semantic information meaningful data? *Philosophy and Phenomenological Research, 70*(2), 351-370.

Forgy, C. (1982) Rete: A fast algorithm for the many patterns/many objects match problem. *Artificial Intelligence, 19*(1), 17-37.

Forgy, C. (n.d.). *OPS5 User's Manual* (Technical Report CMU-CS-81-135). Carnegie Mellon University.

Forgy, C. L. (1982). Rete: A fast algorithm for the many pattern/many object pattern match problem. *Artificial Intelligence, 19*(1), 17-37.

Forgy, C. L., & Shepard S. J. (1987). Rete: A fast match algorithm. AI Expert archive, *2*(1), 34-40).

Forgy, C.L. (1982). Rete: A fast algorithm for the many pattern/many object pattern match problem. *Artificial Intelligence, 19*(1), 17-37.

Forgy, C.L. (1982). Rete: A fast algorithm for the many pattern/many object pattern match problem. *Artificial Intelligence, 19*(1), 17-37.

Fournet, C., & Gonthier, G. (2000). The Join Calculus: A language for distributed mobile programming. In *Proceedings of the Applied Semantics Summer School (APPSEM).*

Fowler, M. (1999). *Refactoring. Improving the design of existing code.* Menlo Park, USA: Addison-Wesley.

Fowler, M. (2004). *Inversion of control containers and the dependency injection pattern.* Retrieved April 3, 2008, from http://www.martinfowler.com/articles/injection.html.

Fowler, M. (2007). *Domain specific languages, 2007. Draft.* Retrieved April 3, 2008, from http://www.martinfowler.com/dslwip/

Fowler, M., Beck, K., Brant, J., Opdyke, W., & Roberts, D. (1999). *Refactoring: Improving the design of existing code.* Addison-Wesley.

Franconi, E. (2002). Tutorial on description logics for conceptual design, information access and ontology integration: Research trends. In *Proc. of 1st International Semantic Web Conference.*

Frank, A. (1992). Qualitative spatial reasoning about distances and directions in geographic space. *Journal of Visual Languages and Computing* (LNCS 639, pp. 343–371).

Franklin, S., Graesser, A. (1997). Is it an agent, or just a program? A taxonomy for autonomous agents. In J. Müller, M. Wooldridge, N. Jennings (Eds.), *Proceedings of the 3rd Workshop on Intelligent Agents III, Agent Theories, Architectures, and Languages (ATAL 1996)* (pp. 21-35). Springer.

Frawley, W., Piatetsky-Shapiro, G. & Matheus, C. (1992). Knowledge discovery in databases: An overview. *AI Magazine*, 213-228.

Freksa, C. (1992). Using orientation information for qualitative spatial reasoning. In *Theories and methods of spatio-temporal reasoning in geographic space* (pp. 162–178).

Friedland, N., Allen, R G., Witbrock, W., Matthews, G., Salay, N., Miraglia, R. et al. (2004). Towards a quantitative platform-independent quantitative analysis of knowledge systems. In *Proceedings of the Ninth International Conference of Knowledge Representation and Reasoning* (pp. 507-515).

Friedman-Hill, E. (2003). *Jess in Action - Java Rule-based Systems.* Manning Publications Co.

Friedman-Hill, E. (2003). *Jess in Action.* Manning, Greenwich.

Fu, A., Wong, M., Sze, S., Wong, W., Wong, W., & Yu, W. (1998). Finding fuzzy sets for the mining of fuzzy association rules for numerical attributes. In *Proceedings of the First International Symposium on Intelligent Data Engineering and Learning (IDEAL'98)* (pp. 263-268).

Fu, F., Jones, C.B., & Abdelmoty, A. (2005). Building a geographical ontology for intelligent spatial search on the Web. In *IASTED - International Conference on Databases and Applications part of the 23rd Multi-Conference on Applied Informatics, Innsbruck, Austria* (pp. 167-172).

Fuchs, N.E., Kaljurand, K., & Kuhn, T. (2008). Attempto controlled English for knowledge representation. In C. Baroglio, P.A. Bonatti, J. Maluszynski, M. Marchiori, A. Polleres, & S. Schaffert (Eds.), *Reasoning Web, Fourth International Summer School 2008* (LNCS 5224). Berlin: Springer.

Fuchs, N.E., Kaljurand, K., & Schneider, G. (2006). *Attempto controlled English meets the challenges of knowledge representation, reasoning, interoperability and user interfaces.* FLAIRS 2006.

Gabbay, D. M., Hogger, C. J., & Robinson, J. A. (Eds.). (1993). Handbook of logic in artificial intelligence and logic programming (Vol. 1). Oxford: Clarendod Press.

Gahegan, M. (1995). Proximity operators for qualitative spatial reasoning. In *Spatial information theory: A theoretical basis for GIS* (LNCS 988, pp. 31-44).

Gallaire, H. & Minker, J. (Eds.) (1978). *Logic and data bases.* New York: Plenum.

Gallaire, H. (1981). Impacts of logic on data bases. In *Proceedings of the International Conference on Very Large Data Bases* (pp. 248-259). New York: IEEE.

Gamma, E., Helm, R., Johnson, R., & Vlissides, J. (1995). *Design patterns: Elements of reusable object-oriented software.* Addison-Wesley Longman Publishing Co.

Gardiner, T., Horrocks, I., & Tsarkov, D. (2006). Automated benchmarking of description logic reasoners. In *Proc. of DL.*

Gasevic, D., Djuric, D., & Devedzic, V. (2005). Bridging MDA and OWL ontologies. *Journal of Web Engineering, 4*(2), 119-134.

Gašević, D., Djurić, D., & Devedžić, V. (2006). *Model driven architecture for Ontology development.* Berlin-Heidelberg: Springer.

Gašević, D., Kaviani, N., & Milanović, M. (2008). Ontologies and software engineering. In S. Staab & R. Studer (Eds.), *Handbook on Ontologies.* Germany: Springer.

Gatziu, S., & Dittrich, K. (1993). *Event in an active object-oriented database system.* Paper presented at Int. Conf. on Rules in Database Systems.

Gehani, N., Jagadish, H.V., & Shmueli, O. (1992). *Event specification in an active object-oriented database.* Paper presented at Int. Conf. on Management of Data.

Gelfond, M., & Lifschitz V. (1993). Representing action and change by logic programs. *Journal of Logic Programming, 17*(2-4), 301-321.

Gelfond, M., & Lifschitz, V. (1991). Classical negation in logic programs and disjunctive databases. *New Generation Computing, 9*(3-4), 365-386.

Genova, G., del Castillo, C.R., Llorens, J. (2003). Mapping UML associations into java code. *Journal of Object Technology, 2*(5),135-162.

Georgeff, M., Lansky, A. (1987). Reactive reasoning and planning: An experiment with a mobile robot. In *Proceedings of the 6th National Conference on Artificial Intelligence (AAAI 1987)* (pp. 677-682).

Georgetown University. (2008). *Data warehouse glossary.* Retrieved February 1, 2008, from http://uis.georgetown.edu/departments/eets/dw/GLOSSARY0816.html

Gerevini, A., & Renz, J. (2002). Combining topological and size information for spatial reasoning. *Artificial Intelligence, 137,* 1-42.

Giacomo, G. D., Lesperance, Y., & Levesque, H. J. (2000). Congolog, a concurrent programming language based on the situation calculus. *Artificial Intelligence, 121*(1-2), 109-169.

Giarratano, J. C., & Riley, G. D. (2005). *Expert systems. Principles and programming.* Thomson Corse Technology.

Gilbo, E. P., & Howard, K.W. (2000). Collaborative optimization of airport arrival and departure traffic flow management strategies for CDM. In *Proceedings of the 3rd USA/Europe Air Traffic Management R&D Seminar*, Naples, Italy, 13-16 June. Retrieved June 5, 2006, from http://www.euro-cdm.org/docs/paper18.pdf

Ginsberg, A. (1988). *Automatic refinement of expert system knowledge bases*. Amsterdam, The Netherlands: Morgan Kaufmann.

Ginsberg, A. (2006). RIF Use cases and requirements. *W3C Working Draft*. Retrieved from http://www.w3.org/TR/rif-ucr/.

Ginsberg, A., Hirtle, D., McCabe, F., & Patranjan, P. (2006). *RIF Use cases and requirements*. W3C Working Draft. Retrieved from http://www.w3.org/TR/rif-ucr.

Ginsberg, A., Hirtle, D., McCabe, F., & Patranjan, P.-L. (2006, July 10). RIF use cases and requirements. *W3C Working Draft* .

Giunchiglia, E., & Lifschitz, V. (1998). An action language based on causal explanation: *Preliminary report in Conf. on Innovative Applications of Artificial Intelligence*. Menlo Park, CA: AAAI Press.

Giunchiglia, E., Kartha, N., & Lifschitz, V. (1997). Representing action: indeterminacy and ramifications. *Artificial Intelligence, 95*, 409-443.

Giunchiglia, E., Lee, J., Lifschitz, V., McCain, N., & Turner, H. (2004). Nonmonotonic causal theories. *Artificial Intelligence, 153*, 49-104.

Giunchiglia, E.m & Lifschitz, V. (1999). Action languages, temporal action logics and the situation calculus. *Linköping Electronic Articles in Computer and Information Science, 4*(040).

Giurca, A. (2007). *Using URML and R2ML in a financial services use case (implemented with JBoss Rules and JenaRules)*. Retrieved Thursday May 1, 2008, http://oxygen.informatik.tu-cottbus.de/rewerse-i1/?q=node/33

Giurca, A., & Wagner, G. (2005). *Towards an abstract syntax and direct-model theoretic semantics for RuleML*. RuleML 2005 (pp. 45-55).

Giurca, A., & Werner, J. (2006). *Translator: R2ML to JBoss rules*. Retrieved from http://oxygen.informatik.tu-cottbus.de/translator/R2MLtoJBossRules/

Giurca, A., Lukichev, S., & Wagner., G. (2006). Modeling Web services with URML. In *Proc. of Workshop Semantics for Business Process Management*.

Gnatovich, R. (2005). Closing the analysis gap. *Imedia Connection*. Retrieved January 10, 2008, from http://www.imediaconnection.com/content/6060.asp

Goasdoue, F., Lattes, V., & Rousset, M.-C. (2000). The use of CARIN Language and algorithms for information integration: The Picsel System. *International Journal on Cooperative Information Systems, 9*, 383-401.

Godik, S., & Moses, T. (2003). eXtensible Access Control Markup Language (XACML). *Specification, OASIS*.

Goldberg, A., & Robson, D. (1983). *Smalltalk-80: The language and its implementation*. Addison-Wesley.

Gómez-Pérez, A., & Rojas-Amaya, D. (1999). Ontological reengineering for reuse. In *Proceedings of the 11th European Workshop on Knowledge Acquisition, Modeling and Management (EKAW 1999)* (pp. 139-156). Berlin, Germany: Springer.

Gonzalez, A. J., & Dankel, D. D.(1993). *The engineering of knowledge-based systems - Theory and practice*. New Jersey, USA: Prentice Hall.

Graco, W., Semenova, T. & Dubossarsky, E. (2007). Toward knowledge-driven data mining. In *Proceedings of the 2007 international workshop on Domain driven data mining: Conference on Knowledge Discovery in Data* (pp. 49-54). New York: ACM.

Graml, T., Bracht, R., & Spies, M. (2007). Patterns of Business rules to enable agile business processes. In *Proceedings of the 11th Enterprise Distributed Object Computing Conference* (pp. 365-378). IEEE Computer Society.

Grant, J., & Beckett, D. (2004). *RDF test cases*. Retrieved April 14, 2008, from http://www.w3.org/TR/rdf-testcases/.

Grau, B.C., Calvanese, D., De Giacomo, G., Horrocks, I., Lutz, C., Motik, B., Parsia, B., et al. (2006). *OWL 1.1 Web ontology language tractable fragments.* W3C Member Submission. Retrieved from, http://www.w3.org/Submission/owl11-tractable/

Greenfield, L. (2008). The data warehousing information center. *LGI International.* Retrieved February 10, 2008, from http://www.dwinfocenter.org/defined.html

Gronau, N, Mueller, C., & Uslar, M. (2004). The kmdl knowledge management approach: Integrating knowledge conversions and business process modeling. In *Proc. 5th international conference on practical aspects of knowledge management, Vienna, Austria* (LNCS 3336, pp. 1-10).

Grønmo, R., Jaeger, C., M., & Hoff, H. (2005). Transformations between UML and OWL-S. In *Proc. of the 1st European Conference Model Driven Architecture - Foundations and Applications* (pp. 269-283).

Gronmo, R., Jaeger, M.C., & Hoff, H. (2005). Transformations between UML and OWL-S. In *Proc. of the 1st European Conf. on Model Driven Architecture - Foundations and Applications* (pp. 269–283).

Groot, P, ten Teije, A, & van Harmelen, F. (2005). A quantitative analysis of the robustness of knowledge-based systems through degradation studies. *Knowledge and information Systems.* Berlin, Germany: Springer

Grosof, B. (2005, December 19). *The production logic programs approach, in a nutshell: Foundations for semantically interoperable Web rules* (Working Paper). MIT Sloan School of Management.

Grosof, B. N. (1997). Prioritized conflict handling for logic programs, In *Proceedings of the International Symposium on Logic Programming* (pp. 197-211). Cambridge: MIT Press.

Grosof, B. N., & Horrocks, I., & Volz, R., & Decker, S. (2003). Description logic programs: combining logic programs with description logic. In *Proceedings of the 14th international conference on World Wide Web 03* (pp. 48-57).

Grosof, B. N., Gandhe, M. D., & Finin, T. W. (2002). SweetJess: Translating DAMLRuleML to JESS. In *Proceedings of the International Workshop on Rule Markup Languages for Business Rules on the Semantic Web.* Italy: Sardinia.

Grosof, B. N., Horrocks, I., Volz, R., & Decker, S. (2003). Description logic programs: combining logic programs with description logic. In *Proceedings of the International Conference on World Wide Web* (pp. 48-57). ACM Press.

Grosof, B. N., Horrocks, I., Volz, R., & Decker, S. (2003). Description logic programs: Combining logic programs with description logic. In *Proc. of the Twelfth International World Wide Web Conference (WWW 2003),* (pp. 48-57). ACM.

Gruber, T. R. (1993). Toward principles for the design of ontologies used for knowledge sharing. In N. Guarino, V.A. Ungheria, & P. Giaretta (Eds.), *Formal ontology in conceptual analysis and knowledge representation.* Kluwer Academic Publishers.

Guarino, N. (1995), Formal ontology, conceptual analysis and knowledge representation. *International Journal of Human-Computer Studies, 43,* 625-640.

Guarino, N. (1997). Understanding, building, and using ontologies: A commentary to "Using Explicit Ontologies in KBS Development", by van Heijst, Schreiber, and Wielinga." *International Journal of Human and Computer Studies, 46,* 293-310. Retrieved from http://citeseer.ist.psu.edu/guarino97understanding.html

Guarino, N., Ungheria, V. A, & Giaretta, P. (1994). Formalizing ontological commitments. *AAAI 1994.*

Gudgin, M., Hadley, M., Mendelsohn, N., Moreau, J., Nielsen, H. F., Karmarkar, A., & Lafon, Y. (2007). *SOAP Ver. 1.2 Part 1: Messaging framework.* W3C Recommendation. Retrieved from http://www.w3.org/TR/soap12-part1/

Guha, S., Rastogi, R., & Shim, K. (2000). ROCK: A robust clustering algorithm for categorical attributes. *Information Systems 25*(5), 345-366.

Guizzardi, G., & Wagner, G. (2004, June 8).Towards ontological foundations for agent modeling concepts using the Unified Foundational Ontology (UFO). In P. Bresciani, P. Giorgini, B. Henderson-Sellers, G. Low & M. Winikoff (Eds.). *Revised Selected Papers of the 6th International Bi-Conference Workshop on Agent-Oriented Information Systems* (LNCS 3508, pp. 110-124).

Guizzardi, R.S.S., Aroyo, L., & Wagner, G. (2003). Agent-oriented knowledge management in learning environments: A peer-to-peer helpdesk case study. In *Agent Mediated Knowledge Management, International Symposium AMKM* (LNCS 2926, pp. 57-72).

Gupta, U. G., & Biegel, J. (1990). A rule-based intelligent test case generator. In *Proc. Menlo Park, USA: AAAI Press–90 Workshop on Knowledge-Based System Verification, Validation and Testing*. Menlo Park, USA: AAAI Press.

Gyenesei, A. (2000). *A fuzzy approach for mining quantitative association rules* (Tech. Rep. No. TUC-STR-336).

Gyenesei, A. (2000). Mining weighted association rules for fuzzy quantitative items. In *Principles of data mining and knowledge discovery* (p. 416-423).

Haarslev, V., & Moller, R. (1999). An empirical evaluation of optimization strategies for abox reasoning in expressive description logics. In *Proceedings of the International Workshop on Description Logics* (Vol. 22). CEUR-WS.

Haarslev, V., & Moller, R. (2001). RACER system description. In R. Gor´e, A. Leitsch, and T. Nipkow (Eds.), *International Joint Conference on Automated Reasoning, IJCAR'2001, June 18-23, Siena, Italy*, (pp. 701–705). Springer-Verlag.

Haarslev, V., & Moller, R. (2003). Racer: A core inference engine for the Semantic Web. In *Proceedings of the International Workshop on Evaluation of Ontology-based Tools* (pp. 27-36).

Haarslev, V., & Muller, R. (2001). RACER system description. In *Proceedings of the First International Joint Conference on Automated Reasoning* (pp. 701-706).

Haarslev, V., Lutz, C., & Moller, R. (1998). Foundations of spatioterminological reasoning with description logics. In *Principles of knowledge representation and reasoning* (pp. 112–123).

Hackathorn, R. (2004). Minimizing action distance. *The Data Administration Newsletter*. Retrieved February 10, 2008, from http://www.tdan.com/i025fe04.htm

Halle, B. von. (2001). *Business rules applied: Building better systems using the business rules approach*. Wiley.

Halpin, T. (2004). Business rule verbalization. In A. Doroshenko, T. Halpin, S. Liddle, & H. Mayr (Eds). *Information Systems Technology and its Applications, 3rd International Conference (ISTA'2004)* (pp. 39-52).

Halpin, H. (2006, May). *Identity, reference, and meaning on the Web*. WWW2006 Workshop on Identity, Reference, and the Web. Retrieved from, http://www.ibiblio.org/hhalpin/irw2006

Halpin, T. (1989). *A logical analysis of information systems: Static aspects of the data-oriented perspective*. PhD thesis, University of Queensland, Brisbane. Australia.

Halpin, T. (2001). *Information modeling and relational databases* (3rd ed).Morgan-Kaufmann.

Halpin, T. (2002). Join Constraints. In *Proceedings of the EMMSAD'02*.

Halpin, T. (2005). Objectification. In *Proceedings of the EMMSAD'05, at CAiSE 2005*.

Halpin, T. (2005a). ORM 2. In *Proceedings of the OTM Workshops 2005*.

Halpin, T. (2005b). FORML Position paper for W3C Workshop on Rule Languages for Interoperability. In *Proceedings of the Workshop Rule Languages for Interoperability*. W3C.

Halpin, T. A., & Curland, M. (2006). Automated verbalization for ORM 2. In R. Meersman, Z. Tari, & P. Herrero (Eds.), *Proceedings of on the Move to Meaningful Internet Systems 2006: OTM 2006 Workshops* (LNCS 4278, pp. 1181-1190).

Halpin, T., & Proper, H. (1995). Subtyping and polymorphism in object-role modeling. *Data & Knowledge Engineering, 15*(3), 251-281

Halpin, T., Evans, K., Hallock, P., & MacLean, W. (2003). *Database modeling with Microsoft® Visio for enterprise architects*. Microsoft® Visio. San Francisco: Morgan Kaufmann.

Halpin, T., Proper, H. (1996). Database schema transformation & optimization. In Proceedings *of the 14th International Conference on Conceptual Modeling* (LNCS 1021).

Halpin, T.A. (2001). *Information modeling and relational databases: From conceptual analysis to logical design*. San Francisco: Morgan Kaufman Publishers.

Halpin, T.A. (2005). ORM 2. In R. Meersman, Z. Tari, & P. Herrero (Eds.), *Proceedings of ORM Workshop'05, On the Move to Meaningful Internet Systems 2005: OTM 2005 Workshops*.

Han, J., & Kamber, M. (2006). *Data mining: Concepts and techniques* (2nd ed.). Morgan Kaufmann.

Han, J., Pei, J., & Yin, Y. (2000). Mining frequent patterns without candidate generation. In W. Chen, J. Naughton & P.A. Bernstein (Eds.), *2000 ACM SIGMOD Intl. Conference on Management of Data* (pp.1-12). ACM Press.

Han, J., Pei, J., Yin, Y., & Mao, R. (2004). Mining frequent patterns without candidate generation: A frequent-pattern tree approach. *Data Mining and Knowledge Discovery, 8*(1), 53-87.

Hanson, E.N., & Widom, J. (1993). An overview of productions rules in database systems. *Knowledge Engineering Review, 8*(2), 21-142.

Happel, H., & Seedorf, S. (2006). Applications of Ontologies in software engineering. In *Proc. of the Int'l WSh. on Semantic Web Enabled Software Engineering*.

Harmelen, F. van, Lifschitz, V., & Porter, B. (Eds.). (2007). *Handbook of knowledge representation*. Elsevier Science.

Hayes, P. (2004). *RDF Semantics*. Retrieved April 14, 2008, from http://www.w3.org/TR/rdf-mt/.

Hayes-Roth, F. (1985). Rule based systems. *ACM Computing Surveys, 28*(9).

Hayes-Roth, F., Waterman, D. A., & Lenat, D.B. (1983). *Building expert systems*. Menlo Park, USA: Addison-Wesley

Heflin, J., Hendler, J., & Luke, S. (1999). *SHOE: A knowledge representation language for Internet applications* (Technical Report CS-TR-4078) (UMIACS TR-99-71).

Herbst, H., Knolmayer, G., Myrach, T., & Schlesinger, M. (1994). The specification of business rules: A comparison of selected methodologies. In A.A. Verijn-Stuart & T.-W. Olle (Eds.), *Methods and associated tools for the information system life cycle*. Amsterdam: Elsevier.

Herre, H., Jaspars, J. O. M., & Wagner, G. (1995). *Partial logics with two kinds of negation as a foundation for knowledge-based reasoning*. (Tech. Rep. CS-R9574). Centrum voor Wiskunde en Informatica (CWI).

Heuer, A. & Sander, P. (1993). The LIVING IN A LATTICE rule language. *Data and Knowledge Engineering, 9*(4), 249-286.

Heymans, S., Predoiu, L., Feier, C., de Bruijn, J., & Van Nieuwenborgh, D. (2006). G-Hybrid knowledge bases. In *Proceedings of the Workshop on Applications of Logic Programming in the Semantic Web and Semantic Web Services* (Vol. 196). CEUR-WS.

Hillston, J. (1996). *A compositional approach to performance modelling*. Cambridge University Press.

Hindriks, K. V., de Boer, F. S., van der Hoek, W., & Meyer, J.-J. C. (2001). Agent programming with declarative goals. In *Proceedings of the 7th International Workshop on Intelligent Agents VII. Agent theories architectures and languages* (pp. 228-243). Springer-Verlag.

Hindriks, K. V., de Boer, F. S., van der Hoek,W., & Meyer, J.-J. C. (1999). Agent programming in 3APL. *Autonomous Agents and Multi-Agent Systems, 2*(4), 357-401.

Hindriks, K.V. (2008). Modules as policy-based intentions: Modular agent programming in GOAL. In *Proceedings of the 5th international workshop on programming multiagent systems (ProMAS'08)* (LNAI 4908).

Hindriks, K.V., de Boer, F.S., van der Hoek, W., & Meyer, J.-J.Ch. (1999). Agent programming in 3APL. *Autonomous Agents and Multi-Agent Systems, 2*(4),357-401.

HIPAA Administrative Simplification: Enforcement; Final Rule. 45 C.F.R., 160 & 164 (2006).

Hirtle, D. (2006). *TRANSLATOR: A TRANSlator from LAnguage TO rules.* Canadian Symposium on Text Analysis (CaSTA), Fredericton, October 2006.

Hirtle, D., Boley, H., Grosof, B., Kifer, M., Sintek, M., Tabet, S., Wagner, G (2006). *Schema specification of RuleML 0.91.* Retrieved from http://www.ruleml.org/spec/

Hitzler, P., & Vrandecic, D. (2005). *The SCREECH OWL reasoner - Scalable approximate ABox reasoning for OWL.* Paper presented at the International Semantic Web Conference, Software Demo.

Hitzler, P., Haase, P., Krtzsch, M., Sure, Y., & Studer, R. (2005). DLP isn't so bad after all. In *Proceedings of the 1st Workshop on OWL: Experiences and Directions* (OWLED-*05). CEUR Workshop Proceedings* (Vol. 188).

Hoare, C.A.R. (1985). *Communication and concurrency.* Prentice Hall.

Hölldobler, S., & J. Schneeberger (1990). A new deductive approach to planning. *New Generation Computing, 8*(3), 225-244.

Holt, J. D., & Chung, S. M. (1999). Efficient mining of association rules in text databases. In *CIKM'99: Proceedings of the eighth international conference on information and knowledge management* (pp.234–242). New York: ACM.

Hong, T.-P., Kuo, C.-S., Chi, S.-C., & Wang, S.-L. (2000). Mining fuzzy rules from quantitative data based on the AprioriTid Algorithm. In *Sac '00: Proceedings of the 2000 ACM symposium on applied computing* (pp. 534-536). New York: ACM Press.

Hori, M., Euzenat, J., & Patel-Schneider, F. P. (2003). OWL Web ontology language, XML presentation syntax. *W3C Note.* Retrieved from, http://www.w3.org/TR/owl-xmlsyntax

Horridge, M., Tsarkov, D., & Redmond, T. (2006). Supporting early adoption of OWL 1.1 with Protégé-OWL and FaCT++. In B. Cuenca Grau, P. Hitzler, C. Shankey, & E. Wallace, (Eds.), *Proceedings of the OWLED*06 Workshop on OWL: Experiences and Directions.* Aachen: Sun SITE Central Europe.

Horrocks, I. (1998). Using an expressive description logic: FaCT or fiction? In *Proceedings of the Sixth International Conference on Principles of Knowledge Representation and Reasoning (KR'98), Trento, Italy, June 2–5, 1998,* (pp. 636–647).

Horrocks, I. (2005). OWL rules, ok? In *W3C Workshop on Rule Languages for Interoperability.*

Horrocks, I. R. (1998). Using an expressive description logic: FaCT or fiction? In *KR'98: Principles of Knowledge Representation and Reasoning* (pp. 636-645).

Horrocks, I., & Patel-Schneider, P. F. (2003). Reducing OWL entailment to description logic satisfiability. In *Proceedings of the International Semantic Web Conference* (pp. 17-29). Springer.

Horrocks, I., & Sattler, U. (2004). Decidability of SHIQ with complex role inclusion axioms. *Artificial Intelligence, 160*(1), 79-104.

Horrocks, I., Kutz, O., & Sattler, U. (2006). The even more irresistible SROIQ. In *Proceedings of the International Conference on Principles of Knowledge Representation and Reasoning* (pp. 57-67). AAAI Press.

Horrocks, I., Li, L., Turi, D. & Bechhofer, S. (2004). The instance store: DL reasoning with large numbers of individuals. In *Proc. of the 2004 Description Logic Workshop (DL~2004),* (pp. 31-40).

Horrocks, I., Li, L., Turi, D., & Bechhofer, S. (2004). The instance store: Description logic reasoning with large numbers of individuals. In *Proceedings of the Workshop on Description Logics.* (Vol. 104) (pp. 31-40). CEUR-WS.

Horrocks, I., Parsia, B., Patel-Schneider, P., F., & Hendler, J. (2005). Semantic Web architecture: Stack or two towers? In *Proc. PPSWR 2005,* (pp. 37-41), Dagstuhl Castle, Germany.

Horrocks, I., Patel-Schneider F.P., Boley, H., Tabet, S., Grosof, B., & Dean, M. (2004). SWRL: A Semantic Web rule language combining OWL and RuleML. *W3C Member Submission.* Retrieved from http://www.w3.org/Submission/SWRL.

Horrocks, I., Patel-Schneider, P. F., Boley, H., Tabet, S., Grosof, B., & Dean, M. (2004). SWRL: *A Semantic Web rule language combining OWL and RuleML* (Tech. rep.). W3C Member Submission.

Horrocks, I., Patel-Schneider, P. F., Boley, H., Tabet, S., Grosof, B., & Dean, M. (2004). *SWRL: A Semantic Web rule language combining OWL and RuleML* (Tech. rep.). W3C. Member submission 21 May 2004. Retrieved April 3, 2008, from http://www.w3.org/Submission/SWRL/

Horrocks, I., Patel-Schneider, P. F., Boley, H., Tabet, S., Grosof, B., & Dean, M. (2004, May 21). Swrl: A Semantic Web rule language combining owl and ruleml. *W3C member submission.*

Horrocks, I., Patel-Schneider, P., Boley, H., Tabet, S., Grosof, B., & Dean, M. (2004, May 21). SWRL: A Semantic Web rule language combining OWL and RuleML. *W3C Member Submission .*

Horrocks, I., Patel-Schneider, P.F., Bechhofer, S., & Tsarkov, D. (2005). OWL rules: A proposal and prototype implementation. *Journal of Web Semantics: Science, Services and Agents on the World Wide Web, 3,* 23-40.

Horrocks, I., Patel-Schneider, P.F., Boley, H., Tabet, S., Grosof, B., and Dean, M. (2004, May 21). *SWRL: A Semantic Web rule language Combining OWL and RuleML.* W3C member submission. Retrieved from, http://www.w3.org/Submission/SWRL/

Horrocks, I., Patel-Schneider, P.F., Boley, H., Tabet, S., Grosof, B., & Dean, M. (2004). *SWRL: A Semantic Web rule language combining owl and RuleML.* W3C Member Submission, 21 May 2004. Retrieved April 22, 2008, from http://www.w3.org/Submission/SWRL/.

Horrocks, I., Patel-Schneider, P.F., Boley, H., Tabet, S., Grosof, B., Dean, M. (2004). *SWRL: A Semantic Web rule language combining OWL and RuleML.* W3C.

Horst, H.J. (2004). Extending the RDFS entailment lemma. In *Proceedings of the International Semantic Web Conference* (pp. 77-91). Springer.

Horst, H.J. (2005). *Completeness, decidability and complexity of entailment for RDF schema and a semantic extension involving the OWL vocabulary. Journal of Web Semantics, 3*(2-3), 79-115.

Howden, N., Rönnquist, R., Hodgson, A., & Lucas, A. (2001). JACK: Summary of an agent infrastructure. In *Proceedings of the 5th International Conference on Autonomous Agents, Montreal, Canada.*

Huntington D. (1999). Knowledge-based systems: A look at rule-based systems. In Liebowitz, J. (Ed.), *Knowledge management handbook* (Section IV, pp. 14-1).

Hustadt, U., Motik, B., & Sattler, U. (2004). Reducing SHIQ-description logic to disjunctive datalog programs. In *Proceedings of the International Conference on Knowledge Representation and Reasoning* (pp. 152-162). AAAI Press.

Hyatt, D. (Ed.) (2001). *XML User Interface Language (XUL) 1.0.* Retrieved April 5, 2008, from http://www.mozilla.org/projects/xul/xul.html

IBM (2001). *IBM Tivoli Enterprise Console - documentation.*

IBM (2001). *IBM Informix Time Series Data Blade Module, User Guide, V.4.0.*

IBM (2003). *Common Event Infrastructure, IBM.*

Ilog (2008). *Ilog Business Rule Management Systems.* Retrieved April 29, 2008, from http://www.ilog.com/products/businessrules

*ILOG, Business Rule Management Systems, Optimization Tools and Engines, Visualization Software Components, Supply Chain Applications* (2008). Retrieved April 22, 2008, from http://www.ilog.com/.

ILOG. (2008). *ILOG Rules*. Retrieved April 19, 2008, from http://www.ilog.com/products/rules/

Infagon. (2006, October). Retrieved from, http://www.mattic.com/Infagon.html

Ingrand, F., Georgeff, M., & Rao, A. (1992). An architecture for real-time reasoning and system control. *IEEE Expert, 7*(6), 34-44.

Inmon, W. (1995). *Tech topic: What is a data warehouse*? New York: Prism Solutions.

Ishida, T. (1994). An optimization algorithm for production systems. *IEEE Transactions on Knownledge and Data Engineering, 6* (4), 549-558.

Ishikawa, H., Suzuki, F., Kozakura, F., Makinouchi, A., Miyagishima, M., Izumida, Y., et al. (1993). The model, language, and implementation of an object-oriented multimedia knowledge base management system. *ACM Trans. on Database Systems, 18*(1), 1-50.

Jaber, M. K., & Voronkov, A. (2006). UNIDOOR: A deductive object-oriented database management system. In *Proceedings of the 22nd International Conference on Data Engineering* (pp. 157). Washington, DC: IEEE Computer Society.

Jackson, P. (1999). *Introduction to expert systems* (3rd ed.). Addison Wesley.

Jackson, P. (1999). *Introduction to Expert Systems*. Addison Wesely Longman (3rd ed).

Jacobs, B. E. (1982). On database logic. *Journal of ACM, 29*(2), 310-332.

Jaffar, J. & Lassez, J.-L. (1987). Constraint logic programming. In *Proceedings of the 14th ACM SIGACT-SIGPLAN Symposium on Principles of programming languages*, (pp. 111-119). New York: ACM.

James F. Allen, J.F. (1983). Maintaining knowledge about temporal intervals. *Communications of the ACM, 26*(11), 832-843.

Jamil, H. M. (1997). Implementing abstract objects with inheritance in datalogneg. In *Proceedings of the International Conference on Very Large Data Bases* (pp. 46–65). Athens, Greece: Morgan Kaufmann.

Jang, M., & Sohn, J-C. (2004). Bossam: An extended rule engine for OWL inferencing. In *Workshop on Rules and Rule Markup Languages for the Semantic Web at the 3rd International Semantic Web Conference* (LNCS 3323, pp. 128-138).

Jarrar, M. (2005). *Towards methodological principles for ontology engineering*. PhD Thesis. Vrije Universiteit Brussel, Belgium.

Jarrar, M. (2006, August). *ORM Markup Language, version 3* (Tech. Rep.) Belgium: Vrije Universiteit Brussel.

Jarrar, M. (2006). Towards the notion of gloss, and the adoption of linguistic resources in formal ontology engineering. In *Proceedings of the WWW2006 Conference*. ACM Press.

Jarrar, M. (2007). Mapping ORM into the SHION/OWL description logic. In *Proceedings of ORM workshop'07, On the Move to Meaningful Internet Systems 2007: OTM 2007 Workshops, Vilamoura, Portugal* (LNCS 4805).

Jarrar, M. (2007). Towards automated reasoning on ORM schemes—Mapping ORM into the DLR_idf description logic. In *Proceedings of the 26th International Conference on Conceptual Modeling (ER 2007)*. (LNCS 4801, pp. 181-197).

Jarrar, M. (2007). Mapping ORM into the SHOIN/OWL description logic—Towards a methodological and expressive graphical notation for ontology engineering. *In OTM workshops, proceeding of the International Workshop on Object-Role Modeling (ORM'07)*. (LNCS 4805, pp. 729-741).

Jarrar, M. (2008). Towards effectiveness and transparency in e-business transactions: An ontology for customer complaint management. In R. Garcia (Ed.), *Semantic Web for Business: Cases and Applications*. Hershey, PA: IGI Global.

Jarrar, M., & Heymans, S. (2008). Towards pattern-based reasoning for friendly ontology debugging. *Journal of Artificial Intelligence Tools, 17*(4).

Jarrar, M., & Meersman, R. (2007). Ontology engineering -The DOGMA approach. In E. Chang, T. Dillon, R.

Meersman & K. Sycara (Eds.), *Advances in Web Semantic. Volume 1, A state-of-the Art Semantic Web advances in Web Semantics IFIP2.12.* Springer-Verlag

Jarrar, M., Demey, J., & Meersman, R. (2003). On reusing conceptual data modeling for ontology engineering. *Journal on Data Semantics, 1*(1) (LNCS 2800, pp. 185-207).

Jarrar, M., Keet, M., & Dongilli, P. (2006). *Multilingual verbalization of ORM conceptual models and axiomatized ontologies* (Tech. Rep.). Belgium: Vrije Universiteit.

*Java Server Pages Technology.* Retrieved April 3, 2008, from http://java.sun.com/products/jsp/.

JBoss Corporation. (2008). *Hibernate – Relational persistence for Java and .NET.* Retrieved November 13, 2008, from http://www.hibernate.org/

JBoss. (2008). *JBoss Drools.* Retrieved April 18, 2008, from http://www.jboss.org/drools/

Jennings, N. R., & Wooldridge, M. J. (1998). *Agent technology - Foundations, applications and markets.* Springer.

*Jess, The Rule Engine For the Java Platform* (2008). Retrieved April 22, 2008, from http://www.jessrules.com/jess/index.shtml

Johnson, W.L. (1994). Agents that learn to explain themselves. In Proceedings *of the National Conference on Artificial Intelligence (AAAI 1996)* (pp. 1257-1263).

Jones, C. B., Alani, H. & Tudhope, D. (2001). Geographical information retrieval with ontologies of place. In *Spatial Information Theory: Foundations of Geographic Information Science, International Conference: COSIT* (LNCS 2205, pp. 322-335).

Jong Soo Park, P. S. Y., Ming-Syan Chen. (1995). An effective hash-based algorithm for mining association rules. In *Proceedings of the 1995 ACM SIGMOD International Conference on Management of Data* (pp.175-186). San Jose, Canada.

Jouault, F., Bézivin, J., Kurtev, I. (2006). TCS: A DSL for the specification of textual concrete syntaxes in model engineering. In *Proc. of the 5th Int'l Conf. on Generative Prog. and Component Eng.* (pp. 249-254).

Jovanović, J., & Gašević, D. (2005). XML/XSLT-based knowledge sharing. *Expert Systems with Applications, 29*(3), 535-553.

JSR 223. *Scripting for the JavaTMPlatform.* Retrieved April 3, 2008, from http://jcp.org/en/jsr/detail?id=223.

JSR 94. (2004, August 04). *JSR94: Java Rule Engine API.* Retrieved April 18, 2008, from http://www.jcp.org/en/jsr/detail?id=94

Kagal, L. (2002). *Rei: A policy language for the Me-Centric Project* (Tech. Rep).HP Labs.

Kakas, A., & Miler, R. (1997). A simple declarative language for describing narratives with actions. *JLP, 31*(1-3), 157200.

Kakas, A., Mancarella, P., Sadri, F., Stathis, K., & Toni, F. (2004). The KGP model of agency. In *The 16th European conference on artificial intelligence* (p. 3337).

Kakas, A., Miler, R., & Toni, F. (1999). An argumentation framework for reasoning about actions and change. In *LPNMR'99.*

Kakas, A.C., Kowalski, R.A., & Toni, F. (1992). Abductive logic programming. *Journal of Logic and Computation, 2*(6), 719-770.

Kalfoglou, Y., & Schorlemmer, M. (2003). Ontology mapping: The state of the art. *The Knowledge Engineering Review, 18*(1), 1-31.

*KAON – The KArlsruhe ONtology and Semantic Web tool suite* (2008). Retrieved April 22, 2008, from http://kaon.semanticweb.org.

*KAON2 – Ontology Management for the Semantic Web* (2008). Retrieved April 22, 2008, from http://kaon2.semanticweb.org.

Kappel, G., Kapsammer, E., Kargl, H., Kramler, G., Reiter, T., Retschitzegger, W., Schwinger, W., Wimmer, M. (2006). Lifting metamodels to ontologies: A step to the semantic integration of modeling languages. In *Proc. of the ACM/IEEE 9th Int'l Conf. on Model Driven Eng. Languages and Sys.* (pp. 528-542).

Karagiannis, D. (1995, August). Bpms: Business process management systems. *SIGOIS Bulletin, 16*(1), 10-13.

Kardasis, P., & Loucopoulos, P. (2004). Expressing and organising business rules. In *Information and software technology, 46*(11), 701-718.

Kattenstroth, H., May, W., & Schenk, F. (2007). Combining OWL with F-Logic rules and defaults. In *Proceedings of the International Workshop on Applications of Logic Programming to the Web, Semantic Web and Semantic Web Services* (Vol. 287) (pp. 60-75). CEUR-WS.

Kaviani, N, Gasevic, D., Hatala, M., Clement, D., & Wagner, G. (2007-2). Integration of rules and policies for Semantic Web services. *International Journal of Advanced Media and Communication (IJAMC), 1*(4), 404-423.

Kaviani, N., Gasevic D., & Hatala, M. (2007-1). Web rule languages to carry policies. In *Proceedings of the 2007 IEEE Workshop on Policies for Distributed Systems and Networks (POLICY 2007), Bologna, Italy* (pp. 188-192).

Keller, G., Nüttgens, M., & Scheer, A.-W. (1992). Semantische Prozeßmodellierung auf der Grundlage Ereignisgesteuerter Prozeßketten (EPK). In A.-W. Scheer (Ed.), *Veröffentlichungen des Instituts für Wirtschaftsinformatik* (Vol. 89). Saarbrücken.

Kelly, R. F. & Pearce, A. R. (2006). Towards high-level programming for distributed problem solving. In *Proceedings of the IEEE/WIC/ACM International Conference on Intelligent Agent Technology* (pp. 490-497). Washington, DC: IEEE Computer Society.

Kemsley, S., Straus, D,. & Sinur, J. (2008) Change the rules without changing your business processes: Using business rules management to turbo-charge your BPM solution. *Intelligent Enterprise Vendor Perspectives Webcast.* Retrieved February 12, 2008 from http://intelligententerprise.com/webinars.jhtml

Kent, S. (2002). Model driven engineering. In *Proceedings of IFM2002* (LNCS 2335).

Kephart, J. O., & Chess, D. M. (2003). The vision of autonomic computing. *Computer, 36*(1), 41-50.

Kiczales, G., Lamping, J., Menhdhekar, A., Maeda, C., Lopes, C., Loingtier, J.-M. Et al. (1997). Aspect-oriented programming. In M. Aksit, & S. Matsuoka (Eds.), *Proceedings European Conference on Object-Oriented Programming* (LNCS 1241, pp. 220-242).

Kietz, J.-U. , Morik, K., Wrobel, S., & Emde, W. (1993). *Knowledge acquisition and machine learning.* London: Academic Press.

Kifer, M. & Wu, J. (1993). A logic for programming with complex objects. *J. Computer and System Sciences, 47*(1), 77-120.

Kifer, M. (1995). Deductive and object data language: A quest for integration. In T. W. Ling, A. O. Mendelzon, & L. Vieille (Eds.), *Proceedings of the International Conference on Deductive and Object-Oriented Databases* (pp. 187-212). (LNCS 1013).

Kifer, M., de Bruijn, J., Boley, H., & Fensel, D. (2005). A realistic architecture for the Semantic Web. In *Proc. RuleML 2005* (pp. 17-29), Galway, Ireland.

Kifer, M., Lausen, D., & Wu, J. (1995). Logical foundations of object-oriented and frame-based languages. *Journal of ACM, 42*(4), 741-843.

Kim, W. (1990). *Introduction to object-oriented databases.* Cambridge, MA: The MIT Press.

Kiryakov, A., Ognyanov, D., & Manov, F. (2005). OWLIM - A pragmatic semantic repository for OWL. In *Proceedings of the International Workshop on Scalable Semantic Web Knowledge Base Systems* (pp. 182-192). Springer.

Klemettinen, M., Mannila, H., Ronkainen, P., Toivonen, H., & Verkamo, A. I. (1994). Finding interesting rules from large sets of discovered association rules. In N. R. Adam, B. K. Bhargava, & Y. Yesha (Eds.), *Third International Conference on Information and Knowledge Management (CIKM'94)* (pp. 401-407). ACM Press.

Klösgen, W., & ytkow, J. M. (Eds.). (2002). *Handbook of data mining and knowledge discovery.* New York: Oxford University Press.

Kluegl, F., & Puppe, F. (1998). The multi-agent simulation environment SeSAm. In *Proceedings of Workshop "Simulation in Knowledge-based Systems"*. Report tr-ri-98-194, Reihe Informatik, Universität Paderborn.

Kluegl, F., Herrler, R., & Oechslein, C. (2004). From simulated to real environments: How to use SeSAm for software development. In *Multi Agent System Technologies* (LNCS 2831, pp. 13-24).

Klyne, G., & Carroll, J.J. (2004). *Resource description framework (RDF): Concepts and abstract syntax*. Recommendation 10 February 2004, W3C. Retrieved April 22, 2008, from http://www.w3.org/TR/rdf-concepts/.

Knauf, R. (2000). *Validating rule-based systems: A complete methodology* [Habilitation Thesis]. Ilmenau, Germany: Ilmenau Technical University, Department of Computer Science and Automation, Chair of Artificial Intelligence.

Knauf, R., Philippow, I., Gonzalez, A., Jantke, K., & Salecker, D. (2002). System refinement in practice - Using a formal method to modify real-life knowledge. In *Proceedings of 15th International Florida Artificial Intelligence Research Society Conference Society (FLAIRS-2002)* (pp. 216-220). Menlo Park, USA: AAAI Press.

Knauf, R., Spreeuwenberg, S., Gerrits, R., & Jendreck, M. (2004). A step out of the ivory tower: Experiences with adapting a test case generation idea to business rules. In *Proc. 17th Intl. FLAIRS Conference* (pp. 343-348). Menlo Park, USA: AAAI Press.

Knauf, R.; Gonzalez, A. J.; and Abel, T. (2002b). A framework for validation of rule-based systems. *IEEE Transactions of Systems, Man and Cybernetics - Part B: Cybernetics 32*(3), 281-295.

Knolmayer, G., Endl, R., Pfahrer, M. (2000). Modeling processes and workflows by business rules. In W. M. Aalst, J. Desel und A. Oberweis (Eds.), *Business process management, models, techniques, and empirical studies* (pp. 16-29). Springer.

KnowGravity Inc. (2002, July 26). Retrieved April 2008, from http://www.knowgravity.com/pdf-e/BRA - Concepts English.pdf

Koenig, A. (1995). Patterns and antipatterns. *Journal of Object-Oriented Programming, 8*(1), 46-48.

Kogut, P., Cranefield, S., Hart, L., Dutra, M., Baclawski, K., Kokar, M., & Smith, J. (2002). UML for Ontology development. *The Knowledge Engineering Review, 17*(1), 61-64.

Kolovski, V., Hendler, J., & Parsia, B. (2007). Analyzing Web access control policies. *WWW* (pp. 677-686).

Kopena, J.B., & Regli, W.C. (2003). *DAMLJessKB: A tool for reasoning with the Semantic Web.* Paper presented at the 2nd Intl. Semantic Web Conference (ISWC2003).

Kowalski, R. (1979). *Logic for problem solving.* Upper Saddle River, NJ: Prentice Hall.

Kowalski, R. (1988). The early years of logic programming. *Communications of the ACM, 31*(1), 38-43.

Kowalski, R. (1992). Database updates in the event calculus. *Journal of Logic Programming, 12*,121-146.

Kowalski, R. A. (1988). The early years of logic programming. *Communications of the ACM, 31(1)*, 38-43.

Kowalski, R., & Kuehner, D. (1971). Linear resolution with selection function. Artificial *Intelligence, 2*, 227-260.

Kowalski, R.A., & Sergot, M.J. (1986). A logic-based calculus of events. *New Generation Computing, 4*, 67-95.

Kozlenkov, A., & Schroeder, M. (2004). Prova: Rule-based java-scripting for a bioinformatics Semantic web. In E. Rahm (Ed.), *DILS* (LNCS 2994, 17-30).

Kozlenkov, A., Paschke, A., & Schroeder, M. (2008). Retrieved April 18, 2008, from http://www.prova.ws/

Kremer R. (1998, April). Visual languages for knowledge representation. *Eleventh Workshop on Knowledge Acquisition, Modeling and Management, Banff, Alberta, Canada.*

Krempels, K.H., & Panchenko, A. (2003). An approach for automated surgery scheduling. In E.K. Burke & H. Rudov (Eds.), *Proceedings of the 6th International Conference on the Practice and Theory of Automated Timetabling (PATAT '06).*

Krötzsch, M, Schaffert, S, & Vrandecic, D. (2007). Reasoning in semantic wikis. In *Reasoning Web* (LNCS 4636).

Krotzsch, M., Hitzler, P., Vrandecic, D., & Sintek, M. (2006). How to reason with OWL in a logic programming system. In *Rules and Rule Markup Languages for the Semantic Web, Second International Conference* (pp. 17–28). IEEE Computer Society.

Krötzsch, M., Rudolph, S., & Hitzler, P. (2008). Description logic rules. In *Proceedings of the 18th European Conference on Artificial Intelligence (ECAI-08)* (pp. 80–84). IOS Press.

Kühne, T. (2006). Matters of (meta-)modeling. *Software and System Modeling, 5*(4), 369-385.

Kulkarni, K., Mattos, N.M., & Cochrane, R. (1999). Active database features in SQL3. *Active rules in database-systems* (pp. 197-219). Springer Verlag.

Kupfer, A., Eckstein, S., Neumann, K., & Mathiak, B.(2006). A coevolution approach for database schemas and related ontologies. In *Proceedings of the 19th IEEE International Symposium on Computer-Based Medical Systems (CBMS'06)* (pp. 605-610). Washington, DC: IEEE Computer Society.

Kurtev, I., Bézivin, J., & Aksit, M. (2002). Technological spaces: An initial appraisal. In *Proceedings of the CoopIS, DOA'2002 Federated Conferences, Industrial track, Irvine, USA.*

Laird, J., Congdon, C.B., & Coulter, K. (2006). *The Soar user's manual. Version 8.6.3.* University of Michigan. Retrieved from http://sitemaker.umich.edu/soar

Lam, D. N., & Barber, K. S. (2005). Comprehending agent software. In F. Dignum, V. Dignum, S. Koenig, S. Kraus, M. P. Singh, & M. Wooldridge (Eds.), *Proceedings of the 4th International Joint Conference on Autonomous Agents and Multiagent Systems (AAMAS 2005)* (pp. 586-593). ACM Press.

Lam, D., & Barber, K. (2004). Debugging agent behaviour in an implemented agent system. In *Proceedings of 2nd International Workshop on Programming Multiagent Systems Languages and tools (PROMAS 2004), New York, USA,*

Lamb, C., Landis, G., Orenstein, J., & Weinreb, D. (1991). The ObjectStore system. *Communications of the ACM, 34*(10), 50-63.

Lamb, N., & Preece, A. (1996). Verification of multi-agent knowledge-based systems. In *ECAI'96 Workshop on Validation, Verification and Refinement of Knowledge-Based Systems* (pp. 114-119).

Lassila, O., & Swick, R. R. (1999). *Resource description framework (rdf) model and syntax specication* (Tech. Rep.). World Wide Web Consortium. Retrieved from http://www.w3.org/TR/REC-rdf-syntax

Lausen, G., Ludascher, B., & May, W. (1998). On logical foundations of active databases. *Logics for databases and information systems* (pp. 389-422).

Lee,W., Stolfo, S. J., & Mok, K.W. (2000). Adaptive intrusion detection: A data mining approach. *Artificial Intelligence Review, 14*(6), 533-567.

Lehman, J. F., Laird, J., & Rosenbloom, P. (1996). A gentle introduction to Soar, an architecture for human cognition. In S. Sternberg, D. Scarborough (Eds.), *Invitation to cognitive science*, 4, 212-249. MIT Press.

Leite, J. A., Alferes, J. J., & Pereira, L. M. (2001). Minerva—A dynamic logic programming agent architecture. In J.-J. Meyer & M. Tambe (Eds.), *Pre-proceedings of the eighth international workshop on agent theories, architectures, and languages (ATAL-2001)* (pp. 133–145).

Leite, J.A. (2003). *Evolving knowledge bases.* IOS-Press.

Leite, J.A., Alferes, J.J., & Pereira, L.M. (2001). On the use of multi-dimensional dynamic logic programming to represent societal agents' viewpoints. In *EPIA01.*

Lester, J.C., & Porter, B.W. (1997). Developing and empirically evaluating robust explanation generators: The knight experiments. *Computational Linguistics, 23*(1), 65-101.

Levesque, H.J., & Brachman, R.J. (1985). A fundamental tradeoff in knowledge representation and reasoning (Revised Version). In R. J. Brachman & H.J. Levesque (Eds.), *Readings in knowledge representation*. San Francisco: Morgan Kaufmann.

Levy, A. Y., & Rousset, M. C. (1995). Combining rules and description logics: An overview of CARIN. In *Proceedings of the International Symposium on Logic Programming* (pp. 635). Cambridge: MIT Press.

Levy, A.A., & Rousset, M.-C. (1998). CARIN: A representation language combining horn rules and description logics. *Articial Intelligence, 104*(1-2),165-209.

Levy, A.Y., & Rousset, M. (1998). Combining horn rules and description logics in CARIN. *Artificial Intelligence, 104*(1-2), 165-209. Elsevier.

Levy, A.Y., & Rousset, M.-C. (1998). Combining Horn rules and description logics in CARIN. *Artificial Intelligence, 104*, 165-209.

Lewis, A. (2007). Web *Services Description Language (WSDL) Version 2.0: Additional MEPs*. W3C Working Group Note. Retrieved from http://www.w3.org/TR/2007/NOTE-wsdl20-additional-meps-20070626/

Li, M., Du X.Y., & Wang, S. (2005). Learning ontology from relational database. In *Proceedings of 2005 International Conference on Machine Learning and Cybernetics, 6*(18-21) (pp. 3410-3415).

Lieberherr, K., Holland, I., & Riel, A. (1988). Object-oriented programming: An objective sense of style. In *Proceedings of OOPSLA '88* (pp. 323-334). ACM.

Liebowitz, J. (2006). Where is the "intelligence" behind "business intelligence"? *Business Intelligence*. Retrieved January 7, 2007, from http://www.businessintelligence.com/ex/asp/code.128/pagenum.1/xe/article.htm

Liebowitz, J. (Ed.). (1998). *The handbook of applied expert systems*. Boca Raton: CRC Press.

Ligęza, A. (1993). Logical foundations for knowledge-based control systems – Knowledge representation, reasoning and theoretical properties. *Scientific Bulletins of AGH, Automatics, 63*(1529).

Ligęza, A. (1993). A note on backward dual resolution and its application to proving completeness of rule-based systems. In *Proceedings of the 13th Int. Joint Conference on Artificial Intelligence (IJCAI), Chambery, France* (pp. 132-137).

Ligęza, A. (2006). *Logical foundations for rule-based systems*. Berlin, Heidelberg: Springer-Verlag.

Ligęza, A., & Nalepa, G. J. (2005). *Visual design and on-line verication of tabular rule-based systems with xtt*. In K. P. Jantke, K.-P. Fähnrich, & W. S. Wittig (Eds.), *Marktplatz internet: Von e-learning bis e-payment: 13. leipziger informatik-tage, lit 2005* (pp. 303312). Bonn: Gesellschaft fur Informatik.

Ligęza, A., & Nalepa, G. J. (2007, May). Knowledge representation with granular attributive logic for XTT-based expert systems. In D. C. Wilson, G. C. J. Sutcli e, & FLAIRS (Eds.), *Flairs-20: Proceedings of the 20th International Florida Artificial Intelligence research society conference : Key West, Florida, May 7-9, 2007* (pp. 530-535). Menlo Park, California: AAAI Press.

Ligęza, A., Wojnicki, I., & Nalepa, G. (2001). Tab-trees: a case tool for design of extended tabular systems. In H. M. et al. (Ed.), *Database and expert systems applications* (Vol. 2113, pp. 422-431). Berlin: Springer-Verlag.

Lillis, D., Collier, R., Toolan, F., & Dunnion, J. (2007). Evaluating communication strategies in a multi-agent information retrieval system. In *Proceedings of the 5th European Workshop on Multi-Agent Systems (EUMAS'07), Hammamet, Tunisia.*

Ling, T. W. & Lee, W. B. T. (1998). DO2: A deductive object-oriented database system. In *Proceedings of the 9th International Conference on Database and Expert System Applications* (pp. 50–59) (LNCS 1460).

Ling, T. W. (1990). The prolog not-predicate and negation as failure. *New Generation Computing, 8*(1), 5-31.

Liu, G, Mok, A.K, & Konana, P. (1998). *A unified approach for specifying timing constraints and composite events in active real-time database systems*. Paper presented at the Real-Time Technology and Applications Symposium, June 1998.

Liu, M. (1996). ROL: A deductive object base language. *Information Systems, 21*(5), 431-457.

Liu, M. (1996). The LOL deductive database programming language. In *Proceedings of International Symposium on Programming Languages, Implementations, Logics and Programs* (pp. 454-468). ( LNCS 1140).

Liu, M. (1998). An overview of rule-based object language. *Journal of Intelligent Information Systems, 10*(1), 5-29.

Liu, M. (1998). Incorporating methods and encapsulation into deductive object-oriented database languages. In *Proceedings of the 9th International Conference on Database and Expert System Applications* (pp. 892-902) ( LNCS 1460).

Liu, M., Dobbie, G., & Ling, T. W. (2002). A logical foundation for deductive object-oriented databases. *ACM Transactions on Database Systems, 27*(1), 117-151.

Lloyd, J. W. (1987). *Foundations of logic programming* (2nd ed.). Springer-Verlag.

Lodha, B., Dinesha, K. V., & Kumar, P. (2005). Need for incorporating a rule-based component in conventional object oriented systems. In *Proceedings of the International Conference on Information Technology: Coding and Computing (ITCC'05).*

Loke, S. W. & Davison, A. (1996). Logic programming with the World Wide Web. In *Proceedings of the 7th ACM Conference on Hypertext* (pp. 235-245). New York: ACM.

Lou, Y., & Ozsoyoglu, M. (1991). LLO: A deductive language with methods and method inheritance. In *Proceedings of the ACM SIGMOD International Conference on Management of Data* (pp. 198-207). Denver, CO: ACM Press.

Luck, M., McBurney, P., Shehory, O., & Willmott, S. (2005). *Agent technology: computing as interaction (A roadmap for agent based computing).* AgentLink.

Luckham, D. (2002). *The power of events: An introduction to complex event processing in distributed enterprise systems.* Addison Wesley Professional.

Luke, S., Balan, G.C., Panait, L., Cioffi-Revilla, C., & Paus, S. (2003). MASON: A Java multi-agent simulation library. In *Proceedings of the Agent 2003 Conference.*

Luke, S., Cioffi-Revilla, C., Panait, L., & Sullivan, K. (2004). MASON: A new multi-agent simulation toolkit. In *Proceedings of the 2004 SwarmFest Workshop.*

Lukichev, S., & Wagner, G. (2005). Visual rules modeling. In *Sixth International Andrei Ershov Memorial Conference Perspectives of System Informatics, Novosibirsk, Russia.* Springer.

Lukichev, S., & Wagner, G. (2006). UML-based rule modeling with Fujaba. In *Proceedings of the 4th International Fujaba Days 2006, University of Bayreuth, Germany* (pp. 31-35).

Lukichev, S., & Wagner, G. (2006). Visual rules modeling. In I. Virbitskaite & A. Voronkov (Eds.), *Proceedings of the 6th International Conference Perspectives of Systems Informatics* (LNCS 4378, pp. 467-673).

Lukichev, S., & Wagner, G. (2006). Visual rules modeling. In *Proceedings of the 6th Inter-national Conference Perspectives of Systems Informatics* (pp. 467-673).

Lukichev, S., & Wagner, G. (2006). Verbalization of the REWERSE I1 rule markup language. REWERSE IST 506779 Report I1-D6.

Lunardhi, A. D., & Passino, K. M. (1995). Verification of qualitative properties of rule-based expert systems. *Applied Artificial Intelligence, 9*, 587-621.

Maier, D. (1986). A logic for objects (Tech. Rep. CS/E-86-012). Beaverton, OR: Oregon Graduate Institute.

Maier, D. (1987). Why database languages are a bad idea. In *Proceedings of the Workshop on Database Programming Languages, Roscoff, France.*

Maier, D. (Ed). (1983). *The theory of relational databases.* Computer Science Press.

Manolescu, I., Brambilla, M., Ceri, S., Comai, S., & Fraternali, P. (2005). Model-driven design and deployment of service-enabled Web applications. *ACM Trans. Inter. Tech., 5*(3), 439-479.

Marsh, D., Tynan, R., O'Kane, D., & O'Hare, G.M.P. (2004). Autonomic wireless sensor networks. *Engineering Applications of Artificial Intelligence Journal, 17*(7), 741-748.

Martincic, C.J. (2001). *Mechanisms for answering "why not" questions in rule- and object-based systems.* PhD thesis, University of Pittsburgh.

Matheus, C., Baclawski, K., & Kokar, M. (2006). BaseVISor: A triples-based inference engine outfitted to process RuleML and R-Entailment rules. In *Proceedings of the International Conference on Rules and Rule Languages for the Semantic Web.*

Matheus, C.J. (2004). SWRLp: An XML-Based SWRL presentation syntax. In *Proceedings of the 3rd International Workshop on Rules and Rule Markup Languages for the Semantic Web,* Hiroshima, Japan (pp. 194-199).

Maybury, M.T. (1992). Communicative acts for explanation generation. *International Journal of Man-Machine Studies, 37*(2), 135-172.

Maybury, M.T. (1998): Planning multimedia explanations using communicative acts. In M.T. Maybury & W. Wahlster (Eds.), *Readings in intelligent user interfaces.* Morgan Kaufmann.

McBride, B. (2001). Jena, Implementing the RDF model and syntax specification. In *Proceedings of the International Workshop on the Semantic Web* (Vol. 40). CEUR-WS.

McBride, B. (2002). Jena: A Semantic Web toolkit, *IEEE Internet Computing, 6(6)*,55-59.

McCabe, T.J., & Butler, C.W. (1989). Design complexity measurement and testing. *Communications of the ACM, 32*(12), 1415-1425.

McCarthy, J. (1959). Programs with common sense. In *Proceedings of the Teddington Conference on the Mechanization of Thought Processes* (pp. 75-91). London: Her Majesty's Stationary Office.

McCarthy, J. (1979). Ascribing mental qualities to machines. In M. Ringle (Ed.), *Philosophical perspectives in artificial intelligence* (pp. 161-195). Humanities Press.

McCarthy, J. (1992). *Elephant 2000: A programming language based on speech acts.* Unpublished Manuscript. Retrieved May 1, 2008 from *http://www-formal.stanford.edu/jmc/*

McCarthy, J. (2007). *Elephant 2000: A programming language based on speech acts.* Invited talk at OOPSLA 2007, Montreal, Canada.

McCarthy, J., & Hayes, P.J. (1969). Some philosophical problems from the standpoint of artificial intelligence. In D. Michie & B. Meltzer (Eds.), *Machine Intelligence* (4th ed.) (pp. 463-502. Edinburgh: Edinburgh University Press.

McCarthy, J., (1979). Ascribing mental qualities to machines. In M. Ringle (Ed.), *Philosophical perspectives in artificial intelligence.* Harvester Press.

McCoy, D. (2002). Business activity monitoring: Calm before the storm. *Gartner Group.* Retrieved May 15, 2007, from http://www.gartner.com/resources/105500/105562/105562.pdf

McGuiness, D.L., & da Silva, P.P. (2003). Infrastructure for Web explanations. In *Proceedings of the Second International Semantic Web Conference (ISWC 2003)* (pp. 113-129).

McGuinness, D. L. & Harmelen, F. (2004). *OWL Web Ontology language overview, W3C Recommendation.* Retrieved April 14, 2008, from http://www.w3.org/TR/owl-features/.

MDA (2008). *Model Driven Architecture,* Retrieved May 2, 2008, from http://www.omg.org/mda/

Meditskos, G., & Bassiliades, N. (2007). *A Semantic Web service discovery and composition prototype framework using production rules.* Workshop on OWL-S: Experiences and Future

Meditskos, G., & Bassiliades, N. (2008). *Combining a DL reasoner and a rule engine for improving entailment-based OWL reasoning.* International Semantic Web Conference (ISWC). Karlsruhe, Germany.

Meditskos, G., & Bassiliades, N. (2008). A rule-based object-oriented OWL reasoner. *IEEE Transactions on Knowledge and Data Engineering, 20,* 397-410.

Meersman, R. (1982). *The RIDL conceptual language* (Research report). Int. Centre for Information Analysis Services, Control Data, Brussels

Meersman, R. (1999). The use of lexicons and other computer-linguistic tools in semantics, design and cooperation of database systems. In Y. Zhang, M. Rusinkiewicz, & Y. Kambayashi (Ed.), *Proceedings of the Second International Symposium on Cooperative Database Systems for Advanced Applications (CODAS99)* (pp. 1-14). Springer.

Meersman, R. (1999), *Semantics ontology tools in information system design.* In Z. Ras & M. Zemankova (Eds.), *Proceedings of the International Symposium on Methodologies for Intelligent Systems (ISMIS99), vol. 1609.* Springer

Meersman, R. (2001). *New frontiers in modeling technology: The promise of ontologies.* In E.J.H. Kerckhoffs & M. Snorek, (Eds.), *Proceedings of the 15th European Multi-Simulation Conference* (pp. 3-6).

Mei, J, Bontas, E.P., & Lin, Z. (2005). OWL2Jess: A transformational implementation of the OWL semantics. In *Proceedings of International Workshops on ISPA* (LNCS 3759, pp. 599-608).

Mei, J., Lin, & Z., Boley, H. (2007). ALC: An integration of description logic and general rules. In *Proceedings of the Web Reasoning and Rule Systems* (pp. 163-177). Springer.

Mei, J., Lin, Z., & Boley, H. (2007). *ALCuP: An integration of description logic and general rules, Web reasoning and rule systems.* In First International Conference, RR 2007 (LNCS 4524, pp. 163-177).

Mei, J., Lin, Z., Boley, H., Li, J., & Bhavsar, V.C. (2007b). The DatalogDL combination of deduction rules and description logics. *Computational Intelligence, 23*(3), 356-372.

Melle, W., Shortliffe, E.H. & Buchanan, B.G. (1984). Emycin: A knowledge engineer's tool for constructing rule-based expert systems. In *Rule-based expert systems.* Addison-Wesley.

Mellor, S. J., & Balcer, M. (2002). *Executable UML: A foundation for model-driven architectures.* Boston: Addison-Wesley Longman Publishing Co., Inc.

Melton, J., & Simon, A.R. (1993). *Understanding the new SQL: A complete guide.* San Mateo: Morgan Kaufmann.

Meo, R., Psaila, G., & Ceri, S. (1996). Composite events in Chimera. In *Advances in Database Technology — EDBT '96.*

Meolic, R., Kapus, T., & Brezonic, Z (2003). An Action computation tree logic with unless operator. In *Proc. of South-East European Workshop on Formal Methods (SEEFM'03).* Thessaloniki, Greece.

Meolic, R., Kapus, T., & Brezonic, Z. (2000). Verification of concurent systems using ACTL. In *IASTED Int. Conf. on Applied Informatics (AI'00).* Anaheim, Calgary: IASTED/ACTA Press.

Meyer, J.-J. C., van der Hoek, W., & van Linder, B. (1999). A logical approach to the dynamics of commitments. *Arificial Intelligence, 113,* 1-40.

Microsoft (2003). Microsoft. Microsoft Visual Studio .NET Enterprise Architect. [CD-ROM]. ASIN: B00005S812.

Microsoft (2007). *The LINQ project.* Retrieved April 3, 2008, from http://msdn.microsoft.com/netframework/future/linq/.

Microsoft. (2007). *SQL Server 2005 books online: Multidimensional expressions (MDX) reference.* Microsoft SQL development center. Retrieved May 10, 2007, from http://msdn2.microsoft.com/en-us/library/ms145506.aspx

Milanovic, M., Gasevic, D., Giurca, A., Wagner, G., & Devedzic, V. (2006). On interchanging between OWL/SWRL and UML/OCL. In *Proc. of the OCLApps Workshop* (pp. 81-95).

Milanović, M., Gašević, D., Giurca, A., Wagner, G., & Devedžić, V. (2007). *Model transformations to share rules between SWRL and R2ML.* Paper presented at the 3rd International Workshop on Semantic Web Enabled

Software Engineering (SWESE 2007), Innsbruck, Austria.

Milanovic, M., Kaviani, N., Gasevic, D., Giurca, A., Wagner, G., Devedzic, V., et al. (2007). Business process integration by using general markup language. In *Proceedings of the 11th IEEE International Enterprise Distributed Object Computing Conference* (pp. 353-364). IEEE Computer Society.

Miller, B.S., & Sprenger, M. (1996). Explanation abilities. In *Computer as assistants: A new generation of support systems* (pp. 68-80). Lawrence Erlbaum Associates.

Miller, J., & Mukerji, J. (2003). MDA guide version 1.0.1. *Object Management Group.*

Milner, R. (1989). *Communication and concurrency.* Prentice-Hall.

Minar, N., Burkhart, R., Langton, C., & Askenazi, M. (1996). *The swarm simulation system: A toolkit for building multi-agent simulations.* SFI Working Paper 96-06-042.

Minsu, J., & Sohn, J.C. (2004). Bossam: An extended rule engine for OWL inferencing. In *Proceedings of Rules and Rule Markup Languages for the Semantic Web* (pp. 128-138). Springer.

Miranker, D. (1989). TREAT: A new and efficient match algorithm for AI production systems. In *Proceedings of National Conference on Artificial Intelligence* (pp. 42-47), Pittman/Morgan Kaufman.

Miranker, D. (1990). Treat*: A new and ecient match algorithme for AI production systems.* Morgan Kaufmann.

MOF (2005). Object Management Group, Meta Object Facility (MOF) Core, v2.0. *OMG Document formal/06-01-01.* Retrieved from http://www.omg.org/cgi-bin/doc?formal/2006-01-01.

MOF. (2008, February 20). *Meta Object Facility (MOF) specification 2.0 facility and object lifecycle.*

Mooney, R.J., & Richards, B.L. (1991). First-order theory revision. In *Machine Learning: Proceedings of the Eighth International Workshop on Machine Learning* (pp. 447-451).

Morgan, T. (2002). *Business rules and information systems. Aligning it with business goals.* Boston, MA: Addison Wesley.

Moss, C. (1994). *Prolog++.* Addison-Wesley.

Motik, B., & Volz, R. (2003). Optimizing query answering in description logics using disjunctive deductive databases. In *Proceedings of the 10th International Workshop on Knowledge Representation meets Databases, Hamburg, Germany* (Vol. 79).

Motik, B., Grau, B.C., Horrocks, I., Wu, Z., Fokoue, A., & Lutz, C. (2008). *OWL 2 Web Ontology Language Profiles.* OWL Working Group.

Motik, B., Horrocks, I., & *Sattler, U. (2007). Adding integrity constraints to OWL. In Proc. of the Third OWL Experiences and Directions Workshop. CEUR.*

Motik, B., Horrocks, I., Rosati, R., & Sattler, U. (2006). Can OWL and logic programming live together happily ever after? In *Proceedings ISWC-2006* (LNCS 4273, pp. 501–514).

Motik, B., Sattler, U., Studer, R. (2005). Query answering for OWL-DL with rules. *Journal of Web Semantics, 3*(1), 41-60.

Motwani, R., Widom, J., Arasu, A., Babcock, B., Babu, S., Datar, M., Manku, G., Olston, C., Rosenstein, J., & Varma, R. (2003). *Query processing, resource management, and approximation in a data stream management system* (Tech. Rep.). Stanford InfoLab.

Mozart (2008). *The Mozart programming system.* Retrieved September 24, 2008, from http://www.mozart-oz.org/

Muggleton, S.H. (1991). Inductive logic programming. *New Generation Computing, 8*(4), 295-318.

Muldoon, C., O'Hare, G.M.P., Phelan, D., Strahan, R., & Collier, R. (2003). ACCESS: An agent system for ubiquitous service delivery. In *Cooperative information agents VII* (LNCS 2782).

Munoz, S., Perez, J., & Gutierrez, C. (2007). Minimal deductive systems for RDF. In E. Franconi, M. Kifer, & W. May (Eds.), *ESWC* (LNCS 4519, pp. 53-67).

MySQL. (2008). *MySQL – The world's most popular open-source database.* Retrieved November 13, 2008, from http://www.mysql.com/

Nagl, C., Rosenberg, F., & Dustdar, S. (2006). VIDRE–A distributed service oriented business rule engine based on RuleML. In *Proc. of the 10th IEEE International Enterprise Distributed Object Computing Conference* (pp. 35-44).

Naish, L. (1992). Declarative diagnosis of missing answers. *New Generation Computing, 10*(3), 255-286.

Nalepa, G. J., & Ligęza, A. (2005). Conceptual modelling and automated implementation of rule-based systems. In T. S. Krzysztof Zieliński (Ed.), *Software engineering: Evolution and emerging technologies* (Vol. 130, pp. 330340). Amsterdam: IOS Press.

Nalepa, G. J., & Ligęza, A. (2005). A graphical tabular model for rule-based logic programming and verification. *Systems Science, 31*(2), 8995.

Nalepa, G. J., & Wojnicki, I. (2008). *Towards formalization of ARD+ conceptual design and refinement method.* In D. C. Wilson & H. C. Lane (Eds.), *Flairs-21: Proceedings of thetwenty-rst international Florida artificial intelligence research society conference: 15-17 May 2008, Coconut Grove, Florida, USA* (pp. 353-358). Menlo Park, California: AAAI Press.

Naqvi, S., & Krishnamurthy, R. (1988). Database updates in logic programming. In *ACM Symposium on Principles of Database Systems.* New York: ACM.

Naqvi, S., & Tsur, S. (1989). *A logical language for data and knowledge bases.* Computer Science Press.

Naqvi, W., & Ibrahim, M. T. (1994). EECA: An active knowledge model. In D. Karagiannis (Ed.), *Proceedings of the 5th international Conference on Database and Expert Systems Applications* (LNCS. 856, pp. 380-389).

Nazareth, D. L. (1989). Issues in the verification of knowledge in rule-based systems. *Int. J. Man-Machine Studies, 30,* 255-271.

Nectaria, T., & Egenhofer M. J. (1996). Multi-resolution spatial databases: Consistency among networks. In *Integrity in Databases-Sixth International Workshop on Foundations of Models and Langauges for Data and Objects* (pp. 119–132).

Negnevitsky, M. (2002). *Artificial intelligence. A guide to intelligent systems.* Harlow, England; London; New York: Addison-Wesley.

Nejdl, W., Olmedilla, D., & Winslett, M. (2004). *PeerTrust: Automated trust negotiation for peers on the Semantic Web.* Paper presented at Workshop on Secure Data Management in a Connected World (SDM'04).

Newell, A. (1990). *Unified theories of cognition.* Harvard University Press.

Newell, A., & Simon, H. A. (1976). Computer science as empirical enquiry. *Communications of the ACM, 19,* 113-126.

Nguyen, T. A., & et al. (1985). Checking an expert systems knowledge base for consistency and completeness. In *Proceedings of the 9-th IJCAI* (pp. 375-378).

Nicholls, C. (2006). BI 2.0: The next generation. *DM Review Magazine.* Retrieved June 7, 2007, from http://www.dmreview.com/issues/20061101/1066763-1.html

Nissanke, N. (1999). *Introductory logic and sets for computer scientists.* Addison Wesley Longman Ltd.

Noble, J. (1997). Basic relationship patterns. In *Proceedings EuroPLOP 97.* Retrieved April 3, 2008, from http://jerry.cs.uiuc.edu/plop/plopd4-submissions/P53.ps

NORMA (n.d.). Retreived October 2006, from http://sourceforge.net/projects/orm

North, K. (1999). Modeling, data semantics, and natural language. *New Architect Magazine.*

North, M.J., Howe, T., Collier, N., & Vos, J.R. (2005). The repast simphony runtime system. In *Proceedings of the Agent 2005 Conference on Generative Social Processes, Models, and Mechanisms, ANL/DIS-06-5* (pp. 1-8). Argonne National Laboratory and The University of Chicago.

North, M.J., Howe, T., Collier, N., & Vos, J.R. (2005). The repast simphony development environment. In

*Proceedings of the Agent 2005 Conference on Generative Social Processes, Models, and Mechanisms, ANL/DIS-06-5* (pp. 159-166). Argonne National Laboratory and The University of Chicago.

Nute, D. (1994). Defeasible logic. In *Handbook of Logic in Artificial Intelligence and Logic Programming, Nonmonotonic Reasoning and Uncertain Reasoning* (Vol. 3). Oxford University Press.

Nwana, H. (1995) Software agents: An overview. *Knowledge Engineering Review, 11*(2), 205-244.

O'Connor, M., Knublauch, H., Samson, T., & Musen, M. (2005). Writing rules for the Semantic Web using SWRL and Jess. *Protégé with Rules WS*. Madrid.

O'Connor, M., Knublauch, H., Tu, S., Grosof, B., Dean, M., Grosso, W., & Musen, M.A. (2005). Supporting rule system interoperability on the Semantic Web with SWRL. In Y. Gil, E. Motta, V.R. Benjamins, & M.A. Musen (Eds.), *Proceedings of the Fourth International Semantic Web Conference – ISWC 2005* (LNCS 3729). Berlin: Springer.

O'Connor, M.J., Shankar, R.D., Tu, S.W., Nyulas, C., Parrish, D.B., Musen, M.A., & Das, A.K. (2007). *Using semantic web technologies for knowledge-driven querying of biomedical data*. Paper presented at the 11th Conference on Artificial Intelligence in Medicine (AIME 07), Amsterdam, Netherlands.

O'Connor, M.J., Tu, S.W., Das, A.K., Musen, M.A. (2007). Querying the Semantic Web with SWRL. *International RuleML Symposium on Rule Interchange and Applications (RuleML2007), Orlando, FL*. Springer Verlag.

O'Dea, D., Geoghegan, S., & Ekins, C. (2005). Dealing with geospatial information in the Semantic Web. In *Proceedings of the 2005 Australasian Ontology Workshop: AOW '05* (pp. 69-73). Australian Computer Society, Inc.

O'Hare G.M.P., Duffy B.R., Collier R.W, Rooney C.F.B., & O'Donoghue R.P.S. (1999). Agent factory: Towards social robots. *First International Workshop of Central and Eastern Europe on Multi - agent Systems (CEEMAS'99), St. Petersburg, Russia.*

O'Hare G.M.P., O'Grady M.J., Marsh D., Ruzzelli A. G. & Tynan R. (2006). Autonomic wireless sensor networks: Intelligent ubiquitous sensing. In *Proceedings of ANIPLA 2006, International Congress on Methodologies for Emerging Technologies in Automation, University La Sapienza, Rome, Italy.*

O'Hare, G.M.P., & O'Grady, M.J. (2003). Gulliver's genie: A multi-agent system for ubiquitous and intelligent content delivery. *Computer Communications, 26*(11), 1177-1187.

O'Hare, G.M.P., Collier, R., Conlon, J., & Abbas, S. (1998). Agent factory: An environment for constructing and visualising agent communities. *9th AICS Conference, Irish Artificial Intelligence and Cognitive Science Conference, UCD, Dublin.*

OASIS (2008). *Organization for the advancement of structured information standards*. Retrieved April 14, 2008, from www.oasis-open.org/.

OASIS. (2007, April 11). *Web services business process execution language version 2.0*. Retrieved April 18, 2008, from http://docs.oasis-open.org/wsbpel/2.0/OS/wsbpel-v2.0-OS.html

Object Management Group (2001). *OMG model driven architecture*. Retrieved April 3, 2008, from http://www.omg.org/mda/

Object Management Group (2006). *Business Process Modeling Notation, V1.1*. Official OMG Specification. Retrieved May 4, 2008 from http://www.bpmn.org/Documents/OMG%20Final%20Adopted%20BPMN%201-0%20Spec%2006-02-01.pdf.

Object Management Group (2007). *MOF QVT final adopted specification*. Retrieved April 3, 2008, from http://www.omg.org/docs/ptc/07-07-07.pdf

Object Management Group (2008). *Business Process Model and Notation (BPMN) specification 2.0*. Retrieved from http://www.omg.org/cgi-bin/doc?bmi/08-02-06

Object Management Group. (2006, May). *Object constraint language version 2.0* (Tech. Rep.). OMG.

*Object Management Group. Object Constraint Language* (OCL). Retrieved February 12, 2008 from http://www.omg.org/technology/documents/modeling_spec_catalog.htm#OCL. Ouyang, C., van der Aalst, W. M. P., Dumas, M., & ter Hofstede, A. H. M. (2006). Translating BPMN to BPEL. *BPM Center.*

ObjectConnections (2008). *Common knowledge.* Retrieved April 29, 2008, from http://www.objectconnections.com

OCL (2006). Object constraint language. *OMG Available Specification, Version 2.0, formal/06-05-01.*

Odell, J., van Dyke Parunak, H., & Bauer, B. (2000). Extending UML for agents. In G. Wagner, Y. Lesperance, & E. Yu (Eds.), *Proceedings of the 2nd International Workshop on Agent-Oriented Information Systems.* Berlin, Germany: iCue Publishing.

Odersky, M., Altherr, P., Cremet, V., Emir, B., Maneth, S., Micheloud, S., et al. (2004). *An overview of the scala programming language* (EPFL Tech. Rep. IC/2004/64).

ODM. (2001, August 5). *ODM: The Object Management Group ontology definition metamodel.* Retrieved from http://www.omg.org/cgi-bin/apps/doc?ad/05-08-01.pdf

ODM. (2003, March 4). Ontology definition metamodel. *Third Revised Submission to OMG/ RFP* (pp. 1-314).

OGC Technical Committee (1999). *OGC, abstract feature specification* (Tech. Rep.) Retrieved 2008, from: http://www.opengis.org/techno/specs.htm.

OMG (1999). *Unified Modeling Language Specification.* Retrieved from http://www.omg.org/uml.

OMG (2003). *Object Constraint Language (OCL), v2.0.* Retrieved from http://www.omg.org/docs/ptc/03-10-14.pdf

OMG (2003). *Unified Modeling Language, v1.3, Section 6: Object Constraint Language Specification.* Retrieved March 2003, from http://www.omg.org/spec/UML/1.3/

OMG (2006). *Business Process Modeling Notation Specification.* Retrieved from http://www.omg.org/docs/dtc/06-02-01.pdf

OMG (2008). *Business process modeling notation.* V1.1, OMG

OMG (2008). *The Object Management Group (OMG).* Retrieved April 14, 2008, from http://www.omg.org/

OMG MOF (2008). *OMG's MetaObject Facility.* Retrieved May 2, 2008, from http://www.omg.org/mof/

OMG Ontology Definition Metamodel (ODM) (2006). Sixth revised submission. *OMG Document ad/2006-05-01.* Retrieved from http://www.omg.org/docs/ad/06-05-01.pdf

OMG PRR (2007). *Production Rule Representation (PRR), Beta 1, OMG Adopted Specification.* Retrieved from http://www.omg.org/docs/dtc/07-11-04.pdf

OMG, Meta Object Facility (MOF) 2.0 XMI Mapping Specification, v2.1. (2005). *OMG Document formal/2005-09-01.* Retrieved from http://www.omg.org/cgi-bin/doc?formal/2005-09-01

OMG, Meta Object Facility (MOF) Core, v2.0. (2006). *OMG Document formal/06-01-01.* Retrieved from http://www.omg.org/cgi-bin/doc?formal/2006-01-01

OMG, MOF QVT Final Adopted Specification. (2005). *OMG document 05-11-01.* Retrieved from http://www.omg.org/docs/ptc/05-11-01.pdf

OMG, Ontology Definition Metamodel Request for Proposal. OMG Document: ad/2003-03-40, (2003). Retrieved from http://www.omg.org/cgi-bin/doc?ad/2003-03-40

OMG. (2006). *Business process modeling notation (bpmn) specification* (Tech. Rep. No. dtc/06-02-01). Object Management Group.

OMG. (2006). *Semantics of business vocabulary and business rules (sbvr)* (Tech. Rep. No. dtc/06-03-02). Object Management Group.

OMG. (2006). *Object Constraint Language, version 2.0.* Object Management Group.

OMG. (2007). *OMG Unified Modeling Language, Infrastructure, V2.1.2.* Object Management Group.

OMG. (2007). *Production rule representation betal* (Tech. Rep.). Object Management Group.

OOjDrew. (2008). Retrieved April 18, 2008, from http://www.jdrew.org/oojdrew/

Oracle (1998). Oracle8 time series data cartridge [White paper].

Oracle (1999). Migrating from Oracle time series to oracle analytic functions: Time series functions [White Paper].

Ordonez, C., Santana, C. A., & Braal, L. de. (2000). Discovering interesting association rules in medical data. In *ACM SIGMOD workshop on research issues in data mining and knowledge discovery* (p. 78-85).

Østerbye, K. (2007). Design of a class library for association relationships. In *Proceedings of the ACM SIGPLAN Symposium on Library-Centric Software Design LCSD'07.* Retrieved April 3, 2008, from http://lcsd.cs.tamu.edu/2007/final/7/7_Paper.pdf.

Ourston, D., & Mooney, R. (1990). Changing the rules: A comprehensive approach to theory refinement. In *Proceedings of the Eighth National Conference on Artificial Intelligence* (pp. 815-820).

Owen, M., & Raj, J. (2006). *Bpmn and business process management. Introduction to the new business process modeling standard.* (Tech. Rep.). OMG.

OWL 2, (2008). *OWL Web ontology language 1.1.* Retrieved November, 10th, 2008 from http://www.w3.org/TR/owl2-profiles/

OWL, (2004). *OWL Web ontology language reference.* Retrieved November, 10th, 2008 from www.w3.org/TR/owl-ref, 2004.

OWL-S (2004). *Semantic markup for Web Services.* Retrieved April 14, 2008, from http://www.w3.org/Submission/OWL-S/

Pan, J. (2005). Benchmarking DL reasoners using realistic ontologies. In *Proc. of the OWL: Experiences and Directions Workshop.*

Pan, J. Z., & Horrocks, I. (2004). *OWL-E: Extending owl with expressive datatype expressions* (Technical report, IMG/2004/KR-SW-01/v1.0). Victoria University of Manchester.

Pan, Y., Xie, G., Ma, L., Yang, Y., Qiu, Z., & Lee, J. (2006). Model-driven ontology engineering. *Journal of Data Semantics VII*, 57-78.

Pang, P., & Raschid, L. (1994). Bottom-up magic evaluation techniques for stratified production rules in a DBMS (Tech. Rep.). University of Maryland Institute for Advanced Computer Studies.

Paolucci, M., & Sycara, K. (2003). Autonomous Semantic Web services. *IEEE Internet Computing, 7*(5), 34-41.

Papazoglou, M., Traverso, P., Dustdar, S., & Leymann, F. (2007, November). Service-oriented computing: State of the art and research challenges. *IEEE Internet Computing*, 64-71.

Parsia, B., Sirin, E., & Kalyanpur, A. (2005). Debugging OWL Ontologies. In *Proceedings of the 14th International World Wide Web Conference (WWW2005), Chiba, Japan, May 2005.* Retrieved from http://www.mindswap.org/papers/debuggingOWL.pdf.

Paschke, A. (2005). *RBSLA: A declarative rule-based service level agreement language based on RuleML.* CIMCA/IAWTIC 2005 (pp. 308 314).

Paschke, A. (2006). *Verification, validation and integrity of distributed and interchanged rule based policies and contracts in the Semantic Web.* Int. Semantic Web and Policy Workshop (SWPW'06), Athens, Georgia.

Paschke, A. (2006). *ECA-RuleML: An approach combining eca rules with temporal interval-based KR event/action logics and transactional update logics.* Retrieved from http://arxiv.org/ftp/cs/papers/0610/0610167.pdf

Paschke, A. (2006). *A typed hybrid description logic programming language with polymorphic order-sorted DL-typed unification for Semantic Web type systems.* OWL-2006 (OWLED'06), Athens, Georgia.

Paschke, A. (2006). *ECA-LP / ECA-RuleML: A homogeneous event-condition-action logic programming language*. Retrieved from http://arxiv.org/ftp/cs/papers/0609/0609143.pdf

Paschke, A. (2006). *ECA-RuleML/ECA-LP*. Paper presented at the Int. Conf. of Rule Markup Languages. (RuleML'06), Athens, Georgia, USA. Retrieved from http://2006.ruleml.org/online-proceedings/poster_submission_54.pdf.

Paschke, A. (2007). *Rule-based service level agreements - Knowledge representation for automated e-contract, SLA and policy management*. München: IDEA Verlag GmbH.

Paschke, A., Boley, H., Kozlenkov, A., & Craig, B. (2007). Rule responder: RuleML-based agents for distributed collaboration on the pragmatic Web. In *Proceedings of the 2nd international Conference on Pragmatic Web* (pp. 17-28). ACM Press.

Paschke, A. (2008). *DoReMoPat - Domain Specific Reference Models for Complex Event Processing*. Paper presented at the EU Consultation workshop on priorities for the FP7 WP2009-10. Retrieved from ftp://ftp.cordis.europa.eu/pub/fp7/ict/docs/ssai/events-20080304-05-paschke_en.pdf

Paschke, A. (2008). *Design patterns for complex event processing*. Paper presented at the 2nd International Conference on Distributed Event-Based Systems (DEBS'08), Rome, Italy, 2008.

Paschke, A. et. al. (2008). *Rif use cases and requirements*. W3C.

Paschke, A. Kozlenkov, A., & Boley, H. (2007). *A homogenous reaction rules language for complex event processing*. Paper presented at the International Workshop on Event Drive Architecture for Complex Event Process (EDA-PS 2007) at VLDB'07, Vienna, Austria.

Paschke, A. Kozlenkov, A., Boley, H., Kifer, M. Tabet, S., Dean, M., & Barrett, K. (2006). *Reaction RuleML*. Retrieved from http://ibis.in.tum.de/research/reaction-ruleml

Paschke, A., & Bichler, M. (2005). SLA representation, management and enforcement. In *Proceedings IEEE International Conference on e-Technology, e-Commerce, e-Service (EEE'05)*.

Paschke, A., Boley, H., & Dietrich, J. (2005). *RIF use case. Rule interchange through test-driven verification and validation*. Retrieved from, http://www.w3.org/2005/rules/wg/wiki/Rule_Interchange_Through_Test-Driven_Verification_and_Validation, 2005.

Paschke, A., Boley, H., Kozlenkov, A., Craig, B. (2007), *Rule responder: A RuleML-based pragmatic agent web for collaborative teams and virtual organizations*. Retrieved from http://responder.ruleml.org

Paschke, A., Boley, H., Kozlenkov, A., Craig, B.L. (2007). *Rule responder: RuleML-based agents for distributed collaboration on the pragmatic Web*. ICPW 2007 (pp. 17-28).

Paschke, A., Dietrich, J., Giurca, A., Wagner, G., Lukichev, S. (2006). *On self-validating rule bases*. Int. Semantic Web Enabled Software Engineering Workshop (SWESE'06), Athens, Georgia.

Paschke, A., Kiss, C., & Al-Hunaty, S. (2006). NPL: Negotiation Pattern Language - A design pattern language for decentralized (agent) coordination and negotiation protocols. In R. Banda (Ed.), *E-negotiation - An introduction*. ICFAI University Press.

Paschke, A., Kozlenkov, A., & Boley, H. (2007) A homogenous reaction rules language for complex event processing. In *Proceedings of International Workshop on Event Drive Architecture for Complex Event Process (EDA-PS 2007)*.

Paschke, A., Kozlenkov, A., Boley, H., Tabet, S., Kifer, M., Dean M. (2008). *Reaction RuleML Version 0.3. Retrieved* from http://ibis.in.tum.de/research/ReactionRuleML/index.htm

Patel-Schneider, P. F., & Horrocks, I. (2006). Position paper: A Comparison of two modelling paradigms in the Semantic Web. In *Proceedings of the International Conference on World Wide Web* (pp. 3-12). ACM Press.

Patel-Schneider, P. F., & Horrocks, I. (2006). Position paper: a comparison of two modelling paradigms in the

Semantic Web. In *Proceedings of the 15th International Conference on World Wide Web (Edinburgh, Scotland, May 23 - 26, 2006). WWW '06, New York, NY* (pp. 3-12). ACM Press.

Patel-Schneider, P. F., Horrocks, I., Tabet, H. B. S., Grosof, B., & Dean, M. (2004). *SWRL: A Semantic Web rule language combining OWL and RuleML* (Internet Report). Retrieved 2008, from http://www.w3.org/Submission/2004/SUBM-SWRL-20040521/.

Paton, N., et al. (1995). Formal specification of active database functionality: A survey. In *2nd Int. Workshop RIDS-95, Athens, Greece.*

Paton, N.W. (1999). Active rules in database systems. *Monographs in computer science.* Springer.

Pătrânjan, P.-L. (2005). *The language XChange: A declarative approach to reactivity on the Web.* PhD Thesis (Dissertation). University of Munich, Germany.

Paulussen, T.O., Zöller, A., Heinzl, A., Braubach, L., Pokahr, A, & Lamersdorf, W. (2006). Agent-based patient scheduling in hospitals. In S. Kirn, O. Herzog, P. Lockemann, & O. Spaniol (Eds.), *Multiagent engineering - Theory and applications in enterprises* (pp. 255-275). Springer.

Pearce, D.J., & Noble, J. (2006). Relationship aspects. In *Proceedings of the 5th international conference on Aspect-oriented software development (AOSD'2006)* (pp. 75-86). ACM.

Pednault, E.P.D. (1994). ADL and the state-transition model of action. *J. Log. Comput. 4*(5), 467-512.

Pei, J., Han, J., Mortazavi-Asl, B., & Zhu, H. (2000). Mining access patterns efficiently from web logs. In *PADKK '00: Proceedings of the 4th Pacific-Asia Conference on Knowledge Discovery and Data Mining, Current Issues and New Applications* (pp. 396-407). London, UK: Springer-Verlag.

Peirce, C. (1955). Abduction and induction. In *Philosophical writings of Peirce.*

Peng, L., Collier, R., Mur, A., Lillis, D., Toolan, F., & Dunnion, J., (2005). A self-configuring agent-based document indexing system. In *Proceedings of the 4th International Central and Eastern European Conference on Multi-Agent Systems, Budapest, Hungary*

Pereira, L.M. (1986). Rational debugging in logic programming. In *Proceedings of the Third International Conference on Logic Programming* (pp. 203-210).

Piatetsky-Shapiro, G. (1991). Discovery, analysis, and presentation of strong rules. In *Knowledge discovery in databases* (pp. 229-248). AAAI/MIT Press.

Plant, R., & Gamble, R. (2003). Methodologies for the development of knowledge-based systems, 1982–2002. *The Knowledge Engineering Review, 18*, 47-81.

Pokahr, A., Braubach, L., & Lamersdorf, W. (2005). Jadex: A BDI reasoning engine. In *Multi-agent programming: Languages, platforms and applications.* Kluwer.

Pokahr, A., Braubach, L., & Lamersdorf, W. (2005) A BDI architecture for goal deliberation. In F. Dignum, V. Dignum, S. Koenig, S. Kraus, M. Singh, & M. Wooldridge (Eds.), *Proceedings of 4th International Joint Conference on Autonomous Agents and Multiagent Systems (AAMAS 2005)* (pp. 1295-1296). ACM.

Pokahr, A., Braubach, L., & Lamersdorf, W. (2005). A goal deliberation strategy for BDI agent systems. In T. Eymann, F. Klügl, F, W. Lamersdorf, M. Klusch, & M. Huhns (Eds.), *Proceedings of the Third German Conference on Multi-Agent System Technologies (MATES-2005)* (pp. 82-94). Springer.

Polleres, A., & Schindlauer, R. (2007). *DLVHEX-SPARQL: A SPARQL Compliant Query engine based on DLVHEX. In Proceedings of 2nd International Workshop on Applications of Logic Programming to the Web, Semantic Web and Semantic Web Services (ALPSWS2007), Porto, Portugal.*

Powell, M.S. (1989). A program development environment based on persistence and abstract data typesIn I. Rosenberg & D. Koch (Eds.), *Persistent object systems.* Springer Verlag.

Preece, A. (1998). Building the right system right. In *Proc. of KAW'98 Eleventh Workshop on Knowledge Acquisition, Modeling and Management.*

Preece, A. D. (1991). *Methods for verifying expert system knowledge bases* (Tech. Rep.).

Preece, A. D. (1993). A new approach to detecting missing knowledge in expert system rule bases. *Int. J. of Man-Machine Studies, 38*, 161-181.

Preece, A. D., & et al. (1991). *Verifying rule-based systems* (Tech. Rep.).

Preece, A., & Shinghal, R. (1994). Foundation and application of knowledge base verification. *International Journal of Intelligent Systems 9*, 683-702.

Proctor, M. (2008). *W3C RIF PRD feature comparison of production rules*. Retrieved October 2008 from http://www.w3.org/2005/rules/wiki/PR_Feature_comparison

Proctor, M., Neale, M., Frandsen, M., Griffith Jr., S., Tirelli, E., Meyer, F., & Verlaenen, K. (n.d.). *Drools documentation*. Retrieved from http://downloads.jboss.com/drools/docs/4.0.7.19894.GA/html/index.html

Proper, H. (1997). Data schema design as a schema evolution process. *Data & Knowledge Engineering, 22*(2), 159-189.

Proper, H., Bleeker, A., Hoppenbrouwers, S. (2004). Object-role modelling as a domain. In *Proceedings of the Workshop on Evaluating Modeling Methods for Systems Analysis and Design.*

Proper, H., van der Weide, T., & ter Hofstede, A. (1993). Formal definition of a conceptual language for the description and manipulation of information models. *Information Systems, 18*(7), 471-495.

Prud'hommeaux, E., & Seaborne, A. (2005). *SPARQL query language for RDF.* http://www.w3.org/TR/2005/WD-rdf-sparql-query-20050217/, 2005.

Prud'hommeaux, E., & Seaborne, A. (2008). *SPARQL query language for RDF.* W3C Recommendation. Retrieved April 14, 2008, from http://www.w3.org/TR/rdf-sparql-query/.

QVT. (2005, June). MOF QVT final adopted specification. *OMG* .

R2ML verbalizer (2006). Retrieved from http://oxygen.informatik.tu-cottbus.de/verbalization/index.jsp

*RacerPro* (2008). Retrieved April 22, 2008, from http://www.racer-systems.com/products/racerpro/index.phtml.

Railsback, S.F., Lytinen, S.L., & Jackson, S.K (2006). Agent-based simulation platforms: Review and development recommendations. *Simulation, 82*(9), 609-623.

Rainer. A. (2005). Web service composition under answer set programming. *KI-Workshop "Planen, Scheduling und Konfigurieren, Entwerfenl" (PuK).*

Rakesh Agrawal, R. S. (1996). Mining quantitative association rules in large relational tables. In H. V. Jagadish & I. S. Mumick (Eds.), *Proceedings of the 1996 ACM SIGMOD international conference on management of data* (pp. 1-12). Montreal, Quebec, Canada.

Ramakrishnan, R. et al. (1998).SQRL: Sorted Relational Query Language. In *SSDBM 1998* (pp. 84-95).

Ramakrishnan, R., Srivastava, D., Sudarshan, S., & Seshadri, P. (1993). Implementation of the CORAL deductive database system. In *Proceedings of the ACM SIGMOD International Conference on Management of Data* (pp. 167-176). Washington, D.C.: ACM Press.

Rao, A. S. (1996). AgentSpeak(L): BDI agents speak out in a logical computable language. In R. van Hoe (Ed.), *Proceedings of the seventh European workshop on modelling autonomous agents in a multi-agent world (MAAMAW'96).* The Netherlands: Eindhoven.

Rao, A. S., & Georgeff, M. P. (1991). Modeling rational agents within a BDI-architecture. In J. Allen, R. Fikes & E. Sandewall (Eds.), *Proceedings of the second international conference on principles of knowledge representation and reasoning (KR'91)* (pp. 473-484). Morgan Kaufmann.

Rao, A., Georgeff, M. (1995). BDI Agents: From theory to practice. In V. Lesser, L. Gasser (Eds.), *Proceedings of the First International Conference on Multiagent Systems* (pp. 312-319). MIT Press.

Rao, A.S., & Georgeff, M.P. (1991). Modeling rational agents within a BDI architecture. In J. Allen, R. Fikes, & E. Sandwall (Eds.), *Proceedings of Second International Conference on Principles of Knowledge Representation and Reasoning* (pp. 473-484). San Mateo, CA: Morgan-Kaufmann.

Rao, A.S., & Georgeff, M.P. (1995). BDI agents: From theory to practice. In *Proceedings of the First International Conference on Multi-Agent Systems (ICMAS'95)* (pp. 312-319). San Francisco, CA.

Raschid, L. (1992). A *semantics for a class of stratified production system programs* (UMIACS-TR-91-114.1). Univ. of Maryland Institute for Advanced Computer Studies, College Park, MD, USA.

Raschid, L., & Lobo, J. (1996). Semantics for update rule programs and implementation in a relational database management system. *ACM Transactions on Database Systems, 22*(4), 526-571.

Reddi, S., Poulovassilis, A., & Small, C. (1999). PFL: An active functional DBPL. In *Active Rules in Database Systems* (pp. 297-308).

Reiter, R. (1984). Towards a logical reconstruction of relational database theory. In M. Brodie, J. Mylopoulos, & J. Schmidt (Eds.), *On conceptual modelling* (pp. 191-233). Springer-Verlag.

Reiter, R. (1995). On specifying database updates. *Journal of Logic Programming, 25*(1), 53-91.

*REWERSE I1 Rule Markup Language (R2ML)* (2006). Retrieved from http://oxygen.informatik.tu-cottbus.de/rewerse-i1/?q=R2ML, 2006.

*REWERSE I1 Rule Markup Language (R2ML)* (2008). Retrieved from http://oxygen.informatik.tu-cottbus.de/rewerse-i1/?q=node/6.

REWERSE I1 Strelka tool (2006). *Strelka – A UML-based Rule Modeling Tool.* Retrieved May 1, 2008, from http://oxygen.informatik.tu-cottbus.de/rewerse-i1/?q=Strelka

Rewerse Working Group I1 - Rule Modeling and Markup (2007). *URML - a UML-based rule modeling language.* Retrieved April 3, 2008, from http://oxygen.informatik.tu-cottbus.de/rewerse-i1/?q=node/7.

Ribaric, M., Gasevic, D., & Milanovic, M. (n.d.). *R2ML: A rule-based approach to modeling of semantically-enriched Web services.*

Richard, B., David, J.-M., & Krivine, J.P. (1993). Building and maintaining a large knowledge based system from a 'knowledge level' perspective: the diva experiment. In *Second Generation Expert Systems* (pp. 376-401). Springer Verlag.

*RIF - Rule Interchange Format Working Group* (2008). Retrieved April 22, 2008, from http://www.w3.org/2005/rules/wiki/RIF_Working_Group

*RIF Working Group* (2005). Retrieved November 10, 2008 from http://www.w3.org/2005/rules/wiki/RIF_Working_Group

Riley, G. (1991). CLIPS: An expert system building tool. In *Proceedings of the Technology 2001Conference, San Jose, CA.*

Riley, G. (2004). *Clips - A tool for building expert systems.* Retrieved May 4, 2008 from http://clipsrules.sourceforge.net/, retrieved on the 4th May 2008.

Robinson, J. A. & Sibert, E. E. (1982). LOGLISP: An alternative to Prolog. In J. E. Hayes, D. Michie, & Y-H. Pao (Eds.), *Machine Intelligence, Vol. 10.* Chichester, England: Ellis Horwood Ltd.

Robinson, J. A. (1965). A machine-oriented logic based on the resolution principle. *Journal of the ACM (JACM), 12(1)*, 23-41.

Roman, D., Keller, U., Lausen, H., de Bruijn, J., Lara, R., Stollberg, M., et al. (2005). Web service modeling ontology. *Applied Ontology, 1*(1), 77-106.

Rooney, C.F.B., Collier, R.W., & O'Hare, G.M.P. (2004). VIPER: VIsual Protocol EditoR. In *Proceedings of the 6th International Conference on Coordination Models and Language (COORDINATION 2004), Pisa, Italy, 24-27 February.*

Rosati, R. (1999). Towards expressive KR systems integrating datalog and description logics: Preliminary report. In *Proceedings of the International Workshop on Description Logics* (Vol. 22) (pp. 160-164). CEUR-WS.

Rosati, R. (2005). On the decidability and complexity of integrating ontologies and rules. *Journal of Web Semantics: Science, Services and Agents on the World Wide Web, 3*, 61-73.

Rosati, R. (2005). Semantic and computational advantages of the safe integration of ontologies and rules. In *Principles and Practice of Semantic Web Reasoning, Third International Workshop, PPSWR 2005, Dagstuhl Castle, Germany* (LNCS 3703, pp. 50-64).

Rosati, R. (2006). DL+log: Tight integration of description logics and disjunctive datalog. In *Proceedings of the Tenth International Conference on Principles of Knowledge Representation and Reasoning* (pp. 68-78). AAAI Press.

Rosati, R. (2006). The limits and possibilities of combining description logics and datalog. In T. Eiter, E. Franconi, R. Hodgson, and S. Stephens (Eds.), *RuleML* (pp. 3-4). IEEE Computer Society.

Rosati, R. (2006). DL+log: Tight integration of description logics and disjunctive datalog. In P. Doherty, J. Mylopoulos, & C.A. Welty (Eds.), *Proceedings of the 10th International Conference on Principles of Knowledge Representation and Reasoning (KR 2006), Lake District, UK* (pp. 68-78). AAAI Press.

Rosati, R. (2006). *Integration ontologies and rules: Semantic and computational issues.* Reasoning Web.

Rosati, R. (2006). Integrating ontologies and rules: Semantic and computational issues. In P. Barahona, F. Bry, E. Franconi, N. Henze, & U. Sattler (Eds.), *Reasoning Web, Second International Summer School 2006, Lissabon, Portugal, September 4_8, 2006, Tutorial Lectures.* (LNCS 4126, pp. 128-151).

Rosca, D., & Wild, C. (2002). Towards a flexible deployment of business rules. *Expert Systems with Applications, 23*(4), 385-394.

Rosca, D., Greenspan, S., & Wild, C. (2004). Enterprise modeling and decision-support for automating the business rules lifecycle. *Journal Automated Software Engineering, 9*(4), 361-404.

Rosca, D., Greenspan, S., Feblowitz, M., & Wild, C. (1997). A decision making methodology in support of the business rules lifecycle. In *Proceedings of the Third IEEE International Symposium on Requirements Engineering* (pp. 236-246).

Rosca, D., Greenspan, S., Wild, C., Reubenstein, H., Maly, K., & Feblowitz, M. (1995). Application of a decision support mechanism to the business rules lifecycle. In *10th Knowledge-Based Software Engineering Conference* (p. 114).

Rosenberg, F., & Dustdar, S. (2005). Business Rules Integration in BPEL – A service-oriented approach. In *Proc. of the 7th International IEEE Conference on E-Commerce Technology* (pp. 476-479).

Rosenberg, F., & Dustdar, S. (2005). Business rules integration in BPEL - A service-oriented approach. In *Proceedings of the 7th International IEEE Conference on E-Commerce Technology (CEC'05)* (pp. 476-479). Munich: IEEE Computer Society.

Rosenberg, F., & Dustdar, S. (2005). Design and implementation of a service-oriented business rules broker. In *Proceedings of the 1st IEEE International Workshop on Service-oriented Solutions for Cooperative Organizations (SoS4CO'05), co-located with the 7th International IEEE Conference on E-Commerce Technology (CEC'05)* (pp. 55-63). Munich: IEEE Computer Society.

Roser, S. & Bauer, B. (2006). An approach to automatically generated model transformations using ontology engineering space. In *Proc. of the 2nd Int'l WSh. on Semantic Web Enabled Software Eng.*

Ross, R. (1997). *The business rule book: Classifying, defining and modeling rules* ($2^{nd}$ ed.) Boston: Database Research Group, Inc..

Ross, R. (2003). *Principles of the business rule approach.* Addison-Wesley.

Ross, R. G. (2003). *Principles of the business rule approach* (1st ed.). Addison-Wesley Professional.

Ross, R. G. (2005). *Business rule concepts: Getting to the point of knowledge* (2nd Ed.).

Ross, R., Collier, R., & O'Hare, G.M.P. (2004). AF-APL – Bridging principles & practice in agent-oriented languages. In *Proceedings of the 2nd International Workshop on Programming Multi-Agent Systems Languages and Tools (PROMAS- 2004), New York.*

Ross, R., Collier, R., O'Hare, G. (2005). AF-APL - Bridging Principles and practice in agent oriented languages. In R. Bordini, M. Dastani, J. Dix, & A. El Fallah Seghrouchni, (Eds.), *Proceedings of Second International Workshop on Programming Multi-Agent Systems (ProMAS 2004)* (pp. 66-88). Springer.

Ross, R.G. (2003). IT systems perspective - The business rule approach. *Computer, 36*(5), 85-87).

Ross., R. G. (1997). *The business rule book: Classifying, defining and modeling rules* (2nd ed.). Boston: Database Research Group, Inc.

Rossetti, L. (2006). *What is business intelligence*? Retrieved January 10, 2008 from http://searchdatamanagement.techtarget.com/sDefinition/0,,sid91_gci213571,00.html

RuleML (2008). *The Rule Markup Initiative*. Retrieved April 14, 2008, from http://www.ruleml.org/.

RuleML Initiative (2004). *The Rule Markup Initiative Web Site*. Retrieved from, http://ruleml.org

*RuleML, Rule Markup Initiative* (2008). Retrieved April 22, 2008, from http://www.ruleml.org/.

Rumbaugh, J. (1987). Relations as semantic constructs in an object-oriented language. In *OOPSLA 87: Proceedings on Object-oriented programming systems, languages and applications* (pp. 466-481). ACM.

Rupp, N.A. (n.d.). *The login of the bottom line: An introduction to The Drools Project*. Retrieved from http://www.theserverside.com/tt/articles/article.tss?l=Drools

Russell, S. & Norvig, P. (1995). *Artificial intelligence: A modern approach*. Englewood Cliffs, NJ: Prentice Hall.

Russell, S., & Norvig, P. (2003). *Artificial intelligence: A modern approach* (2nd ed.). Prentice-Hall.

Sadri, F., & Kowalski, R. (1995). Variants of the event calculus. In *Int. Conf. on Logic Programming, Kanagawa, Japan.*

Sadri, F., & Ullman, J. D. (1982). Template dependencies: A large class of dependencies in relational databases and its complete axiomatization. *Journal of the ACM (JACM), 29*(2), 363-372.

Sadri, R. (2001). *Optimization of sequence queries in database systems*. PhD Thesis, UCLA.

Sadri, R., Zaniolo, C., Zarkesh, A., & Adibi, J. (2001). *Optimization of sequence queries in database systems.*

Sagonas, K., Swift, T., & Warren, D. (1993). *XSB: An overview of its use and implementation* (Tech. Rep). Retrieved 2006, from http://citeseer.ist.psu.edu/195039.html

Sagonas, K., Swift, T., & Warren, D.S. (1994). XSB as an efficient deductive database engine. *ACM SIGMOD Record, 23*(2), 442-453.

Saitta, L., Botta, M., & Neri, F. (1993). Multistrategy learning and theory revision. *Machine Learning, 11*(2), 153-172.

Sanderson, M., & Kohler, J. (2004). Analyzing geographic queries. In *Workshop in Geographic Information Retrieval.*

Sandewall, E. (1989). Combining logic and differential equations for describing real world systems In *KR89*. Morgan Kaufman.

Sandia National Laboratories. (2008). *Jess, the Rule Engine for the Java Platform*. Retrieved November 13, 2008 from http://herzberg.ca.sandia.gov/

Sandifer, A, & Von Halle, B. (2000). Business rules: Capturing the most elusive information asset. In S. Purba (Ed.), *High-performance databases: Design, develop-*

*ment and deployment* (pp. 111-128). CRC Press, Taylor & Francis Group.

Sangiorgi, D. & Walker, D. (2001). *The Pi-calculus: A theory of mobile processes*. Cambridge University Press.

Saraswat, V.A., Jagadeesan, R., & V. Gupta (1996). Timed default concurrency constraint programming. *Journal of Symbolic Computation, 22*(5/6).

Sarbanes-Oxley Act of 2002, H.R. 3763, 107th Cong., 2nd Sess. (2002).

Sardina, S., Giacomo, G. D., Lespérance, Y., & Levesque, H. J. (2004). On the semantics of deliberation in indigolog ù from theory to implementation. *Annals of Mathematics and Artificial Intelligence, 41*(2-4), 259-299.

Savasere, A., Omiecinski, E., & Navathe, S. B. (1995). An efficient algorithm for mining association rules in large databases. In U. Dayal, P. M. D. Gray, & S. Nishio (Eds.), *Proceedings of 21th International Conference on Very Large Data Bases (VLDB'95)* (p. 432-444). Morgan Kaufmann.

SBVR (2006). Object Management Group. *Semantics of business vocabulary and business rules (SBVR) – Interim specification*. Retrieved from http://www.omg.org/docs/bei/05-08-01.pdf

SBVR (2008). *Semantics of business vocabulary and business rules*. Formal specification by OMG, http://www.omg.org/spec/SBVR/1.0/

SBVR Specification (2008). *Object Management Group*. Retrieved May 4, 2008 from http://www.omg.org/docs/formal/08-01-02.pdf

Scheer, A.-W. (1999). *ARIS - Business process frameworks* (3rd ed.). Springer.

Schmidt, D. (2006). Model-driven engineering. *IEEE Computer, 39*(2), 25-31.

Schmidt, R. (2002). Web services based execution of business rules. *International Workshop on Rule Markup Languages for Business Rules on the Semantic Web.*

Schmidt-Schauß, M. (1989). Subsumption in KL-ONE is undecidable. In *Proceedings of the 1st International Conference on Principles of Knowledge Representation and Reasoning* (pp. 421-431). Toronto, Canada.

Schneier, B. (1992). *The Rete matching algorithm: A description of the pattern-matching algorithm underlying Drools.*

Schockaert, S., & Cock, M. (2007). Reasoning about vague topological information. In *Proceedings of the Sixteenth ACM Conference on Information and Knowledge Management* (pp. 593-602).

Schockaert, S., Smart P. D., Abdelmoty A.I., & Jones C.B. (2008). Mining topological relations from the Web. In *DEXA Workshops* (pp. 652-656). IEEE Computer Society.

Schreiber, G., Akkermans, H., Anjewierden, A., de Hoog, R., Shadbolt, N., Van de Velde, W., & Wielinga, B. (2001). *Knowledge engineering and management - The CommonKADS methodology* (2nd ed.). Cambridge, USA: MIT Press.

Schroeder, M., & Wagner, G. (1998) Vivid agents: theory, architecture, and applications. *Applied Artificial Intelligence, 14*, 645-675.

Scott, A.C., Clancey, W.J., Davis, R., & Shortliffe, E. (1984). Methods for generating explanations. In *Rule-based expert systems*. Addison-Wesley.

Searle, J.R. (1969). *Speech acts*. Cambridge, UK: Cambridge University Press.

Seidewitz, E. (2003). What models mean. *IEEE Software, 20*(5), 26-32.

Seiriö, M., & Berndtsson, M. (2005). Design and implementation of an ECA rule markup language. In *Proceedings RuleML 2005* (pp. 98-112).

Selic, B. (2003). The pragmatics of model-driven development. *IEEE Software, 20*(5), 19-25.

Sellis, T., Lin, C.C., & Raschid, L. (1993). Coupling production systems and database systems. *In ACM Sigmond Int. Conf. on the Management of Data.*

SeRQL (2008). *The SeRQL query language*. User Manual. Retrieved April 14, 2008, from http://www.openrdf.org/doc/sesame/users/ch06.html

Servigne, S., Ubeda, T., Puricelli, A. & Laurini, R. (2000). Methodology for spatial consistency improvement of geographic databases. *GeoInformatica*, 7-34.

Shadri, P., Livny, M., & Ramakrishnan, R. (1994). Sequence query processing. In *SIGMOD 1994* (pp. 430-441).

Shadri, P., Livny, M., Ramakrishnan, R. (1995). SEQ: A Model for sequence databases. In *ICDE 1995* (pp. 232-239).

Shahar, Y. (1999). *Timing is everything: Temporal reasoning and temporal data maintenance in medicine*. Paper presented at the Seventh Joint European Conference on Artificial Intelligence in Medicine and Medical Decision Making (AIMDM'99), Aalborg, Denmark.

Shanahan, M. (1990). Representing continuous change in the event calculus. In *ECAI'90*.

Shanahan, M. (1997). *Solving the frame problem*. London: MIT Press.

Shanahan, M. (1997). Event calculus planning revisited. In *4th European Conference on Planning (ECP97)* (LNAI 1348).

Shapiro, E.Y. (1982). *Algorithmic program debugging*. PhD thesis, Yale University.

Sheers, KR. (1996). HP OpenView event correlation services. *Hewlett Packard Journal, 47*(5), 31-42.

Sheth, A., Verma, K., & Gomadam, K. (2006). Semantics to energize the full services spectrum. *Comm. of the ACM, 49*(7), 55-61.

Shoham, Y. (1993) Agent-oriented programming. *Artificial Intelligence, 60*(1), 51-92.

Shortliffe, R.J. (1974). *MYCIN: A rule-based computer program for advising physicians regarding antimicrobial therapy selection*. PhD thesis, Stanford University.

Shortliffe, R.J., Davis, R., Axline, S.G., Buchanan, B.G., Green, C.C. & Cohen, S.N. (1975): Computer-based consultations in clinical therapeutics: Explanation and rule acquisition capabilities of the mycin system. In *Computers and Biomedical Research, 8*, 303-320.

Silver, G. A. & Silver, M. L. (1989). *Systems analysis and design*. MA: Addison Wesley.

Sintek, M., & Decker, S. (2002). TRIPLE – A query, inference, and transformation language for the Semantic Web. In I. Horrocks & J.A. Hendler (Eds.), *Proceedings of the First International Semantic Web Conference – ISWC 2002* (LNCS 2342). Berlin: Springer.

Sirin, E., Parsia, B., Cuenca Grau, B., Kalyanpur, A., & Katz, Y. (2007). Pellet: A practical OWL-DL reasoner. *Journal of Web Semantics: Science, Services and Agents on the World Wide Web, 5*, 51-53.

Sirin, E., Parsia, B., Grau, B. C., Kalyanpur, A., & Katz, Y. (2007). Pellet: A Practical OWL-DL Reasoner. *Web Semantics: Science, Services and Agents on the World Wide Web, 5*(2), 51-53.

Sirin, E., Parsia, B., Grau, B., C., Kalyanpur A., & Katz, Y. (2007). Pellet: A practical OWL-DL reasoner. *Journal of Web Semantics, 5*(2).

Smart, P.D., Abdelmoty, A.I., Jones, C.B., & El-Geresy, B.A. (2007). A framework for combining rules and geo-ontologies. In *Web Reasoning and Rule Systems: First International Conference, RR 2007, Innsbruck, Austria* (LNCS 4524, pp. 133-147).

Smith, H., & Fingar, P. (2003). *Business process management: The third wave*.

Smith, M., Welty, C., & McGuinness, D. (2004, February 10). OWL Web ontology language guide. *W3C Recommendation* .

Smith, T.R., & Frew, J. (1995). Alexandria digital library. *Communications of the ACM, 38*(4), 61-62.

Snodgrass, R.T. (1995). *The TSQL2 temporal query language*. Boston: Kluwer Academic Publishers.

Snoeck, M. (2003). Sequence constraints in business modeling and business process modeling. In *Enterprise information system IV* (pp. 194-201). Hingham, MA: Kluwer Academic Publishers.

*Software and Systems Modeling (SoSyM), 1*(2), 142-156.

Soloviev, V. (1992). An overview of three commercial object-oriented database management systems: ONTOS, ObjectStore, O2. *SIGMOD Record, 21*(1), 93-104.

Sommerville, I. (1996). *Software engineering* (5th ed.). Menlo Park, USA: Addison-Wesley.

Sommerville, I. (2004). *Software engineering* (7th ed.). Pearson Education Limited.

Son, T. C., & Baral, C. (2001). Formalizing sensing actions—a transition function based approach. *Artif. Intell., 125*(1-2), 19-91.

Sowa, J.F (2000). *Knowledge representation: Logical, philosophical, and computational foundations*. Addison-Wesley.

Spinellis, D. (2001). Notable design patterns for domain-specific languages. *Journal of Systems and Software, 56*(1), 91-99.

Spyns P., Meersman R., & Jarrar M. (2002). Data modeling versus Ontology engineering. *SIGMOD Record: Special Issue on Semantic Web and Data Management, 31*(4), 12-17.

Spyns, P., Tang, Y., & Meersman, R. (2007). A model theory inspired collaborative ontology engineering methodology. *Journal of Applied Ontology, special issue on "Ontological Foundations for Conceptual Modeling," 4*(1), 1-23

SQWRL (2008). *SQWRL*. Retrieved November 10, 2008 from http://protege.cim3.net/cgi-bin/wiki.pl?SQWRL

Srivastava, D., Ramakrishnan, R., Srivastava, D., & Sudarshan, S. (1993). CORAL++: Adding object-orientation to a logic database language. In *Proceedings of the International Conference on Very Large Data Bases* (pp. 158-170). Dublin, Ireland: Morgan Kaufmann.

Stefik, M. J. (1995). *Introduction to knowledge systems*. San Francisco: Morgan Kaufmann.

Stein, S., & Ivanov, K. (2007). EPK nach BPEL Transformation als Voraussetzung für praktische Umsetzung einer SOA Software Engineering 2007. In W.K. Bleek, J. Raasch, & H. Züllighoven, H. (Eds.), *Gesellschaft für Informatik (GI) Vol. 105* (pp.75-80).

Sterling, L. & Shapiro, E. (1986). *The art of Prolog*. MIT Press.

Sterling, L. & Taveter, K. (in press). Event-based optimization of air-to-air business processes. In *Proceedings of the AAAI Spring Symposium on Intelligent Event Processing, Stanford, CA, USA, March 23-25*. Menlo Park, CA: AAAI Press.

Sterling, L., & Shapiro, E. (1994). *The art of Prolog* (2nd ed.) Cambridge, MA; London,: The MIT Press.

Sterling, L., & Taveter, K. (2009). *The art of agent-oriented modeling*. Cambridge, MA; London: The MIT Press.

Studer, R., Benjamins, V. R., & Fensel, D. (1998). Knowledge engineering: Principles and methods. *Data and Knowledge Engineering, 25*(1-2), 161-197.

Stumptner, M., & Wotawa, F. (1998). A survey of intelligent debugging. AI *Communications, 11*(1), 35-51.

Sun Microsystems (1998). *JavaBeans*. Retrieved April 3, 2008, from http://java.sun.com/products/javabeans/

Sun Microsystems (2007). *JavaServer Pages Technology version 2.1*. Retrieved April 3, 2008, from http://java.sun.com/products/jsp/

Sun Microsystems. (2008). *Java Persistence API*. Retrieved November 13, 2008, from http://java.sun.com/javaee/overview/faq/persistence.jsp

Swanson, E.B. (1976). The dimensions of maintenance. In *Proceedings of the 2nd International Conference of Software Engineering*.

Swartout, W.R. (1983). Xplain: A system for creating and explaining expert consulting programs. *Artificial Intelligence, 21*(3), 285-325

Swift, T. (2004). Deduction in Ontologies via ASP. In *Proceedings of Logic Programming and Nonmonotonic Reasoning* (pp. 275-288). Springer.

SWRL (2004). *SWRL: A Semantic Web rule language cCombining OWL and RuleML*. Retrieved November 10, 2008 from http://www.w3.org/Submission/SWRL/

SWRL Built-ins (2004). *Built-ins for SWRL*. Retrieved November 10, 2008 from http://www.daml.org/2004/04/swrl/builtins.html

Tabet, S., Wagner, G., Spreeuwenberg, S., Vincent, P., Jacques, G., de Sainte Marie, C., Pellant, J., Frank, J., & Durand, J. (2007). *Object Management Group: Production Rule Representation (PRR)*. Retrieved from http://www.w3.org/2004/12/rules-ws/paper/53/

Tambe, M., & Rosenbloom, P. (1989). Eliminating expensive chunks by restricting expressiveness. In *Proceedings of the International Joint Conference on Artificial Intelligence* (pp. 431-456).

Tan, P.-N., Steinbach, M., & Kumar, V. (2005). *Introduction to data mining*. Addison Wesley.

Tang, Y., & Meersman, R. (2008). Use semantic decision tables to improve meaning evolution support systems. In F.E. Sandnes, Y. Zhang, C. Rong, L.T. Yang, & J. Ma (Eds.), *Proc. of UIC'08 (The 5th International Conference on Ubiquitous Intelligence and Computing), Building Smart Worlds in Real and Cyber Spaces* (LNCS 5181).

Tang, Y., & Trog, D. (in press). Model ontological commitments using ORM$^+$ in T-Lex. Accepted at *ORM workshop 2008, OTM 2008*.

Tang, Y., Spyns, P. & Meersman, R. (2007). Towards semantically grounded decision rules using ORM$^+$. In A. Paschke & Y. Biletskiy (Eds.), *Proc. of International RuleML Symposium on Rule Interchange and Applications (RuleML'07)* (LNCS 4824).

Taniguchi, K., Ishio, T., Kamiya, T., Kusumoto, S., & Inoue, K. (2005). Extracting sequence diagram from execution trace of Java program. In *Proc. of 8$^{th}$ workshop on Principles of Software Evolution* (pp. 148-151).

Tatara, Eric & North, Michael J. & Collier, Nick & Vos, Jerry R. (2006). An Introduction to Repast Modeling by Using a Simple Predator-Prey Example. In *Proceedings of the Agent 2006 Conference on Social Agents: Results and Prospects* (pp. 1-11). Argonne National Laboratory and The University of Chicago.

Taveter, K. & Wagner, G. (2001). Agent-oriented enterprise modeling based on business rules. In H. S. Kunii, S. Jajodia, & A. Sølvberg (Eds.), *Conceptual Modeling – ER 2001* (LNCS 2224, pp. 527-540). Berlin, Germany: Springer-Verlag.

Taveter, K. & Wagner, G. (2005). Towards radical agent-oriented software engineering processes based on AOR modelling. In B. Henderson-Sellers & P. Giorgini (Eds.), *Agent-oriented methodologies* (pp. 277-316). Hershey, PA: Idea Group.

Taveter, K. (2001). Agent-oriented business modeling and simulation. Unpublished PhD thesis, Tallinn Technical University.

Taveter, K. (2004). *A multi-perspective methodology for agent-oriented business modeling and simulation*. Ph.D. thesis, Tallinn University of Technology, Estonia.

Taveter, K., & Wagner, G. (2001). Agent-oriented enterprise modeling based on business rules. In *Proceedings of the 20th International Conference on Conceptual Modeling (ER'01)* (pp. 527-540). London: Springer.

Taveter, K., & Wagner, G. (2001, September). Agent-oriented enterprise modeling based on business rules. In *Proceedings of International Workshop on Open Enterprise Solutions: Systems, Experiences and Organizations (OES-SEO2001)*. Rome.

Taveter, K., & Wagner, G. (2001, November). Agent-oriented enterprise modeling based on business rules. In *Proceedings of the 20th International Conference on Conceptual Modeling (ER2001)* (LNCS 2224, pp. 527-540).

Taveter, K., Wagner, G. (2001). Agent-Oriented enterprise modeling based on business rules. In *Proceedings of 20th International Conference on Conceptual*

*Modeling (ER2001)*, Yokohama, Japan (LNCS 2224, pp. 527-540).

Temporal Built-ins (2008). *SWRL Temporal Ontology*. Retrieved November 10, 2008 from http://protege.cim3.net/cgi-bin/wiki.pl?SWRLTemporalBuiltIns

Temporal Ontology (2008). *SWRL Temporal Ontology*. Retrieved November 10, 2008 from http://swrl.stanford.edu/ontologies/built-ins/3.3/temporal.owl

Th´eret, P. (1994). *De l'ecacit'e des interpr'eteurs de syst ems de r egles de production dans les syst emes a base de connaissances*. PhD thesis, Universit'e Paris XIII.

The Buiness Rules Group. (2000, July). *Defining business rules - What are they really?* Retrieved April 10, 2008, from Business Rules Group: http://www.businessrules-group.org/first_paper/BRG-whatisBR_3ed.pdf

*The ISO 17799 IT security standard*. (n.d.) Retrieved May 4, 2008 from http://www.iso.org/iso/iso_catalogue/catalogue_tc/catalogue_detail.htm?csnumber=39612.

*The ISO 27002:2005 IT security standard*. (n.d.). Retrieved May 4, 2008 from http://www.iso.org/iso/iso_catalogue/catalogue_tc/catalogue_detail.htm?csnumber=50297.

The Object Management Group (2004). *Unified Modeling Language: Superstructure, Version 2.0*. Retrieved April 3, 2008, from http://www.omg.org/docs/formal/05-07-04.pdf

*The Rule Markup Initiative*. (2001). Retrieved April 18, 2008, from http://www.ruleml.org

Thielscher, M. (1999). From situation calculus to fluent calculus: State update axioms as a solution to the inferential frame problem. *Artificial Intelligence, 111*, 277-299.

Thomas, D., & Hunt, A. (2000). *Programming Ruby: The pragmatic programmer's guide*. Addison-Wesley Longman Publishing.

Thomas, S. R. (1995). The PLACA agent programming language. In M. Wooldridge, N. Jennings (Eds.), *Proceedings of the 1st International Workshop Intelligent Agents I, Agent Theories, Architectures and Languages (ATAL 1994)* (pp. 355-370). Springer.

Timm, J., & Gannod, G. (2005). A model-driven approach for specifying Semantic Web services. In *Proc. of IEEE International Conference on Web Services* (pp. 313-320).

Tinnemeier, N., Dastani, M., & Meyer, J.-J. C. (2008). Orwell's nightmare for agents? Programming multi-agent organisations. In *Proceedings of the Sixth International Workshop on Programming Multi-Agent Systems (ProMAS'08)*, Estoril, Portugal.

Toninelli, A., Bradshaw, J., Kagal, L., & Montanari, R. (2005). *Rule-based and Ontology-based policies: Toward a hybrid approach to control agents in pervasive environments*. Paper preseneted at the Semantic Web and Policy Workshop, Ireland.

Tonti, G., Bradshaw, J., Jeffers, R., Montanari, R., Suri, N., & Uszok, A. (2003). Semantic Web languages for policy representation and reasoning: A comparison of KAoS, Rei, and Ponder. In *The International Semantic Web Conferencel (ISWC)* (pp. 419-437).

Torsun, I. S. (1995). *Foundations of intelligent knowledge-based systems*. London, San Diego, New York, Boston, Sydney, Tokyo, Toronto: Academic Press.

Trog, D., Vereecken, J., Christiaens, S., De Leenheer, P., & Meersman, R. (2006). T-Lex: A Role-Based Ontology Engineering Tool. In R. Meersman, Z. Tari, & P. Herrero (Eds.), *Proc. of on the Move to Meaningful Internet Systems 2006: OTM 2006 Workshops* (LNCS 4278, pp. 1191-1200).

Tsarkov, D., & Horrocks, I. (2006). FaCT++ description logic reasoner: System description. In U. Furbach & N. Shankar (Eds.), *Proceedings of the International Joint Conference on Automated Reasoning, IJCAR 2006 (LNAI* 4130). Berlin: Springer.

Tsur, S. & Zaniolo, C. (1986). LDL: A logic-based data language. In *Proceedings of the International Conference on Very Large Data Bases* (pp. 33-41). Kyoto, Japan: Morgan Kaufmann.

Ullman, J. (1982). *Principles of database systems*. Computer Science Press.

Ullman, J. D. (1989). *Principles of database and knowledge-base systems* (Vols. 1 & 2). Computer Science Press.

UML (2005). Object Management Group, Unified Modeling Language 2.0. Docs. *formal/05-07-04 & formal/05-07-05.*

URML (2008). *UML-based rule modeling language.* Retrieved from http://oxygen.informatik.tu-cottbus.de/rewerse-i1/?q=URML

URML Metamodel (2008). Retrieved May 1, 2008, from http://oxygen.informatik.tu-cottbus.de/strelka/URML-Metamodel.htm

Uszok, A., & Bradshaw, J. (n.d.). IHMC ontology and policy management. Retrieved from http://ontology.ihmc.us/

Uszok, A., Bradshaw, J., Jeffers, R., Suri, N., Hayes, P., Breedy, M., et al. (2003). KAoS policy and domain services: Toward a description-logic approach to policy representation, deconfliction, and enforcement. In *IEEE 4th International Workshop on Policies for Distributed Systems and Networks* (pp. 93-96).

Van Belleghem, K., Denecker, M., & De Schreye, D. (1997). A strong correspondence between description logics and open logic programming. In *Proceedings of the International Conference on Logic Programming* (pp. 346-360). MIT Press.

van Hoeve, W. J., Pesant, G., Rousseau, L-M., & Sabharwal, A (2006). Revisiting the sequence constraint. In F. Benhamou (Ed.), *Proc. of Principles and Practice of Constraint Programming - CP 2006, 12th International Conference (CP 2006)* (LNCS 4204, pp. 620-634).

van Luin, J., Tulba, F., & Wagner, G. (2004, May). Remodeling the beer game as an agent-object-relationship simulation. In C. Espinasse (Ed.). *Proceedings of the 5th workshop on Agent-Based Simulation (ABS2004)* (pp. 5-10). Lisbon, Portugal.

van Riemsdijk, M. B., Dastani, M. Meyer, J.J., & de Boer, F.S. (2006). Goal-oriented modularity in agent programming. In *Proceedings of the Fifth International Joint Conference on Autonomous Agents and Multiagent Systems (AAMAS'06)*, (pp. 1271-1278).

Van Roy, P. & Haridi, S. (2004). *Concepts, models and techniques of computer programming.* Cambridge, MA; London: The MIT Press.

Vanthienen, J., Mues, C., & G.Wets. (1997). Inter-tabular verification in an interactive environment. In F. H. J. Vanthienen (Ed.), *Eurovav-97, 4th European symposium on the validation and verification of knowledge based systems* (Vol. II, p. 155-165). Leuven, Belgium: Katholieke Universiteit Leuven.

Vara, J., Castro, V., & Marcos, E. (2005). WSDL Automatic Generation from UML Models in a MDA Framework. *International Journal of Web Services Practices, 1*(1-2), 1-12.

Vasilecas, O., & Smaizys, A. (2006). The framework: an approach to support business rule based data analysis. *In Proceedings of the 7th International Baltic Conference on Databases and Information Systems* (pp. 141-147).

Vergunst, N., Steunebrink, B., Dastani, M., Dignum, F., & Meyer, J. (2007). Towards programming multimodal dialogues. In *Proceedings of the workshop on Communication Between Human and Artificial Agents (CHAA'07).* IEEE Computer Society Press.

Veritas (1999). VERITAS NerveCentertm. *VERITAS Software.* Retrieved from, http://eval.veritas.com/Webfiles/docs/NCOverview.pdf.

Vermeir, D. (1983). Semantic hierarchies and abstraction in conceptual schemata. *Journal of Information Systems, 8*(2), 117-124

Vermesan, A. (1998). Foundation and application of expert system verification and validation. In J. Liebowitz (Ed.), *The handbook of applied expert systems.* CRC Press.

Vermesan, A., & Coenen, F. (Eds.). (1999). *Validation and verification of knowledge based systems. theory, tools and practice.* Boston: Kluwer Academic Publisher.

VisioModeler (2006). Retrieved October 2006, from http://www.microsoft.com/downloads/ results.aspx?displaylang=en\&freeText=VisioModeler

Volz, R. (2004). *Web Ontology reasoning with logic databases*. PhD thesis, AIFB, University of Karlsruhe.

Vretanos, P. (2005). *Open geospatial consortium filter encoding implementation specification* (Tech. Rep.). Retrieved 2008, from http://www.opengis.org/techno/specs.htm.

*W3C – Rule Interchange Format Working Group*. (2008). Retrieved November 2, 2008 from http://www.w3.org/2005/rules/wiki/RIF_Working_Group

W3C (2000). W3C Resource Description Framework. Candidate Recommendation. Retrieved from, http://www.w3.org/TR/2000/CR-rdf-schema-20000327/

W3C (2007). Semantic Web Layer Cake. Retrieved from, http://www.w3.org/2007/03/layerCake.png

W3C (2008). *The Semantic Web activity*. Retrieved April 14, 2008, from http://www.w3.org/2001/sw/.

W3C 2006. *RIF use cases and requirements*. W3C Working Draft. Retrieved July 10, 2006, from http://www.w3.org/TR/rif-ucr/

*W3C Semantic Web Activity* (2008). Retrieved April 22, 2008, from http://www.w3.org/2001/sw/.

W3C. (1998). *Extensible markup language (xml) 1.0* (Tech. Rep.). W3C recommendation.

Wagner, G. (1997). Vivid agents - How they deliberate, how they react, how they are verified. (Extended version of G. Wagner, Logical and operational model of scalable knowledge and perception-based agents). In W. Van de Velde and J.W. Perram (Eds.), *Agents breaking away* (LNCS 1038).

Wagner, G. (2000, May). Agent-oriented analysis and design of organizational information systems. In J. Barzdins & A. Caplinskas (Eds.). *Proceedings of 4th International Workshop on Databases and Information Systems*. Vilnius, Lituania.

Wagner, G. (2002, July). A UML profile for external AOR models. In *Proceedings of Third International Workshop on Agent-Oriented Software Engineering (AOSE-2002)* (Vol. 2585). Bologna, Italy: Springer-Verlag.

Wagner, G. (2003). Web rules need two kinds of negation (LNCS 2901, pp. 33-50).

Wagner, G. (2003, July). The agent-object-relationship meta-model: Towards a unified view of state and behavior. In *Information Systems, 28*(5), 475-504.

Wagner, G. (2004). AOR modeling and simulation: Towards a general architecture for

Wagner, G., A.Giurca, & Lukichev, S. (2005). R2ml: A general approach for marking up rules. In F. Bry, F. Fages, M. Marchiori, & H. Ohlbach (Eds.), *Principles and practices of Semantic Web reasoning, Dagstuhl seminar proceedings 05371*.

Wagner, G., Antoniou, G., Tabet, S., & Boley, H. (2004). The abstract syntax of RuleML - Towards a general Web rule language framework. In *Proceedings of the International Conference on Web Intelligence* (pp. 628-631). IEEE Computer Society.

Wagner, G., Damasio, C., & Antoniou, G. (2005). Towards a general Web rule language. *International Journal of Web Engineering and Technology, 2*(2), 181-206.

Wagner, G., Damásio, C.V., & Antoniou, G. (2005). Towards a general Web rule language. *Int. J. Web Eng. Technol, 2*(2/3): 181-206 (2005)

Wagner, G., Giurca, A., & Lukichev, S. (2005). A general markup framework for integrity and derivation rules. In *Principles and practices of Semantic Web reasoning*. Dagstuhl.

Wagner, G., Giurca, A., & Lukichev, S. (2005). R2ML: A general approach for marking-up rules. In *principles and practices of Semantic Web reasoning*.

Wagner, G., Giurca, A., & Lukichev, S. (2006). A general markup framework for integrity and derivation rules. In F. Bry, F. Fages, M. Marchiori, & H.-J. Ohlbach (Eds.). *Principles and practices of Semantic Web reasoning* (Number 05371 in Dagstuhl Seminar Proceedings). IBFI Schloss Dagstuhl.

Wagner, G., Giurca, A., & Lukichev, S. (2006). A general markup framework for integrity and derivation rules. In F. Bry et al. (Eds.), *Principles and Practices of semantic web reasoning*. Dagstuhl Seminar Proceedings. Internationales Begegnungs- und Forschungszentrum fuer Informatik (IBFI), Schloss Dagstuhl, Germany.

Wagner, G., Giurca, A., & Lukichev, S. (2006). A usable interchange format for rich syntax rules integrating OCL, RuleML and SWRL. In *Proceedings of the Workshop of Reasoning on the Web*. Edinburgh, UK.

Wagner, G., Giurca, A., & Lukichev, S. (2006). A *usable interchange format for rich syntax rules integrating OCL, RuleML and SWRL*. Paper presented at Reasoning on the Web Workshop at WWW2006, Edinburgh, UK.

Wagner, G., Giurca, A., & Lukichev, S. (2006). R2ML: A general approach for marking up rules. In *Dagstuhl Seminar Proc. 05371*.

Wagner, G., Giurca, A., Lukichev, S., Nicolae, O., & Daconescu, M. (2007). Case Study 1: Userv product Derby 2005. REWERSE IST 506779 Report I1-D9.

Wagner, G.; Giurca, A. & Lukichev, S. (2006). A usable interchange format for rich syntax rules integrating OCL, RuleML and SWRL. In *Proceedings of the Reasoning on the Web (RoW2006) Workshop at WWW2006*.

Wagner, K., & Klückmann, J. (2006). Prozessdesign als Grundlage von Compliance Management, Enterprise Architecture und Business Rules. In A.-W. Scheer, H. Kruppke, W. Jost & H. Kindermann (Eds.), *Agilität durch ARIS Geschäfprozessmanagement* (pp. 125-136). Springer Berlin/Heidelberg.

Wang, K., Billington, D., Blee, J., & Antoniou, G. (2004). Combining description logic and defeasible logic for the Semantic Web. In *Proceedings of Rules and Rule Markup Languages for the Semantic Web* (pp. 170-181). Springer.

Waterman, D. A. (1968). *Machine learning of heuristics*. PhD thesis, Stanford University.

Weerawarana, S., Curbera, F., Leymann, F., Storey, T., & Ferguson, F. D. (2005). *Web services platform architecture*. Prentice Hall.

Welty, C., & Guarino, N (2001). Supporting ontological analysis of taxonomic relationships. *Data and Knowledge Engineering, 39*(1), 51-74.

Weske, M. (2007). *Business process management: Concepts, languages, architectures*. Springer.

Wessel, M. (2001). Obstacles on the way to spatial reasoning with description logics – Some undecidability results. *Working Notes of the 2001 International Description Logics Workshop (DL-2001) Vol. 49*, Stanford, CA, USA.

Wessel, M. (2003). Some practical issues in building a hybrid deductive geographic information system with a DL component. In *Proceedings of the 10th International Workshop on Knowledge Representation meets Databases (KRDB 2003)* (Vol. 79). Hamburg, Germany.

Westman, R., & Fritzson, P. (1993). Graphical user interfaces for algorithmic debugging. In *Automated and algorithmic debugging* (pp. 273-286).

White, C. (2004). Building the smart business: In-Line real-time BI. *DM Review Magazine*. Retrieved June 10, 2007, from http://www.dmreview.com/issues/20040901/1009281-1.html

Whiting, R. (2002). Analysis gap. *Information Week*. Retrieved January 10, 2008, from http://www.informationweek.com/story/IWK20020418S0007

Widom, J. (1992). A denotational semantics for starbust production rule language. *SIGMOD record, 21*(3), 4-9.

Widom, J., & Ceri, S. (1996). *Active database systems: Triggers and rules for advanced database processing*. Morgan Kaufmann.

Widom, J., & Finkelstein, S.J. (1990). *Set-oriented production rules in relational database systems*. Paper presented at ACM Sigmond Int. Conf. on the Management of Data.

Wielemaker, J. (2003). *An overview of the SWI-Prolog programming environment (Tech. Rep.)*. University of Amsterdam, The Netherlands.

Wilensky, U. (1999). *NetLogo: Center for connected learning and computer based modeling*. Northwestern University. Evanston, IL, USA.

Wilkins, D.C. (1990). Knowledge base refinement as improving and incorrect, inconsistent and incomplete domain theory. *Machine Learning*, 3, 493-513.

Winikoff, M. (2005). JACK intelligent agents: An industrial strength platform. In R. Bordini, M. Dastani, J. Dix, & A. El Fallah Seghrouchni (Eds.), *Multi-agent programming: Languages, platforms and applications* (pp. 175-193). Springer.

Winikoff, M. (2005). JACK$^{TM}$ intelligent agents: An industrial strength platform. In *Multi-agent programming: Languages, platforms and applications*. Kluwer.

Wintraecken, J. J. V. R. (1990). *The NIAM information analysis method: theory and practice*. Deventer, The Nederlands: Kluwer.

Wise, L. (2008). With broader BI Usage, who owns the data? *IT Business Edge*. Retrieved January 09, 2008, from http://www.itbusinessedge.com/item/?ci=37391

Wogulis, J., & Pazzani, M (1993). A methodology for evaluating theory revision systems: results with Audrey ii. In *Proceedings of the Sixth International Workshop on Machine Learning* (pp. 332-337).

Wong, S. & Tojo, S. (1996). A deductive object-oriented database system for situated inference in law. *IEEE Transactions on Knowledge and Data Engineering, 8*(3), 496-503.

Wooldridge, M. (2000). *Reasoning about rational agents*. MIT Press.

Wooldridge, M. (2001). *An introduction to multiagent systems*. John Wiley & Sons.

Wooldridge, M., & Jennings, N. (1995) Agent theories, architectures, and languages: A survey. In M. Wooldridge, N. Jennings (Eds.), *Proceedings of International Workshop on Intelligent Agents I, Agent Theories, Architectures, and Languages (ATAL'94)* (pp. 1-39). Springer.

Wooldridge, M., & Jennings, N.R. (1995). Intelligent agents: Theory and practice. *Knowledge Engineering Review, 10*(2).

Wray, R., Jones, R. (2006) Considering Soar as an agent architecture. In R. Sun (Ed.), *Cognition and multi-agent interaction - From cognitive modeling to social simulation*. Cambridge University Press.

Wright, I., & Marshall, J. (2003). The execution kernel of RC++: RETE*, A faster RETE with treat as a special case. *International Journal of Intelligent Games and Simulation, 2*(1), 36-48.

WSDL2R2ML (2006). *The WSDL to R2ML transformation scenario*. Retrieved from http://www.eclipse.org/m2m/atl/atlTransformations/#WSDL2R2ML

WSDM (2006). *OASIS Web Services Distributed Management WSDM V1.1 as an OASIS Standard*. Retrieved from http://www.oasis-open.org/committees/wsdm/, 2006

*XML metamodel* (2006). Retrieved from http://mdr.netbeans.org/mdrxml.html

XMLA council. (2007). *XML for analysis*. Retrieved May 11, 2007 from http://www.xmlforanalysis.com/index.htm

XMLMapper (2008). *SWRLTab XMLMapper Tool*. Retrieved November 10, 2008 from http://protege.cim3.net/cgi-bin/wiki.pl?SWRLTabXMLBuiltIns

XPath (2007). *XML Path Language (XPath) 2.0*. Retrieved November 10, 2008 from http://www.w3.org/TR/xpath20/

XQuery (2007). *XQuery 1.0: An XML Query Language*. Retrieved November 10, 2008 from http://www.w3.org/TR/xquery/

XSB. (2007, October 6). *XSB Prolog*. Retrieved from http://xsb.sourceforge.net/

Yang, D., Kifer, M., & Zhao, C. (2003). FLORA-2: A rule-based knowledge representation and inference infrastructure for the Semantic Web. In *Proceedings of the International Conference on Ontologies, Databases and Applications of Semantics* (pp. 671-688). Springer.

Yemini, S.A., Kliger, S., Mozes E., Yemini, Y., & Ohsie, D. (1996). High speed and robust event correlation. *IEEE Communications Magazine, 34*(5), 82-90.

Yu, V. L., Buchanan, D.G., Shortliffe, E.H., Wraith, E.H., Davis, S.D., Scott, R. et al. (1979). Evaluating the performance of a computer-based consultant. *Computer Programs in Biomedicine, 95*(1).

Zacharias, V. (2008). Development and verification of rule based systems – A survey of developers. In *Proceedings of the International RuleML symposium on Rule Interchange and Application (RuleML 2008), Orlando, USA.*

Zacharias, V. (2008). Rules as a simple way to model knowledge – Closing the gap between promise and reality. In *Proceedings of the 10th International Conference on Enterprise Information Systems (ICEIS 2008), Barcelona, Spain.*

Zacharias, V. (2008). *Trie! Rules*. Retrieved April 28, 2008, from http://code.google.com/p/trie-rules/, accessed 2008-28-04.

Zacharias, V., & Abecker, A. (2007). On modern debugging for rule-based systems In *Proceedings of the Nineteenth International Conference on Software Engineering and Knowledge Engineering (SEKE 2007)*, Boston, USA, July 9-11.

Zacharias, V., & Abecker, A. (2007). explorative debugging for rapid rule base development. In *Proceedings of the 3rd Workshop on Scripting for the Semantic Web, 4th European Semantic Web Conference (ESWC 2007), Innsbruck, Austria.*

Zadeh, L. (1965). Fuzzy sets. *Information and Control, 8*(3), 338-353.

Zarri, G.P. (1986). The use of inference mechanisms to improve the retrieval facilities from large relational databases. In F. Rabitti, *Proceedings of the Ninth International ACM Conference on Research and Development in Information Retrieval*. New York: ACM.

Zarri, G.P. (1998). Representation of temporal knowledge in events: The formalism, and its potential for legal narratives. *Information & Communications Technology Law – Special Issue on Models of Time, Action, and Situations 7*, 213-241.

Zarri, G.P. (2003). A conceptual model for representing narratives. In R. Jain, A. Abraham, C. Faucher, & B.J. van der Zwaag (Eds.), *Innovations in Knowledge Engineering*. Adelaide: Advanced Knowledge International.

Zarri, G.P. (2005). Integrating the two main inference modes of NKRL, Transformations and hypotheses. *Journal on Data Semantics (JoDS), 4*, 304-340.

Zarri, G.P. (2007). An implemented representation and reasoning system for creating and exploiting large knowledge bases of 'narrative' information. In Z. Ma (Ed.), *Intelligent Databases, Technologies and Applications*. Hershey, PA: IGI Global.

Zeng, L., Flaxer, D., Chang, H., & Jun-Jang, J. (2002). PLM flow—Dynamic business process composition and execution by rule inference. In *Technologies for e-services* (LNCS 2444, pp. 51-95).

Zhao, D., & Lu, Y.-s. (2004). Association rule mining in intrusion detection systems. In S. J. B. Yoo, K.-w. Cheung, Y.-C. Chung, & G. Li (Eds.), *Network Architectures, Management*, (Vol. 5282, p. 577-581).

Zhao, G., Gao, Y., & Meersman, R. (2004). An Ontology-based approach to business modeling. In *Proceedings of the International Conference of Knowledge Engineering and Decision Support (ICKEDS2004)* (pp. 213–221).

Zimmer, D., & Unland, R. (1999). On the semantics of complex events in active database management systems. In *ICDE'99* (pp. 392-399).

Zimmermann, J., & Buchmann, A.P. (1999). *REACH. Active rules in database systems* (pp. 263-277).

Zou, Y., Finin, T., & Chen, H. (2004). F-OWL: An inference engine for Semantic Web. In *Proceedings of the International Workshop on Formal Approaches to Agent-Based Systems* (pp. 238-248). Springer.

# About the Contributors

**Adrian Giurca**, is currently investigating methods and applications for information systems of the next generation especially (agent based) reasoning on social media/software and reasoning on the Web and Semantic Web. Formerly, he worked on knowledge representation (particularly with uncertainty representation) and logic programming. Since 2005, he is a senior researcher at the Institute for Informatics of the Brandenburg University of Technology, Germany. From 1990 through 2004, he was with the Faculty of Mathematics and Computer Science from University of Craiova, Romania. During this time he was also consultant for a number of small size enterprises. He received its PhD in 2004 from the University of Bucharest, Romania. He has been or is involved in research projects founded by the European Commission (FP6). He has been member of the Network of Excellence REWERSE (http://rewerse.net 2004-2008) of the 6th Framework Program of the European Commission He is regularly serving as a reviewer for international conferences and journals, and he actively contributes to current research themes in web technologies.

**Dragan Gašević** is a Canada Research Chair in Semantic Technologies and an assistant professor in the School of Computing and Information Systems at Athabasca University. He is also an Adjunct professor in the School of Interactive Arts and Technology at Simon Fraser University Surrey. His research interests include semantic technologies, software language engineering, technology-enhanced learning, and service-oriented architectures. He is a recipient of Alberta Ingenuity's 2008 New Faculty Award and has been a principle investigator of several large scale research projects in Canada and Europe. He has (co-)authored around 200 research papers. He is the lead author of the book monograph entitled "Model Driven Engineering and Ontology Development" whose second edition is published by Springer in 2009. He has been serving on editorial boards of five international journals and has edited special issues in journals such as IET Software, IEEE Transactions on Software Engineering, and Elsevier's Information Systems. Being organizer, chair, and member of program committees of many international conferences and workshops, he is currently serving on the steering committee of the International Conference on Software Language Engineering. Dragan's home page is available at http://dgasevic.athabascau.ca/.

**Kuldar Taveter** (PhD in Engineering) is a professor of Software Engineering in the Department of Informatics at Tallinn University of Technology, Estonia. Before that he was employed by the University of Melbourne, Australia, and in 1997-2005 by the Technical Research Centre of Finland, which acts as a mediator between academy and industry. His main research interests are agent-oriented software engineering and agent-based business process automation and simulation. He has also conducted research on business rules and ontologies

* * *

**Alia I. Abdelmoty** is a lecturer of Computer Science in Cardiff University. Her research interests are in the area of Geoinformatics with particular focus on qualitative spatial and spatiotemporal reasoning, visual query languages and geographic information retrieval. Her goal is to realise the theoretical developments in the area of qualitative spatial reasoning in practical engines to solve problems in the geographic domain and to extend those results to other spatial application domains. She is also working in the area of understanding geospatial semantics to enhance the personalisation of web search. She has been involved in several UK funded projects in this area as well as the EC FP 5 project SPIRIT (Spatially-aware information retrieval on the internet) and the current FP 6 project TRIPOD, for automatic generation of photographic captions, using the geographic content of images.

**Nick Bassiliades** received his PhD degree in parallel knowledge base systems in 1998 from the Department of Informatics, Aristotle University, Thessaloniki, Greece, where he is currently an assistant professor. His research interests include knowledge based systems, rule systems, ontologies and the semantic web. He has published over 90 scientific papers and articles at international journals, conferences and edited volumes and co-authored an international book on Parallel, Object-Oriented, and Active Knowledge Base Systems and a Greek book on Artificial Intelligence. He has been involved in Greek and European projects concerning knowledge based systems, intelligent agents, e-learning, web services, semantic web, etc. He is a member of the Board of the Greek Artificial Intelligence Society, a director of the RuleML initiative and a member of IEEE and ACM.

**Joachim Baumeister** is senior researcher at the University of Würzburg, Germany, where he also received his PhD in 2004. His research focuses on semantic technologies and practical knowledge engineering such as continuous knowledge formalization, advanced evaluation and refactoring methods. He is regularly serving as a reviewer for international conferences and journals, and he actively contributes to current research themes which are documented by more than 60 referred publications. Dr. Baumeister gives lectures for example on Knowledge-Based Systems and the Semantic Web. For about 10 years he is working in and coordinating industrial research projects, mainly in the medical area, but also in the biological and technical domain.

**Harold Boley** is adjunct professor at the Faculty of Computer Science, University of New Brunswick, and leader of the Semantic Web Laboratory at NRC-IIT, Fredericton, Canada. He developed the Relational-Functional Markup Language before starting and co-leading the Rule Markup (RuleML)Initiative. He co-designed the Semantic Web Rule Language (SWRL), which combines the W3C-recommended Web Ontology Language OWL and RuleML. He led the design of a First-Order Logic Web language (FOL RuleML) and contributed to the design of the Semantic Web Services Language (SWSL). He is co-editor of the W3C Rule Interchange Format (RIF). His focus is on Semantic Web knowledge representation combining ontologies and rules. He is currently exploring this for person/organization-centric profiles supporting FOAF-extending social networking, e-tourism, and expert finding, as realized, e.g., with RuleML FOAF in FindXpRT, eTourPlan, and Rule Responder.

**Lars Braubach** is, since 2007, a project leader within the VSIS group at Hamburg University. He earned his Diploma in Informatics in 2003 and his Dr. rer.nat. in Informatics in the area of distributed (multi-agent) systems with focus on architectures and methodologies for realizing intelligent systems in 2007. Since 2003, he worked in the DFG funded project MedPAge as part of the priority research program (SPP) 1083. As part of MedPAge, he co-develops the Jadex agent framework since 2003 together with Alexander Pokahr. Since 2007, he co-acquired and manages two DFG-funded research projects. The SodekoVS project deals with fundamental research in the area of self-organizing systems, whereby in the technology transfer project, Go4Flex, in cooperation with the industry partner Daimler, agile workflow descriptions are explored. He has published over 45 peer-reviewed scientific publications and has co-organized a number of workshops and conferences – mostly in the area of multi-agent systems.

**Rem Collier** is a lecturer in the School of Computer Science and Informatics at University College Dublin (UCD). He has a BSc (Hons) in Mathematics from the University of Bristol (1994), a MSc and M.Phil. in Computer Science from the University of Manchester Institute of Science and Technology (1995/96), and a PhD in Computer Science from UCD (2001). His research interests include Multi-Agent Systems, Agent-Oriented Software Engineering, and Information Retrieval. He has published over 50 refereed papers/chapters and has won grant income in the region of €365K. In 2003, Dr Collier received the prestigious Cooperative Information Agents, System Innovation Award for ACCESS: An Agent Architecture for Ubiquitous Service Delivery.

**Amar Das** is assistant professor of Medicine (Biomedical Informatics) and of Psychiatry & Behavioral Science at Stanford University. He holds a BA from Northwestern University and an MD and PhD from Stanford University. His research focuses on the advancement of model-driven architectures and Semantic Web technologies (XML, OWL, SWRL) to support temporal reasoning, data integration, and collaborative systems in health care and the life sciences.

**Mehdi Dastani** is a lecturer in computer science at the Utrecht University. He is working in the area of multi-agent systems and multi-agent programming for the last ten years and has published many papers on these subjects. Currently, he is the general chair of the European Agent Systems Summer School (EASSS), a co-organizer of ProMAS (The International Workshop on Programming Multi-Agent Systems) and multi-agent programming contest. He has been the PC member of various international conferences and workshops such as AAMAS (International Joint Conference on Autonomous Agents and Multi-Agent Systems), AAAI (Association for the Advancement of Artificial Intelligence), IJCAI (International Joint Conference on Artificial Intelligence), and JELIA (European Conference on Logics in Artificial Intelligence).

**Vladan Devedžić** is a professor of computer science at the Department of Software Engineering, FON - School of Business Administration, University of Belgrade, Serbia. His main research interests include software engineering, intelligent systems, knowledge representation, ontologies, Semantic Web, intelligent reasoning, and applications of artificial intelligence techniques to education and medicine. He can be reached at devedzic@etf.rs.

**Jens Fietrich** is a senior lecturer in Computer Science and Software Engineering at the School of Engineering and Advanced Technology (SEAT http://engineering.massey.ac.nz/) at the Turitea (Palmerston North) Campus of Massey University in New Zealand <http://www.massey.ac.nz/>. Jens has a Master in Mathematics and a PhD in Computer Science, both from the University of Leipzig <http://www.uni-leipzig.de/> in Germany. After completing his PhD in 1996, Jens worked for a couple of years in industry as an IT consultant in Germany, Namibia, Switzerland and the UK, and returned in 2003 to academia. Jens' research interests are in the areas of design pattern formalisation, software componentry, agile software engineering techniques and business rule automation. He is a member of the IEEE.

**Dragan Djuric**, PhD, is assistant professor at the Department of Software Engineering, FON - School of Business Administration, University of Belgrade, Serbia. His interests mostly include modeling, enterprise software architecture, object-oriented development, Java platform, Semantic Web, and intelligent information systems. He can be reached at dragan@dragandjuric.com.

**Schahram Dustdar** is full professor of Computer Science with a focus on Internet Technologies at the Distributed Systems Group, Information Systems Institute, Vienna University of Technology (TU Wien) where he is director of the Vita Lab. He is also Honorary professor of Information Systems at the Department of Computing Science at the University of Groningen (RuG), The Netherlands. More information on: http://www.infosys.tuwien.ac.at/Staff/sd

**Marwane El Kharbili** is working in the ARIS research department of IDS Scheer AG, working on its business rule management solution for business process management (BPM). As part of his research activities, He is involved in various research projects, such as the European project SUPER (www.ip-super.org). He is also a PhD student currently working on his thesis tackling semantic compliance management issues in BPM. His topics of interest are business rule management, BPM, semantic web, compliance management and complex event processing. Marwane has an engineering degree (Diplom-Inform. Eq. Master) from the university of Karlsruhe in Germany and a Masters degree of engineering (ingenieur informatique) from the ENSIMAG school of applied mathematics and informatics at the Grenoble INP university in France.

**Baher El-Geresy** is a senior research fellow in the Faculty of Advanced Technology at the University of Glamorgan and a member of the Geographic Information Research Centre there. His research work is focussed on the area of qualitative spatial and spatiotemporal representation and reasoning, incomplete spatial reasoning, and constraint satisfaction problems in the spatial domain. He holds a PhD in Engineering and in Computer science. His research goal is to develop a general unified theory for qualitative space and time.

**Mihai Gabroveanu** is currently working as assistant professor in the Department of Informatics at the University of Craiova, Romania. He received his MS degree in Artificial Intelligence from the Faculty of Mathematics and Informatics, University of Craiova, Romania in 1999 and is now PhD student at the Faculty of Mathematics and Informatics from the University of Bucharest, Romania. His current research is focused on data mining models and algorithms, data warehouse and grid computing. He is also interested in web programming, algorithms and data structures.

**Stathes Hadjiefthymiades** received his BSc, MSc and PhD in Informatics and Telecommmunications from the University of Athens (NKUA), in 1993, 1996, and 1999, respectively, and an MSc in engineering-economics from the National Technical University of Athens, in 2002. In 1992, he joined the consulting firm Advanced Services Group as an analyst/developer of telematic systems. In 1995, he joined the Communication Networks Laboratory, NKUA. From September 2001 to July 2002, he was a visiting assistant professor at the Department of Information and Communication Systems Engineering, University of Aegean. In July 2002, he joined the faculty of the Hellenic Open University as an assistant professor. Since December 2003, he has been with the faculty of the department of Informatics and Telecommunications, NKUA, where he currently is an assistant professor. He has participated in numerous EU and national projects. His research interests include the areas of mobile, pervasive computing, web engineering, and networked multimedia.

**Marek Hatala** is an associate professor at the School of Interactive Arts and Technology at Simon Fraser University. Dr. Hatala has obtained his PhD in Artificial Intelligence from the Technical University of Kosice in Slovakia. At SFU, he is the Director of Laboratory for Ontological Research where he leads research that applies AI methods from the semantic web toward solving difficult problems in other domains. The lab's recent projects include: supporting learners in interactive learning environments, user modeling in ambient intelligence systems such as museum guides, interoperability research in service oriented architectures and security policies, and modeling of software engineering languages to achieve interoperability in the software engineering environments. Dr. Hatala's areas of interest include information and knowledge management, artificial intelligence, user modeling, distributed systems and interoperability, security and trust policies, e-learning, and collaborative systems.

**Mustafa Jarrar** is a Marie Curie fellow at the University of Cyprus. Before that he worked as a senior research scientist at STARLab, Vrije Universiteit Brussel, Belgium. After graduating in 1995 in Computer Engineering, he received his Masters in 2000, and his PhD in Computer Science in early 2005 from the Vrije Universiteit Brussel. Jarrar published more than 50 refereed articles in the areas of database, conceptual modeling, ontologies, automated reasoning, and lexical semantics. He chaired 12 international events and workshops, a PC members of more than 50 conferences and journals. Dr. jarrar served as a project leader for STARLab in 8 national and international project, and has recently co-founded the Ontology Outreach Advisory (ontology-advisory.org), which is an international not-for-profit association with the aim of developing strategies for ontology recommendation and standardization, and promoting the ontology technology to industry. Dr. Jarrar is a full member of the IFIP2.6 on Database Semantics, the IFIP2.12 on Web Semantics, the IEEE Learning Standards Committee, and the CEN/ISSS ICT Skills and Curricula. http://www.jarrar.info

**Christopher B. Jones** has been professor of Geographical Information Systems (GIS) in the School of Computer Science at Cardiff University, UK, since 2000. He has held previous academic positions at the University of Glamorgan and the University of Cambridge. Prior to his academic career, he was employed on geological computing applications at BP Exploration and at the British Geological Survey. Current research interests focus on the area of geographical information retrieval and include the acquisition and modelling of knowledge of geographical places and place names and the development of methods to index documents and photographic images for spatially-aware web search engines. He has also worked in the areas of automated cartographic design with regard to map generalisation and automated text placement; multiscale spatial databases; environmental change detection; spatio-temporal data modelling; and the computer reconstruction and visualisation of fossils.

**Isambo Karali** is an assistant professor in the Department of Informatics and Telecommunications, University of Athens, since November 2007. From September 1999 to October 2007, she was Lecturer in the same Department. From 1989 to 1999, she was research associate in the Department and was employed in European Union research projects. She obtained her first degree in Mathematics in 1986 (Department of Mathematics, University of Athens), her MSc in Computer Science in 1988 (University College, University of London) and her PhD in Informatics in 1995 (Department of Informatics, University of Athens). Her research interests include knowledge representation, logic programming as well as Web information extraction and representation. She has published a number of papers and has been reviewer in various journals and conferences related to the above areas. She is a member of scientific societies, such as ACM and LMS.

**Nima Kaviani** is a PhD student at the department of Electrical and Computer Engineering (EECE) in the University of British Columbia. He received his MSc on Information Technology from Simon Fraser University, where he primarily focused on policy modeling and policy engineering using Web rule languages. His research interests are these days mainly in the area of semantic enabled middleware systems with considerations for policy enforcement and behavior management. He has also an extensive background in the areas of Robotic and Multi-Agent Systems.

**Antoni Ligęza** graduated from Faculty of Electrical Engineering, Automatics and Electronics (present: Faculty of Electrical Engineering, Automatics, Informatics and Electronics, EAIiE), AGH – University of Science and Technology in Cracow, Poland; received M.Sc. in electronics/automatic control in 1980. After completing Doctors Studies he received his PhD degree in Computer Science (1983), and the habilitation (docent degree; polish Dr habilitowany) in 1994 in Computer Science/Artificial Intelligence, both from the EAIiE Faculty at AGH. In 2006 he received the professor title from the President of Poland. His main research concern Knowledge Engineering (Artificial Intelligence) including knowledge representation and inference methods, rule-based systems, automated plan generation, technical diagnostics, logics and systems science. Some most important original research results include development of backward plan generation model (1983), independent discovery of dual resolution method for automated inference (1991), the concepts of granular sets and relations (2000), granular attributive logic (2003) , diagnostic inference models in the form of logical AND/OR/NOT causal graphs (1995) and Potential Conflict Structures (1996). He was visiting reseracher/professor at Technical University of Denmark, Lyngby (1988), LAAS, Toulouse, France (1992, 1996), Universite de Nancy I, France (1994), University of Balearic Islands, Spain (1994, 1995, 2005), University of Girona, Spain (1996, 1997), and Universite de Caen, France (2004, 2005, 2007). He published (as author and co-author) more than 200 research papers, including recent monograph Logical Foundations for Rule-Based Systems Springer, 2006 (author), and Knowledge-Driven Computing, Springer, 2008 (co-editor and co-author). Member of IEEE Computer Society and ACM.

**Mengchi Liu**; obtained his PhD in Computer Science from the University of Calgary in Canada in 1992. Currently, he is a full professor in Computer Science. His research interests include rule-based languages, advanced database systems, intelligent information systems. XML, and web technologies.

**Sergey Lukichev** received the BSc degree in mathematics in 2000 and the MSc degree in computer science in 2002 from the Novosibirsk State University, Russia. Since 2004, he has been a PhD student at the Brandenburg Technical University in Cottbus, Germany, chair of Internet Technologies. His research interests include Semantic Web, rule languages, rule interchange, rule modeling, and rule verification. He has been also involved in the European research project REWERSE concerning rule modeling and rule markup.

**Georgios Meditskos** received the BSc degree in computer science in 2004 and the MSc degree in computer science in 2007 from the Aristotle University of Thessaloniki, Greece. Since 2004, he has been a PhD student in computer science at the Aristotle University of Thessaloniki. His research interests include semantic Web, rule languages, semantic Web services, and artificial intelligence. He has been also involved in Greek and European projects concerning ontologies, knowledge based systems, and semantic web.

**Robert Meersman** received his PhD in Mathematics at the Vrije Universiteit Brussel (VUB) in 1976. He founded the first InfoLab at University of Hasselt (Belgium, 1983-86), the second at University of Tilburg (The Netherlands, 1986-95) and VUB STARLab in 1995. While he was R&D project manager at Control Data Corp (1978-83), he worked on the definition of the NIAM (now ORM) method as well as on its query and constraint languages (RIDL) and on the first tools for this methodology. He is now a full professor at VUB since 1995. His current research is focused on ontologies and their relationship and use in databases, semantic web and community-driven semantic design methodologies and tools.

**Anton Michlmayr** received the MSc degree in computer science from Vienna University of Technology in 2005. He is currently a research assistant and PhD student at the Distributed Systems Group, Information Systems Institute, Vienna University of Technology. His research interests include software architectures for distributed systems with an emphasis on distributed event-based systems, service-oriented architectures and Web service technologies. More information can be found at: http://www.infosys.tuwien.ac.at/Staff/michlmayr

**Milan Milanović** is a PhD candidate at Department of Software Engineering, FON - School of Business Administration, University of Belgrade, Serbia. His interests include, Modeling and metamodeling, Model transformations, Rule and policy-based process modeling, Rule languages, Business modeling, Business rules and Service Oriented Architectures. He can be reached at milan@milanovic.org.

**Mark Musen** is professor of Medicine (Biomedical Informatics) and Computer Science (by courtesy) at Stanford University, where he is head of the Stanford Center for Biomedical Informatics Research. He holds an MD from Brown University and a PhD from Stanford. Dr. Musen conducts research related to intelligent systems, the Semantic Web, reusable ontologies and knowledge representations, and biomedical decision support. His long-standing work on a system known as Protégé has led to an open-source technology now used by thousands of developers around the world to build intelligent computer systems and new computer applications for e-science and the Semantic Web. He is known for his research on the application of intelligent computer systems to assist health-care workers in guideline-directed therapy and in management of clinical trials. He is principal investigator of the National Center for Biomedical Ontology, one of the seven National Centers for Biomedical Computing supported by the NIH Roadmap.

**Christoph Nagl** received his MSc degree in computer science from the Vienna University of Technology in 2006. During his study he focused his research on service-oriented architectures and business rules, especially the design and implementation of ViDRE. He is now working as an IT consultant in industrial projects.

**Grzegorz J. Nalepa** holds a position of assistant professor in the Institute of Automatics at AGH University of Science and Technology in Krakow, Poland. He is interested mainly in practical applications of Artificial Intelligence, Knowledge Engineering, rule-based expert systems. His research also includes the Semantic Web, software engineering, and computer security. Current research interests include applications of rule-based systems and business rules design and analysis, in the field of software engineering and control systems. He is the author of over fifty scientific papers. He lectures Introduction to Unix/GNU/Linux, AI Languages and Computer Security at the Faculty of Electrical Engineering of AGH UST. He is the author of the XTT rule-based systems design, implementation and analysis method. Since 2008 he has been coordinating the HeKatE project, led by prof. Antoni Ligeza. Tho project develops new design frameworks for rule-base intelligent systems.

**Martin J. O'Connor** is a systems software developer in the Stanford Center for Biomedical Informatics Research at Stanford University. He received his MSc in computer science from the University of Dublin, Trinity College. His research interests include the development of data integration tools and query languages for the Semantic Web.

**Gregory O'Hare** is a senior lecturer in the School of Computer Science and informatics at University College Dublin (UCD). He is a Principle Investigator in CLARITY: The Center for Sensor Web Technologies, a €16.8M SFI-funded CSET (2008-2013), and is Director of the Practice and Research in Intelligent Systems & Media (PRISM) Laboratory. Prior to joining UCD he was on the faculty of the Universities of Central Lancashire and Manchester. His research interests include Distributed Artificial Intelligence, Multi-Agent Systems, Agent-Oriented Software Engineering and Mobile & Ubiquitous Computing. He has published over 270 refereed papers/chapters, 4 books and has won significant grant income (ca ¤11.5M). In 2003 he received the prestigious Cooperative Information Agents, System Innovation Award for ACCESS: An Agent Architecture for Ubiquitous Service Delivery. Dr O'Hare is a Fellow of the British Computer Society and the Irish Computer Society, a member of the ACM, AAAI and a Chartered Engineer. He currently serves on the Editorial Review Board of 2 International Journals.

**Vassilis Papataxiarhis** received his BSc in Informatics from the Department of Informatics & Telecommunications at the National and Kapodistrian University of Athens (NKUA), Greece in 2006 and his M.Sc. in the area of "Advanced Information Systems" from the same Department in 2008. Nowadays, he is a reseacher in the areas of Artificial Intelligence and Autonomic Computing in the department. He is also a member of the Pervasive Computing Research Group (p-comp) and the Communication Networks Laboratory (CNL) of the NKUA. He has participated in several European and national research projects. His research interests are mainly focused on the fields of Semantic Web technologies, Knowledge Representation and Reasoning, Machine Learning and Ambient Intelligence.

**Emilian Pascalau**, was born in 1982, in Romania. He's current activities involve research and development of methods and enterprise systems comprising on one hand browser based rule engines, rules in the context of mashups, and on the other hand agent based systems that follow human like reasoning. Since October 2008 he is a research assistant and PhD student at Hasso Plattner Institut for IT Systems Engineering, at University of Potsdam. From 2007 - 2008 he was a research assistant at the Institute for Informatics of the Brandenburg University of Technology, Germany, with a DAAD scholarship. During this period he has been involved in research projects founded by the European Commission (FP6). He received in 2006, his Master diploma in Computer Science - Distributive Systems in Internet and Intranet from Babes-Bolyai University, Cluj-Napoca. During his master studies Emilian Pascalau worked for Softvision Inc. Cluj-Napoca office and San Jose, California office, as a Java Enterprise developer.

**Adrian Paschke** leads the chair for Corporate Semantic Web at the FU Berlin (FUB) and is vice-director of the Semantics Technologies Institute Berlin (STI Berlin), director of RuleML Inc., and research director of the Centre for Information Technology Transfer (CITT) GmbH. He is project leader of several open-source projects such as the Prova Semantic Web rule engine, Rule Responder, national projects such as the BMBF funded project Corporate Semantic Web, and is involved in several international and EU projects such as the EU Network of Excellence REWERSE, EU Strep Sealife, EU IP DoReMoPat. He is steering-committee chair of the RuleML Web Rule Standardization Initiative, Co-Chair of the Reaction RuleML technical group, founding member of the Event Processing Technology Society (EPTS), voting member of OMG, and invited member of several W3C groups such as the W3C Rule Interchange Format (W3C RIF), where he edits several Web rule standards documents.

**Alexander Pokahr** is, since 2007, a project leader within the VSIS group at Hamburg University. He earned his Diploma in Informatics in 2003 and his Dr. rer. nat. in Informatics in the area of programming languages and development tools for distributed intelligent software systems. Together with Lars Braubach he worked on the MedPAge project and in the DFG priority research program SPP 1083 "Intelligent Agents in Real-World Business Applications" for the last 5 years and co-developed the Jadex agent framework. Currently he manages two co-acquired DFG-funded research projects. In the SodekoVS project it is aimed at making self-organizing mechanisms generically software technically usable, whereas in the 2009 starting Go4Flex project workflows will be made more flexible by using goal-oriented concepts. He has published over 45 peer-reviewed scientific publications and has co-organized a number of workshops and conferences – mostly in the area of multi-agent systems.

**Frank Puppe** has worked full time in the area of computer science and artificial intelligence since 1983, when he earned his master (diplom) degree at Bonn university. Further academic degrees were his PhD at Kaiserslautern university in 1986 and his habilitation at Karlsruhe university in 1991. Since 1992 he is appointed as full professor at University of Würzburg, Germany. He worked as consultant for knowledge based systems for many companies including Siemens AG, Daimler-Chrysler AG, Renk AG, German Telekom AG, Koenig&Bauer AG, and was project leader in 4 large national joint projects on intelligent tutoring&training systems, diagnostic problem solving, cooperating problem solvers and intelligent assistance for scheduling tasks. He published 6 books (incl. 2 edited volumes) and more than 100 articles in the field of knowledge based systems. His interests encompass all aspects of knowledge-based systems.

**Marko Ribaric** is a research staff member at Institute Mihailo Pupin in Belgrade, Serbia, and member of a Good Old AI research group. His research interest are (but not limited) in model-driven engineering, model transformations, and the Semantic Web. He received a M.Sc. degree in computer science from School of Electrical Engineering, University of Belgrade.

**Florian Rosenberg** is research assistant and PhD student at the Distributed Systems Group, Information Systems Institute, Vienna University of Technology. His research areas include software engineering in general, service-oriented architectures and Web service technologies with an emphasis on service composition and Quality of Service (QoS) aspects. More information can be found at: http://www.infosys.tuwien.ac.at/Staff/rosenberg

**Dietmar Seipel** is working in the areas of databases and artificial intelligence. He received his diploma (1986) and PhD (1989) degree in the fields of relational database design and recursive query processing in deductive databases, respectively. Since his habilitation at the University of Tübingen in 1995, which covered disjunctive deductive databases and non-monotonic reasoning, he works as a professor at the University of Würzburg. His current research interests include databases for semi-structured data, knowledge-based systems, Semantic Web, declarative programming and software engineering issues. He is cooperating in research projects with industrial partners and colleagues from university on applications in bioinformatics, computational linguistics, network diagnosis, and medicine.

**Shahin Sheidaei**, born in 1983 in Tehran, Iran, received his BSc in 2006 from the Department of Computer Engineering, Iran University of Science and Technology, Iran. As of January 2009 he is doing his M.Sc. in the school of interactive arts and technology, Simon Fraser University, Canada. His main area of research is policies, security and model driven engineering.

**Philip Smart** is a completing PhD student in the School of Computing at Cardiff University. Philip's PhD involves the development of a geospatial rule language for the representation and reasoning over geospatial ontology bases. Philip is also a research associate on the EC FP 6 project TRIPOD where he is responsible for the development of a toponym ontology to facilitate the resolution of place names in image captions as well as place identification from 3D models. Philip's research interests include the development and maintenance of geographic ontologies, integration of spatial logics and rule languages, mediation based ontology architectures and development of Java based geospatial web services.

**Panagiotis Stamatopoulos** received the BSc degree in physics (1982) and the PhD degree in computer science (1988) from the National and Kapodistrian University of Athens, Greece. After receiving his PhD, he joined, as a post-doctoral researcher the Institute of Informatics and Telecommunications of the NCSR "Democritos", Greece. Since 1993, he is a faculty member of the Department of Informatics and Telecommunications, of the National and Kapodistrian University of Athens. Currently, he is holding an assistant professor position at this department. Dr. Stamatopoulos is working in the area of artificial intelligence and his main research interests include constraint programming, machine learning, logic programming and natural language processing. He is a member of ACM, AAAI and IEEE Computer Society.

**Leon Sterling** is a professor of Software Innovation and Engineering in the Department of Computer Science and Software Engineering and Director of eResearch at the University of Melbourne, Australia. He received a BSc(Hons) from the University of Melbourne in 1976 and a PhD in Pure Mathematics from the Australian National University in 1981. He has worked in universities in the UK, Israel, and USA. His teaching and research specialities are software engineering, agent technology, logic programming, especially the language Prolog, and artificial intelligence.

**Yan Tang** is a researcher at the Semantic Technology and Application Research Laboratory (STARLab), Department of Science, Vrije Universiteit Brussel (VUB, Free University of Brussels). Her research interests include semantic decision making, decision support and decision table, ontology based application architecture, ontology creation methodology, ontology modeling and business rule modeling. Her PhD thesis is about the theory and practices of Semantic Decision Tables (SDT), which is under the supervision of prof. Dr. Robert Meersman. During her PhD period, she has published her work in several international journals, conference proceedings as well as book chapters, one of which has been edited by World Scientific.

**Bojan Tomić** is a teaching assistant and PhD student at the Department of Information Systems and Technologies, Faculty of Organizational Sciences, University of Belgrade (Serbia). In 2005 he received his BS degree and the award for the "Best Student Research Work in Technical Sciences", granted by the University of Belgrade for his paper titled "JavaDON: An Open-source Expert System Shell". Besides being author and coauthor of several papers, he authored one book and took part in several AI related projects. Bojan Tomić is currently pursuing his PhD thesis in the area of expert system application in the business domain. His other research interests include software engineering and software testing. He is a member of the Good Old AI (GOAI) research group.

**Vassileios Tsetsos** received his BSc in Informatics from the Department of Informatics & Telecommunications at the University of Athens, Greece in 2003 and his M.Sc. in "Communication Systems and Data Networks" from the same Department in 2005. Nowadays, he is a PhD candidate in the department. He is a member of the Communication Networks Laboratory (CNL) of the University of Athens and the Pervasive Computing Research Group (p-comp). He has participated in several European, funded by the EU IST & ICT programmes, and national research projects. His research interests are in the areas of pervasive and mobile computing, Semantic Web technologies and middleware for context-aware and sensor-based services.

**Gerd Wagner** is professor of Internet Technology at Brandenburg University of Technology at Cottbus, Germany. He has studied Mathematics, Philosophy and Informatics in Heidelberg, San Francisco and Berlin. His research interests include agent-oriented modeling and agent-based simulation, foundational ontologies, (business) rule technologies and the Semantic Web. He has more than 90 publications in these areas in international journals and books and has published two monographies.

**Valentin Zacharias** works as a senior researcher in the Information Process Engineering group at the FZI Research Center for Information Technology in Karlsruhe, Germany. He wrote his PhD about tool support for the correct creation of rule bases at the University of Karlsruhe. Before that, he studied computer science and psychology at the Free University in Berlin and the University of Massachusetts, Dartmouth. His main research interest is the verification and validation of knowledge based systems - a topic he worked on for multiple years in Project Halo. In the past years he has also worked on multiple public and private funded projects bringing new rule technologies and semantic technologies to practical applications.

**Gian Piero Zarri** is known internationally as the author of NKRL ("Narrative Knowledge Representation Language"), a standard representation of the meaning of the so-called 'narrative' documents. His record of research is extensive (more than one hundred and fifty refereed papers); he is the editor of the book "Operational Expert Systems Applications in Europe" (Pergamon Press, 1991), co-author of "Intelligent Database Systems" (Addison-Wesley/ACM Press, 2001), and author of a recent book on NKRL, "Representation and Management of Narrative Information" (Springer, 2009). Prof. Zarri has been on the editorial board of important scientific journals and on the Program Committee of international conferences. He has worked as an expert for the European Commission, some French agencies, the US National Foundation, the Italian Ministry of Research, the Dutch Jacquard Program, the Austrian "Vienna Science and Technology Fund", etc. He has also co-operated intensively with industry, both with large industrial groups and with SMEs.

# Index

## Symbols



## A



## B

## C



## D



## E

## S

2 370003 345452

Lightning Source UK Ltd.
Milton Keynes UK
UKOW03n1909221214

243410UK00005BA/14/P